JOHN A. THORPE
PAUL G. KUMPEL

State University of New York at Stony Brook

Elementary Linear Algebra

 SAUNDERS COLLEGE PUBLISHING

Philadelphia   New York   Chicago
San Francisco   Montreal   Toronto
London   Sydney   Tokyo   Mexico City
Rio de Janeiro   Madrid

Address orders to:
383 Madison Avenue
New York, NY 10017

Address editorial correspondence to:
West Washington Square
Philadelphia, PA 19105

Text Typeface: 10/12 Times Roman
Compositor: Waldman Graphics, Inc.
Acquisitions Editor: Leslie Hawke
Project Editor: Patrice L. Smith
Copyeditor: Robin Bonner
Managing Editor & Art Director: Richard L. Moore
Art/Design Assistant: Virginia A. Bollard
Text Design: Emily Harste
Cover Design: Lawrence R. Didona
Text Artwork: Vantage Art, Inc.
Production Manager: Tim Frelick
Assistant Production Manager: Maureen Iannuzzi

Cover credit: Frank Whitney/The Image Bank

Library of Congress Cataloging in Publication Data

Thorpe, John A.
  Elementary linear algebra.

  Includes index.

  1. Algebras, Linear. I. Kumpel, P. G. (Paul G.)
II. Title.
QA184.T48 1983   512'.5   83-14442
ISBN 0-03-061249-7

ELEMENTARY LINEAR ALGEBRA   ISBN 0-03-061249-7
© 1984 by CBS College Publishing. All rights reserved. Printed in the United States of
   America.
Library of Congress catalog card number 83-14442

4567   032   987654321

CBS COLLEGE PUBLISHING
Saunders College Publishing
Holt, Rinehart and Winston
The Dryden Press

To my sons
Ken and Steve
   J.T.

To my son and my daughter
Ted and Lisa
   P.K.

This text is designed for a first course in linear algebra at the college level. The material is accessible to any student who has mastered high school algebra, geometry, and elementary trigonometry.

We begin in Chapter 1 with a study of the problem of solving systems of linear equations using Gaussian elimination. This is a natural starting point for the study of linear algebra since it provides a smooth transition from material that the student has learned in high school algebra. Solutions of systems of linear equations lead naturally to the idea of vectors in $\mathbb{R}^n$ and then to the structure of $\mathbb{R}^n$ as a vector space. After studying the algebra of $\mathbb{R}^n$, we look at the geometry of $\mathbb{R}^2$, and then of $\mathbb{R}^n$. We describe lines in $\mathbb{R}^n$ using vector equations. The concept of dot product is then introduced as a way of finding the angle between two vectors and is used to describe lines in $\mathbb{R}^2$, planes in $\mathbb{R}^3$, and hyperplanes in $\mathbb{R}^n$. The cross product is introduced as a way to find a vector perpendicular to two given vectors in $\mathbb{R}^3$. Its properties are explained, and it is used to find volumes of parallelepipeds and of tetrahedra and to prove a generalized Pythagorean Theorem for tetrahedra in $\mathbb{R}^3$. The chapter concludes with an application, the vertex method for solving linear programming problems in $\mathbb{R}^2$.

Chapter 2 introduces the idea of the linear map associated with a matrix. We begin with a discussion of linear maps from the plane to itself. The geometry is emphasized here, using a multitude of explicit examples. We then move on to the more general case of linear maps from $\mathbb{R}^n$ to $\mathbb{R}^m$. We discuss the algebra of linear maps and the algebra of matrices. We motivate matrix multiplication using the correspondence between matrices and linear maps: Matrix multiplication is the matrix operation that corresponds to composition of linear maps. The chapter continues with a discussion of inverse maps and inverse matrices. It concludes with an application to Markov processes.

Determinants are introduced in Chapter 3. The determinant of a 2×2 matrix is defined by the usual formula. Then we observe how row operations affect 2×2 determinants and use these properties to *define* the determinant of an $n \times n$ matrix for arbitrary n. This approach leads the student quickly to a computationally superior algorithm for computing determinants, without the necessity of developing extensive theory first. Then we show that determinants can also be evaluated either using expansion by minors or using permutations. We derive the cofactor formula for the inverse of an invertible matrix, and use it to validate Cramer's Rule. Proofs of the theorems are outlined for the 3×3 case, where they are easy to follow. We conclude the chapter with a proof that the determinant of a product is a product of determinants and with a discussion of the determinant as a measure of area and volume magnification under a linear map.

Chapter 4 is devoted to the study of subspaces, spanning sets, and linear independence in $\mathbb{R}^n$. Here we discuss the coordinates of a vector with respect to an ordered basis, orthonormal bases, and the Gram-Schmidt orthogonalization process. This chapter closes with a discussion of least squares approximations.

In Chapter 5 we explore the special properties of linear maps from $\mathbb{R}^n$ to itself. We discuss eigenvectors and eigenvalues, the diagonalization of square matrices, and isometries. We include here a thorough discussion of rotations in 3-space. We conclude the chapter with an application of diagonalization techniques to the study of conic sections.

The text concludes with a chapter on general vector spaces. The concepts of linear maps, linear independence, bases, and dimension for finite dimensional vector spaces are discussed here. These abstract ideas are copiously illustrated, using examples from the earlier chapters. As with each of the preceding chapters, we close this chapter with an application, this time to linear difference equations.

In writing this book, we have been guided by several principles.

1. *A textbook should be organized in such a way that the easier material comes early and the more abstract and difficult concepts come later, after the student has been given a good grounding in the concrete aspects of the subject*. In this spirit, we develop geometric and computational linear algebra in the first three chapters, postponing until Chapter 4 the more sophisticated notions of linear independence and bases, and postponing until Chapter 6 the abstract concept of a general vector space.

2. *The art of good teaching consists of guiding students in a smooth learning process that constantly builds on the present state of their knowledge*. Our adherence to this principle should be evident throughout the book, beginning with page 1 where we draw upon and extend the students' knowledge of high school algebra by studying systems of linear equations.

3. *A new idea should be introduced only when the need for it is clear to the student*. In particular, unmotivated definitions must be avoided. We introduce, for example, the dot product to measure the angle between vectors, the cross product to find a vector perpendicular to two given vectors, and matrix multiplication to solve the problem of finding a matrix operation that corresponds to composing linear maps.

4. *Students learn better from examples than from abstract discussions*. For this reason, we often discuss a carefully chosen example to illustrate a theorem before discussing its proof. This makes the discussion of the proof more meaningful.

5. *Students at this level learn best by doing computations*. Abstraction can be understood, but not without constant drill in the computational techniques that illustrate and validate the abstraction. We include many computational examples and exercises in every section so that students will never find themselves with only theory to think about.

6. *Students rarely learn by studying rigorous mathematical proofs*. Although such proofs are satisfying to mathematicians and are essential for serious students of more advanced mathematics, these proofs are usually inappropriate for beginning students. So our approach to rigor is as follows:

(i) Each theorem is stated precisely and in as much generality as is appropriate.

(ii) Careful, rigorous proofs are included whenever they are fairly transparent or whenever they add significantly to the students' understanding of the theorem.

(iii) We do not hesitate to present a proof only for low dimensions whenever the low dimensional proof captures all of the essential ideas of the general case. The increase in clarity from this procedure is often dramatic.

(iv) Occasionally, a particularly difficult proof is omitted entirely and replaced by a discussion designed to enable the student to see why the stated result is reasonable.

7. *Linear algebra is best understood in a geometric, rather than an algebraic, context.* We take every opportunity to call the student's attention to the geometric content of linear algebra. Thus, immediately after defining linear maps of the plane, we devote an entire section to a careful discussion of geometric examples: expansions, contractions, rotations, and reflections. We have prepared more than 200 figures, most in two colors, illustrating the geometric meaning of concepts as they arise. In fact, a student can get a very good idea of what the subject is about merely by studying the figures and their detailed captions.

8. *Applications should be treated seriously if they are treated at all.* We have included an application at the end of each chapter. Each of these applications is treated in some depth. We avoid teasing references to the possibility of applying linear algebra to "real world situations," preferring to cover a few applications meaningfully and to leave the others to specialized courses.

We have included at the end of each chapter a collection of review exercises that should help students focus on any areas of weakness.

The result of our efforts is, we hope, a book from which students can learn. It is clear that linear algebra is a subject that college freshmen and sophomores, and even high school seniors, can study. The challenge is to create a learning instrument that will be appropriate for this audience. This has been our goal.

John A. Thorpe
Paul G. Kumpel

ACKNOWLEDGMENTS

We would like to thank the following reviewers whose thoughtful comments have added significantly to the quality of this work:

Professor Joe Albree of Auburn University

Professor Michael Burke of the College of San Mateo

Professor Albert G. Fadell of SUNY at Buffalo

Professor William R. Fuller of Purdue University

Professor Joseph A. Gallian of the University of Minnesota at Duluth

Professor David R. Hill of Temple University

Professor Richard Porter of Northeastern University

Thanks also are due to Professor Richard M. Koch of the University of Oregon whose copious comments on an earlier work of ours strongly influenced the presentation in several parts of this work and to Professor Gary Simon of New York University who generously read part of our manuscript and provided us with many helpful comments.

On the production end, we owe considerable debts to our typist, Wanda Mocarski, for her patience and for the high quality of her work, and to the staff of Saunders College Publishing, especially Don Jackson and Jim Porterfield who provided the initial encouragement that got us started on this project, to Leslie Hawke whose continual enthusiastic support has seen us through the writing stages, and to Patrice Smith who has guided us so professionally through the publication process.

John A. Thorpe
Paul G. Kumpel

CONTENTS OVERVIEW

CONTENTS

Linear Equations, Matrices, and Vectors

This book is devoted to the study of linear algebra and some of its applications. The simplest linear algebra problem involves finding the point or points where two straight lines in the plane meet. You have undoubtedly learned how to solve this problem in a previous algebra course. It amounts to solving two simultaneous linear equations. This first chapter presents a systematic study of this problem and its generalizations.

1.1 Solving Systems of Linear Equations

We begin by developing a systematic procedure for solving a system of linear equations. We start with an example.

EXAMPLE 1

Consider the system

$$x_1 + 2x_2 + x_3 = 3$$
$$2x_1 + 4x_2 + x_3 = 3$$
$$3x_1 + 7x_2 \quad\quad = 2$$

of three equations in three unknowns. We shall find all solutions of this system; that is, we shall find all triples (x_1, x_2, x_3) of real numbers that satisfy all three equations simultaneously.

The first step in the solution procedure is to eliminate x_1 from all equations except the first. This is done by adding -2 times the first equation to the second equation and -3 times the first equation to the third. The system then becomes

$$x_1 + 2x_2 + \quad x_3 = \quad 3$$
$$- \quad x_3 = -3$$
$$x_2 - 3x_3 = -7.$$

The second step eliminates x_2 from all equations except the second. Since the second equation contains no x_2, we first interchange the second and third equations to get

$$x_1 + 2x_2 + \quad x_3 = \quad 3$$
$$x_2 - 3x_3 = -7$$
$$- \quad x_3 = -3.$$

We can now eliminate x_2 from all equations except the second by adding -2 times the second equation to the first:

$$x_1 \qquad + 7x_3 = \quad 17$$
$$x_2 - 3x_3 = \quad -7$$
$$- \quad x_3 = \quad -3.$$

Next, we multiply the third equation by -1 so that the coefficient of x_3 in that equation is $+1$:

$$x_1 \qquad + 7x_3 = \quad 17$$
$$x_2 - 3x_3 = \quad -7$$
$$x_3 = \quad 3.$$

Finally, we eliminate x_3 from all equations but the third by adding appropriate multiples of the third equation to the other two:

$$x_1 \qquad\qquad = \quad -4$$
$$x_2 \qquad = \quad 2$$
$$x_3 = \quad 3.$$

We find, then, that the only solution of the original system is $(x_1, x_2, x_3) = (-4, 2, 3)$. ■

The work involved in solving a system of linear equations can be simplified considerably by the use of a shorthand notation. We can delete the unknowns and the plus and equal signs and still retain all the basic information about the system. The numbers that remain form a rectangular array called a **matrix**. The matrix of the system

$$x_1 + 2x_2 + x_3 = 3$$
$$2x_1 + 4x_2 + x_3 = 3$$
$$3x_1 + 7x_2 \qquad = 2$$

of Example 1 is

$$\begin{pmatrix} 1 & 2 & 1 & 3 \\ 2 & 4 & 1 & 3 \\ 3 & 7 & 0 & 2 \end{pmatrix}.$$

Notice that the last column in this matrix contains the numbers on the right-hand side of the system and the other columns contain the coefficients of the unknowns. The rows of the matrix represent the equations of the system. These equations can be recovered by putting back in their proper places the unknowns x_1, x_2, and x_3 together with the plus and equal signs.

Notice further that multiplying an equation in the system by a real number corresponds to multiplying a row of the matrix by that number. Interchanging two equations corresponds to interchanging two rows of the matrix. Adding one equation to another corresponds to adding one row of the matrix to another.

Therefore, the operations used to solve a system of linear equations correspond to three *row operations* on the matrix:

(i) multiply a row by a nonzero real number

(ii) interchange two rows

(iii) add a multiple of one row to another.

The operations we used to solve the system of Example 1 can be represented in the following compact way:

$$\begin{pmatrix} 1 & 2 & 1 & 3 \\ 2 & 4 & 1 & 3 \\ 3 & 7 & 0 & 2 \end{pmatrix} \xrightarrow{\text{(iii)}} \begin{pmatrix} 1 & 2 & 1 & 3 \\ 0 & 0 & -1 & -3 \\ 0 & 1 & -3 & -7 \end{pmatrix} \xrightarrow{\text{(ii)}} \begin{pmatrix} 1 & 2 & 1 & 3 \\ 0 & 1 & -3 & -7 \\ 0 & 0 & -1 & -3 \end{pmatrix}$$

$$\xrightarrow{\text{(iii)}} \begin{pmatrix} 1 & 0 & 7 & 17 \\ 0 & 1 & -3 & -7 \\ 0 & 0 & -1 & -3 \end{pmatrix} \xrightarrow{\text{(i)}} \begin{pmatrix} 1 & 0 & 7 & 17 \\ 0 & 1 & -3 & -7 \\ 0 & 0 & 1 & 3 \end{pmatrix}$$

$$\xrightarrow{\text{(iii)}} \begin{pmatrix} 1 & 0 & 0 & -4 \\ 0 & 1 & 0 & 2 \\ 0 & 0 & 1 & 3 \end{pmatrix}$$

The number above each arrow indicates the row operation used in that step. Notice that the last matrix obtained is just the matrix of the system

$$\begin{aligned} x_1 & & & = -4 \\ & x_2 & & = 2 \\ & & x_3 & = 3 \end{aligned}$$

from which the solution was evident.

Let us look at another example. This time we do not write out the equations at each stage of the solution procedure. Instead, we use the shorthand notation. In studying this example, do not attempt to understand the rationale for each step on the first reading. Just follow the steps and check that they accomplish the stated goals. The rationale will be discussed immediately following the example.

EXAMPLE 2

To solve the system

$$\begin{aligned} x_3 + x_4 & = 0 \\ -2x_1 - 4x_2 + x_3 \qquad & = -3 \\ 3x_1 + 6x_2 - x_3 + x_4 & = 5 \end{aligned}$$

we apply a sequence of row operations to the matrix

$$\begin{pmatrix} 0 & 0 & 1 & 1 & 0 \\ -2 & -4 & 1 & 0 & -3 \\ 3 & 6 & -1 & 1 & 5 \end{pmatrix}.$$

First, interchange the first row with the third row to make the entry in the first row, first column, different from zero:

$$(ii) \longrightarrow \begin{pmatrix} 3 & 6 & -1 & 1 & 5 \\ -2 & -4 & 1 & 0 & -3 \\ 0 & 0 & 1 & 1 & 0 \end{pmatrix}.$$

Then add the second row to the first row to transform the first row, first column, entry into a 1:

$$(iii) \longrightarrow \begin{pmatrix} 1 & 2 & 0 & 1 & 2 \\ -2 & -4 & 1 & 0 & -3 \\ 0 & 0 & 1 & 1 & 0 \end{pmatrix}.$$

Next add twice the first row to the second row to make all other entries in the first column equal to zero; this has the effect of eliminating x_1 from all equations but the first:

$$(iii) \longrightarrow \begin{pmatrix} 1 & 2 & 0 & 1 & 2 \\ 0 & 0 & 1 & 2 & 1 \\ 0 & 0 & 1 & 1 & 0 \end{pmatrix}.$$

Add -1 times the second row to the third row to make all entries in the third column, other than the 1 in the second row, equal to zero; this has the effect of eliminating x_3 from all equations but the second:

$$(iii) \longrightarrow \begin{pmatrix} 1 & 2 & 0 & 1 & 2 \\ 0 & 0 & 1 & 2 & 1 \\ 0 & 0 & 0 & -1 & -1 \end{pmatrix}.$$

Multiply the third row by -1 to transform the third row, fourth column entry into a 1:

$$(i) \longrightarrow \begin{pmatrix} 1 & 2 & 0 & 1 & 2 \\ 0 & 0 & 1 & 2 & 1 \\ 0 & 0 & 0 & 1 & 1 \end{pmatrix}.$$

Finally, add -1 times the third row to the first row and add -2 times the third row to the second row to make all the other entries in the fourth column equal to zero:

$$(iii) \longrightarrow \begin{pmatrix} 1 & 2 & 0 & 0 & 1 \\ 0 & 0 & 1 & 0 & -1 \\ 0 & 0 & 0 & 1 & 1 \end{pmatrix}.$$

The system corresponding to this last matrix is

$$
\begin{aligned}
x_1 + 2x_2 \qquad\qquad &= \quad 1 \\
x_3 \qquad &= \ -1 \\
x_4 &= \quad 1.
\end{aligned}
$$

From these equations it is clear that $x_3 = -1$ and $x_4 = 1$. But the first equation still contains both x_1 and x_2. The fact that x_2 can't be eliminated means simply that we can choose any value for x_2. If we set $x_2 = c$, then $x_1 = 1 - 2c$, and the solutions of the system are

$$x_1 = 1 - 2c$$

$$x_2 = c$$

$$x_3 = -1$$

$$x_4 = 1,$$

where c is any real number. In other words, the set of solutions is the set of all ordered quadruples $(1 - 2c, c, -1, 1)$, where c is any real number. ■

Look back now at the matrices obtained in these two examples:

$$\begin{pmatrix} 1 & 0 & 0 & -4 \\ 0 & 1 & 0 & 2 \\ 0 & 0 & 1 & 3 \end{pmatrix} \quad \text{and} \quad \begin{pmatrix} 1 & 2 & 0 & 0 & 1 \\ 0 & 0 & 1 & 0 & -1 \\ 0 & 0 & 0 & 1 & 1 \end{pmatrix}.$$

You will see that they have several features in common.

A "descending staircase" can be drawn in each matrix so that

(i) each stair has height one row
(ii) below the staircase all entries are zero
(iii) in each corner of the staircase is the number 1
(iv) each column containing one of these corner 1's has all other entries zero.

A matrix in which a descending staircase can be drawn having properties (i), (ii), (iii), and (iv) is said to be a *row echelon matrix.*

We can now describe in general how to solve a system of linear equations using what is often called the method of *Gaussian elimination.*

Consider the system

$$a_{11}x_1 + a_{12}x_2 + \cdots + a_{1n}x_n = b_1$$

$$a_{21}x_1 + a_{22}x_2 + \cdots + a_{2n}x_n = b_2$$

(∗)

$$\cdots$$

$$a_{m1}x_1 + a_{m2}x_2 + \cdots + a_{mn}x_n = b_m$$

of m equations in n unknowns. Equations of this type are called *linear equations.* A system of linear equations is called a *linear system.* A *solution* of the linear system (∗) is an ordered n-tuple $(x_1, \ldots, x_n)$ of real numbers that satisfies all the equations in the system simultaneously. The *matrix of the linear system* (∗) is the matrix

$$\begin{pmatrix} a_{11} & a_{12} & \cdots & a_{1n} & b_1 \\ a_{21} & a_{22} & \cdots & a_{2n} & b_2 \\ & & \cdots & & \\ a_{m1} & a_{m2} & \cdots & a_{mn} & b_m \end{pmatrix}.$$

> To find all solutions of a system of linear equations
>
> 1. Write down the matrix of the system.
> 2. Use row operations to convert this matrix to a row echelon matrix. (It is best to work from left to right on one column at a time.)
> 3. Write down the system of equations that corresponds to this row echelon matrix, set the unknowns that do not correspond to corner 1's equal to c_1, c_2, . . . , and read off the solutions by solving for the unknowns corresponding to the corner 1's.

The process of using row operations to convert a matrix to a row echelon matrix is called *row reduction*.

REMARK 1   Often, during the row reduction process, there is more than one reasonable candidate for the next row operation to be performed. A thoughtful choice among these can sometimes make your work easier. For example, in the second step of Example 2 we added the second row to the first row to get

$$\begin{pmatrix} 3 & 6 & -1 & 1 & 5 \\ -2 & -4 & 1 & 0 & -3 \\ 0 & 0 & 1 & 1 & 0 \end{pmatrix} \xrightarrow{\text{(iii)}} \begin{pmatrix} 1 & 2 & 0 & 1 & 2 \\ -2 & -4 & 1 & 0 & -3 \\ 0 & 0 & 1 & 1 & 0 \end{pmatrix}.$$

Alternatively, we could have multiplied the first row by $\frac{1}{3}$ to get

$$\begin{pmatrix} 3 & 6 & -1 & 1 & 5 \\ -2 & -4 & 1 & 0 & -3 \\ 0 & 0 & 1 & 1 & 0 \end{pmatrix} \xrightarrow{\text{(i)}} \begin{pmatrix} 1 & 2 & -\frac{1}{3} & \frac{1}{3} & \frac{5}{3} \\ -2 & -4 & 1 & 0 & -3 \\ 0 & 0 & 1 & 1 & 0 \end{pmatrix}.$$

Either of these operations is fine for introducing a 1 in the first row, first column. But the first choice is better in this case because it avoids introducing fractions into the matrix. As you gain experience in reducing matrices, you should pause for a second at each stage to choose judiciously among the options available.

REMARK 2   Every matrix can be reduced to a row echelon matrix by a sequence of row operations. It can be shown that the row echelon matrix obtained is uniquely determined by the original matrix. Thus, if two people use different sequences of row operations to reduce a given matrix to a row echelon matrix, they will both attain the same result (provided, of course, that they do the arithmetic correctly).

REMARK 3   The reason that Gaussian elimination works is that row operations are reversible. Using a reverse sequence of row operations it is possible to convert the row echelon matrix back to the original matrix. It follows that not only is every solution of the original system a solution of the simplified system but also every solution of the simplified system is a solution of the original system. In other words, *the two systems have the same solution set*.

EXAMPLE 3

Let us solve the system

$$x_1 + x_2 + x_3 - x_4 = 2$$

$$2x_1 + 2x_2 + 3x_3 + x_4 = 5$$

$$x_1 - x_2 + 2x_3 + 3x_4 + x_5 = 4.$$

First we write down the matrix of the system. Then we use row operations to reduce the matrix to a row echelon matrix. It is always best to work on only one column at a time, starting with the first column and working to the right. In studying this example, you should be able to describe explicitly the operations used in each step.

$$\begin{pmatrix} 1 & 1 & 1 & -1 & 0 & 2 \\ 2 & 2 & 3 & 1 & 0 & 5 \\ 1 & -1 & 2 & 3 & 1 & 4 \end{pmatrix}$$

$$\xrightarrow{\text{(iii)}} \begin{pmatrix} 1 & 1 & 1 & -1 & 0 & 2 \\ 0 & 0 & 1 & 3 & 0 & 1 \\ 0 & -2 & 1 & 4 & 1 & 2 \end{pmatrix}$$

$$\xrightarrow{\text{(ii)}} \begin{pmatrix} 1 & 1 & 1 & -1 & 0 & 2 \\ 0 & -2 & 1 & 4 & 1 & 2 \\ 0 & 0 & 1 & 3 & 0 & 1 \end{pmatrix}$$

$$\xrightarrow{\text{(i)}} \begin{pmatrix} 1 & 1 & 1 & -1 & 0 & 2 \\ 0 & 1 & -\frac{1}{2} & -2 & -\frac{1}{2} & -1 \\ 0 & 0 & 1 & 3 & 0 & 1 \end{pmatrix}$$

$$\xrightarrow{\text{(iii)}} \begin{pmatrix} 1 & 0 & \frac{3}{2} & 1 & \frac{1}{2} & 3 \\ 0 & 1 & -\frac{1}{2} & -2 & -\frac{1}{2} & -1 \\ 0 & 0 & 1 & 3 & 0 & 1 \end{pmatrix}$$

$$\xrightarrow{\text{(iii)}} \begin{pmatrix} 1 & 0 & 0 & -\frac{7}{2} & \frac{1}{2} & \frac{3}{2} \\ 0 & 1 & 0 & -\frac{1}{2} & -\frac{1}{2} & -\frac{1}{2} \\ 0 & 0 & 1 & 3 & 0 & 1 \end{pmatrix}$$

Now we write the system corresponding to the row echelon matrix:

$$x_1 \qquad\qquad - \tfrac{7}{2}x_4 + \tfrac{1}{2}x_5 = \tfrac{3}{2}$$

$$x_2 \qquad - \tfrac{1}{2}x_4 - \tfrac{1}{2}x_5 = -\tfrac{1}{2}$$

$$x_3 + 3x_4 \qquad\qquad = 1$$

The unknowns x_4 and x_5 can be assigned any values, so we set $x_4 = c_1$ and $x_5 = c_2$. Then

$$x_1 = \tfrac{3}{2} + \tfrac{7}{2}c_1 - \tfrac{1}{2}c_2$$

$$x_2 = -\tfrac{1}{2} + \tfrac{1}{2}c_1 + \tfrac{1}{2}c_2$$

$$x_3 = 1 - 3c_1$$

$$x_4 = c_1$$

$$x_5 = c_2.$$

Hence the solution set of the given system is the set of all 5-tuples of the form

$$(\tfrac{3}{2} + \tfrac{7}{2}c_1 - \tfrac{1}{2}c_2, \ -\tfrac{1}{2} + \tfrac{1}{2}c_1 + \tfrac{1}{2}c_2, \ 1 - 3c_1, \ c_1, \ c_2)$$

where c_1 and c_2 are any two real numbers. ■

You have seen examples of linear systems that have exactly one solution and of linear systems that have infinitely many solutions. It should not surprise you to find that some linear systems have no solutions at all.

EXAMPLE 4

To solve the system

$$x_1 - x_2 + x_3 = 1$$

$$x_1 + 2x_2 - x_3 = 7$$

$$-x_1 + 4x_2 - 3x_3 = 4$$

we reduce the matrix of the system as follows:

$$\begin{pmatrix} 1 & -1 & 1 & 1 \\ 1 & 2 & -1 & 7 \\ -1 & 4 & -3 & 4 \end{pmatrix} \xrightarrow{\text{(iii)}} \begin{pmatrix} 1 & -1 & 1 & 1 \\ 0 & 3 & -2 & 6 \\ 0 & 3 & -2 & 5 \end{pmatrix}$$

$$\xrightarrow{\text{(iii)}} \begin{pmatrix} 1 & -1 & 1 & 1 \\ 0 & 3 & -2 & 6 \\ 0 & 0 & 0 & -1 \end{pmatrix} \xrightarrow{\text{(i)}} \begin{pmatrix} 1 & -1 & 1 & 1 \\ 0 & 1 & -\tfrac{2}{3} & 2 \\ 0 & 0 & 0 & 1 \end{pmatrix}$$

$$\xrightarrow{\text{(iii)}} \begin{pmatrix} 1 & 0 & \tfrac{1}{3} & 3 \\ 0 & 1 & -\tfrac{2}{3} & 2 \\ 0 & 0 & 0 & 1 \end{pmatrix} \xrightarrow{\text{(iii)}} \begin{pmatrix} 1 & 0 & \tfrac{1}{3} & 0 \\ 0 & 1 & -\tfrac{2}{3} & 0 \\ 0 & 0 & 0 & 1 \end{pmatrix}.$$

This last matrix corresponds to the system

$$x_1 \qquad + \tfrac{1}{3}x_3 = 0$$

$$x_2 - \tfrac{2}{3}x_3 = 0$$

$$0 = 1$$

which, of course, has *no solution,* since $0 \neq 1$. ■

What occurred in the last example is typical of the case in which a system has no solution: There was a corner 1 in the last column of the row echelon matrix. In general, it is easy to see how many solutions a linear system has by looking at the location of the corner 1's.

If a corner 1 occurs in the last column of the row echelon matrix, then the corresponding system has no solution. If there is no corner 1 in the last column, then the system has one or more solutions (see Exercises 11 and 12).

Furthermore, if every column except the last in a row echelon matrix has a corner 1, then the system has exactly one solution. If there is no corner 1 in the last column and in at least one additional column then the system has infinitely many solutions (see Exercises 13 and 14).

EXERCISES

1. Find all solutions (if there are any) of each of the following linear systems using Gaussian elimination.

   (a) $\begin{aligned} x_1 + x_2 &= 1 \\ x_1 - x_2 &= -1 \end{aligned}$ (b) $\begin{aligned} x_1 + x_2 &= 1 \\ -x_1 + x_2 &= -1 \end{aligned}$ (c) $\begin{aligned} x_1 + x_2 &= 1 \\ -x_1 - x_2 &= 0 \end{aligned}$

2. Use row operations to reduce the following matrices to row echelon matrices.

   (a) $\begin{pmatrix} 0 & 0 & 1 \\ 0 & 1 & 0 \\ 1 & 0 & 0 \end{pmatrix}$ (b) $\begin{pmatrix} 2 & -1 & 3 \\ 1 & 0 & -1 \\ 3 & -1 & 1 \\ -1 & 1 & -1 \end{pmatrix}$

   (c) $\begin{pmatrix} 1 & 1 & -1 & -1 & 2 \\ 2 & 2 & 0 & 3 & 1 \\ -1 & -1 & 0 & 1 & 1 \end{pmatrix}$

3. Solve:

$$\begin{aligned} 2x_1 + x_2 - x_3 &= 6 \\ x_1 - 2x_2 - 2x_3 &= 1 \\ -x_1 + 12x_2 + 8x_3 &= 7 \end{aligned}$$

4. Solve:

$$\begin{aligned} x_1 + 2x_2 + 4x_3 &= 1 \\ x_1 + x_2 + 3x_3 &= 2 \\ 2x_1 + 5x_2 + 9x_3 &= 1 \end{aligned}$$

5. Solve:

$$\begin{aligned} -x_1 + 2x_2 + x_3 - 2x_4 &= 2 \\ x_1 - x_2 - 2x_3 + 2x_4 &= 0 \\ -2x_1 + 4x_2 \qquad - 2x_4 &= -2 \end{aligned}$$

6. Solve:

$$\begin{aligned} 3x_1 + 2x_2 - x_3 &= 1 \\ 2x_1 - 2x_2 + 5x_3 &= -3 \\ 7x_1 - 2x_2 + 9x_3 &= -5 \end{aligned}$$

7. Solve:

$$\begin{aligned} 3x_1 + x_2 - x_3 + 2x_4 &= 7 \\ 2x_1 - 2x_2 + 5x_3 - 7x_4 &= 1 \\ -4x_1 - 4x_2 + 7x_3 - 11x_4 &= -13 \end{aligned}$$

8. Solve:

$$
\begin{aligned}
x_1 + x_2 - 3x_3 &= 1 \\
x_2 + x_3 - x_4 &= 2 \\
-x_1 + x_2 \qquad\quad + x_4 &= -1 \\
x_1 + 6x_2 - 8x_3 + x_4 &= 3
\end{aligned}
$$

9. Solve:

$$
\begin{aligned}
2x_2 - x_3 - 2x_4 \qquad\quad &= -1 \\
-2x_1 + 2x_2 - 3x_3 - 4x_4 - 8x_5 &= 0 \\
x_3 + 3x_4 + 5x_5 &= 1 \\
x_1 \qquad\quad + x_3 + 2x_4 + 6x_5 &= 0 \\
2x_2 - x_3 - x_4 + 3x_5 &= -1
\end{aligned}
$$

10. Let $a_{11}, a_{12}, a_{21},$ and a_{22} be real numbers. Show that the system

$$
\begin{aligned}
a_{11}x_1 + a_{12}x_2 &= 0 \\
a_{21}x_1 + a_{22}x_2 &= 0
\end{aligned}
$$

has a nontrivial solution (one different from $(0, 0)$) if and only if

$$
a_{11}a_{22} - a_{12}a_{21} = 0.
$$

11. Explain why a system of linear equations whose row echelon matrix has a corner 1 in the last column must have no solutions.

12. Explain why a system of linear equations whose row echelon matrix has *no* corner 1 in the last column must have at least one solution.

13. Explain why a system of linear equations whose row echelon matrix has a corner 1 in every column except the last must have exactly one solution.

14. Explain why a system of linear equations whose row echelon matrix has no corner 1 in the last column and in at least one additional column must have infinitely many solutions.

1.2
Solution Vectors, Row Vectors, and $\mathbb{R}^n$

In this section we shall study the set of all ordered n-tuples $(a_1, a_2, \ldots, a_n)$ of real numbers. This set is denoted by the symbol $\mathbb{R}^n$. Elements of $\mathbb{R}^n$ are called *vectors*. We saw in the previous section that vectors occur as solutions of systems of linear equations. We shall see in this and future sections that vectors also occur in many other settings and that they are useful in solving many different kinds of problems.

Vectors are usually denoted by boldface letters. Thus a vector in $\mathbb{R}^n$ is an ordered n-tuple,

$$
\mathbf{a} = (a_1, a_2, \ldots, a_n)
$$

where $a_1, a_2, \ldots, a_n$ are real numbers. The number a_1 is called the *first entry* of $\mathbf{a}$, a_2 is the *second entry* of $\mathbf{a}$, and so on. Two vectors, $\mathbf{a} = (a_1, a_2, \ldots, a_n)$ and $\mathbf{b} = (b_1, b_2, \ldots, b_n)$, in $\mathbb{R}^n$ are said to be *equal* if and only if their corresponding entries are equal. Thus $\mathbf{a} = \mathbf{b}$ if and only if $a_1 = b_1, a_2 = b_2, \ldots, a_n = b_n$.

Vectors that occur as solutions of linear systems are called ***solution vectors***. Vectors also occur as rows in matrices. They are called ***row vectors***. For example, we find in the matrix

$$\begin{pmatrix} 1 & -1 & 3 & 2 \\ 0 & 5 & 4 & 3 \\ -1 & 1 & -1 & 1 \end{pmatrix}$$

the three row vectors

$$\mathbf{a} = (1, -1, 3, 2)$$

$$\mathbf{b} = (0, 5, 4, 3)$$

and

$$\mathbf{c} = (-1, 1, -1, 1).$$

These three vectors are elements of $\mathbb{R}^4$.

The row operations on matrices that were used in Section 1 to help solve systems of linear equations suggest a way to add vectors and a way to multiply vectors by real numbers.

Given two vectors $\mathbf{a}$ and $\mathbf{b}$ in $\mathbb{R}^n$, their ***sum*** is the vector obtained by adding corresponding entries: if $\mathbf{a} = (a_1, a_2, \ldots, a_n)$ and $\mathbf{b} = (b_1, b_2, \ldots, b_n)$ then

$$\mathbf{a} + \mathbf{b} = (a_1 + b_1, a_2 + b_2, \ldots, a_n + b_n).$$

Notice that vectors can be added only if they have the same size. You cannot add a vector in $\mathbb{R}^2$ to a vector in $\mathbb{R}^3$.

Given a vector $\mathbf{a}$ and a real number c, the ***scalar multiple*** $c\mathbf{a}$ of the vector $\mathbf{a}$ is the vector obtained by multiplying each entry in $\mathbf{a}$ by c: if $\mathbf{a} = (a_1, a_2, \ldots, a_n)$ then

$$c\mathbf{a} = (ca_1, ca_2, \ldots, ca_n).$$

The number c is called a ***scalar*** here, to emphasize that it is a real number and not a vector.

EXAMPLE 1

In $\mathbb{R}^3$, we have

$$(1, -1, 0) + (1, 1, 2) = (2, 0, 2),$$

$$3(-3, 2, 1) = (-9, 6, 3),$$

and

$$2(1, 3, 2) + (-1)(1, 1, 1) = (2, 6, 4) + (-1, -1, -1) = (1, 5, 3).$$

In $\mathbb{R}^4$, if $\mathbf{a} = (1, 0, -1, 2)$ and $\mathbf{b} = (2, 3, -1, 7)$, then

$$3\mathbf{a} + 2\mathbf{b} = 3(1, 0, -1, 2) + 2(2, 3 -1, 7)$$

$$= (3, 0, -3, 6) + (4, 6, -2, 14)$$

$$= (7, 6, -5, 20). \quad \blacksquare$$

The operation of vector addition has the following four properties: for $\mathbf{a}$, $\mathbf{b}$, and $\mathbf{c}$ in $\mathbb{R}^n$,

(A_1) $\mathbf{a} + \mathbf{b} = \mathbf{b} + \mathbf{a}$

(A_2) $(\mathbf{a} + \mathbf{b}) + \mathbf{c} = \mathbf{a} + (\mathbf{b} + \mathbf{c})$

(A_3) $\mathbf{0} + \mathbf{a} = \mathbf{a} + \mathbf{0} = \mathbf{a}$, where $\mathbf{0}$ is the vector $(0, 0, \ldots, 0)$

(A_4) $\mathbf{a} + (-\mathbf{a}) = (-\mathbf{a}) + \mathbf{a} = \mathbf{0}$, where $-\mathbf{a}$ is the vector $(-1)\mathbf{a}$.

These properties follow directly from the corresponding properties of real numbers. For example, if $\mathbf{a} = (a_1, a_2, \ldots, a_n)$ and $\mathbf{b} = (b_1, b_2, \ldots, b_n)$ then

$$\mathbf{a} + \mathbf{b} = (a_1, a_2, \ldots, a_n) + (b_1, b_2, \ldots, b_n)$$

$$= (a_1 + b_1, a_2 + b_2, \ldots, a_n + b_n)$$

$$= (b_1 + a_1, b_2 + a_2, \ldots, b_n + a_n)$$

$$= (b_1, b_2, \ldots, b_n) + (a_1, a_2, \ldots, a_n)$$

$$= \mathbf{b} + \mathbf{a}.$$

This computation establishes property A_1. The remaining three properties can be verified in a similar way.

The operation of scalar multiplication has the following four properties: for $\mathbf{a}$ and $\mathbf{b}$ vectors in $\mathbb{R}^n$, and c and d real numbers,

(S_1) $c(\mathbf{a} + \mathbf{b}) = c\mathbf{a} + c\mathbf{b}$

(S_2) $(c + d)\mathbf{a} = c\mathbf{a} + d\mathbf{a}$

(S_3) $c(d\mathbf{a}) = (cd)\mathbf{a}$

(S_4) $1\mathbf{a} = \mathbf{a}$.

These properties are also easy to check.

Vector subtraction in $\mathbb{R}^n$ also makes sense. For $\mathbf{a}$ and $\mathbf{b}$ vectors in $\mathbb{R}^n$ we define $\mathbf{a} - \mathbf{b}$ to be the vector $\mathbf{a} + (-1)\mathbf{b}$.

The operations of vector addition and scalar multiplication give us a nice way to write solutions of systems of linear equations.

EXAMPLE 2

The solution set of the system

$$x_3 + x_4 = 0$$

$$-2x_1 - 4x_2 + x_3 = -3$$

$$3x_1 + 6x_2 - x_3 + x_4 = 5$$

was found in Example 2 of Section 1 to be the set of all vectors in $\mathbb{R}^4$ of the form

$$\mathbf{x} = (1 - 2c, c, -1, 1),$$

where c is any real number. This $\mathbf{x}$ can be rewritten as

$$\mathbf{x} = (1, 0, -1, 1) + (-2c, c, 0, 0),$$

or

$$\mathbf{x} = (1, 0, -1, 1) + c(-2, 1, 0, 0).$$

This shows that all solutions of the given system can be obtained by adding scalar multiples of the vector $(-2, 1, 0, 0)$ to the vector $(1, 0, -1, 1)$. The vector $(1, 0, -1, 1)$ is itself a solution of the system. The vector $(-2, 1, 0, 0)$ is not. It is, however, a solution to a different system; namely, the system

$$x_3 + x_4 = 0$$

$$-2x_1 - 4x_2 + x_3 \quad\quad = 0$$

$$3x_1 + 6x_2 - x_3 + x_4 = 0,$$

which is obtained from the original system by replacing each number on the right side of that system with zero. You can easily check that this is so. ∎

A system of linear equations is called *__homogeneous__* if the numbers on the right side of the equations are all zero. The system

$$x_3 + x_4 = 0$$

$$-2x_1 - 4x_2 + x_3 \quad\quad = 0$$

$$3x_1 + 6x_2 - x_3 + x_4 = 0$$

is homogeneous, but the system

$$x_3 + x_4 = \quad 0$$

$$-2x_1 - 4x_2 + x_3 \quad\quad = -3$$

$$3x_1 + 6x_2 - x_3 + x_4 = \quad 5$$

is not.

Given any system of linear equations, a homogeneous system can be obtained from it by replacing all of the numbers on the right with zeros. The resulting homogeneous system is called the *__associated homogeneous system__*. For example, the system

$$2x_1 + x_2 = 3$$

$$x_1 - x_2 = 1$$

has the associated homogeneous system

$$2x_1 + x_2 = 0$$

$$x_1 - x_2 = 0.$$

For the system discussed in Example 2 we see that the method of Gaussian elimination leads to a formula exhibiting every solution of the given system as

$$\mathbf{x} = \mathbf{u} + c\mathbf{v},$$

where **u** is one solution of the given system and where **v** is a solution of the associated homogeneous system. A similar formula appears in the next example.

EXAMPLE 3
Let us solve the system

$$x_1 + 2x_2 + 5x_3 + 5x_4 = 3$$

$$x_1 + 3x_2 + 8x_3 + 7x_4 = 2$$

$$2x_1 + 3x_2 + 7x_3 + 8x_4 = 7.$$

The matrix of this system reduces as follows:

$$\begin{pmatrix} 1 & 2 & 5 & 5 & 3 \\ 1 & 3 & 8 & 7 & 2 \\ 2 & 3 & 7 & 8 & 7 \end{pmatrix} \xrightarrow{\text{(iii)}} \begin{pmatrix} 1 & 2 & 5 & 5 & 3 \\ 0 & 1 & 3 & 2 & -1 \\ 0 & -1 & -3 & -2 & 1 \end{pmatrix}$$

$$\xrightarrow{\text{(iii)}} \begin{pmatrix} 1 & 0 & -1 & 1 & 5 \\ 0 & 1 & 3 & 2 & -1 \\ 0 & 0 & 0 & 0 & 0 \end{pmatrix}.$$

This row echelon matrix corresponds to the system

$$x_1 \quad - x_3 + x_4 = 5$$

$$x_2 + 3x_3 + 2x_4 = -1$$

$$0 = 0.$$

Taking $x_3 = c_1$ and $x_4 = c_2$ we see that $x_1 = 5 + c_1 - c_2$ and $x_2 = -1 - 3c_1 - 2c_2$, so the solution set is the set of all vectors of the form

$$\mathbf{x} = (5 + c_1 - c_2, -1 - 3c_1 - 2c_2, c_1, c_2),$$

where c_1 and c_2 are arbitrary real numbers. We can write this **x** as

$$\mathbf{x} = (5, -1, 0, 0) + (c_1, -3c_1, c_1, 0) + (-c_2, -2c_2, 0, c_2)$$

or as

$$\mathbf{x} = (5, -1, 0, 0) + c_1(1, -3, 1, 0) + c_2(-1, -2, 0, 1).$$

This formula exhibits every solution of the given system in the form

$$\mathbf{x} = \mathbf{u} + c_1\mathbf{v}_1 + c_2\mathbf{v}_2.$$

The vector $\mathbf{u} = (5, -1, 0, 0)$ is a solution of the given system. The vectors $\mathbf{v}_1 = (1, -3, 1, 0)$ and $\mathbf{v}_2 = (-1, -2, 0, 1)$ are solutions of the associated homogeneous system. ■

We have seen here two special cases of a very general phenomenon: *Gaussian elimination always leads to vectors* **v**, $\mathbf{u}_1, \mathbf{u}_2, \ldots, \mathbf{u}_k$ *such that the solution set of the given linear system consists of all vectors of the form*

$$\mathbf{x} = \mathbf{v} + c_1\mathbf{u}_1 + c_2\mathbf{u}_2 + \cdots + c_k\mathbf{u}_k,$$

where $c_1, c_2, \ldots, c_k$ are arbitrary real numbers. The vector $\mathbf{v}$ is a solution of the given system. The vectors $\mathbf{u}_1, \mathbf{u}_2, \ldots, \mathbf{u}_k$ are solutions of the associated homogeneous system.

The formula

$$\mathbf{x} = \mathbf{v} + c_1\mathbf{u}_1 + c_2\mathbf{u}_2 + \cdots + c_k\mathbf{u}_k$$

is said to describe the ***general solution*** of the given linear system. We call the vector $\mathbf{v}$ a ***particular solution*** of this system.

EXAMPLE 4

Let us find the general solution of the homogeneous linear system

$$x_1 - x_2 + x_3 - x_4 + x_5 = 0$$
$$2x_1 + x_2 - x_3 \qquad - x_5 = 0.$$

The matrix of this system reduces as follows:

$$\begin{pmatrix} 1 & -1 & 1 & -1 & 1 & 0 \\ 2 & 1 & -1 & 0 & -1 & 0 \end{pmatrix}$$

$$\xrightarrow{\text{(iii)}} \begin{pmatrix} 1 & -1 & 1 & -1 & 1 & 0 \\ 0 & 3 & -3 & 2 & -3 & 0 \end{pmatrix}$$

$$\xrightarrow{\text{(i)}} \begin{pmatrix} 1 & -1 & 1 & -1 & 1 & 0 \\ 0 & 1 & -1 & \frac{2}{3} & -1 & 0 \end{pmatrix}$$

$$\xrightarrow{\text{(iii)}} \begin{pmatrix} 1 & 0 & 0 & -\frac{1}{3} & 0 & 0 \\ 0 & 1 & -1 & \frac{2}{3} & 1 & 0 \end{pmatrix}.$$

This row echelon matrix corresponds to the system

$$x_1 \qquad\qquad - \tfrac{1}{3}x_4 \qquad = 0$$
$$x_2 - x_3 + \tfrac{2}{3}x_4 - x_5 = 0.$$

Setting $x_3 = c_1$, $x_4 = c_2$, and $x_5 = c_3$ we get that $x_1 = \tfrac{1}{3}c_2$ and $x_2 = c_1 - \tfrac{2}{3}c_2 + c_3$ so the general solution is

$$\mathbf{x} = (\tfrac{1}{3}c_2, \; c_1 - \tfrac{2}{3}c_2 + c_3, \; c_1, \; c_2, \; c_3)$$

or

$$\mathbf{x} = c_1(0, 1, 1, 0, 0) + c_2(\tfrac{1}{3}, -\tfrac{2}{3}, 0, 1, 0) + c_3(0, 1, 0, 0, 1).$$

This solution is of the form

$$\mathbf{x} = \mathbf{v} + c_1\mathbf{u}_1 + c_2\mathbf{u}_2 + c_3\mathbf{u}_3,$$

where $\mathbf{v} = \mathbf{0}$. The particular solution $\mathbf{v}$ will always be zero when the system is homogeneous. ∎

EXERCISES

1. Find:

   (a) $(1, -1, 1, -1, 0) + (-1, 1, -1, 1, 0)$

   (b) $(2, 3, 5, 7) + (-1, 3, -5, 7)$

   (c) $2(1, 1, -1)$

   (d) $-1(1, 0, 0, -1)$

   (e) $3(1, 2, 3, 4) + 4(-1, -1, 0, 3)$

   (f) $(1, 3, -2, 1) - (1, -1, -3, 0)$

2. Solve the following vector equations:

   (a) $2\mathbf{x} = (-2, 4, 2, 6)$

   (b) $3\mathbf{x} + (-1, 1, 1) = (2, 0, 0)$

   (c) $\begin{cases} \mathbf{x} + \mathbf{y} = (1, 1) \\ \mathbf{x} - \mathbf{y} = (1, -1) \end{cases}$

3. Express each of the solutions to Exercises 3–9 of Section 1 in the form

$$\mathbf{u} + c_1\mathbf{v}_1 + \cdots + c_k\mathbf{v}_k.$$

4. Solve:

$$x_1 + 2x_2 - x_3 = 1$$
$$2x_1 + 4x_2 - 2x_3 = 2$$
$$-x_1 - 2x_2 + x_3 = -1$$

5. Solve:

$$2x_1 + x_2 = 2$$
$$7x_1 + 4x_2 = 3$$
$$3x_1 + 2x_2 = -1$$

6. Solve:

$$x_1 - 2x_2 + x_4 - x_6 = 0$$
$$-x_1 + 2x_2 + x_3 - x_5 = 1$$
$$2x_1 - 4x_2 + 3x_3 - 3x_6 = 4$$

7. Verify properties A_2–A_4 for vector addition in $\mathbb{R}^n$.

8. Verify properties S_1–S_4 for scalar multiplication in $\mathbb{R}^n$.

9. Show that if $\mathbf{u}_1$ and $\mathbf{u}_2$ are any two solutions of a system of linear equations, then $\mathbf{u}_1 - \mathbf{u}_2$ is a solution of the associated homogeneous system. Conclude that, given any particular solution $\mathbf{u}$, every solution is of the form $\mathbf{u} + \mathbf{v}$ where $\mathbf{v}$ is a solution of the associated homogeneous system.

10. (a) Show that if both $\mathbf{v}_1$ and $\mathbf{v}_2$ are solutions of a given homogeneous system of linear equations, then so is $\mathbf{v}_1 + \mathbf{v}_2$.

   (b) Show that if $\mathbf{v}$ is a solution of a homogeneous system of linear equations, then so is $c\mathbf{v}$ for every real number c.

   (c) Conclude that if $\mathbf{v}_1, \ldots, \mathbf{v}_k$ are any k solutions of a homogeneous system of linear equations and $c_1, \ldots, c_k$ are any k real numbers, then $c_1\mathbf{v}_1 + \cdots + c_k\mathbf{v}_k$ is also a solution.

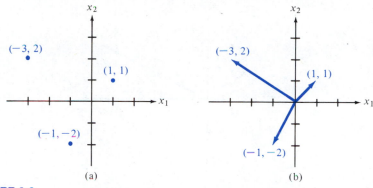

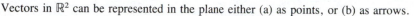

FIGURE 1.1
Vectors in $\mathbb{R}^2$ can be represented in the plane either (a) as points, or (b) as arrows.

1.3
The Geometry of $\mathbb{R}^2$

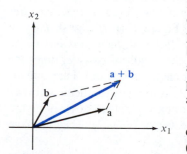

FIGURE 1.2
Vector addition in $\mathbb{R}^2$.

You have now seen some of the *algebra* of $\mathbb{R}^n$, the space of all ordered n-tuples of real numbers. This section and the next will present some of the *geometry* of $\mathbb{R}^n$.

Let us begin by looking at $\mathbb{R}^2$. A vector in $\mathbb{R}^2$ is an ordered pair of real numbers. You have seen ordered pairs of real numbers before, as coordinates of points in the Cartesian coordinate plane (see Figure 1.1a). Viewing ordered pairs of real numbers as points in the plane is one way to represent elements of $\mathbb{R}^2$ geometrically. Another way to represent elements of $\mathbb{R}^2$ geometrically is with arrows (see Figure 1.1b). Thus, each vector in $\mathbb{R}^2$ can be viewed either as a point in the plane or as the arrow whose tail is at the origin and whose head is at that point.

Viewing vectors as arrows leads to a useful interpretation of vector addition. If $\mathbf{a} = (a_1, a_2)$ and $\mathbf{b} = (b_1, b_2)$ are represented as arrows with tails at $(0, 0)$, then $\mathbf{a} + \mathbf{b} = (a_1 + b_1, a_2 + b_2)$ *is the arrow along the diagonal of the parallelogram whose sides are* $\mathbf{a}$ *and* $\mathbf{b}$ (see Figure 1.2). The reason that this **parallelogram law of addition** holds can be seen in Figure 1.3. The shaded

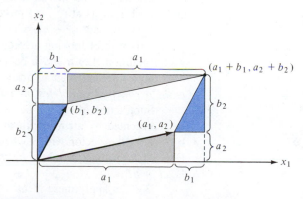

FIGURE 1.3
Proof of the parallelogram law of addition.

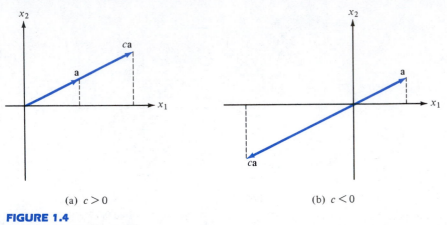

(a) $c > 0$ (b) $c < 0$

FIGURE 1.4
Scalar multiplication in $\mathbb{R}^2$.

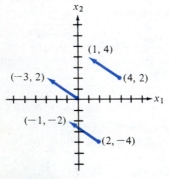

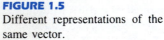

FIGURE 1.5
Different representations of the same vector.

right triangles in this figure are congruent in pairs. Therefore, the quadrilateral with vertices $(0, 0)$, (a_1, a_2), $(a_1 + b_1, a_2 + b_2)$, and (b_1, b_2) has opposite sides equal and hence is a parallelogram.

Scalar multiplication can also be interpreted using arrows. If $\mathbf{a} = (a_1, a_2)$ is a vector in $\mathbb{R}^2$ and if c is a real number, then *the vector* $c\mathbf{a} = (ca_1, ca_2)$ *lies on the same line through the origin as* $\mathbf{a}$ (see Figure 1.4). *If* $c > 0$, *then* $c\mathbf{a}$ *points in the same direction as* $\mathbf{a}$ (Figure 1.4a). *If* $c < 0$, *then* $c\mathbf{a}$ *points in the opposite direction* (Figure 1.4b). In each case the length of $c\mathbf{a}$ is $|c|$ times the length of $\mathbf{a}$. This follows from the fact that the triangles in Figure 1.4 are similar.

Another useful way to picture vectors in $\mathbb{R}^2$ is to visualize them as arrows with tails at some point other than the origin. For example, the arrows in Figure 1.5 can *all* be considered to represent the same vector $(-3, 2)$. They all have the same length and the same direction.

In general, when visualizing vectors as arrows with tails at points other than the origin, *two arrows represent the same vector if and only if they have the same length and the same direction*. The arrow whose tail is at (b_1, b_2) and whose head is at (c_1, c_2) represents the vector $\mathbf{a} = (a_1, a_2)$ if and only if $c_1 - b_1 = a_1$ and $c_2 - b_2 = a_2$ (see Figure 1.6).

This way of looking at vectors yields another geometric interpretation of vector addition. Let $\mathbf{a}$ and $\mathbf{b}$ be vectors in $\mathbb{R}^2$. View $\mathbf{a}$ as an arrow whose tail is at the origin and $\mathbf{b}$ as an arrow whose tail is at the head of $\mathbf{a}$. Then $\mathbf{a} + \mathbf{b}$ *is the arrow that completes the triangle* (see Figure 1.7).

Subtraction of vectors has a similar geometric interpretation. Let $\mathbf{a}$ and $\mathbf{b}$ be represented by arrows with their tails at the origin (see Figure 1.8). Then $\mathbf{b} - \mathbf{a}$ is the vector which when added to $\mathbf{a}$ gives $\mathbf{b}$: $\mathbf{a} + (\mathbf{b} - \mathbf{a}) = \mathbf{b}$. Hence $\mathbf{b} - \mathbf{a}$ *is represented by the arrow that goes from the head of* $\mathbf{a}$ *to the head of* $\mathbf{b}$.

We can apply these ideas to describe straight lines in the plane. Suppose $\mathbf{a}$ and $\mathbf{d}$ are vectors in $\mathbb{R}^2$ with $\mathbf{d} \neq \mathbf{0}$. Let us view $\mathbf{a}$ as an arrow with its tail at the origin and $\mathbf{d}$ as an arrow with its tail at the head of $\mathbf{a}$. We will find an

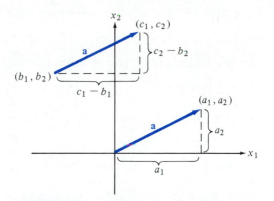

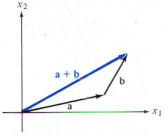

FIGURE 1.6
Two representations of the vector **a**.

FIGURE 1.7
Another interpretation of vector addition in $\mathbb{R}^2$.

equation that describes the line ℓ through the head of **a** in the direction of **d** (see Figure 1.9). A point **x** in the plane lies on the line ℓ if and only if the vector $\mathbf{x} - \mathbf{a}$ is a scalar multiple of **d** (see Figure 1.10). In other words, **x** is on ℓ if and only if $\mathbf{x} - \mathbf{a} = t\mathbf{d}$ for some real number t. Said yet another way, **x** is on ℓ if and only if

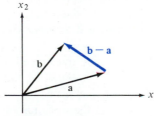

FIGURE 1.8
Vector subtraction in $\mathbb{R}^2$.

$$\mathbf{x} = \mathbf{a} + t\mathbf{d}$$

for some real number t.

This vector equation $\mathbf{x} = \mathbf{a} + t\mathbf{d}$ is a ***parametric representation*** of the line through **a** in the direction of **d**. The vector **d** is a ***direction vector*** for ℓ. The scalar t is the ***parameter***.

EXAMPLE 1

Let ℓ be the line through $(-1, 1)$ in the direction of the vector $(2, 3)$ (see Figure 1.11). A parametric representation of ℓ is given by the equation

$$\mathbf{x} = (-1, 1) + t(2, 3).$$

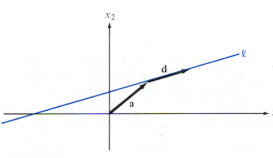

FIGURE 1.9
The line in $\mathbb{R}^2$ through **a** in the direction of **d**.

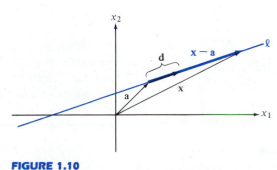

FIGURE 1.10
$\mathbf{x} - \mathbf{a} = t\mathbf{d}$.

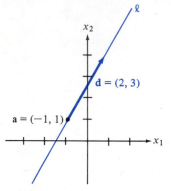

FIGURE 1.11
The line in $\mathbb{R}^2$ through $\mathbf{a} = (-1, 1)$ in the direction of the vector $\mathbf{d} = (2, 3)$.

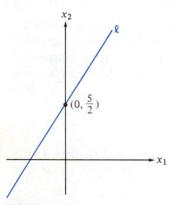

FIGURE 1.12
ℓ is the line with slope $\frac{3}{2}$ and x_2-intercept $\frac{5}{2}$.

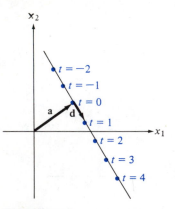

FIGURE 1.13
The vector equation $\mathbf{x} = \mathbf{a} + t\mathbf{d}$ describes the motion of the point $\mathbf{x}$ along a line.

If we write $\mathbf{x} = (x_1, x_2)$, then this vector equation becomes

$$(x_1, x_2) = (-1, 1) + t(2, 3).$$

This is equivalent to the pair of scalar equations

$$x_1 = -1 + 2t$$

and

$$x_2 = 1 + 3t.$$

We can, if we wish, eliminate t from this pair of scalar equations to get the single scalar equation $3x_1 - 2x_2 = -5$, or $x_2 = \frac{3}{2}x_1 + \frac{5}{2}$. We see, then, that ℓ is the line with slope $\frac{3}{2}$ and x_2-intercept $\frac{5}{2}$ (see Figure 1.12). ∎

REMARK   Every line ℓ has many parametric representations. We can describe ℓ by the equation $\mathbf{x} = \mathbf{a} + t\mathbf{d}$ where $\mathbf{a}$ is any point on ℓ and $\mathbf{d}$ is any vector that points in the direction of ℓ. For example, among the many vector equations that describe the line ℓ discussed above are: $\mathbf{x} = (-1, 1) + t(2, 3)$, $\mathbf{x} = (-1, 1) + t(-2, -3)$, $\mathbf{x} = (0, \frac{5}{2}) + t(2, 3)$, and $\mathbf{x} = (0, \frac{5}{2}) + t(-4, -6)$.

Notice that we use the various geometric interpretations of vectors in $\mathbb{R}^2$ interchangeably. The way in which we visualize a vector depends on the situation. For example, in describing a line ℓ by a vector equation $\mathbf{x} = \mathbf{a} + t\mathbf{d}$, we visualize $\mathbf{x}$ and $\mathbf{a}$ as *points* on ℓ and we visualize $\mathbf{d}$ as an *arrow* whose tail is at the point $\mathbf{a}$.

It is often helpful to think of the vector equation of a line dynamically. Think of the parameter t as measuring time. Think of $\mathbf{x} = \mathbf{a} + t\mathbf{d}$ as the position at time t of a point moving in the plane. Then, as t runs from $-\infty$ to $+\infty$, the point $\mathbf{x}$ travels along the line through the point $\mathbf{a}$ in the direction of the vector $\mathbf{d}$ (see Figure 1.13). In this dynamic picture, the vector $\mathbf{d}$ represents the *velocity* of the moving point.

We say that two nonzero vectors are *parallel* if each is a scalar multiple of the other. Notice that two lines in $\mathbb{R}^2$ are parallel if and only if their direction vectors are parallel.

The fact that a line can be described parametrically in many different ways makes the job of finding the point where two nonparallel lines intersect somewhat delicate.

EXAMPLE 2
The lines $\mathbf{x} = (-1, -4) + t(1, 3)$ and $\mathbf{x} = (2, 3) + t(1, 1)$ are not parallel, since their direction vectors $(1, 3)$ and $(1, 1)$ are not parallel. Hence these lines must intersect. But there is no reason to expect that the point of intersection occurs for the same value of the parameter t in each equation. Therefore, to find the point of intersection, we must look for real numbers t_1 and t_2 such that

$$(-1, -4) + t_1(1, 3) = (2, 3) + t_2(1, 1).$$

We can rewrite this equation as the vector equation

$$(t_1 - t_2, 3t_1 - t_2) = (3, 7)$$

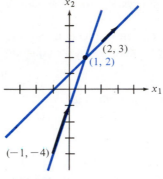

FIGURE 1.14
The lines $\mathbf{x} = (-1, -4) + t(1, 3)$ and $\mathbf{x} = (2, 3) + t(1, 1)$ intersect at the point $(1, 2)$.

or, equivalently, as the pair of scalar equations

$$t_1 - t_2 = 3$$
$$3t_1 - t_2 = 7.$$

The solution of this pair of linear equations can be found by reducing the matrix

$$\begin{pmatrix} 1 & -1 & 3 \\ 3 & -1 & 7 \end{pmatrix} \quad \text{to} \quad \begin{pmatrix} 1 & 0 & 2 \\ 0 & 1 & -1 \end{pmatrix}.$$

We find that $t_1 = 2$ and $t_2 = -1$. Substituting the value $t = t_1 = 2$ into the equation $\mathbf{x} = (-1, -4) + t(1, 3)$ gives

$$\mathbf{x} = (-1, -4) + 2(1, 3) = (1, 2).$$

Substituting the value $t = t_2 = -1$ into the equation $\mathbf{x} = (2, 3) + t(1, 1)$ gives

$$\mathbf{x} = (2, 3) + (-1)(1, 1) = (1, 2).$$

So the point $(1, 2)$ lies on both lines (see Figure 1.14). ■

Associated with each arrow in the plane is a direction and a length. The length of the arrow from $(0, 0)$ to (a_1, a_2) can be computed using the Pythagorean Theorem. It is equal to $\sqrt{a_1^2 + a_2^2}$ (see Figure 1.15). This number is the **length** of the vector $\mathbf{a} = (a_1, a_2)$ and it is denoted by the symbol $\|\mathbf{a}\|$. Thus,

$$\|\mathbf{a}\| = \|(a_1, a_2)\| = \sqrt{a_1^2 + a_2^2}.$$

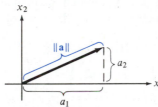

FIGURE 1.15
The length of the vector $\mathbf{a} = (a_1, a_2)$ is $\|\mathbf{a}\| = \sqrt{a_1^2 + a_2^2}$.

If $\mathbf{a} = (a_1, a_2)$ and $\mathbf{b} = (b_1, b_2)$ are any two points in the plane, then the arrow from $\mathbf{a}$ to $\mathbf{b}$ is $\mathbf{b} - \mathbf{a}$. Its length,

$$\|\mathbf{b} - \mathbf{a}\| = \sqrt{(b_1 - a_1)^2 + (b_2 - a_2)^2},$$

is equal to the **distance** from $\mathbf{a}$ to $\mathbf{b}$ (see Figure 1.16).

EXAMPLE 3
If $\mathbf{a} = (3, -4)$ and $\mathbf{b} = (-1, 2)$ then

$$\|\mathbf{a}\| = \sqrt{9 + 16} = \sqrt{25} = 5$$
$$\|\mathbf{b}\| = \sqrt{1 + 4} = \sqrt{5}$$

and the distance from $\mathbf{a}$ to $\mathbf{b}$ is

$$\|\mathbf{b} - \mathbf{a}\| = \sqrt{(-1 - 3)^2 + (2 - (-4))^2} = \sqrt{16 + 36} = \sqrt{52}. \quad ■$$

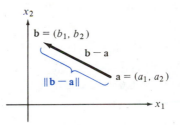

FIGURE 1.16
The distance from $\mathbf{a}$ to $\mathbf{b}$ is equal to $\|\mathbf{b} - \mathbf{a}\|$.

Notice that if $\mathbf{a}$ and $\mathbf{b}$ are vectors in $\mathbb{R}^2$ and t is any real number, then

$$\|t\mathbf{a}\| = |t|\,\|\mathbf{a}\|$$

and

$$\|\mathbf{a} + \mathbf{b}\| \leq \|\mathbf{a}\| + \|\mathbf{b}\|.$$

The inequality $\|\mathbf{a} + \mathbf{b}\| \leq \|\mathbf{a}\| + \|\mathbf{b}\|$ is called the **triangle inequality** because

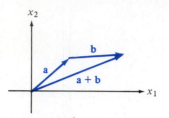

FIGURE 1.17
The triangle inequality states that $\|\mathbf{a} + \mathbf{b}\| \leq \|\mathbf{a}\| + \|\mathbf{b}\|$.

it states that the length of any side of a triangle is always less than the sum of the lengths of the other two sides (see Figure 1.17).

Consider now the line ℓ passing through a given pair of points $\mathbf{a}$ and $\mathbf{b}$ in $\mathbb{R}^2$. Since the vector $\mathbf{b} - \mathbf{a}$ points along this line (see Figure 1.18), the equation

$$\mathbf{x} = \mathbf{a} + t(\mathbf{b} - \mathbf{a})$$

is a parametric representation of ℓ. This equation can be written as

$$\mathbf{x} = (1 - t)\mathbf{a} + t\mathbf{b}.$$

Notice that $\mathbf{x} = \mathbf{a}$ when $t = 0$ and that $\mathbf{x} = \mathbf{b}$ when $t = 1$.

When $0 \leq t \leq 1$, the point $\mathbf{x} = (1 - t)\mathbf{a} + t\mathbf{b}$ lies on the ***line segment*** joining $\mathbf{a}$ to $\mathbf{b}$ (see Figure 1.19). The parameter t measures the ratio $\|\mathbf{x} - \mathbf{a}\|/\|\mathbf{b} - \mathbf{a}\|$ of the distance from $\mathbf{a}$ to $\mathbf{x}$ to the distance from $\mathbf{a}$ to $\mathbf{b}$. This is true because $\mathbf{x} - \mathbf{a} = t(\mathbf{b} - \mathbf{a})$, hence $\|\mathbf{x} - \mathbf{a}\| = |t| \|\mathbf{b} - \mathbf{a}\| = t\|\mathbf{b} - \mathbf{a}\|$, and so $t = \|\mathbf{x} - \mathbf{a}\|/\|\mathbf{b} - \mathbf{a}\|$. When $t = \frac{1}{2}$ we obtain the ***midpoint*** $\frac{1}{2}\mathbf{a} + \frac{1}{2}\mathbf{b}$ of the line segment from $\mathbf{a}$ to $\mathbf{b}$. When $t = \frac{3}{4}$ we obtain the point $\frac{1}{4}\mathbf{a} + \frac{3}{4}\mathbf{b}$ which lies three fourths of the way from $\mathbf{a}$ to $\mathbf{b}$.

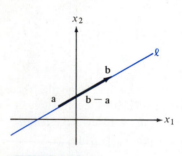

FIGURE 1.18
The line ℓ passing through $\mathbf{a}$ and $\mathbf{b}$ is described by the parametric equation $\mathbf{x} = \mathbf{a} + t(\mathbf{b} - \mathbf{a})$.

EXAMPLE 4

Let $\mathbf{a} = (1, 3)$ and $\mathbf{b} = (3, -2)$.

To find a *vector equation* describing the line ℓ passing through the points $\mathbf{a}$ and $\mathbf{b}$ we compute $\mathbf{b} - \mathbf{a} = (2, -5)$. The equation $\mathbf{x} = \mathbf{a} + t(\mathbf{b} - \mathbf{a})$, or $\mathbf{x} = (1, 3) + t(2, -5)$, is a parametric representation of ℓ.

To find the *midpoint* of the line segment from $\mathbf{a}$ to $\mathbf{b}$, we compute

$$\mathbf{a} + \tfrac{1}{2}(\mathbf{b} - \mathbf{a}) = \tfrac{1}{2}\mathbf{a} + \tfrac{1}{2}\mathbf{b} = (\tfrac{1}{2}, \tfrac{3}{2}) + (\tfrac{3}{2}, -1) = (2, \tfrac{1}{2}).$$

To find *the points that divide the line segment from* $\mathbf{a}$ *to* $\mathbf{b}$ *into three equal parts* we compute

$$\mathbf{a} + \tfrac{1}{3}(\mathbf{b} - \mathbf{a}) = \tfrac{2}{3}\mathbf{a} + \tfrac{1}{3}\mathbf{b} = (\tfrac{2}{3}, 2) + (1, -\tfrac{2}{3}) = (\tfrac{5}{3}, \tfrac{4}{3})$$

and

$$\mathbf{a} + \tfrac{2}{3}(\mathbf{b} - \mathbf{a}) = \tfrac{1}{3}\mathbf{a} + \tfrac{2}{3}\mathbf{b} = (\tfrac{1}{3}, 1) + (2, -\tfrac{4}{3}) = (\tfrac{7}{3}, -\tfrac{1}{3}).$$

The point $\frac{2}{3}\mathbf{a} + \frac{1}{3}\mathbf{b} = (\frac{5}{3}, \frac{4}{3})$ lies one third of the way from $\mathbf{a}$ to $\mathbf{b}$. The point $\frac{1}{3}\mathbf{a} + \frac{2}{3}\mathbf{b} = (\frac{7}{3}, -\frac{1}{3})$ lies two thirds of the way from $\mathbf{a}$ to $\mathbf{b}$. ∎

The *line segment S* from $\mathbf{a}$ to $\mathbf{b}$, where $\mathbf{a}$ and $\mathbf{b}$ are two points in $\mathbb{R}^2$, can be described as the set of all vectors in $\mathbb{R}^2$ of the form $(1 - t)\mathbf{a} + t\mathbf{b}$ where $0 \leq t \leq 1$, that is

$$S = \{(1 - t)\mathbf{a} + t\mathbf{b} \,|\, 0 \leq t \leq 1\}$$

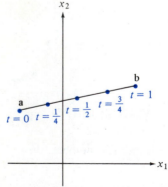

FIGURE 1.19
The point $(1 - t)\mathbf{a} + t\mathbf{b}$ $(0 \leq t \leq 1)$ lies on the line segment from $\mathbf{a}$ to $\mathbf{b}$.

(Figure 1.19). *Parallelograms* in $\mathbb{R}^2$ can be described in a similar way.

Suppose $\mathbf{a}$ and $\mathbf{b}$ are two nonzero vectors in $\mathbb{R}^2$, with $\mathbf{b}$ not a scalar multiple of $\mathbf{a}$. The ***parallelogram spanned by*** $\mathbf{a}$ ***and*** $\mathbf{b}$ is the set

$$P = \{s\mathbf{a} + t\mathbf{b} \,|\, 0 \leq s \leq 1, 0 \leq t \leq 1\}$$

(see Figure 1.20). If $\mathbf{c}$ is any point in $\mathbb{R}^2$, the ***parallelogram with sides*** $\mathbf{a}$ ***and*** $\mathbf{b}$ ***and with initial vertex*** $\mathbf{c}$ is the set

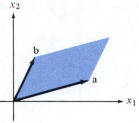

FIGURE 1.20
The parallelogram spanned by **a** and **b**.

$$P = \{\mathbf{c} + s\mathbf{a} + t\mathbf{b} | 0 \le s \le 1, 0 \le t \le 1\}$$

(see Figure 1.21).

EXAMPLE 5

The set

$$\{s(1, 0) + t(0, 2) | 0 \le s \le 1, 0 \le t \le 1\}$$

represents the parallelogram sketched in Figure 1.22a. This parallelogram is a rectangle.

The parallelogram

$$\{s(1, 0) + t(0, 1) | 0 \le s \le 1, 0 \le t \le 1\}$$

is sketched in Figure 1.22b. This parallelogram is a square.

The parallelogram

$$\{(-1, -1) + s(2, 1) + t(1, 2) | 0 \le s \le 1, 0 \le t \le 1\}$$

is sketched in Figure 1.22c. ■

EXERCISES

1. Let $\mathbf{a} = (1, 2)$ and $\mathbf{b} = (1, -3)$. Sketch each of the following vectors as arrows in the plane.

   (a) **a**   (b) **b**
   (c) 2**a**   (d) −**b**
   (e) **a** + **b**   (f) **b** − **a**
   (g) 2**a** + 3**b** (h) **a** − 2**b**

2. Let $\mathbf{a} = (-3, 3)$ and $\mathbf{b} = (2, 1)$. Sketch and find vector equations for each of the following lines. In each case, find the slope and x_2-intercept of the line by eliminating the parameter.

   (a) The line through **a** in the direction of **b**.
   (b) The line through **a** in the direction of 2**b**.
   (c) The line through 2**a** in the direction of **b**.
   (d) The line through **a** and **b**.

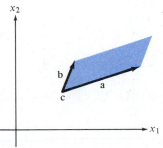

FIGURE 1.21
The parallelogram with sides **a** and **b** and with initial vertex **c**.

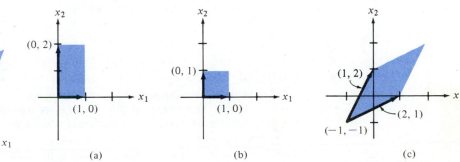

FIGURE 1.22
Parallelograms in $\mathbb{R}^2$. The parallelogram in (a) is a rectangle. The parallelogram in (b) is a square.

3. Let $\mathbf{a} = (4, 1)$ and $\mathbf{b} = (-1, 1)$. Sketch and find vector equations for each of the following lines. In each case find the slope and x_2-intercept of the line by eliminating the parameter.

    (a)   The line through $\mathbf{a}$ in the direction of $\mathbf{b}$.

    (b)   The line through $\mathbf{b}$ in the direction of $\mathbf{a}$.

    (c)   The line through $\mathbf{a}$ and $\mathbf{b}$.   (d)   The line through $-\mathbf{a}$ and $2\mathbf{b}$.

4. Find a vector equation for each of the following lines.
    [*Hint:* First find two points on the line.]

    (a) $2x_1 + x_2 = 1$ (b) $x_1 - x_2 = 2$

    (c) $x_2 = 5x_1 + 2$ (d) $x_2 - 3x_1 = 4$

5. Find the point of intersection of each of the following pairs of lines.

    (a) $\mathbf{x} = (2, 1) + t(-1, 1)$ and $\mathbf{x} = (1, 0) + t(2, 1)$

    (b) $\mathbf{x} = (1, -1) + t(4, 2)$ and $\mathbf{x} = (-1, 0) + t(-1, 3)$

    (c) $\mathbf{x} = (3, 2) + t(1, 1)$ and $\mathbf{x} = (2, 1) + t(3, -2)$

6. Let $\mathbf{a} = (-1, 3)$ and $\mathbf{b} = (4, 9)$. Find

    (a) $\|\mathbf{a}\|$ (b) $\|\mathbf{b}\|$ (c) $\|\mathbf{a} + \mathbf{b}\|$

    (d) $\|\mathbf{a} - \mathbf{b}\|$ (e) $\|\mathbf{a}\| + \|\mathbf{b}\|$

    (f)   The midpoint of the line segment from $\mathbf{a}$ to $\mathbf{b}$.

    (g)   The point on the line segment from $\mathbf{a}$ to $\mathbf{b}$ that lies two thirds of the way from $\mathbf{a}$ to $\mathbf{b}$.

    (h)   The point on the line segment from $\mathbf{a}$ to $\mathbf{b}$ that lies three tenths of the way from $\mathbf{a}$ to $\mathbf{b}$.

7. Let S be the line segment from $(1, 10)$ to $(2, 20)$.

    (a)   Find the midpoint of S.

    (b)   Find the points that divide S into three equal parts.

    (c)   Find the points that divide S into five equal parts.

8. Repeat Exercise 7 for the line segment from $(-1, 5)$ to $(2, -1)$.

9. Sketch the parallelogram spanned by $\mathbf{a}$ and $\mathbf{b}$, where

    (a) $\mathbf{a} = (1, 1), \mathbf{b} = (1, 2)$ (b) $\mathbf{a} = (-1, 1), \mathbf{b} = (1, 1)$

    (c) $\mathbf{a} = (1, 2), \mathbf{b} = (1, -2)$ (d) $\mathbf{a} = (-1, -3), \mathbf{b} = (-2, -5)$

10. Sketch the parallelograms with sides $\mathbf{a}$ and $\mathbf{b}$ and with initial vertex $\mathbf{c} = (4, 5)$, where $\mathbf{a}$ and $\mathbf{b}$ are as in Exercise 9.

11. Sketch each of the following parallelograms.

    (a) $\{s(3, 1) + t(1, 2) | 0 \le s \le 1, 0 \le t \le 1\}$

    (b) $\{(0, 1) + s(-2, -3) + t(-3, 1) | 0 \le s \le 1, 0 \le t \le 1\}$

    (c) $\{(3, 3) + s(1, 4) + t(1, 1) | 0 \le s \le 1, 0 \le t \le 1\}$

    (d) $\{(-4, -2) + s(4, 1) + t(1, -1) | 0 \le s \le 1, 0 \le t \le 1\}$

12. Show that the diagonals of a parallelogram bisect each other. [*Hint:* Let the parallelogram have one vertex at $\mathbf{0}$, choose $\mathbf{a}$ and $\mathbf{b}$ as in Figure 1.23, and compute the midpoints of the diagonals in terms of $\mathbf{a}$ and $\mathbf{b}$.]

13. Show that the medians of a triangle meet at a point that is located, on each median, two thirds of the way from the vertex to the midpoint of the opposite side. [*Hint:* Let $\mathbf{a}'$ be the point that is two thirds of the way along the median from $\mathbf{a}$ to the midpoint of the opposite side (see Figure 1.24). Find a formula expressing $\mathbf{a}'$ in terms of $\mathbf{a}$, $\mathbf{b}$, and $\mathbf{c}$. Then do the same for the corresponding points $\mathbf{b}'$ and $\mathbf{c}'$ on the medians from $\mathbf{b}$ and from $\mathbf{c}$.]

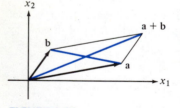

FIGURE 1.23

The diagonals of a parallelogram bisect each other.

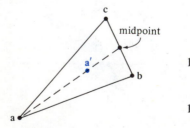

FIGURE 1.24

$\mathbf{a}'$ is two thirds of the way along the line segment joining $\mathbf{a}$ to the midpoint of the opposite side.

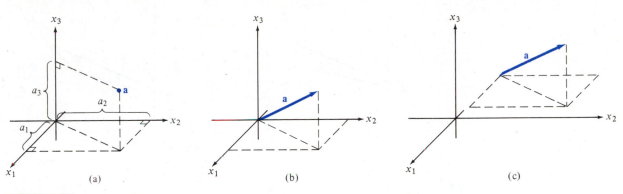

FIGURE 1.25
Three geometric representations of the same vector **a** in $\mathbb{R}^3$.

1.4
The Geometry of $\mathbb{R}^n$

The ideas of the previous section generalize to the space $\mathbb{R}^n$ of ordered n-tuples of real numbers, although the figures are somewhat more difficult to draw when n equals three, and they are impossible to draw when n is greater than three.

Let us first consider the case $n = 3$, where the concepts are easiest to visualize. A vector $\mathbf{a} = (a_1, a_2, a_3)$ in $\mathbb{R}^3$ can be viewed as a point in 3-space (see Figure 1.25a), as an arrow in 3-space with tail at the origin (see Figure 1.25b), or as an arrow in 3-space with tail at any arbitrary point (see Figure 1.25c). An arrow in 3-space with tail at the point (b_1, b_2, b_3) and head at the point (c_1, c_2, c_3) represents the vector $\mathbf{a} = (a_1, a_2, a_3)$ if and only if $c_1 - b_1 = a_1, c_2 - b_2 = a_2,$ and $c_3 - b_3 = a_3$ (see Figure 1.26).

If two vectors **a** and **b** are viewed as arrows with tails at the same point, then their sum $\mathbf{a} + \mathbf{b}$ is represented by the arrow along the diagonal of the parallelogram with sides **a** and **b** (see Figure 1.27a). If **b** is viewed as an arrow

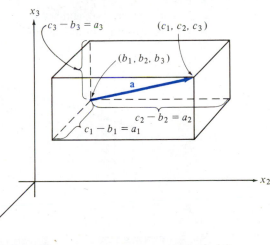

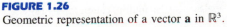

FIGURE 1.26
Geometric representation of a vector **a** in $\mathbb{R}^3$.

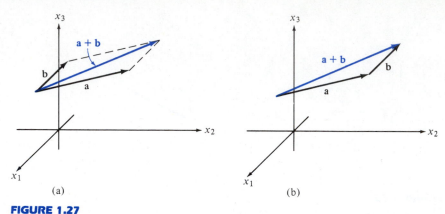

FIGURE 1.27

Vector addition in $\mathbb{R}^3$.

whose tail is at the head of **a**, then **a** + **b** is represented by the arrow from the tail of **a** to the head of **b** (see Figure 1.27b). If c is a scalar greater than zero, then the vector $c\mathbf{a}$ points in the same direction as **a** but is c times as long (see Figure 1.28a). If c is less than zero, then $c\mathbf{a}$ points in the direction opposite to **a** and is $|c|$ times as long as **a** (see Figure 1.28b).

The sum **a** + **b** + **c** of three vectors in $\mathbb{R}^3$ can also be described geometrically. If **a**, **b**, and **c** are viewed as arrows with tails at the same point, then **a** + **b** + **c** is the arrow along the diagonal of the parallelepiped with edges **a**, **b**, and **c** (see Figure 1.29). To see this, move **c** so that its tail is at the head of **a** + **b**.

A vector equation for a line in $\mathbb{R}^3$ can be found in the same way as for a line in $\mathbb{R}^2$. Given vectors **a** and **d** in $\mathbb{R}^3$ with $\mathbf{d} \neq \mathbf{0}$, a point **x** is on the line ℓ through **a** in the direction of **d** if and only if $\mathbf{x} - \mathbf{a} = t\mathbf{d}$, for some real number t (see Figure 1.30). In other words, **x** is on ℓ if and only if

$$\mathbf{x} = \mathbf{a} + t\mathbf{d},$$

for some real number t. The vector equation $\mathbf{x} = \mathbf{a} + t\mathbf{d}$ is a ***parametric representation*** for the line ℓ.

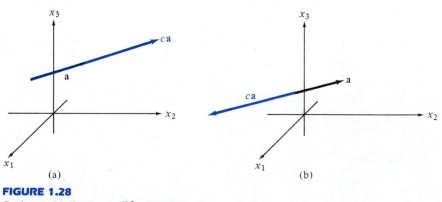

FIGURE 1.28

Scalar multiplication in $\mathbb{R}^3$: (a) $c > 0$, (b) $c < 0$.

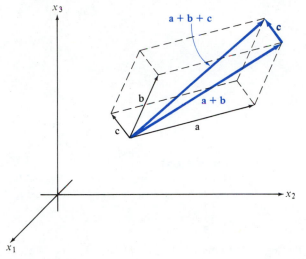

FIGURE 1.29
The sum of three vectors in $\mathbb{R}^3$.

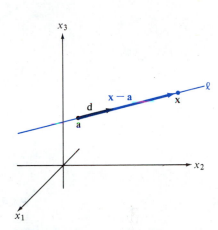

FIGURE 1.30
The line in $\mathbb{R}^3$ through **a** in the direction of **d**: **x** is on ℓ if and only if $\mathbf{x} - \mathbf{a} = t\mathbf{d}$ for some t.

EXAMPLE 1

Let ℓ be the line in $\mathbb{R}^3$ through $(1, 2, 1)$ in the direction of the vector $\mathbf{d} = (2, 1, 3)$. Then

$$\mathbf{x} = (1, 2, 1) + t(2, 1, 3)$$

is a vector equation for ℓ. If $\mathbf{x} = (x_1, x_2, x_3)$ then this vector equation can be rewritten as the system of three scalar equations

$$x_1 = 1 + 2t,$$

$$x_2 = 2 + t,$$

$$x_3 = 1 + 3t.$$

Notice that there is no way to eliminate t from these three equations to obtain a single scalar equation describing ℓ. *In $\mathbb{R}^3$, and in $\mathbb{R}^n$ whenever $n > 2$, more than one scalar equation will always be needed to describe a line.* ■

If **a** and **b** are two points in $\mathbb{R}^3$, the *line through* **a** *and* **b** can be described by the vector equation

$$\mathbf{x} = \mathbf{a} + t(\mathbf{b} - \mathbf{a}),$$

or

$$\mathbf{x} = (1 - t)\mathbf{a} + t\mathbf{b}$$

(see Figure 1.31). Just as in $\mathbb{R}^2$, $\mathbf{x} = \mathbf{a}$ when $t = 0$, $\mathbf{x} = \mathbf{b}$ when $t = 1$, and **x** lies on the *line segment* joining **a** to **b** whenever $0 \le t \le 1$.

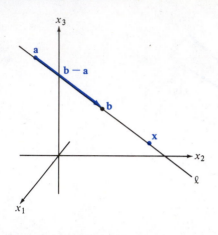

FIGURE 1.31

A point $\mathbf{x}$ is on the line ℓ through $\mathbf{a}$ and $\mathbf{b}$ if and only if $\mathbf{x} = \mathbf{a} + t(\mathbf{b} - \mathbf{a})$ for some t.

EXAMPLE 2

Let ℓ be the line in $\mathbb{R}^3$ through the points $(2, 1, 2)$ and $(3, -1, 1)$. We can take $\mathbf{a} = (2, 1, 2)$ and $\mathbf{b} = (3, -1, 1)$ in the equation

$$\mathbf{x} = \mathbf{a} + t(\mathbf{b} - \mathbf{a})$$

to get a vector equation for ℓ. We find that

$$\mathbf{x} = (2, 1, 2) + t(1, -2, -1).$$

Of course, this is only one of the many vector equations that can be used to describe ℓ. We can find another by taking $\mathbf{a} = (3, -1, 1)$ and $\mathbf{b} = (2, 1, 2)$ to obtain

$$\mathbf{x} = (3, -1, 1) + t(-1, 2, 1). \quad \blacksquare$$

The previous section demonstrated how to use vector methods to find the point of intersection of any pair of nonparallel lines in $\mathbb{R}^2$. In $\mathbb{R}^3$, lines that are not parallel need not intersect, but the same technique can be used to find out if the given lines do intersect and, if they do, to find the point of intersection.

EXAMPLE 3

Let ℓ_1 be the line in $\mathbb{R}^3$ whose equation is

$$\mathbf{x} = (2, 3, 2) + t(1, 4, 1)$$

and let ℓ_2 be the line whose equation is

$$\mathbf{x} = (-1, 2, 2) + t(2, -3, -1).$$

These lines intersect if and only if there are real numbers t_1 and t_2 such that

$$(2, 3, 2) + t_1(1, 4, 1) = (-1, 2, 2) + t_2(2, -3, -1).$$

This equation can be rewritten as the vector equation

$$t_1(1, 4, 1) - t_2(2, -3, -1) = (-3, -1, 0)$$

or as the system of scalar equations

$$t_1 - 2t_2 = -3$$

$$4t_1 + 3t_2 = -1$$

$$t_1 + t_2 = 0.$$

Row reduction of the matrix of this system yields

$$\begin{pmatrix} 1 & -2 & -3 \\ 4 & 3 & -1 \\ 1 & 1 & 0 \end{pmatrix} \longrightarrow \begin{pmatrix} 1 & 0 & -1 \\ 0 & 1 & 1 \\ 0 & 0 & 0 \end{pmatrix}$$

so (t_1, t_2) satisfies the original vector equation if and only if it satisfies the system

$$t_1 \qquad = -1$$

$$t_2 = 1$$

$$0 = 0.$$

Since there is a solution of this system, $(t_1, t_2) = (-1, 1)$, the lines ℓ_1 and ℓ_2 intersect. The point of intersection can be found either by substituting $t_1 = -1$ in the equation of ℓ_1 to get

$$\mathbf{x} = (2, 3, 2) + (-1)(1, 4, 1) = (1, -1, 1)$$

or by substituting $t_2 = 1$ into the equation of ℓ_2 to get

$$\mathbf{x} = (-1, 2, 2) + 1(2, -3, -1) = (1, -1, 1).$$

The point $(1, -1, 1)$ is the point where ℓ_1 and ℓ_2 meet. ■

The *length* $\|\mathbf{a}\|$ of a vector $\mathbf{a} = (a_1, a_2, a_3)$ in $\mathbb{R}^3$ is given by the formula

$$\|\mathbf{a}\| = \sqrt{a_1^2 + a_2^2 + a_3^2}$$

(see Figure 1.32). The *distance* D in $\mathbb{R}^3$ between the points $\mathbf{a} = (a_1, a_2, a_3)$

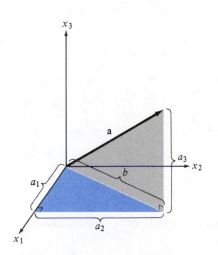

FIGURE 1.32
Using the Pythagorean Theorem twice, note that the length $\|\mathbf{a}\|$ of the vector $\mathbf{a}$ is given by the formula $\|\mathbf{a}\| = \sqrt{b^2 + a_3^2} = \sqrt{a_1^2 + a_2^2 + a_3^2}$.

and $\mathbf{b} = (b_1, b_2, b_3)$ is equal to the length of the arrow that extends from $\mathbf{a}$ to $\mathbf{b}$. It can be calculated from the formula

$$D = \|\mathbf{b} - \mathbf{a}\| = \sqrt{(b_1 - a_1)^2 + (b_2 - a_2)^2 + (b_3 - a_3)^2}.$$

EXAMPLE 4

Let $\mathbf{a}$ and $\mathbf{b}$ be two points in $\mathbb{R}^3$ and let $\mathbf{c} = \frac{1}{3}\mathbf{a} + \frac{2}{3}\mathbf{b}$. Then $\mathbf{c} = (1 - t)\mathbf{a} + t\mathbf{b}$, where $t = \frac{2}{3}$, and hence $\mathbf{c}$ lies on the line segment from $\mathbf{a}$ to $\mathbf{b}$. The distance from $\mathbf{a}$ to $\mathbf{c}$ is given by the formula

$$\|\mathbf{c} - \mathbf{a}\| = \|\tfrac{1}{3}\mathbf{a} + \tfrac{2}{3}\mathbf{b} - \mathbf{a}\| = \|\tfrac{2}{3}\mathbf{b} - \tfrac{2}{3}\mathbf{a}\| = \tfrac{2}{3}\|\mathbf{b} - \mathbf{a}\|.$$

Therefore the distance from $\mathbf{a}$ to $\mathbf{c}$ is equal to two thirds of the distance from $\mathbf{a}$ to $\mathbf{b}$. In other words, *the point* $\mathbf{c} = \frac{1}{3}\mathbf{a} + \frac{2}{3}\mathbf{b}$ *lies on the line segment from* $\mathbf{a}$ *to* $\mathbf{b}$ *and is located two thirds of the way from* $\mathbf{a}$ *to* $\mathbf{b}$. ■

Although we cannot draw pictures in $\mathbb{R}^n$ when $n > 3$, we can still describe lines, lengths of vectors, and distances between points in exactly the same way we did in $\mathbb{R}^2$ and in $\mathbb{R}^3$.

Given the vectors $\mathbf{a}$ and $\mathbf{d}$ in $\mathbb{R}^n$ with $\mathbf{d} \neq \mathbf{0}$, the **line through $\mathbf{a}$ in the direction of** $\mathbf{d}$ is the set of all $\mathbf{x}$ in $\mathbb{R}^n$ of the form

$$\mathbf{x} = \mathbf{a} + t\mathbf{d}$$

where t is any real number. If $\mathbf{a}$ and $\mathbf{b}$ are two points in $\mathbb{R}^n$, the **line through $\mathbf{a}$ and $\mathbf{b}$** is the set of all $\mathbf{x}$ in $\mathbb{R}^n$ of the form

$$\mathbf{x} = \mathbf{a} + t(\mathbf{b} - \mathbf{a}).$$

If $0 \leq t \leq 1$ then the point

$$\mathbf{x} = \mathbf{a} + t(\mathbf{b} - \mathbf{a}) = (1 - t)\mathbf{a} + t\mathbf{b}$$

lies on the **line segment** from $\mathbf{a}$ to $\mathbf{b}$. The point $\mathbf{m} = \frac{1}{2}\mathbf{a} + \frac{1}{2}\mathbf{b}$ is the **midpoint** of the line segment from $\mathbf{a}$ to $\mathbf{b}$. The point $\frac{2}{5}\mathbf{a} + \frac{3}{5}\mathbf{b}$ lies on this segment and is three fifths of the way from $\mathbf{a}$ to $\mathbf{b}$.

The **length** $\|\mathbf{a}\|$ of a vector $\mathbf{a} = (a_1, a_2, \ldots, a_n)$ in $\mathbb{R}^n$ is given by the formula

$$\|\mathbf{a}\| = \sqrt{a_1^2 + a_2^2 + \cdots + a_n^2}.$$

The **distance** D between points $\mathbf{a}$ and $\mathbf{b}$ is given by the formula

$$D = \|\mathbf{b} - \mathbf{a}\|.$$

Notice that, if

$$\mathbf{c} = (1 - t)\mathbf{a} + t\mathbf{b} \qquad (0 \leq t \leq 1)$$

is on the line segment from $\mathbf{a}$ to $\mathbf{b}$, then

$$\|\mathbf{c} - \mathbf{a}\| = \|t(\mathbf{b} - \mathbf{a})\| = |t|\,\|\mathbf{b} - \mathbf{a}\|$$

so *t measures the ratio of the distance between* $\mathbf{a}$ *and* $\mathbf{c}$ *to the distance between* $\mathbf{a}$ *and* $\mathbf{b}$.

EXAMPLE 5

Let $\mathbf{a} = (1, 1, 0, 1)$ and $\mathbf{b} = (1, -1, 2, 4)$. Then $\mathbf{a}$ and $\mathbf{b}$ are vectors in $\mathbb{R}^4$. The distance from $\mathbf{a}$ to $\mathbf{b}$ is

$$D = \|\mathbf{b} - \mathbf{a}\| = \|(0, -2, 2, 3)\| = \sqrt{0^2 + (-2)^2 + 2^2 + 3^2} = \sqrt{17}.$$

The line through $\mathbf{a}$ and $\mathbf{b}$ is described by the equation

$$\mathbf{x} = \mathbf{a} + t(\mathbf{b} - \mathbf{a})$$

or

$$\mathbf{x} = (1, 1, 0, 1) + t(0, -2, 2, 3).$$

The midpoint of the line segment from $\mathbf{a}$ to $\mathbf{b}$ is

$$\mathbf{m} = \tfrac{1}{2}\mathbf{a} + \tfrac{1}{2}\mathbf{b} = (1, 0, 1, \tfrac{5}{2}).$$

The points $\mathbf{x}_1$ and $\mathbf{x}_2$ that divide the line segment from $\mathbf{a}$ to $\mathbf{b}$ into three equal parts are

$$\mathbf{x}_1 = \tfrac{2}{3}\mathbf{a} + \tfrac{1}{3}\mathbf{b} = (1, \tfrac{1}{3}, \tfrac{2}{3}, 2)$$

and

$$\mathbf{x}_2 = \tfrac{1}{3}\mathbf{a} + \tfrac{2}{3}\mathbf{b} = (1, -\tfrac{1}{3}, \tfrac{4}{3}, 3). \quad \blacksquare$$

EXAMPLE 6

Let us determine if the lines

$$\mathbf{x} = (1, -1, 1, 0) + t(2, 3, 2, -1) \text{ and } \mathbf{x} = (1, 0, 1, 2) + t(-1, 3, 4, 2)$$

in $\mathbb{R}^4$ intersect. We seek t_1 and t_2 such that

$$(1, -1, 1, 0) + t_1(2, 3, 2, -1) = (1, 0, 1, 2) + t_2(-1, 3, 4, 2),$$

that is, such that

$$t_1(2, 3, 2, -1) - t_2(-1, 3, 4, 2) = (0, 1, 0, 2).$$

Therefore, we want to solve the linear system

$$2t_1 + t_2 = 0$$
$$3t_1 - 3t_2 = 1$$
$$2t_1 - 4t_2 = 0$$
$$-t_1 - 2t_2 = 2.$$

The matrix of this system reduces as follows:

$$\begin{pmatrix} 2 & 1 & 0 \\ 3 & -3 & 1 \\ 2 & -4 & 0 \\ -1 & -2 & 2 \end{pmatrix} \longrightarrow \begin{pmatrix} 1 & 0 & 0 \\ 0 & 1 & 0 \\ 0 & 0 & 1 \\ 0 & 0 & 0 \end{pmatrix}.$$

There is a corner 1 in the last column of the reduced matrix, so no solution exists, hence the lines do not intersect. ■

Two lines in $\mathbb{R}^n$ are **parallel** if their direction vectors are parallel, that is, if the direction vectors are scalar multiples of one another. Thus the line $\mathbf{x} = \mathbf{a} + t\mathbf{d}$ is parallel to the line $\mathbf{x} = \mathbf{b} + t\mathbf{e}$ if and only if $\mathbf{e} = c\mathbf{d}$ for some $c \neq 0$.

EXAMPLE 7

The lines

$$\mathbf{x} = (2, 1, 3) + t(-1, 1, 1) \quad \text{and} \quad \mathbf{x} = (-1, 2, 1) + t(2, -2, -2)$$

are parallel because $(2, -2, -2) = (-2)(-1, 1, 1)$. The lines

$$\mathbf{x} = (2, 3, 2) + t(1, 4, 1) \quad \text{and} \quad \mathbf{x} = (-1, 2, 2) + t(2, -3, -1)$$

are not parallel because there is no scalar c for which $(2, -3, -1) = c(1, 4, 1)$.

Notice that parallel lines may coincide. For example, consider the lines ℓ_1 and ℓ_2 with equations

$$\mathbf{x} = (-1, 1, 2) + t(1, 3, 1) \text{ and } \mathbf{x} = (2, 10, 5) + t(2, 6, 2).$$

These lines are parallel because the direction vector $(1, 3, 1)$ for ℓ_1 is a scalar multiple of the direction vector $(2, 6, 2)$ for ℓ_2: $(1, 3, 1) = \frac{1}{2}(2, 6, 2)$. Therefore, ℓ_1 and ℓ_2 coincide if and only if they have a point in common. The point $(-1, 1, 2)$ is on ℓ_1. Is it also on ℓ_2? It is if there is a real number t such that

$$(-1, 1, 2) = (2, 10, 5) + t(2, 6, 2).$$

This number t must satisfy the three scalar equations

$$-1 = 2 + 2t$$

$$1 = 10 + 6t$$

$$2 = 5 + 2t.$$

Solving these equations yields the solution $t = -\frac{3}{2}$. Hence the point $(-1, 1, 2) = (2, 10, 5) - \frac{3}{2}(2, 6, 2)$ is also on ℓ_2. It follows that ℓ_1 and ℓ_2 are the same line. ■

EXERCISES

1. Sketch the following vectors as arrows in $\mathbb{R}^3$.
    (a) (1, 0, 0)   (b) (0, 1, 0)   (c) (0, 0, 1)
    (d) (1, 1, 0)   (e) (1, 0, 1)   (f) (0, 1, 1)
    (g) (-1, 1, 0)   (h) (1, 1, 2)   (i) (1, 1, -1)

2. Find a vector equation for each of the following lines in $\mathbb{R}^3$:
    (a) the line through (1, 2, 1) in the direction of $(-3, 1, 4)$
    (b) the line through $(-1, 1, -1)$ in the direction of (2, 1, 2)
    (c) the line through $(1, -2, 3)$ in the direction of $(-4, 1, 1)$

    (d)   the line through the points $(1, 2, 1)$ and $(-1, 1, -1)$

    (e)   the line through the points $(1, -2, 3)$ and $(-4, 1, -1)$

    (f)   the line through the points $(2, -3, 2)$ and $(2, 3, 2)$

3.   Let ℓ be the line $\mathbf{x} = (1, -1, 1) + t(1, 0, 1)$. Which of the following points are on ℓ?

    (a) $(1, -1, 1)$ (b) $(1, 0, 1)$

    (c) $(3, -1, 1)$ (d) $(-1, -1, -1)$

4.   For each of the following pairs of lines in $\mathbb{R}^3$, determine if they intersect. If they intersect, find the point (or points) of intersection.

    (a) $\mathbf{x} = (-2, -3, 0) + t(5, 3, 1), \mathbf{x} = (5, -6, 8) + t(-1, -3, 2)$

    (b) $\mathbf{x} = (1, 0, 0) + t(0, 0, 1), \mathbf{x} = t(0, 1, 0)$

    (c) $\mathbf{x} = (1, 1, 1) + t(-1, 0, 1), \mathbf{x} = (0, 1, -1) + t(2, 0, -2)$

    (d) $\mathbf{x} = (1, 1, 1) + t(-1, 0, 1), \mathbf{x} = (0, 1, 2) + t(-1, 0, 1)$

    (e) $\mathbf{x} = (3, 4, -2) + t(0, 2, -1), \mathbf{x} = (5, 2, -3) + t(1, 0, -1)$

5.   Find the length of each of the following vectors.

    (a) $(1, -1, 1)$ (b) $(3, -2, 5)$

    (c) $(1, 0, 1)$ (d) $(2, 1, 2)$

    (e) $(\frac{1}{3}, \frac{2}{3}, \frac{2}{3})$

6.   Find the distance between each of the following pairs of points in $\mathbb{R}^3$.

    (a) $(1, 2, 4)$ and $(-1, 1, 2)$ (b) $(0, 3, 2)$ and $(7, -1, 6)$

    (c) $(0, 0, 0)$ and $(5, 4, 3)$ (d) $(-1, -1, 2)$ and $(3, -1, 2)$

    (e) $(1, 3, 3)$ and $(1, 5, 3)$

7.   Let $\mathbf{a} = (-1, 3, 7)$ and $\mathbf{b} = (2, 1, 3)$.

    (a) Find the distance between $\mathbf{a}$ and $\mathbf{b}$.

    (b) Find an equation describing the line through $\mathbf{a}$ and $\mathbf{b}$.

    (c) Find the midpoint of the line segment from $\mathbf{a}$ to $\mathbf{b}$.

    (d) Find the points that divide the line segment from $\mathbf{a}$ to $\mathbf{b}$ into three equal parts.

8.   Find a vector equation for each of the following lines:

    (a) the line in $\mathbb{R}^4$ through $(1, 1, -1, 0)$ in the direction of $(1, 2, 3, 4)$

    (b) the line in $\mathbb{R}^6$ through $(1, 0, 1, 0, 1, 0)$ in the direction of $(-2, -1, 0, 1, 2, 3)$

    (c) the line in $\mathbb{R}^4$ through the points $(1, 1, -1, 0)$ and $(1, 2, 3, 4)$

    (d) the line in $\mathbb{R}^5$ through the points $(1, 2, 3, 4, 5)$ and $(5, 4, 3, 2, 1)$

    (e) the line in $\mathbb{R}^n$ through the points $(1, 2, 3, \ldots, n)$ and $(0, 1, 2, \ldots, n - 1)$

9.   Let $\mathbf{a} = (-1, 3, 0, 2)$, $\mathbf{b} = (2, 3, 7, 4)$, and $\mathbf{c} = (0, 1, -1, 2)$. Find

    (a) $\|\mathbf{a}\|$ (b) $\|\mathbf{b}\|$

    (c) $\|\mathbf{c}\|$ (d) the distance from $\mathbf{a}$ to $\mathbf{b}$

    (e) the distance from $\mathbf{b}$ to $\mathbf{c}$ (f) the distance from $\mathbf{a}$ to $\mathbf{c}$

10.   Find the midpoint of the line segment from $\mathbf{a}$ to $\mathbf{b}$ where

    (a) $\mathbf{a} = (1, 3, 2)$ and $\mathbf{b} = (-1, -3, -2)$

    (b) $\mathbf{a} = (-1, 0, 4)$ and $\mathbf{b} = (1, 1, 1)$

    (c) $\mathbf{a} = (4, 3, 2, 1)$ and $\mathbf{b} = (6, -1, -2, 7)$

    (d) $\mathbf{a} = (1, 2, 3, \ldots, n)$ and $\mathbf{b} = (n, n - 1, \ldots, 1)$

11. Let $\mathbf{a} = (-1, 2, -1)$ and $\mathbf{b} = (2, -7, 8)$.

(a) Find the points that divide the line segment from $\mathbf{a}$ to $\mathbf{b}$ into three equal parts.

(b) Find the points that divide the line segment from $\mathbf{a}$ to $\mathbf{b}$ into four equal parts.

(c) Find the point that lies one fifth of the way from $\mathbf{a}$ to $\mathbf{b}$.

(d) Find the point that lies three fifths of the way from $\mathbf{a}$ to $\mathbf{b}$.

12. Find a vector equation for the line connecting the point $(1, -1, 0)$ to the midpoint of the segment from $(-2, 2, -2)$ to $(3, -1, 4)$.

13. Find a vector equation for the line connecting the point $(-2, 1, 3)$ to the midpoint of the segment from $(-1, 3, 7)$ to $(-3, 3, 3)$

14. Find vector equations for the medians of the triangle whose vertices are $(1, 0, 0)$, $(0, 1, 0)$, and $(0, 0, 1)$.

15. For each of the following triangles, find the point where the medians intersect.

(a) The triangle with vertices $(1, 2, 1)$, $(3, 6, 2)$, and $(-1, 0, 2)$

(b) The triangle with vertices $(-1, 0, 2)$, $(3, 1, 3)$, and $(5, 5, 1)$

(c) The triangle with vertices $(1, 1, 1)$, $(1, -1, 1)$, and $(9, 7, 15)$

16. Determine which of the following pairs of lines are parallel.

(a) $\mathbf{x} = (-1, 1) + t(-3, 4)$, $\mathbf{x} = (1, 0) + t(4, -3)$

(b) $\mathbf{x} = (-1, 1) + t(-3, 4)$, $\mathbf{x} = (1, 0) + t(3, -4)$

(c) $\mathbf{x} = (1, 0, 5) + t(1, 2, -1)$, $\mathbf{x} = (5, 5, 6) + t(-2, -4, 2)$

(d) $\mathbf{x} = (1, 1, 2, 0) + t(-1, 3, 0, 6)$, $\mathbf{x} = (3, 7, -4, 0) + t(-\frac{1}{6}, \frac{1}{2}, 0, 1)$

(e) $\mathbf{x} = (3, 2, 1, 0) + t(0, 1, 2, 3)$, $\mathbf{x} = (0, 1, 2, 3) + t(3, 2, 1, 0)$

17. Find a vector equation for the line in $\mathbb{R}^3$ passing through $(1, 3, 5)$ and parallel to the line $\mathbf{x} = (1, -3, 2) + t(-1, 1, 1)$.

18. Let ℓ be the line in $\mathbb{R}^3$ that passes through the points $(1, 2, 3)$ and $(2, -3, 1)$. Find a vector equation for the line through the origin parallel to ℓ.

19. Two particles are moving in 3-space according to the following equations:

first particle: $\mathbf{x}(t) = (-2, -3, 0) + t(5, 3, 1)$

second particle: $\mathbf{x}(t) = (5, -6, 8) + t(-1, -3, 2)$.

Here, t denotes time and $\mathbf{x}(t)$ denotes the position of the particle at time t. Do these particles collide? If so, when?

20. Show that the solution set of a system of n linear equations in two unknowns, viewed as a subset of $\mathbb{R}^2$, is one of the following: (i) the empty set, (ii) a single point, (iii) a line, or (iv) the whole plane.

21. Show that, if $\mathbf{a}$ is any vector in $\mathbb{R}^n$ and t is any real number, then $\|t\mathbf{a}\| = |t|\,\|\mathbf{a}\|$.

22. Let $\mathbf{a} = (a_1, \ldots, a_n)$, let $\mathbf{d} = (d_1, \ldots, d_n)$, $\mathbf{d} \neq \mathbf{0}$, and let ℓ be the line $\mathbf{x} = \mathbf{a} + t\mathbf{d}$. Show that, if $d_i \neq 0$ for all i ($1 \leq i \leq n$) then ℓ is the solution set of the system of $n - 1$ equations

$$\frac{x_1 - a_1}{d_1} = \frac{x_2 - a_2}{d_2} = \cdots = \frac{x_n - a_n}{d_n}.$$

These equations are called the **symmetric equations** for the line ℓ.

1.5
The Dot Product

An important problem in the geometry of $\mathbb{R}^n$ is that of finding the angle between two vectors. This section begins with a discussion of this problem in $\mathbb{R}^2$.

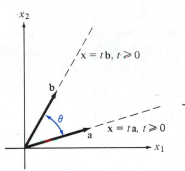

FIGURE 1.33
The angle between two vectors.

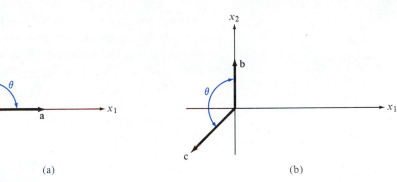

(a)   (b)

FIGURE 1.34
The angle θ in (a) is $\pi/2$ and in (b) is $3\pi/4$.

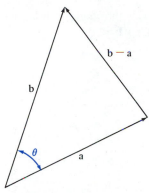

FIGURE 1.35
The angle can be found by using the law of cosines.

Let **a** and **b** be nonzero vectors in $\mathbb{R}^2$. The *angle between* **a** *and* **b** is defined to be the angle θ, $0 \le \theta \le \pi$, formed by the half-lines $\mathbf{x} = t\mathbf{a}$ ($t \ge 0$) and $\mathbf{x} = t\mathbf{b}$ ($t \ge 0$) (see Figure 1.33). Since θ is required to be between 0 and π, the angle between the vectors **a** and **b** is uniquely determined.

EXAMPLE 1
The angle between the vectors $\mathbf{a} = (1, 0)$ and $\mathbf{b} = (0, 1)$ is $\pi/2$ (see Figure 1.34a). The angle between the vectors $\mathbf{b} = (0, 1)$ and $\mathbf{c} = (-1, -1)$ is $3\pi/4$ (see Figure 1.34b). ■

We can find a formula for the angle between two vectors by applying the law of cosines to the triangle with sides **a**, **b**, and $\mathbf{b} - \mathbf{a}$ (see Figure 1.35). The law of cosines states that $C^2 = A^2 + B^2 - 2AB \cos \theta$, where A, B, and C are the lengths of the sides of a triangle and where θ is the angle opposite the side of length C. If $A = \|\mathbf{a}\|$, $B = \|\mathbf{b}\|$, and $C = \|\mathbf{b} - \mathbf{a}\|$, then this formula becomes

$$\|\mathbf{b} - \mathbf{a}\|^2 = \|\mathbf{a}\|^2 + \|\mathbf{b}\|^2 - 2\|\mathbf{a}\|\,\|\mathbf{b}\| \cos \theta$$

where θ is the angle between **a** and **b**. If $\mathbf{a} = (a_1, a_2)$ and $\mathbf{b} = (b_1, b_2)$, then

$$\|\mathbf{a}\|^2 = a_1^2 + a_2^2, \quad \|\mathbf{b}\|^2 = b_1^2 + b_2^2,$$

and

$$\|\mathbf{b} - \mathbf{a}\|^2 = \|(b_1 - a_1, b_2 - a_2)\|^2 = (b_1 - a_1)^2 + (b_2 - a_2)^2$$

so

$$(b_1 - a_1)^2 + (b_2 - a_2)^2 = a_1^2 + a_2^2 + b_1^2 + b_2^2 - 2\|\mathbf{a}\|\,\|\mathbf{b}\| \cos \theta.$$

Expanding the left side and combining terms yields

$$-2a_1 b_1 - 2a_2 b_2 = -2\|\mathbf{a}\|\,\|\mathbf{b}\| \cos \theta,$$

or

$$\cos \theta = \frac{a_1 b_1 + a_2 b_2}{\|\mathbf{a}\|\,\|\mathbf{b}\|}.$$

The expression $a_1b_1 + a_2b_2$, which appears in the numerator of this formula, occurs frequently enough in vector calculations to justify giving it a name of its own. It is called the **dot product** of $\mathbf{a} = (a_1, a_2)$ and $\mathbf{b} = (b_1, b_2)$ and we write

$$\mathbf{a} \cdot \mathbf{b} = a_1b_1 + a_2b_2.$$

The formula for the angle θ between nonzero vectors $\mathbf{a}$ and $\mathbf{b}$ in $\mathbb{R}^2$ can then be written as

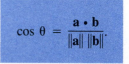

$$\cos \theta = \frac{\mathbf{a} \cdot \mathbf{b}}{\|\mathbf{a}\| \, \|\mathbf{b}\|}.$$

Notice that $\cos \theta$ has the same sign as $\mathbf{a} \cdot \mathbf{b}$. Hence the angle θ is **acute** ($\theta < \pi/2$) if $\mathbf{a} \cdot \mathbf{b} > 0$ and it is **obtuse** ($\theta > \pi/2$) if $\mathbf{a} \cdot \mathbf{b} < 0$.

EXAMPLE 2

To find the angle θ between the vectors $\mathbf{a} = (\sqrt{3}, 1)$ and $\mathbf{b} = (-\sqrt{3}, 1)$, compute

$$\|\mathbf{a}\| = \sqrt{(\sqrt{3})^2 + 1^2} = 2$$

$$\|\mathbf{b}\| = \sqrt{(-\sqrt{3})^2 + 1^2} = 2$$

and

$$\mathbf{a} \cdot \mathbf{b} = (\sqrt{3})(-\sqrt{3}) + (1)(1) = -3 + 1 = -2.$$

Then

$$\cos \theta = \frac{\mathbf{a} \cdot \mathbf{b}}{\|\mathbf{a}\| \, \|\mathbf{b}\|} = \frac{-2}{(2)(2)} = -\frac{1}{2}.$$

This angle θ is obtuse. In fact,

$$\theta = \frac{2\pi}{3}. \quad \blacksquare$$

EXAMPLE 3

To find the angle θ between the vectors $\mathbf{a} = (-1, 3)$ and $\mathbf{b} = (2, 1)$, compute

$$\|\mathbf{a}\| = \sqrt{(-1)^2 + 3^2} = \sqrt{10}$$

$$\|\mathbf{b}\| = \sqrt{2^2 + 1^2} = \sqrt{5}$$

and

$$\mathbf{a} \cdot \mathbf{b} = (-1)(2) + (3)(1) = 1.$$

Then

$$\cos \theta = \frac{1}{\sqrt{10}\sqrt{5}} = \frac{1}{5\sqrt{2}} \approx 0.14.$$

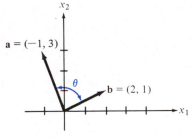

FIGURE 1.36
The angle θ is approximately 1.43 radians.

This angle θ is an acute angle. If we need to know an approximate value for θ, we can use a calculator or a table to find that

$$\theta \approx 1.43 \text{ radians,}$$

which seems reasonable if we sketch the vectors (see Figure 1.36) and recall that $\pi/2 \approx 1.57$. ■

If **a** and **b** are nonzero vectors in $\mathbb{R}^2$ with $\mathbf{a} \cdot \mathbf{b} = 0$ then $\cos \theta = 0$ and hence $\theta = \pi/2$. Conversely, if $\theta = \pi/2$, then $\cos \theta = 0$ and therefore $\mathbf{a} \cdot \mathbf{b} = \|\mathbf{a}\| \, \|\mathbf{b}\| \cos \theta = 0$. This shows that *two nonzero vectors in $\mathbb{R}^2$ are perpendicular if and only if their dot product is zero.*

EXAMPLE 4

To see if two vectors in $\mathbb{R}^2$ are perpendicular, compute their dot product. If $\mathbf{a} = (1, 2)$ and $\mathbf{b} = (-2, 1)$ then $\mathbf{a} \cdot \mathbf{b} = 0$, so **a** and **b** are perpendicular. If $\mathbf{c} = (1, 3)$ and $\mathbf{d} = (3, 1)$, then $\mathbf{c} \cdot \mathbf{d} \neq 0$, so **c** and **d** are not perpendicular. ■

The dot product has many useful properties. Among them are the following. If **a**, **b**, and **c** are any vectors in $\mathbb{R}^2$, then $(\mathbf{a} + \mathbf{b}) \cdot \mathbf{c} = \mathbf{a} \cdot \mathbf{c} + \mathbf{b} \cdot \mathbf{c}$, $\mathbf{a} \cdot (\mathbf{b} + \mathbf{c}) = \mathbf{a} \cdot \mathbf{b} + \mathbf{a} \cdot \mathbf{c}$, $\mathbf{a} \cdot \mathbf{b} = \mathbf{b} \cdot \mathbf{a}$, and $\mathbf{a} \cdot \mathbf{a} = \|\mathbf{a}\|^2$. These properties of the dot product are immediately apparent from the definition. With the help of these properties we can use the dot product to prove theorems in geometry.

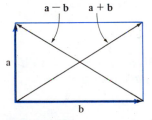

FIGURE 1.37
Vectors can be used to study the geometry of a rectangle.

EXAMPLE 5

You learned in plane geometry that the diagonals of a rectangle are perpendicular if and only if the rectangle is a square. This can be verified using vector methods.

Let **a** and **b** be vectors along adjacent sides of the rectangle. Then the vectors $\mathbf{a} + \mathbf{b}$ and $\mathbf{a} - \mathbf{b}$ are along the diagonals (see Figure 1.37). Since

$$(\mathbf{a} + \mathbf{b}) \cdot (\mathbf{a} - \mathbf{b}) = \mathbf{a} \cdot (\mathbf{a} - \mathbf{b}) + \mathbf{b} \cdot (\mathbf{a} - \mathbf{b})$$

$$= \mathbf{a} \cdot \mathbf{a} - \mathbf{a} \cdot \mathbf{b} + \mathbf{b} \cdot \mathbf{a} - \mathbf{b} \cdot \mathbf{b}$$

$$= \|\mathbf{a}\|^2 - \|\mathbf{b}\|^2$$

we see that the diagonals are perpendicular (that is, $(\mathbf{a} + \mathbf{b}) \cdot (\mathbf{a} - \mathbf{b}) = 0$) if and only if $\|\mathbf{a}\|^2 = \|\mathbf{b}\|^2$. But the condition $\|\mathbf{a}\|^2 = \|\mathbf{b}\|^2$ holds if and only if adjacent sides of the rectangle have equal length, that is, if and only if the rectangle is a square. ■

The dot product idea extends easily to $\mathbb{R}^n$. If $\mathbf{a} = (a_1, a_2, \ldots, a_n)$ and $\mathbf{b} = (b_1, b_2, \ldots, b_n)$ are two vectors in $\mathbb{R}^n$, we define the *dot product* of **a** and **b** to be the real number

$$\mathbf{a} \cdot \mathbf{b} = a_1 b_1 + a_2 b_2 + \cdots + a_n b_n.$$

EXAMPLE 6

The dot product of the vectors $\mathbf{a} = (1, 2, -1)$ and $\mathbf{b} = (3, -1, 2)$ in $\mathbb{R}^3$ is

$$\mathbf{a} \cdot \mathbf{b} = (1, 2, -1) \cdot (3, -1, 2) = (1)(3) + (2)(-1) + (-1)(2) = -1.$$

The dot product of the vectors $\mathbf{c} = (1, -1, 0, 2)$ and $\mathbf{d} = (2, -1, 1, -3)$ in $\mathbb{R}^4$ is

$$\mathbf{c} \cdot \mathbf{d} = (1, -1, 0, 2) \cdot (2, -1, 1, -3)$$
$$= (1)(2) + (-1)(-1) + (0)(1) + 2(-3)$$
$$= -3. \quad \blacksquare$$

The dot product has the following useful properties. For any vectors $\mathbf{a}$, $\mathbf{b}$, and $\mathbf{c}$ in $\mathbb{R}^n$ and any real number t,

(D$_1$) $\mathbf{a} \cdot \mathbf{b} = \mathbf{b} \cdot \mathbf{a}$

(D$_2$) $\mathbf{a} \cdot (\mathbf{b} + \mathbf{c}) = \mathbf{a} \cdot \mathbf{b} + \mathbf{a} \cdot \mathbf{c}$

(D$_3$) $(t\mathbf{a}) \cdot \mathbf{b} = t(\mathbf{a} \cdot \mathbf{b})$

(D$_4$) $\mathbf{a} \cdot \mathbf{a} = \|\mathbf{a}\|^2 \geq 0$, and $\mathbf{a} \cdot \mathbf{a} = 0$ if and only if $\mathbf{a} = \mathbf{0}$.

These properties are easy to verify (see Exercise 10).

The dot product can be used to define the angle between two vectors in $\mathbb{R}^n$. Let $\mathbf{a}$ and $\mathbf{b}$ be nonzero vectors in $\mathbb{R}^n$. The **angle between a and b** is the unique θ with $0 \leq \theta \leq \pi$ that satisfies the equation

$$\cos \theta = \frac{\mathbf{a} \cdot \mathbf{b}}{\|\mathbf{a}\| \, \|\mathbf{b}\|}.$$

The angle θ is **acute** ($\theta < \pi/2$) if $\mathbf{a} \cdot \mathbf{b} > 0$. It is **obtuse** ($\theta > \pi/2$) if $\mathbf{a} \cdot \mathbf{b} < 0$. It is a **right angle** (that is, $\mathbf{a}$ and $\mathbf{b}$ are **perpendicular**) if $\mathbf{a} \cdot \mathbf{b} = 0$.

EXAMPLE 7

To find the angle θ between the vectors $\mathbf{a} = (-1, 2, -3)$ and $\mathbf{b} = (1, 1, 1)$ in $\mathbb{R}^3$, compute

$$\|\mathbf{a}\| = \sqrt{(-1)^2 + (2)^2 + (-3)^2} = \sqrt{14}, \|\mathbf{b}\| = \sqrt{1^2 + 1^2 + 1^2} = \sqrt{3}$$

and

$$\mathbf{a} \cdot \mathbf{b} = (-1)(1) + (2)(1) + (-3)(1) = -2.$$

Then

$$\cos \theta = \frac{-2}{\sqrt{14}\sqrt{3}}.$$

This angle θ is obtuse, since $\mathbf{a} \cdot \mathbf{b} < 0$. A short computation with a calculator shows that $\theta \approx 1.88$ radians. $\quad \blacksquare$

REMARK 1   The angle between nonzero vectors $\mathbf{a}$ and $\mathbf{b}$ in $\mathbb{R}^n$ has been defined as the number θ that satisfies $\cos \theta = \mathbf{a} \cdot \mathbf{b}/\|\mathbf{a}\| \, \|\mathbf{b}\|$. But how do we know that there is always such a θ? Since $-1 \leq \cos \theta \leq 1$, there will be such a θ if and only if $-1 \leq \mathbf{a} \cdot \mathbf{b}/\|\mathbf{a}\| \, \|\mathbf{b}\| \leq 1$. It turns out that there is a famous inequality, called the *Cauchy-Schwarz inequality*, which states that $|\mathbf{a} \cdot \mathbf{b}| \leq \|\mathbf{a}\| \|\mathbf{b}\|$ whenever $\mathbf{a}$ and $\mathbf{b}$ are two vectors in $\mathbb{R}^n$. An outline of a proof of this inequality can be found in Exercise 12. This inequality ensures that $-1 \leq \mathbf{a} \cdot \mathbf{b}/\|\mathbf{a}\| \|\mathbf{b}\| \leq 1$, and therefore that the equation $\cos \theta = \mathbf{a} \cdot \mathbf{b}/\|\mathbf{a}\| \|\mathbf{b}\|$ has a solution θ, for all nonzero vectors $\mathbf{a}$ and $\mathbf{b}$ in $\mathbb{R}^n$.

REMARK 2   It is natural to ask if the angle between nonzero vectors $\mathbf{a}$ and $\mathbf{b}$ in $\mathbb{R}^3$ as defined earlier is really the same as the angle between the arrows that represent $\mathbf{a}$ and $\mathbf{b}$. Using the definition of θ, it is easy to verify that the law of cosines still holds in $\mathbb{R}^3$ (and in $\mathbb{R}^n$ for all n; see Exercise 13). It follows that θ is indeed the angle between the arrows that represent $\mathbf{a}$ and $\mathbf{b}$ (see Figure 1.38).

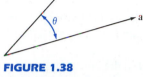

FIGURE 1.38
The angle between $\mathbf{a}$ and $\mathbf{b}$.

Now suppose $\mathbf{d}$ is any nonzero vector in $\mathbb{R}^n$. Using the dot product, any vector $\mathbf{a}$ in $\mathbb{R}^n$ can be decomposed into a sum of two vectors,

$$\mathbf{a} = \mathbf{v} + \mathbf{w},$$

where $\mathbf{v}$ is a scalar multiple of $\mathbf{d}$ and $\mathbf{w}$ is perpendicular to $\mathbf{d}$ (see Figure 1.39). The vector $\mathbf{v}$ will be equal to $c\mathbf{d}$ for some real number c, so we will have

$$\mathbf{a} = c\mathbf{d} + \mathbf{w}.$$

To find c, we need only take the dot product of both sides of this equation with $\mathbf{d}$. Since $\mathbf{w}$ is required to be perpendicular to $\mathbf{d}$ we will have $\mathbf{w} \cdot \mathbf{d} = 0$ and hence

$$\mathbf{a} \cdot \mathbf{d} = c\mathbf{d} \cdot \mathbf{d} + \mathbf{w} \cdot \mathbf{d} = c\mathbf{d} \cdot \mathbf{d}$$

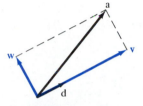

FIGURE 1.39
Decomposing a vector $\mathbf{a}$ into components $\mathbf{v}$ along $\mathbf{d}$, and $\mathbf{w}$ perpendicular to $\mathbf{d}$.

or

$$c = \frac{\mathbf{a} \cdot \mathbf{d}}{\mathbf{d} \cdot \mathbf{d}}.$$

The equation $\mathbf{a} = c\mathbf{d} + \mathbf{w}$ can now be solved for $\mathbf{w}$:

$$\mathbf{w} = \mathbf{a} - c\mathbf{d} = \mathbf{a} - \frac{\mathbf{a} \cdot \mathbf{d}}{\mathbf{d} \cdot \mathbf{d}} \mathbf{d}.$$

Notice that if $\mathbf{w}$ is given by this formula, then

$$\mathbf{w} \cdot \mathbf{d} = \left(\mathbf{a} - \frac{\mathbf{a} \cdot \mathbf{d}}{\mathbf{d} \cdot \mathbf{d}} \right) \cdot \mathbf{d} = \mathbf{a} \cdot \mathbf{d} - \frac{\mathbf{a} \cdot \mathbf{d}}{\mathbf{d} \cdot \mathbf{d}} \mathbf{d} \cdot \mathbf{d} = 0,$$

so $\mathbf{w}$ *is* perpendicular to $\mathbf{d}$, as required.

The vector

$$\mathbf{v} = \frac{\mathbf{a} \cdot \mathbf{d}}{\mathbf{d} \cdot \mathbf{d}} \mathbf{d}$$

is *the component of* **a** *along* **d**. The vector

$$w = a - v = a - \frac{a \cdot d}{d \cdot d} d$$

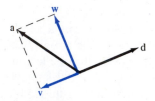

FIGURE 1.40
Components of **a** along **d** and perpendicular to **d**, when **a** · **d** < 0.

is *the component of* **a** *perpendicular to* **d**. *These vectors are the unique vectors* **v** *and* **w** *in* $\mathbb{R}^n$ *with* **v** *a scalar multiple of* **d** *and* **w** *perpendicular to* **d** *such that* **a** = **v** + **w** (see Figure 1.39).

Notice that the component of **a** along **d** may point in the opposite direction from **d** (see Figure 1.40). This is the case when **a** · **d** is negative; that is, when the angle between **a** and **d** is obtuse.

EXAMPLE 8

Let **a** = (1, 0) and let **d** = (1, 1). The component of **a** along **d** is

$$v = \frac{a \cdot d}{d \cdot d} d = \frac{(1, 0) \cdot (1, 1)}{(1, 1) \cdot (1, 1)} (1, 1) = \tfrac{1}{2}(1,1) = (\tfrac{1}{2}, \tfrac{1}{2}).$$

The component of **a** = (1, 0) perpendicular to **d** is

$$w = a - v = (1, 0) - (\tfrac{1}{2}, \tfrac{1}{2}) = (\tfrac{1}{2}, -\tfrac{1}{2})$$

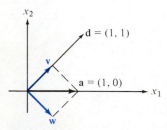

FIGURE 1.41
Decomposing **a** = (1, 0) with respect to **d** = (1, 1).

(see Figure 1.41).

If **a** = (3, 7, −1, 4) and **d** = (1, −1, 0, 2), then the component of **a** along **d** is

$$v = \frac{a \cdot d}{d \cdot d} d = \tfrac{4}{6}(1, -1, 0, 2) = (\tfrac{2}{3}, -\tfrac{2}{3}, 0, \tfrac{4}{3})$$

and the component of **a** perpendicular to **d** is

$$w = a - v = (\tfrac{7}{3}, \tfrac{23}{3}, -1, \tfrac{8}{3}). \quad \blacksquare$$

A vector **u** is a ***unit vector*** if its length $\|u\|$ is equal to 1. Given any nonzero vector **d** we can find a unit vector **u** pointing in the same direction as **d** by dividing **d** by its length:

$$u = \frac{d}{\|d\|} = \frac{1}{\|d\|} d.$$

The formula for the component of a vector **a** along a unit vector **u** is especially simple. Since **u** · **u** = $\|u\|^2$ = 1, the component of **a** along **u** is given by the formula **v** = (**a** · **u**)**u**.

EXAMPLE 9

Let **d** = (3, −1, 2). Then

$$u = \frac{1}{\|d\|} d = \frac{1}{\sqrt{14}} (3, -1, 2) = \left(\frac{3}{\sqrt{14}}, \frac{-1}{\sqrt{14}}, \frac{2}{\sqrt{14}} \right)$$

is a unit vector in the direction of **d**. The component of **a** = (2, 3, 5) in the direction of **u** is

$$\mathbf{v} = (\mathbf{a} \cdot \mathbf{u})\mathbf{u} = \frac{13}{\sqrt{14}}\left(\frac{3}{\sqrt{14}}, \frac{-1}{\sqrt{14}}, \frac{2}{\sqrt{14}}\right) = \left(\frac{39}{14}, -\frac{13}{14}, \frac{13}{7}\right).$$

Notice that this is the same as the component of **a** in the direction of **d**:

$$\mathbf{v} = \frac{\mathbf{a} \cdot \mathbf{d}}{\mathbf{d} \cdot \mathbf{d}}\mathbf{d} = \frac{13}{14}(3, -1, 2) = \left(\frac{39}{14}, -\frac{13}{14}, \frac{13}{7}\right). \quad \blacksquare$$

EXERCISES

1. Compute the dot product of the vectors **a** and **b** and find the cosine of the angle θ between them. In each case, decide if θ is 0, $\pi/2$, π, acute, or obtuse. Finally, use a calculator when necessary to find the angle θ.
   (a) **a** = (1, 2), **b** = (-2, 3)
   (b) **a** = (-1, 1), **b** = (1, 1)
   (c) **a** = (1, 1), **b** = (2, 2)
   (d) **a** = (2, 1), **b** = (-4, -2)
   (e) **a** = (3, 4), **b** = (1, -2)
   (f) **a** = (1, 2, 1), **b** = (-1, 1, 1)
   (g) **a** = (1, 2, -1), **b** = (3, 1, 1)
   (h) **a** = (-1, 2, 3), **b** = (4, -2, -1)
   (i) **a** = (-1, 1, 2, -1), **b** = (1, -1, 1, 0)
   (j) **a** = (1, 1, 0, 1, 1), **b** = (2, 2, 3, 2, 2)

2. Show that if **a** = (a_1, a_2), **a** $\neq$ (0, 0), then the vector **b** defined by **b** = $(-a_2, a_1)$ is perpendicular to **a**.

3. Use the result of Exercise 2 to find a vector equation for the line in $\mathbb{R}^2$ through the point (5, -2) and perpendicular to the line
   (a) **x** = (0, 1) + t(1, 1)   (b) **x** = (-1, 1) + $t(-\frac{1}{2}, \frac{1}{2})$
   (c) **x** = (2, 3) + t(-3, 4)   (d) **x** = (-7, 6) + t(1, 3)
   (e) **x** = (2, 2) + t(2, 1)   (f) **x** = (-1, 1) + t(-1, 1)

4. Let **a** = (2, -3). Find the component of **a** along **d** and find the component of **a** perpendicular to **d**, where
   (a) **d** = (1, 0)   (b) **d** = (0, 1)
   (c) **d** = (-1, 1)   (d) **d** = (3, 2)
   (e) **d** = (-4, 6)   (f) **d** = (2, 3).

5. Let **d** = (1, -1, 1). Find the component of **a** along **d** and find the component of **a** perpendicular to **d**, where
   (a) **a** = (1, 0, 0)   (b) **a** = (0, 1, 0)
   (c) **a** = (0, 0, 1)   (d) **a** = (-1, 1, -1)
   (e) **a** = (1, 2, 1)   (f) **a** = (2, 3, 4)

6. Let $\mathbf{d}_1$ and $\mathbf{d}_2$ be nonzero vectors in $\mathbb{R}^n$ that are perpendicular ($\mathbf{d}_1 \cdot \mathbf{d}_2 = 0$) and let **a** be any vector in $\mathbb{R}^n$. Let $\mathbf{v}_1$ be the component of **a** along $\mathbf{d}_1$ and let $\mathbf{v}_2$ be the component of **a** along $\mathbf{d}_2$. Show that the vector

$$\mathbf{w} = \mathbf{a} - \mathbf{v}_1 - \mathbf{v}_2$$

is perpendicular to both $\mathbf{d}_1$ and $\mathbf{d}_2$.

7. Find the unit vector **u** in the direction of **d** and sketch both **d** and **u**, where

   (a) **d** = (2, 3)   (b) **d** = (−1, 1)

   (c) **d** = (0, 7)   (d) **d** = (2, 1, 2)

   (e) **d** = (1, −1, −1)   (f) **d** = ($\frac{1}{2}$, $\frac{1}{2}$, $\frac{1}{2}$)

8. Use vector methods to show that a parallelogram has perpendicular diagonals if and only if it is a rhombus. [*Hint:* Imitate the argument in Example 5. Recall that a parallelogram is a *rhombus* if adjacent sides have equal length.]

9. Use vector methods to show that if an altitude of a triangle intersects the opposite side in its midpoint, then the triangle is isosceles.

10. Verify properties D_1–D_4 for the dot product of vectors in $\mathbb{R}^n$.

11. Show that for any two vectors **a** and **b** in $\mathbb{R}^n$

$$\mathbf{a} \cdot \mathbf{b} = \tfrac{1}{4}(\|\mathbf{a} + \mathbf{b}\|^2 - \|\mathbf{a} - \mathbf{b}\|^2).$$

12. (a) Show that the quadratic polynomial $At^2 + Bt + C$ attains its smallest value when $t = -B/2A$. [*Hint:* Complete the square.]

    (b) Let **a** and **b** be nonzero vectors in $\mathbb{R}^n$ and consider the quadratic polynomial $\|t\mathbf{b} + \mathbf{a}\|^2 = (t\mathbf{b} + \mathbf{a}) \cdot (t\mathbf{b} + \mathbf{a})$. Show that the smallest value of this polynomial occurs when $t = -\mathbf{a} \cdot \mathbf{b}/\|\mathbf{b}\|^2$ and calculate this smallest value.

    (c) Use the fact that $\|t\mathbf{b} + \mathbf{a}\|^2 \geq 0$ for all t to show that $|\mathbf{a} \cdot \mathbf{b}| \leq \|\mathbf{a}\|\,\|\mathbf{b}\|$.

    REMARK. The inequality in (c) is the *Cauchy-Schwarz inequality*.

13. Let **a** and **b** be any two nonzero vectors in $\mathbb{R}^n$ and let θ be the angle between **a** and **b**. Show that

$$\|\mathbf{b} - \mathbf{a}\|^2 = \|\mathbf{a}\|^2 + \|\mathbf{b}\|^2 - 2\|\mathbf{a}\|\,\|\mathbf{b}\| \cos \theta.$$

[This is the *law of cosines* in $\mathbb{R}^n$ (see Figure 1.42). If θ is a right angle then this formula reduces to the *Pythagorean Theorem:* $\|\mathbf{b} - \mathbf{a}\|^2 = \|\mathbf{a}\|^2 + \|\mathbf{b}\|^2$.]

14. Let **a** and **b** be any two vectors in $\mathbb{R}^n$. Prove the **triangle inequality**

$$\|\mathbf{a} + \mathbf{b}\| \leq \|\mathbf{a}\| + \|\mathbf{b}\|.$$

[*Hint:* Use the Cauchy-Schwarz inequality (Exercise 12) to show that

$$\|\mathbf{a} + \mathbf{b}\|^2 \leq (\|\mathbf{a}\| + \|\mathbf{b}\|)^2.]$$

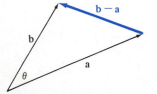

FIGURE 1.42
The law of cosines in $\mathbb{R}^n$ states that $\|\mathbf{b} - \mathbf{a}\|^2 = \|\mathbf{a}\|^2 + \|\mathbf{b}\|^2 - 2\|\mathbf{a}\|\,\|\mathbf{b}\| \cos \theta$.

1.6
Lines, Planes, and Hyperplanes

In Section 3 vector methods were used to describe lines in $\mathbb{R}^2$ and, in Section 4, these methods were generalized to describe lines in $\mathbb{R}^n$ for all n. This section will introduce an alternate method for describing lines in $\mathbb{R}^2$ and show how this method generalizes to describe planes in $\mathbb{R}^3$ and "hyperplanes" in $\mathbb{R}^n$ for all $n > 3$.

Let ℓ be a line in $\mathbb{R}^2$. In Section 3 we described the line ℓ in terms of a point **a** on ℓ and a vector **d** *parallel* to ℓ (see Figure 1.43a). Now we shall describe ℓ in terms of a point **a** on ℓ and a vector **b** *perpendicular* to ℓ (see Figure 1.43b).

Suppose then, that **a** is a point in $\mathbb{R}^2$ and that **b** is a nonzero vector in $\mathbb{R}^2$. Let ℓ be the line in $\mathbb{R}^2$ that passes through **a** and is perpendicular to **b**. Then a point **x** in $\mathbb{R}^2$ is on ℓ if and only if the vector **x** − **a** is perpendicular to **b** (see Figure 1.44). That is, **x** is on ℓ if and only if

$$\mathbf{b} \cdot (\mathbf{x} - \mathbf{a}) = 0.$$

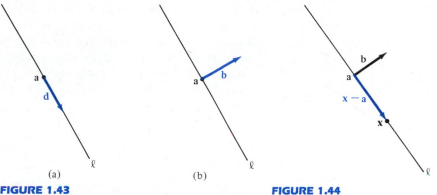

(a)   (b)

FIGURE 1.43
A line ℓ in $\mathbb{R}^2$ is uniquely determined by specifying a point **a** on ℓ and either (a) a vector **d** parallel to ℓ, or (b) a vector **b** perpendicular to ℓ.

FIGURE 1.44
x is on ℓ if and only if $\mathbf{x} - \mathbf{a}$ is perpendicular to **b**.

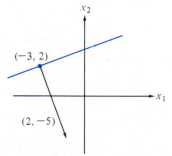

FIGURE 1.45
The line ℓ passes through $(-3, 2)$ and is perpendicular to the vector $(2, -5)$.

This equation is a ***scalar equation for the line*** ℓ. Notice that this equation can be rewritten as $\mathbf{b} \cdot \mathbf{x} = \mathbf{b} \cdot \mathbf{a}$ or as

$$b_1 x_1 + b_2 x_2 = c,$$

where $\mathbf{b} = (b_1, b_2)$, $\mathbf{x} = (x_1, x_2)$, and $c = \mathbf{b} \cdot \mathbf{a}$.

EXAMPLE 1

To find a scalar equation for the line ℓ in $\mathbb{R}^2$ that passes through the point $(-3, 2)$ and is perpendicular to the vector $(2, -5)$ (see Figure 1.45), let $\mathbf{a} = (-3, 2)$ and $\mathbf{b} = (2, -5)$. Then the equation

$$\mathbf{b} \cdot (\mathbf{x} - \mathbf{a}) = 0$$

or

$$(2, -5) \cdot (\mathbf{x} - (-3, 2)) = 0$$

describes ℓ. This equation can be rewritten as

$$(2, -5) \cdot \mathbf{x} = (2, -5) \cdot (-3, 2)$$

or as

$$2x_1 - 5x_2 = -16$$

where $\mathbf{x} = (x_1, x_2)$. ∎

Theorem 1   *The equation*

$$b_1 x_1 + b_2 x_2 = c$$

describes a line ℓ in $\mathbb{R}^2$ whenever $(b_1, b_2) \neq (0, 0)$. Moreover, the coefficients b_1 and b_2 of x_1 and x_2 are the entries of a vector $\mathbf{b} = (b_1, b_2)$ that is perpendicular to ℓ.

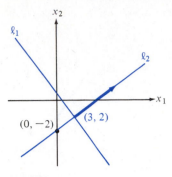

FIGURE 1.46

The line ℓ_2 is perpendicular to the line ℓ_1 and passes through the point $(0, -2)$.

Proof To see this, rewrite the equation in the form $\mathbf{b} \cdot \mathbf{x} = c$ where $\mathbf{b} = (b_1, b_2)$ and $\mathbf{x} = (x_1, x_2)$. Let $\mathbf{a}$ be any solution of this equation. Then $\mathbf{b} \cdot \mathbf{a} = c$. Hence the equation $\mathbf{b} \cdot \mathbf{x} = c$ is the same as the equation $\mathbf{b} \cdot \mathbf{x} = \mathbf{b} \cdot \mathbf{a}$, or $\mathbf{b} \cdot (\mathbf{x} - \mathbf{a}) = 0$. In other words, $\mathbf{x}$ satisfies the equation $\mathbf{b} \cdot \mathbf{x} = c$ if and only if $\mathbf{x} - \mathbf{a}$ is perpendicular to $\mathbf{b}$. Thus (see Figure 1.44), *the solution set of the equation* $\mathbf{b} \cdot \mathbf{x} = c$ *is precisely the line in* $\mathbb{R}^2$ *through* $\mathbf{a}$ *perpendicular to* $\mathbf{b} = (b_1, b_2)$. ■

EXAMPLE 2

Let ℓ_1 be the line

$$3x_1 + 2x_2 = 1$$

in $\mathbb{R}^2$. Let us find an equation for the line ℓ_2 that is perpendicular to ℓ_1 and passes through the point $(0, -2)$ (see Figure 1.46). Since the vector $(3, 2)$ is perpendicular to ℓ_1, it is parallel to ℓ_2. Hence a vector equation for ℓ_2 is

$$\mathbf{x} = (0, -2) + t(3, 2).$$

If we want a scalar equation for ℓ_2 we can write this vector equation as the pair of scalar equations

$$x_1 = 3t$$

$$x_2 = -2 + 2t$$

and eliminate t to get

$$2x_1 - 3x_2 = 6. \quad ■$$

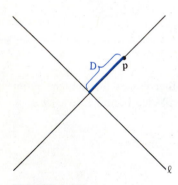

FIGURE 1.47

D is the distance from $\mathbf{p}$ to ℓ.

Now let us find a formula for the distance in $\mathbb{R}^2$ from a point $\mathbf{p}$ to a line ℓ. The **distance D from $\mathbf{p}$ to ℓ** is the length of the line segment that connects $\mathbf{p}$ to ℓ and is perpendicular to ℓ (see Figure 1.47). If the line ℓ has equation $\mathbf{b} \cdot \mathbf{x} = c$ then the vector $\mathbf{b}$ is perpendicular to ℓ so $D = \|t\mathbf{b}\|$ where t is chosen so that $\mathbf{p} + t\mathbf{b}$ is on ℓ (see Figure 1.48). To find t, we substitute $\mathbf{x} = \mathbf{p} + t\mathbf{b}$ into the equation $\mathbf{b} \cdot \mathbf{x} = c$ for ℓ to get

$$\mathbf{b} \cdot (\mathbf{p} + t\mathbf{b}) = c$$

or

$$\mathbf{b} \cdot \mathbf{p} + t\mathbf{b} \cdot \mathbf{b} = c.$$

Solving for t yields

$$t = \frac{c - \mathbf{b} \cdot \mathbf{p}}{\mathbf{b} \cdot \mathbf{b}}.$$

It follows that

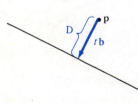

FIGURE 1.48

The distance D is equal to $\|t\mathbf{b}\|$.

$$D = \|t\mathbf{b}\| = \left\| \frac{c - \mathbf{b} \cdot \mathbf{p}}{\mathbf{b} \cdot \mathbf{b}} \mathbf{b} \right\| = \frac{|c - \mathbf{b} \cdot \mathbf{p}|}{\|\mathbf{b}\|^2} \|\mathbf{b}\| = \frac{|\mathbf{b} \cdot \mathbf{p} - c|}{\|\mathbf{b}\|}.$$

Thus, *the distance D from the point* **p** *to the line* **b** • **x** = *c is given by the formula*

$$D = \frac{|\mathbf{b} \cdot \mathbf{p} - c|}{\|\mathbf{b}\|}.$$

EXAMPLE 3

To find the distance D from the point $(-1, 2)$ to the line $3x_1 - 2x_2 = 4$, take
$\mathbf{b} = (3, -2)$, $\mathbf{p} = (-1, 2)$, $c = 4$, and calculate

$$D = \frac{|\mathbf{b} \cdot \mathbf{p} - c|}{\|\mathbf{b}\|} = \frac{|-11|}{\sqrt{13}} = \frac{11}{\sqrt{13}}. \quad \blacksquare$$

EXAMPLE 4

To find the distance D from the point $(-1, 2)$ to the line ℓ with vector equation
$\mathbf{x} = (1, 1) + t(7, -6)$, we must first find a scalar equation for ℓ. We can do
that either by writing this vector equation as a pair of scalar equations and then
eliminating t, or by observing that the point $(1, 1)$ is on ℓ and that the vector
$(6, 7)$ is perpendicular to ℓ. [The latter fact is immediately apparent, because
the vector $(7, -6)$ is parallel to ℓ and $(6, 7) \cdot (7, -6) = 0$.] Either way, we
find the equation

$$6x_1 + 7x_2 = 13$$

for ℓ. The distance from $\mathbf{p} = (-1, 2)$ to ℓ is therefore

$$D = \frac{|6(-1) + 7(2) - 13|}{\|(6, 7)\|} = \frac{5}{\sqrt{85}}. \quad \blacksquare$$

We can find a scalar equation describing a *plane* in $\mathbb{R}^3$ in exactly the same
way that we find a scalar equation describing a *line* in $\mathbb{R}^2$.

Suppose **a** is a point in $\mathbb{R}^3$ and **b** is a nonzero vector in $\mathbb{R}^3$. Let P be the
plane in $\mathbb{R}^3$ that passes through the point **a** and is perpendicular to **b** (see Figure
1.49). Then a point **x** in $\mathbb{R}^3$ is on P if and only if **x** − **a** is perpendicular to **b**
(see Figure 1.50). That is, **x** is on P if and only if $\mathbf{b} \cdot (\mathbf{x} - \mathbf{a}) = 0$. Thus, in
$\mathbb{R}^3$, *the equation*

$$\mathbf{b} \cdot (\mathbf{x} - \mathbf{a}) = 0$$

describes the plane through **a** *perpendicular to* **b**.

EXAMPLE 5

Let us find an equation for the plane P in $\mathbb{R}^3$ that passes through the point
$(1, -1, 2)$ and is perpendicular to the vector $(2, 3, -1)$. Taking $\mathbf{a} =
(1, -1, 2)$ and $\mathbf{b} = (2, 3, -1)$, we find that the equation

$$\mathbf{b} \cdot (\mathbf{x} - \mathbf{a}) = 0,$$

or

$$(2, 3, -1) \cdot (\mathbf{x} - (1, -1, 2)) = 0,$$

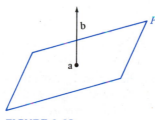

FIGURE 1.49

P is the plane through **a** perpendicular to **b**.

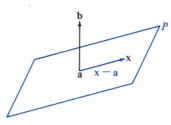

FIGURE 1.50

x is on P if and only if **x** − **a** is perpendicular to **b**.

describes P. This equation can be rewritten as

$$(2, 3, -1) \cdot \mathbf{x} = (2, 3, -1) \cdot (1, -1, 2)$$

or as

$$2x_1 + 3x_2 - x_3 = -3,$$

where $\mathbf{x} = (x_1, x_2, x_3)$. ∎

We know that the equation $b_1x_1 + b_2x_2 = c$ describes a line in $\mathbb{R}^2$ whenever $(b_1, b_2) \neq (0, 0)$ and that the vector $\mathbf{b} = (b_1, b_2)$ is perpendicular to that line. This result generalizes to $\mathbb{R}^3$, as is shown in the following theorem.

Theorem 2 *The equation*

$$b_1x_1 + b_2x_2 + b_3x_3 = c$$

describes a plane P in $\mathbb{R}^3$ whenever $(b_1, b_2, b_3) \neq (0, 0, 0)$. Moreover, the coefficients b_1, b_2 and b_3 of x_1, x_2 and x_3 are the entries of a vector $\mathbf{b} = (b_1, b_2, b_3)$ that is perpendicular to P.

Proof The proof of Theorem 2 is nearly identical to the proof of Theorem 1. In that proof, simply replace (b_1, b_2) with (b_1, b_2, b_3), replace (x_1, x_2) with (x_1, x_2, x_3) and, in the last sentence, substitute the words "*line in $\mathbb{R}^2$*" for the words "*plane in $\mathbb{R}^3$*." ∎

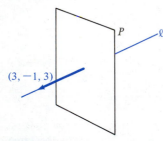

FIGURE 1.51
The line ℓ is perpendicular to the plane P.

EXAMPLE 6
Let P be the plane

$$3x_1 - x_2 + 3x_3 = 5.$$

Let us find an equation for the line ℓ that is perpendicular to P and passes through the point $(2, 1, -3)$ (see Figure 1.51). Since the vector $(3, -1, 3)$ is perpendicular to P, it is parallel to ℓ. Hence a vector equation for ℓ is

$$\mathbf{x} = (2, 1, -3) + t(3, -1, 3). \quad ∎$$

Two planes $\mathbf{b}_1 \cdot \mathbf{x} = c_1$ and $\mathbf{b}_2 \cdot \mathbf{x} = c_2$ in $\mathbb{R}^3$ are *parallel* if the vectors $\mathbf{b}_1$ and $\mathbf{b}_2$ are parallel, that is, if $\mathbf{b}_1$ and $\mathbf{b}_2$ are scalar multiples of one another (see Figure 1.52).

$\mathbf{b}_2$

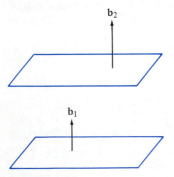

$\mathbf{b}_1$

FIGURE 1.52
The planes $\mathbf{b}_1 \cdot \mathbf{x} = c_1$ and $\mathbf{b}_2 \cdot \mathbf{x} = c_2$ are parallel if $\mathbf{b}_1$ and $\mathbf{b}_2$ are scalar multiples of one another.

EXAMPLE 7
The planes $2x_1 - x_2 + 3x_3 = 1$ and $x_1 + 2x_2 - 5x_3 = -2$ are not parallel because the vectors $(2, -1, 3)$ and $(1, 2, -5)$ are not scalar multiples of one another. To find the intersection of these planes, we must find all points (x_1, x_2, x_3) in $\mathbb{R}^3$ that lie on both planes. Therefore, we must find all (x_1, x_2, x_3) that simultaneously satisfy the equations

$$2x_1 - x_2 + 3x_3 = 1$$

$$x_1 + 2x_2 - 5x_3 = -2.$$

To solve this system, we reduce the matrix of the system to its row echelon matrix:

$$\begin{pmatrix} 2 & -1 & 3 & 1 \\ 1 & 2 & -5 & -2 \end{pmatrix} \longrightarrow \begin{pmatrix} 1 & 0 & \frac{1}{5} & 0 \\ 0 & 1 & -\frac{13}{5} & -1 \end{pmatrix}$$

The system corresponding to the row echelon matrix is

$$x_1 \quad\quad + \quad \tfrac{1}{5}x_3 = \quad 0$$
$$x_2 - \tfrac{13}{5}x_3 = -1.$$

If we set $x_3 = t$, we find the general solution

$$(x_1, x_2, x_3) = (-\tfrac{1}{5}t, -1 + \tfrac{13}{5}t, t)$$

or

$$\mathbf{x} = (0, -1, 0) + t(-\tfrac{1}{5}, \tfrac{13}{5}, 1).$$

Thus, these two planes intersect in the line that passes through the point $(0, -1, 0)$ and is in the direction of the vector $(-\tfrac{1}{5}, \tfrac{13}{5}, 1)$. ■

Now consider the equation

$$b_1x_1 + b_2x_2 + \cdots + b_nx_n = c,$$

where $\mathbf{b} = (b_1, b_2, \ldots, b_n)$ is a nonzero vector in $\mathbb{R}^n$ and c is a real number. When $n = 2$, the solution set of this equation is a *line* in $\mathbb{R}^2$. When $n = 3$, the solution set is a *plane* in $\mathbb{R}^3$. When n is greater than 3, the solution set of this equation is called a **hyperplane** in $\mathbb{R}^n$. In particular, given a point $\mathbf{a}$ in $\mathbb{R}^n$ and a nonzero vector $\mathbf{b}$ in $\mathbb{R}^n$, the **hyperplane through a perpendicular to b** is the solution set of the equation $\mathbf{b} \cdot (\mathbf{x} - \mathbf{a}) = 0$, or $\mathbf{b} \cdot \mathbf{x} = c$ where $c = \mathbf{b} \cdot \mathbf{a}$.

EXAMPLE 8

Let us find an equation for the hyperplane H in $\mathbb{R}^4$ through the point $(1, 3, 4, 2)$ perpendicular to the vector $(-1, 2, 2, 1)$. Taking $\mathbf{a} = (1, 3, 4, 2)$ and $\mathbf{b} = (-1, 2, 2, 1)$, we find that the equation

$$\mathbf{b} \cdot (\mathbf{x} - \mathbf{a}) = 0$$

or

$$(-1, 2, 2, 1) \cdot (\mathbf{x} - (1, 3, 4, 2)) = 0$$

describes H. This equation can be rewritten as

$$(-1, 2, 2, 1) \cdot \mathbf{x} = (-1, 2, 2, 1) \cdot (1, 3, 4, 2)$$

or as

$$-x_1 + 2x_2 + 2x_3 + x_4 = 15,$$

where $\mathbf{x} = (x_1, x_2, x_3, x_4)$. ■

EXAMPLE 9

Let ℓ be the line

$$\mathbf{x} = (2, -1, 0, 3) + t(3, -2, -1, 3)$$

in $\mathbb{R}^4$ and let H be the hyperplane

$$3x_1 - 7x_2 + 2x_3 + 4x_4 = 14.$$

Let us find the intersection of ℓ and H.

Each point $\mathbf{x}$ on ℓ is of the form

$$\mathbf{x} = (2 + 3t, -1 - 2t, -t, 3 + 3t),$$

for some $t \in \mathbb{R}$. This point $\mathbf{x}$ will be on H if and only if

$$3(2 + 3t) - 7(-1 - 2t) + 2(-t) + 4(3 + 3t) = 14,$$

that is, if and only if $33t = -11$ or $t = -\frac{1}{3}$. Hence ℓ and H intersect at the point

$$\mathbf{x} = (2, -1, 0, 3) - \tfrac{1}{3}(3, -2, -1, 3)$$

or

$$\mathbf{x} = (1, -\tfrac{1}{3}, \tfrac{1}{3}, 2). \quad \blacksquare$$

Recall that the formula

$$D = \frac{|\mathbf{b} \cdot \mathbf{p} - c|}{\|\mathbf{b}\|}$$

gives the distance from the point $\mathbf{p}$ to the line $\mathbf{b} \cdot \mathbf{x} = c$ when $\mathbf{p}$ and $\mathbf{b}$ are in $\mathbb{R}^2$. *This same formula gives the distance from* $\mathbf{p}$ *to the plane* $\mathbf{b} \cdot \mathbf{x} = c$ *when* $\mathbf{p}$ *and* $\mathbf{b}$ *are in* $\mathbb{R}^3$, *and it gives the distance from* $\mathbf{p}$ *to the hyperplane* $\mathbf{b} \cdot \mathbf{x} = c$ *when* $\mathbf{p}$ *and* $\mathbf{b}$ *are in* $\mathbb{R}^n$ *for* $n > 3$. The proof that this formula is valid for arbitrary n is almost identical to the proof when $n = 2$, so we will not reproduce it here.

EXAMPLE 10

To find the distance D from the point $(3, 0, -1)$ to the plane

$$-5x_1 + 7x_2 + 4x_3 = -3$$

we take $\mathbf{p} = (3, 0, -1)$, $\mathbf{b} = (-5, 7, 4)$, and $c = -3$ to get

$$D = \frac{|\mathbf{b} \cdot \mathbf{p} - c|}{\|\mathbf{b}\|} = \frac{|-19 + 3|}{\sqrt{90}} = \frac{16}{3\sqrt{10}}. \quad \blacksquare$$

The ideas of this section can be used to give a geometric interpretation of the solution set of a system of linear equations. It has been shown that the solution set of an equation of the form

$$a_{11}x_1 + a_{12}x_2 + \cdots + a_{1n}x_n = b_1$$

is a hyperplane in $\mathbb{R}^n$. Therefore, *the solution set of any linear system*

$$a_{11}x_1 + a_{12}x_2 + \cdots + a_{1n}x_n = b_1$$

$$\cdots$$

$$a_{m1}x_1 + a_{m2}x_2 + \cdots + a_{mn}x_n = b_m$$

is the intersection of m hyperplanes in $\mathbb{R}^n$. If $b_1 = b_2 = \cdots = b_m = 0$, that is, if the system is homogeneous, then these equations can be rewritten as

$$\mathbf{a}_1 \cdot \mathbf{x} = 0$$

$$\cdots$$

$$\mathbf{a}_m \cdot \mathbf{x} = 0,$$

where $\mathbf{a}_1 = (a_{11}, \ldots, a_{1n}), \ldots, \mathbf{a}_m = (a_{m1}, \ldots, a_{mn})$. In this case, the solution set is the set of all vectors $\mathbf{x}$ in $\mathbb{R}^n$ that are perpendicular to *all* of the vectors $\mathbf{a}_1, \ldots, \mathbf{a}_m$.

EXERCISES

1. Find a scalar equation for the line in $\mathbb{R}^2$ that passes through $\mathbf{a}$ and is perpendicular to $\mathbf{b}$, where
   (a) $\mathbf{a} = (1, 2)$, $\mathbf{b} = (1, -1)$ (b) $\mathbf{a} = (3, -2)$, $\mathbf{b} = (-\frac{1}{2}, \frac{1}{4})$
   (c) $\mathbf{a} = (-5, 1)$, $\mathbf{b} = (2, -2)$ (d) $\mathbf{a} = (0, 0)$, $\mathbf{b} = (\frac{1}{2}, 1)$
   (e) $\mathbf{a} = (-2, 4)$, $\mathbf{b} = (0, 1)$

2. Find a scalar equation for each of the following lines in $\mathbb{R}^2$.
   (a) $\mathbf{x} = (1, -1) + t(2, -1)$ (b) $\mathbf{x} = (1, 2) + t(3, 4)$
   (c) $\mathbf{x} = (-1, 0) + t(1, 0)$ (d) $\mathbf{x} = (10, 20) + t(5, -1)$
   (e) $\mathbf{x} = (-2, 4) + t(0, 1)$

3. Find a scalar equation for the line in $\mathbb{R}^2$ that passes through the point $(2, 3)$ and is perpendicular to the given line.
   (a) $\mathbf{x} = (1, 1) + t(-1, 2)$ (b) $\mathbf{x} = t(3, 4)$
   (c) $\mathbf{x} = (1, 0) + t(0, 1)$ (d) $\mathbf{x} = (5, -2) + t(\frac{3}{2}, 1)$
   (e) $\mathbf{x} = (-1, -2) + t(-3, -4)$

4. Find a vector equation for the line in $\mathbb{R}^2$ that passes through the given point and is perpendicular to the given line.
   (a) $(0, 0)$, $x_1 + x_2 = 5$ (b) $(3, 7)$, $-x_1 + 2x_2 = 0$
   (c) $(-1, -4)$, $2x_1 + 4x_2 = 1$ (d) $(2, 2)$, $3x_1 + 4x_2 = 5$
   (e) $(0, 0)$, $x_1 = 1$

5. Find a scalar equation for each of the lines whose vector equations were found in Exercise 4.

6. Find the distance from the point $(-1, 3)$ to each of the following lines.
   (a) $x_1 - x_2 = 3$ (b) $2x_1 - 3x_2 = 0$
   (c) $x_1 + 3x_2 = 8$ (d) $3x_1 - 2x_2 = 1$
   (e) $x_1 = 0$ (f) $x_2 = 7$

7. Find the distance from the point $(1, 1)$ to each of the following lines.
   (a) $\mathbf{x} = (0, 1) + t(1, 1)$ (b) $\mathbf{x} = (1, -1) + t(2, -1)$

(c) $\mathbf{x} = (-1, 0) + t(1, 0)$ (d) $\mathbf{x} = (1, 1) + t(-3, 2)$

(e) $\mathbf{x} = t(-1, 1)$.

8. Find an equation for the plane in $\mathbb{R}^3$ that passes through $\mathbf{a}$ and is perpendicular to $\mathbf{d}$, where

(a) $\mathbf{a} = (1, 2, 1)$, $\mathbf{d} = (-3, 1, 4)$

(b) $\mathbf{a} = (-1, 1, -1)$, $\mathbf{d} = (2, 1, 2)$

(c) $\mathbf{a} = (1, -2, 3)$, $\mathbf{d} = (-4, 1, 1)$

(d) $\mathbf{a} = (3, 3, 1)$, $\mathbf{d} = (3, 3, 1)$

(e) $\mathbf{a} = (-1, 1, 2)$, $\mathbf{d} = (0, 0, 1)$

9. Find an equation for the plane that passes through the given point and is perpendicular to the given line.

(a) $(1, 1, -1)$, $\mathbf{x} = (2, 1, 2) + t(2, 3, -1)$

(b) $(4, 4, 2)$, $\mathbf{x} = (-1, 0, 1) + t(-1, 2, 3)$

(c) $(1, -1, 0)$, $\mathbf{x} = (-1, 0, 1) + t(1, 6, -2)$

(d) $(2, 1, 2)$, $\mathbf{x} = (2, 2, 1) + t(-1, 3, 2)$

(e) $(0, 0, 0)$, $\mathbf{x} = t(1, 0, -1)$

10. Find a vector equation for the line that passes through the given point and is perpendicular to the given plane.

(a) $(1, -1, 0)$, $x_1 - x_2 + x_3 = 5$

(b) $(3, 2, 1)$, $3x_1 - x_2 + 3x_3 = 14$

(c) $(-1, 1, 1)$, $2x_1 + 2x_2 - x_3 = 7$

(d) $(1, 2, 3)$, $4x_1 + 5x_2 + 6x_3 = 10$

(e) $(5, 1, 3)$, $x_2 = 0$

11. Determine which of the following pairs of planes are parallel. For each pair that is not parallel, find a vector equation for the line of intersection.

(a) $2x_1 - 3x_2 + x_3 = 1$ and $3x_1 - 2x_2 = 0$

(b) $x_1 + x_2 - 2x_3 = 4$ and $-x_1 - 2x_2 + x_3 = -1$

(c) $3x_1 - 2x_2 + 2x_3 = -1$ and $-3x_1 + 2x_2 - 2x_3 = 0$

(d) $x_1 + x_2 + x_3 = 1$ and $x_3 = 0$

(e) $2x_1 - 3x_2 - 2x_3 = 5$ and $-4x_1 + 6x_2 + 4x_3 = 3$

12. Find a vector equation for the line of intersection of the given pair of planes in $\mathbb{R}^3$.

(a) $x_1 + x_2 + x_3 = 5$ and $x_1 - x_2 + x_3 = 1$

(b) $2x_1 + x_3 = 0$ and $3x_1 + 4x_2 = -1$

(c) $x_1 = 0$ and $x_2 = 0$

(d) $x_1 - 2x_2 + 3x_3 = 4$ and $x_1 + x_2 - x_3 = 2$

13. Find the point where the given line and the given plane intersect.

(a) $\mathbf{x} = (1, -1, 3) + t(3, 5, 0)$, $2x_1 + 3x_2 - x_3 = 7$

(b) $\mathbf{x} = (0, 1, 1) + t(-3, 1, -2)$, $-3x_1 + 3x_2 + x_3 = 8$

(c) $\mathbf{x} = t(1, 1, 1)$, $x_1 + x_2 + x_3 = 1$

(d) $\mathbf{x} = (5, 7, 3) + t(0, 1, -1)$, $x_3 = -5$

(e) $\mathbf{x} = (1, 0, 1) + t(-1, -1, 4)$, $x_1 + x_2 + 2x_3 = -1$

14. Find the distance from the given point to the given plane.

(a) $(1, -1, 1)$, $x_1 - x_2 + x_3 = 3$ (b) $(1, 2, 3)$, $3x_1 - x_2 - x_3 = 4$

(c) $(3, -2, 1)$, $2x_1 - x_2 + x_3 = 2$

(d) $(0, 0, 0)$, $x_1 - x_2 + x_3 = 7$

(e) $(0, 0, 1)$, $x_1 + x_2 + x_3 = 10$

15. Find an equation for the hyperplane through **a** perpendicular to **b**, where

   (a) $\mathbf{a} = (-1, 0, 1, -1)$, $\mathbf{b} = (2, -1, 3, 4)$

   (b) $\mathbf{a} = (3, 3, 1, 3)$, $\mathbf{b} = (6, 5, 5, 0)$

   (c) $\mathbf{a} = (0, 0, 0, 0, 0)$, $\mathbf{b} = (-2, -1, 0, 2, 5)$

   (d) $\mathbf{a} = (1, \frac{1}{2}, \frac{1}{3}, \frac{1}{4}, \frac{1}{5})$, $\mathbf{b} = (2, 4, 6, 8, 10)$

   (e) $\mathbf{a} = (1, \frac{1}{2}, \frac{1}{3}, \ldots, \frac{1}{n})$, $\mathbf{b} = (2, 4, 6, \ldots, 2n)$

16. Find a vector equation for the line through the given point perpendicular to the given hyperplane.

   (a) $(1, -1, -2, 3, 4)$, $x_1 - x_2 + 3x_3 - x_4 + x_5 = 2$

   (b) $(2, 1, 3, 4, -1)$, $x_1 - x_5 = 3$

   (c) $(-1, 2, 4, -1)$, $x_1 - x_3 + x_4 = 0$

17. Find an equation for the hyperplane through the given point perpendicular to the given line.

   (a) $(1, 2, 3, 4)$, $\mathbf{x} = (\frac{1}{5}, \frac{2}{5}, \frac{3}{5}, \frac{4}{5}) + t(-1, 0, 0, -1)$

   (b) $(1, -1, 1, -1)$, $\mathbf{x} = (1, -1, 1, -1) + t(1, 2, 1, 2)$

   (c) $(1, 2, 1, 2, 1)$, $\mathbf{x} = (1, 0, 0, 0, 1) + t(2, -1, 2, -1, 2)$

18. The three hyperplanes $x_1 + x_2 + x_3 + x_4 = 1$, $x_1 - x_2 + x_3 - x_4 = 0$, and $x_1 - x_3 + x_4 = 2$ intersect in a line in $\mathbb{R}^4$. Find a vector equation for that line.

19. Two hyperplanes $\mathbf{b}_1 \cdot \mathbf{x} = c_1$ and $\mathbf{b}_2 \cdot \mathbf{x} = c_2$ in $\mathbb{R}^n$ are **parallel** if the vectors $\mathbf{b}_1$ and $\mathbf{b}_2$ are scalar multiples of one another. They are **perpendicular** if the vectors $\mathbf{b}_1$ and $\mathbf{b}_2$ are perpendicular. For each of the following pairs of hyperplanes, determine if they are parallel, perpendicular, or neither.

   (a) $x_1 - x_2 + x_3 - x_4 = 1$ and $x_1 + x_2 + x_3 + x_4 = 0$

   (b) $x_1 - x_2 + x_3 - x_4 = 1$ and $-x_1 + x_2 - x_3 + x_4 = 0$

   (c) $x_1 + 2x_2 + x_4 = 0$ and $x_1 - 2x_2 + x_3 - x_4 = 3$

   (d) $2x_1 - 2x_2 + 3x_3 - x_4 = 5$ and $6x_1 - 6x_2 + 9x_3 - 3x_4 = 5$

   (e) $2x_1 - 3x_2 + 3x_3 - 2x_4 = 1$ and $x_1 + x_2 - x_3 - 2x_4 = 4$

20. (a) Show that the distance D between the parallel lines $b_1 x_1 + b_2 x_2 = c_1$ and $b_1 x_1 + b_2 x_2 = c_2$ in $\mathbb{R}^2$ is given by the formula $D = |c_1 - c_2|/\sqrt{b_1^2 + b_2^2}$.

   (b) Find a similar formula for the distance between parallel planes in $\mathbb{R}^3$.

   (c) Does this formula generalize to cover parallel hyperplanes in $\mathbb{R}^n$?

21. The equation $\mathbf{b} \cdot \mathbf{x} = c$ for a line, plane, or hyperplane is said to be in **normal form** if $\|\mathbf{b}\| = 1$ and $c \geq 0$.

   (a) Show that each line in $\mathbb{R}^2$, plane in $\mathbb{R}^3$, and hyperplane in $\mathbb{R}^n (n > 3)$ can be described by a unique normal form equation.

   (b) Show that if $\|\mathbf{b}\| = 1$ and $c \geq 0$ then c is equal to the distance from the origin **0** to the line, plane, or hyperplane $\mathbf{b} \cdot \mathbf{x} = c$.

   (c) Show that the normal form equation for a line ℓ in $\mathbb{R}^2$ can be written as

$$x_1 \cos \theta + x_2 \sin \theta = c$$

where c and θ are as in Figure 1.53.

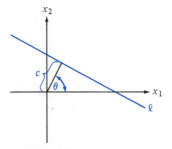

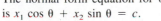

FIGURE 1.53
The normal form equation for ℓ is $x_1 \cos \theta + x_2 \sin \theta = c$.

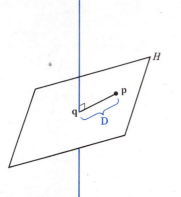

FIGURE 1.54

The distance D from **p** to ℓ is equal to the distance from **p** to **q**.

22. Show that the line $\mathbf{x} = \mathbf{a} + t\mathbf{d}$ and the hyperplane $\mathbf{b} \cdot \mathbf{x} = c$ intersect at the point

$$\mathbf{a} + \frac{c - \mathbf{b} \cdot \mathbf{a}}{\mathbf{b} \cdot \mathbf{d}} \mathbf{d}$$

provided that $\mathbf{b} \cdot \mathbf{d} \neq 0$. What happens if $\mathbf{b} \cdot \mathbf{d} = 0$?

23. Let ℓ be the line in $\mathbb{R}^n$ with vector equation $\mathbf{x} = \mathbf{a} + t\mathbf{d}$ and let **p** be any point in $\mathbb{R}^n$. Let H be the hyperplane (line if $n = 2$, plane if $n = 3$) through **p**, perpendicular to ℓ.

    (a) Show that ℓ and H intersect at the point

    $$\mathbf{q} = \mathbf{a} - \frac{(\mathbf{a} - \mathbf{p}) \cdot \mathbf{d}}{\|\mathbf{d}\|^2} \mathbf{d}.$$

    (b) The ***distance from* p *to* ℓ** is given by the formula

    $$D = \|\mathbf{q} - \mathbf{p}\|$$

    where **q** is the point found in part (a) (see Figure 1.54). (Do you recognize $\mathbf{q} - \mathbf{p}$ as a component of $\mathbf{a} - \mathbf{p}$?) Find D when ℓ is the line $\mathbf{x} = (-1, 1, 3) + t(2, 0, 2)$ in $\mathbb{R}^3$ and **p** is the point $(0, 0, 0)$.

24. The ***angle*** between two hyperplanes, $\mathbf{a} \cdot \mathbf{x} = c$ and $\mathbf{b} \cdot \mathbf{x} = d$, in $\mathbb{R}^n$ (planes if $n = 3$, lines if $n = 2$) is defined to be the angle between the vectors **a** and **b**. Find the angle between each of the following:

    (a) the lines $3x_1 + x_2 = 5$ and $x_1 + 2x_2 = 7$, in $\mathbb{R}^2$

    (b) the planes $x_1 + x_2 + x_3 = 2$ and $2x_1 - x_2 - x_3 = 1$, in $\mathbb{R}^3$

    (c) the hyperplanes $x_1 = 0$ and $3x_1 + x_2 + x_3 + x_4 = 0$, in $\mathbb{R}^4$

1.7
The Cross Product

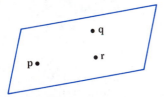

FIGURE 1.55

The plane through **p**, **q**, and **r**.

In the previous section we learned that in $\mathbb{R}^3$ the equation $\mathbf{b} \cdot (\mathbf{x} - \mathbf{a}) = 0$ describes the plane through **a** perpendicular to **b**. The vector **b**, however, is often not given explicitly. Suppose, for example, that we wish to find an equation for the plane in $\mathbb{R}^3$ passing through three noncollinear points **p**, **q**, and **r** (see Figure 1.55). The problem of finding a vector perpendicular to this plane is the same as the problem of finding a vector **b** perpendicular to the vectors $\mathbf{q} - \mathbf{p}$ and $\mathbf{r} - \mathbf{p}$ (see Figure 1.56). We will now develop a formula for finding a vector perpendicular to any given pair of vectors in $\mathbb{R}^3$.

Suppose $\mathbf{a} = (a_1, a_2, a_3)$ and $\mathbf{b} = (b_1, b_2, b_3)$ are any two vectors $\mathbb{R}^3$, neither of which is a scalar multiple of the other. In order to find a vector **x** perpendicular to both **a** and **b** we must solve the equations $\mathbf{a} \cdot \mathbf{x} = 0$ and $\mathbf{b} \cdot \mathbf{x} = 0$; that is, we must solve the linear system

$$a_1 x_1 + a_2 x_2 + a_3 x_3 = 0$$

$$b_1 x_1 + b_2 x_2 + b_3 x_3 = 0.$$

The general solution of this system is

$$\mathbf{x} = c(a_2 b_3 - a_3 b_2, \, a_3 b_1 - a_1 b_3, \, a_1 b_2 - a_2 b_1),$$

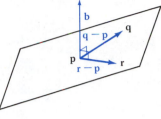

FIGURE 1.56
The vector **b** is perpendicular to the plane through **p**, **q**, and **r** if and only if it is perpendicular to the vectors **q** − **p** and **r** − **p**.

where c is an arbitrary real number (see Exercise 10). If we take $c = 1$, we obtain the particular solution

$$(a_2b_3 - a_3b_2, \ a_3b_1 - a_1b_3, \ a_1b_2 - a_2b_1).$$

This vector is perpendicular to both **a** and **b**.

Given any two vectors $\mathbf{a} = (a_1, a_2, a_3)$ and $\mathbf{b} = (b_1, b_2, b_3)$ in $\mathbb{R}^3$, the vector $(a_2b_3 - a_3b_2, \ a_3b_1 - a_1b_3, \ a_1b_2 - a_2b_1)$ is called the **cross product** of **a** and **b** and is denoted $\mathbf{a} \times \mathbf{b}$. To calculate $\mathbf{a} \times \mathbf{b}$, it is probably easiest to remember the formula in the following way. Given real numbers a, b, c, and d, the number $ad - bc$ is called the **determinant** of the 2×2 matrix

$$\begin{pmatrix} a & b \\ c & d \end{pmatrix}$$

and we write

$$\det \begin{pmatrix} a & b \\ c & d \end{pmatrix} = ad - bc.$$

Notice that

$$a_2b_3 - a_3b_2 = \det \begin{pmatrix} a_2 & a_3 \\ b_2 & b_3 \end{pmatrix},$$

$$a_3b_1 - a_1b_3 = -\det \begin{pmatrix} a_1 & a_3 \\ b_1 & b_3 \end{pmatrix},$$

and

$$a_1b_2 - a_2b_1 = \det \begin{pmatrix} a_1 & a_2 \\ b_1 & b_2 \end{pmatrix}.$$

Therefore

$$\mathbf{a} \times \mathbf{b} = \left(\det \begin{pmatrix} a_2 & a_3 \\ b_2 & b_3 \end{pmatrix}, \ -\det \begin{pmatrix} a_1 & a_3 \\ b_1 & b_3 \end{pmatrix}, \ \det \begin{pmatrix} a_1 & a_2 \\ b_1 & b_2 \end{pmatrix} \right).$$

Notice further that these three 2×2 matrices can be obtained simply by successively deleting columns of the 2×3 matrix

$$\begin{pmatrix} \mathbf{a} \\ \mathbf{b} \end{pmatrix} = \begin{pmatrix} a_1 & a_2 & a_3 \\ b_1 & b_2 & b_3 \end{pmatrix}$$

whose rows are the vectors **a** and **b**. In other words,

$$\mathbf{a} \times \mathbf{b} = \left(\det \begin{pmatrix} \boxed{a_1} & a_2 & a_3 \\ \boxed{b_1} & b_2 & b_3 \end{pmatrix}, \ -\det \begin{pmatrix} a_1 & \boxed{a_2} & a_3 \\ b_1 & \boxed{b_2} & b_3 \end{pmatrix}, \ \det \begin{pmatrix} a_1 & a_2 & \boxed{a_3} \\ b_1 & b_2 & \boxed{b_3} \end{pmatrix} \right).$$

The shaded columns here and elsewhere in the text indicate columns that have been deleted. Don't forget the minus sign before the second entry in this formula!

EXAMPLE 1

If $\mathbf{a} = (1, -1, 2)$ and $\mathbf{b} = (3, 0, 1)$, then

$$\mathbf{a} \times \mathbf{b} =$$

$$\left(\det\begin{pmatrix} 1 & -1 & 2 \\ 3 & 0 & 1 \end{pmatrix}, \ -\det\begin{pmatrix} 1 & -1 & 2 \\ 3 & 0 & 1 \end{pmatrix}, \ \det\begin{pmatrix} 1 & -1 & 2 \\ 3 & 0 & 1 \end{pmatrix} \right)$$

$$= \left(\det\begin{pmatrix} -1 & 2 \\ 0 & 1 \end{pmatrix}, \ -\det\begin{pmatrix} 1 & 2 \\ 3 & 1 \end{pmatrix}, \ \det\begin{pmatrix} 1 & -1 \\ 3 & 0 \end{pmatrix} \right),$$

$$= ((-1)(1) - (2)(0), \ -[(1)(1) - (2)(3)], \ (1)(0) - (-1)(3))$$

$$= (-1, 5, 3).$$

This answer can be checked by verifying that the dot product of this vector with both $\mathbf{a} = (1, -1, 2)$ and $\mathbf{b} = (3, 0, 1)$ is zero. ∎

EXAMPLE 2

Let us find an equation for the plane P in $\mathbb{R}^3$ through the three points $\mathbf{p} = (1, -1, 1)$, $\mathbf{q} = (2, 3, 1)$, and $\mathbf{r} = (-1, 2, 3)$. The point $\mathbf{p} = (1, -1, 1)$ is on P. The vector

$$(\mathbf{q} - \mathbf{p}) \times (\mathbf{r} - \mathbf{p}) = (1, 4, 0) \times (-2, 3, 2)$$

$$= \left(\det\begin{pmatrix} 1 & 4 & 0 \\ -2 & 3 & 2 \end{pmatrix}, \ -\det\begin{pmatrix} 1 & 4 & 0 \\ -2 & 3 & 2 \end{pmatrix}, \ \det\begin{pmatrix} 1 & 4 & 0 \\ -2 & 3 & 2 \end{pmatrix} \right)$$

$$= (8, -2, 11)$$

is perpendicular to P. Therefore an equation for P is

$$(8, -2, 11) \cdot (\mathbf{x} - (1, -1, 1)) = 0$$

or

$$8x_1 - 2x_2 + 11x_3 = 21. \quad ∎$$

REMARK 1   In Example 2, there were many choices, so the form of the answer can vary. For example, $\mathbf{q} = (2, 3, 1)$ could have been used as the point on P and $(\mathbf{p} - \mathbf{q}) \times (\mathbf{r} - \mathbf{q})$ could have been used as the vector perpendicular to P. However, since

$$(\mathbf{p} - \mathbf{q}) \times (\mathbf{r} - \mathbf{q}) = (-1, -4, 0) \times (-3, -1, 2) = (-8, 2, -11),$$

these choices would have led to the equation

$$(-8, 2, -11) \cdot (\mathbf{x} - (2, 3, 1)) = 0$$

or

$$-8x_1 + 2x_2 - 11x_3 = -21$$

for P which is equivalent to the equation obtained in Example 2.

The cross product is also useful for finding a line perpendicular to a given pair of lines.

EXAMPLE 3

The lines $\mathbf{x} = (1, -1, 1) + t(1, 2, 3)$ and $\mathbf{x} = (0, -3, -2) + t(3, 2, 1)$ intersect. Let us find the point where they intersect and a vector equation for the line perpendicular to both of the given lines passing through their point of intersection.

To find the point of intersection we solve the equation

$$(1, -1, 1) + t_1(1, 2, 3) = (0, -3, -2) + t_2(3, 2, 1)$$

or, equivalently, the linear system

$$t_1 - 3t_2 = -1$$
$$2t_1 - 2t_2 = -2$$
$$3t_1 - t_2 = -3.$$

The matrix of this system reduces as follows:

$$\begin{pmatrix} 1 & -3 & -1 \\ 2 & -2 & -2 \\ 3 & -1 & -3 \end{pmatrix} \longrightarrow \begin{pmatrix} 1 & 0 & -1 \\ 0 & 1 & 0 \\ 0 & 0 & 0 \end{pmatrix}.$$

We see that $t_1 = -1$ and $t_2 = 0$, so the point of intersection of the two lines is $(0, -3, -2)$.

Now we must find a vector that is perpendicular to both of the given lines. Since $(1, 2, 3)$ is parallel to the first line and $(3, 2, 1)$ is parallel to the second, the vector

$$(1, 2, 3) \times (3, 2, 1) = (-4, 8, -4)$$

is perpendicular to both. Hence the line

$$\mathbf{x} = (0, -3, -2) + t(-4, 8, -4)$$

is perpendicular to both of the given lines and it passes through their point of intersection. ■

There is a nice geometric description of the cross product of two vectors. We already know that $\mathbf{a} \times \mathbf{b}$ is perpendicular to both $\mathbf{a}$ and $\mathbf{b}$. Let us now compute a formula for the length $\|\mathbf{a} \times \mathbf{b}\|$ of $\mathbf{a} \times \mathbf{b}$.

Note that

$$\|\mathbf{a} \times \mathbf{b}\|^2 = \|(a_2b_3 - a_3b_2, -(a_1b_3 - a_3b_1), a_1b_2 - a_2b_1)\|^2$$
$$= (a_2b_3 - a_3b_2)^2 + (a_1b_3 - a_3b_1)^2 + (a_1b_2 - a_2b_1)^2$$
$$= a_2^2b_3^2 + a_3^2b_2^2 + a_1^2b_3^2 + a_3^2b_1^2 + a_1^2b_2^2 + a_2^2b_1^2$$
$$- 2(a_2b_3a_3b_2 + a_1b_3a_3b_1 + a_1b_2a_2b_1).$$

If we compare this with

$$\|\mathbf{a}\|^2\|\mathbf{b}\|^2 = (a_1^2 + a_2^2 + a_3^2)(b_1^2 + b_2^2 + b_3^2)$$

$$= a_1^2b_1^2 + a_1^2b_2^2 + a_1^2b_3^2 + a_2^2b_1^2 + a_2^2b_2^2 + a_2^2b_3^2$$

$$+ a_3^2b_1^2 + a_3^2b_2^2 + a_3^2b_3^2$$

we see that

$$\|\mathbf{a}\|^2\|\mathbf{b}\|^2 - \|\mathbf{a} \times \mathbf{b}\|^2 = a_1^2b_1^2 + a_2^2b_2^2 + a_3^2b_3^2$$

$$+ 2(a_2b_3a_3b_2 + a_1b_3a_3b_1 + a_1b_2a_2b_1)$$

$$= (a_1b_1 + a_2b_2 + a_3b_3)^2$$

$$= (\mathbf{a} \cdot \mathbf{b})^2.$$

Solving for $\|\mathbf{a} \times \mathbf{b}\|^2$ yields the formula

$$\|\mathbf{a} \times \mathbf{b}\|^2 = \|\mathbf{a}\|^2\|\mathbf{b}\|^2 - (\mathbf{a} \cdot \mathbf{b})^2.$$

But, if $\mathbf{a}$ and $\mathbf{b}$ are both nonzero, then $(\mathbf{a} \cdot \mathbf{b})^2 = \|\mathbf{a}\|^2\|\mathbf{b}\|^2 \cos^2 \theta$, where θ is the angle between $\mathbf{a}$ and $\mathbf{b}$; hence

$$\|\mathbf{a}\|^2\|\mathbf{b}\|^2 - (\mathbf{a} \cdot \mathbf{b})^2 = \|\mathbf{a}\|^2\|\mathbf{b}\|^2 - \|\mathbf{a}\|^2\|\mathbf{b}\|^2 \cos^2 \theta$$

$$= \|\mathbf{a}\|^2\|\mathbf{b}\|^2(1 - \cos^2 \theta) = \|\mathbf{a}\|^2\|\mathbf{b}\|^2 \sin^2 \theta$$

or

$$\|\mathbf{a} \times \mathbf{b}\| = \|\mathbf{a}\| \|\mathbf{b}\| \sin \theta,$$

where θ is the angle between $\mathbf{a}$ and $\mathbf{b}$. (Note that $\sin \theta \geq 0$ since $0 \leq \theta \leq \pi$.)

This formula tells us that if $\mathbf{a}$ and $\mathbf{b}$ are both nonzero, then *the length* $\|\mathbf{a} \times \mathbf{b}\|$ *of the cross product* $\mathbf{a} \times \mathbf{b}$ *is equal to the area of the parallelogram spanned by* $\mathbf{a}$ *and* $\mathbf{b}$ (see Figure 1.57). Of course, if either $\mathbf{a}$ or $\mathbf{b}$ is equal to zero, then $\|\mathbf{a} \times \mathbf{b}\| = 0$.

The relationship between cross product and area can be used to compute areas of parallelograms and of triangles in $\mathbb{R}^3$.

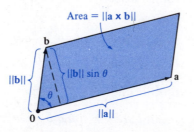

FIGURE 1.57

The parallelogram spanned by $\mathbf{a}$ and $\mathbf{b}$ has base length $\|\mathbf{a}\|$ and height $\|\mathbf{b}\| \sin \theta$. Its area is $\|\mathbf{a}\| \|\mathbf{b}\| \sin \theta = \|\mathbf{a} \times \mathbf{b}\|$.

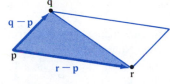

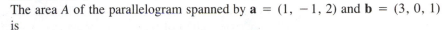

FIGURE 1.58
The area of the triangle with vertices **p**, **q**, and **r** is one half the area of the parallelogram spanned by **q** − **p** and **r** − **p**.

EXAMPLE 4

The area A of the parallelogram spanned by $\mathbf{a} = (1, -1, 2)$ and $\mathbf{b} = (3, 0, 1)$ is

$$A = \|\mathbf{a} \times \mathbf{b}\| = \|(1, -1, 2) \times (3, 0, 1)\| = \|(-1, 5, 3)\|$$

$$= \sqrt{1 + 25 + 9} = \sqrt{35} \approx 5.9.$$

The area B of the triangle whose vertices are $\mathbf{p} = (-1, 1, 3)$, $\mathbf{q} = (2, -2, 1)$, and $\mathbf{r} = (5, 4, 3)$ is one-half the area of the parallelogram spanned by $\mathbf{q} - \mathbf{p}$ and $\mathbf{r} - \mathbf{p}$ (see Figure 1.58). Hence

$$B = \tfrac{1}{2}\|(\mathbf{q} - \mathbf{p}) \times (\mathbf{r} - \mathbf{p})\| = \tfrac{1}{2}\|(3, -3, -2) \times (6, 3, 0)\|$$

$$= \tfrac{1}{2}\|(6, -12, 27)\|$$

$$= \tfrac{1}{2}\sqrt{909} \approx 15.07. \quad \blacksquare$$

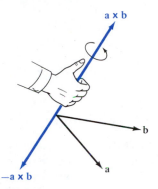

FIGURE 1.59
There are only two vectors in $\mathbb{R}^3$ of length $\|\mathbf{a} \times \mathbf{b}\|$ that are perpendicular to both **a** and **b**. The one determined by the right hand rule is $\mathbf{a} \times \mathbf{b}$.

We now know that if **a** and **b** are both nonzero, then $\mathbf{a} \times \mathbf{b}$ points in a direction perpendicular to **a** and **b** and that its length is equal to the area of the parallelogram spanned by **a** and **b**. But there are only two vectors in $\mathbb{R}^3$ with this length that are perpendicular to both **a** and **b**. These two vectors point in opposite directions (see Figure 1.59). It turns out that $\mathbf{a} \times \mathbf{b}$ is the one determined by the ***right hand rule:*** *the direction of $\mathbf{a} \times \mathbf{b}$ is the direction in which your right thumb points when the fingers of your right hand are curled in the direction from **a** to **b*** (Figure 1.59). No formal proof of this will be given here, but it can be seen intuitively why it is so. Move **a** and **b** continuously until **a** points in the direction of $\mathbf{e}_1 = (1, 0, 0)$ and **b** points in the direction of $\mathbf{e}_2 = (0, 1, 0)$, being careful to keep the angle between **a** and **b** always between 0 and π. At the end of this motion, the cross product must point in the direction of $\mathbf{e}_1 \times \mathbf{e}_2 = (0, 0, 1) = \mathbf{e}_3$; that is, it must point in the direction determined by the right hand rule (see Figure 1.60). But, throughout the motion, the cross product $\mathbf{a} \times \mathbf{b}$ moved continuously so it must have always pointed to the side of the plane of **a** and **b** determined by the right hand rule.

We have come, then, to a complete geometric description of $\mathbf{a} \times \mathbf{b}$.

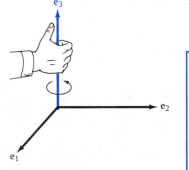

FIGURE 1.60
The direction of $\mathbf{e}_3$ is determined by the right hand rule.

The cross product $\mathbf{a} \times \mathbf{b}$ of two vectors **a** and **b** in $\mathbb{R}^3$, neither of which is a scalar multiple of the other, is the vector

(1) that is perpendicular to both **a** and **b**,

(2) whose length is the area of the parallelogram spanned by **a** and **b**, and

(3) whose direction is determined by the right hand rule.

If $\mathbf{a} = \mathbf{0}$ or $\mathbf{b} = \mathbf{0}$, or if **b** is a scalar multiple of **a**, then $\mathbf{a} \times \mathbf{b} = \mathbf{0}$.

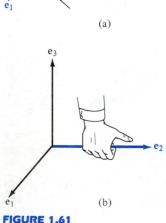

FIGURE 1.61
The right hand rule applied (a) to $\mathbf{e}_2 \times \mathbf{e}_3$, and (b) to $\mathbf{e}_3 \times \mathbf{e}_1$.

EXAMPLE 5

$\mathbf{e}_2 \times \mathbf{e}_3 = \mathbf{e}_1$ and $\mathbf{e}_3 \times \mathbf{e}_1 = \mathbf{e}_2$. These formulas are clear using the geometric characterization of the cross product (see Figure 1.61). Note that these vectors are mutually perpendicular, each pair of these vectors spans a square of area 1, and their directions are related by the right hand rule. ■

The cross product has several useful algebraic properties. We list some of them in the following theorem. Most of these can be proved directly from the definition of the cross product.

Theorem 1   *For all vectors* $\mathbf{a}$, $\mathbf{b}$, *and* $\mathbf{c}$ *in* $\mathbb{R}^3$,

(a) $\mathbf{a} \times \mathbf{b} = -\mathbf{b} \times \mathbf{a}$

(b) $(\mathbf{a} \times \mathbf{b}) \cdot \mathbf{a} = 0$ *and* $(\mathbf{a} \times \mathbf{b}) \cdot \mathbf{b} = 0$

(c) $(\mathbf{a} + \mathbf{b}) \times \mathbf{c} = \mathbf{a} \times \mathbf{c} + \mathbf{b} \times \mathbf{c}$

(d) $\mathbf{a} \cdot (\mathbf{b} \times \mathbf{c}) = \mathbf{b} \cdot (\mathbf{c} \times \mathbf{a}) = \mathbf{c} \cdot (\mathbf{a} \times \mathbf{b})$

(e) $\mathbf{a} \times (\mathbf{b} \times \mathbf{c}) = (\mathbf{a} \cdot \mathbf{c})\mathbf{b} - (\mathbf{a} \cdot \mathbf{b})\mathbf{c}$

(f) $\|\mathbf{a} \times \mathbf{b}\|^2 = \|\mathbf{a}\|^2\|\mathbf{b}\|^2 - (\mathbf{a} \cdot \mathbf{b})^2$.

Moreover,

(g)   *if* $\mathbf{a}$ *and* $\mathbf{b}$ *are nonzero, then* $\mathbf{a} \times \mathbf{b} = \mathbf{0}$ *if and only if* $\mathbf{a}$ *and* $\mathbf{b}$ *are parallel.*

REMARK 2   Property (a) says that the cross product is not commutative. It is not associative either since, for example, $\mathbf{e}_1 \times (\mathbf{e}_1 \times \mathbf{e}_2) = \mathbf{e}_1 \times \mathbf{e}_3 = -\mathbf{e}_3 \times \mathbf{e}_1 = -\mathbf{e}_2$, whereas $(\mathbf{e}_1 \times \mathbf{e}_1) \times \mathbf{e}_2 = \mathbf{0}$ from (g), so $\mathbf{e}_1 \times (\mathbf{e}_1 \times \mathbf{e}_2) \neq (\mathbf{e}_1 \times \mathbf{e}_1) \times \mathbf{e}_2$. Property (b) expresses the fact that $\mathbf{a} \times \mathbf{b}$ is perpendicular to both $\mathbf{a}$ and $\mathbf{b}$. Property (c) is a distributive law.

To verify property (g), note first that $\mathbf{a} \times \mathbf{b} = \mathbf{0}$ if and only if $\|\mathbf{a} \times \mathbf{b}\| = 0$. But $\|\mathbf{a} \times \mathbf{b}\| = \|\mathbf{a}\| \|\mathbf{b}\| \sin \theta$ so, if $\mathbf{a} \neq \mathbf{0}$ and $\mathbf{b} \neq \mathbf{0}$, then $\mathbf{a} \times \mathbf{b} = \mathbf{0}$ if and only if $\sin \theta = 0$. In other words, $\mathbf{a} \times \mathbf{b} = \mathbf{0}$ if and only if $\theta = 0$ or $\theta = \pi$, that is, if and only if $\mathbf{a}$ and $\mathbf{b}$ are parallel.

Property (f) was proved above, in the derivation of the formula $\|\mathbf{a} \times \mathbf{b}\| = \|\mathbf{a}\| \|\mathbf{b}\| \sin \theta$.

You are asked to verify the remaining properties in the exercises.

The product $\mathbf{a} \cdot (\mathbf{b} \times \mathbf{c})$ that appears in (d) of Theorem 1 is called the ***triple scalar product*** of $\mathbf{a}$, $\mathbf{b}$ and $\mathbf{c}$. It is, of course, a real number, and it has an interesting geometric interpretation. We know that

$$\mathbf{a} \cdot (\mathbf{b} \times \mathbf{c}) = \|\mathbf{a}\| \|\mathbf{b} \times \mathbf{c}\| \cos \theta,$$

where θ is the angle between $\mathbf{a}$ and $\mathbf{b} \times \mathbf{c}$. Also, we know that $\|\mathbf{b} \times \mathbf{c}\|$ is the area of the parallelogram spanned by $\mathbf{b}$ and $\mathbf{c}$. That parallelogram is the base of the parallelepiped spanned by $\mathbf{a}$, $\mathbf{b}$ and $\mathbf{c}$ (see Figure 1.62). Clearly $\|\mathbf{a}\| \, |\cos \theta|$ is the altitude of this parallelepiped; hence $\|\mathbf{b} \times \mathbf{c}\| \, \|\mathbf{a}\| \, |\cos \theta| = |\mathbf{a} \cdot (\mathbf{b} \times \mathbf{c})|$ is its volume. Therefore, we have proved the following theorem.

Theorem 2   *The volume V of the parallelepiped in* $\mathbb{R}^3$ *spanned by vectors* $\mathbf{a}$, $\mathbf{b}$ *and* $\mathbf{c}$ *is equal to the absolute value of the triple scalar product of* $\mathbf{a}$, $\mathbf{b}$ *and* $\mathbf{c}$: $V = |\mathbf{a} \cdot (\mathbf{b} \times \mathbf{c})|$.

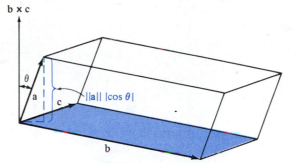

FIGURE 1.62

The parallelepiped spanned by **a**, **b**, and **c** has base area $\|\mathbf{b} \times \mathbf{c}\|$ and height $\|\mathbf{a}\| \, |\cos \theta|$. Its volume is therefore equal to $\|\mathbf{a}\| \, \|\mathbf{b} \times \mathbf{c}\| \, |\cos \theta| = |\mathbf{a} \cdot (\mathbf{b} \times \mathbf{c})|$.

EXAMPLE 6

Let us compute the volume V of the parallelepiped spanned by the vectors $\mathbf{a} = (3, 0, 1)$, $\mathbf{b} = (3, 3, 0)$, and $\mathbf{c} = (1, 2, 3)$. Since

$$\mathbf{b} \times \mathbf{c} = (3, 3, 0) \times (1, 2, 3) = \left(\det\begin{pmatrix} 3 & 0 \\ 2 & 3 \end{pmatrix}, \; -\det\begin{pmatrix} 3 & 0 \\ 1 & 3 \end{pmatrix}, \; \det\begin{pmatrix} 3 & 3 \\ 1 & 2 \end{pmatrix} \right),$$

$$= (9, -9, 3),$$

we find that

$$V = |\mathbf{a} \cdot (\mathbf{b} \times \mathbf{c})| = |(3, 0, 1) \cdot (9, -9, 3)| = 30. \quad\blacksquare$$

We can also use cross products to compute volumes of tetrahedra. The volume of a tetrahedron is equal to one-third the area of its base times its height. The base of the tetrahedron spanned by **a**, **b**, and **c** is the triangle spanned by $\mathbf{b} - \mathbf{a}$ and $\mathbf{c} - \mathbf{a}$ (see Figure 1.63). Its area is $\frac{1}{2}\|(\mathbf{b} - \mathbf{a}) \times (\mathbf{c} - \mathbf{a})\|$. The height of this tetrahedron is equal to the length of the component of **a** in the direction perpendicular to the base. Therefore, the height is

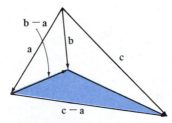

FIGURE 1.63

The tetrahedron spanned by **a**, **b**, and **c**.

$$\left\| \frac{\mathbf{a} \cdot [(\mathbf{b} - \mathbf{a}) \times (\mathbf{c} - \mathbf{a})]}{\|(\mathbf{b} - \mathbf{a}) \times (\mathbf{c} - \mathbf{a})\|^2} \, (\mathbf{b} - \mathbf{a}) \times (\mathbf{c} - \mathbf{a}) \right\|$$

or

$$\frac{|\mathbf{a} \cdot [(\mathbf{b} - \mathbf{a}) \times (\mathbf{c} - \mathbf{a})]|}{\|(\mathbf{b} - \mathbf{a}) \times (\mathbf{c} - \mathbf{a})\|}.$$

The volume of the tetrahedron is

$$V = \frac{1}{3} \cdot \frac{1}{2} \|(\mathbf{b} - \mathbf{a}) \times (\mathbf{c} - \mathbf{a})\| \cdot \frac{|\mathbf{a} \cdot (\mathbf{b} - \mathbf{a}) \times (\mathbf{c} - \mathbf{a})|}{\|(\mathbf{b} - \mathbf{a}) \times (\mathbf{c} - \mathbf{a})\|}$$

$$= \frac{1}{6} |\mathbf{a} \cdot (\mathbf{b} - \mathbf{a}) \times (\mathbf{c} - \mathbf{a})|$$

$$= \frac{1}{6} |\mathbf{a} \cdot (\mathbf{b} \times \mathbf{c} - \mathbf{b} \times \mathbf{a} - \mathbf{a} \times \mathbf{c} + \mathbf{a} \times \mathbf{a})|$$

$$= \frac{1}{6} |\mathbf{a} \cdot (\mathbf{b} \times \mathbf{c})|$$

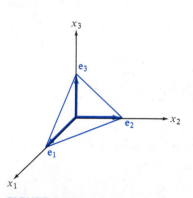

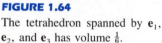

FIGURE 1.64
The tetrahedron spanned by $\mathbf{e}_1$, $\mathbf{e}_2$, and $\mathbf{e}_3$ has volume $\frac{1}{6}$.

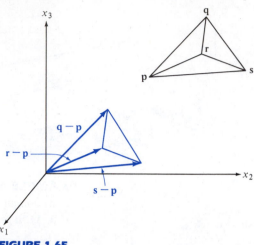

FIGURE 1.65
The tetrahedron with vertices $\mathbf{p}$, $\mathbf{q}$, $\mathbf{r}$, and $\mathbf{s}$ has the same volume as the tetrahedron spanned by $\mathbf{q} - \mathbf{p}$, $\mathbf{r} - \mathbf{p}$, and $\mathbf{s} - \mathbf{p}$.

since $\mathbf{a}$ is perpendicular to both $\mathbf{b} \times \mathbf{a}$ and $\mathbf{a} \times \mathbf{c}$, and $\mathbf{a} \times \mathbf{a} = \mathbf{0}$. Hence, *the volume of the tetrahedron spanned by $\mathbf{a}$, $\mathbf{b}$, and $\mathbf{c}$ is given by the formula*

$$V = \tfrac{1}{6}|\mathbf{a} \cdot (\mathbf{b} \times \mathbf{c})|.$$

EXAMPLE 7

The volume of the tetrahedron spanned by $\mathbf{e}_1 = (1, 0, 0)$, $\mathbf{e}_2 = (0, 1, 0)$, and $\mathbf{e}_3 = (0, 0, 1)$ (see Figure 1.64) is

$$V = \tfrac{1}{6}|\mathbf{e}_1 \cdot (\mathbf{e}_2 \times \mathbf{e}_3)| = \tfrac{1}{6}|\mathbf{e}_1 \cdot \mathbf{e}_1| = \tfrac{1}{6}.$$

The volume of the tetrahedron with vertices $\mathbf{p} = (1, 5, 5)$, $\mathbf{q} = (-1, 7, 8)$, $\mathbf{r} = (2, 7, 7)$, and $\mathbf{s} = (0, 8, 6)$ is the same as that of the tetrahedron spanned by the vectors $\mathbf{q} - \mathbf{p}$, $\mathbf{r} - \mathbf{p}$, and $\mathbf{s} - \mathbf{p}$ (see Figure 1.65). Its volume is

$$V = \tfrac{1}{6}|(\mathbf{q} - \mathbf{p}) \cdot ((\mathbf{r} - \mathbf{p}) \times (\mathbf{s} - \mathbf{p}))|$$

$$= \tfrac{1}{6}|(-2, 2, 3) \cdot ((1, 2, 2) \times (-1, 3, 1))|$$

$$= \tfrac{1}{6}|(-2, 2, 3) \cdot (-4, -3, 5)|$$

$$= \tfrac{17}{6}. \quad \blacksquare$$

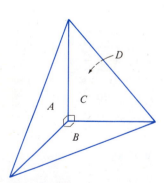

FIGURE 1.66
A right tetrahedron.

Theorem 3 (The 3-dimensional Pythagorean theorem). *Let A, B, C, and D be the areas of the four faces of a right tetrahedron, with D the area of the face opposite the vertex where the edges meet at right angles (see Figure 1.66). Then $D^2 = A^2 + B^2 + C^2$.*

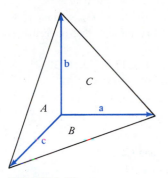

FIGURE 1.67
The mutually perpendicular vectors **a**, **b**, and **c** span a right tetrahedron.

Proof Let **a**, **b**, and **c** span the given tetrahedron, with **a** perpendicular to the face with area A, **b** perpendicular to the face with area B, and **c** perpendicular to the face with area C (see Figure 1.67).

By the formula for the area of a triangle in terms of the cross product we see that

$$D = \tfrac{1}{2}\|(\mathbf{b} - \mathbf{a}) \times (\mathbf{c} - \mathbf{a})\|$$

and hence

$$D^2 = \tfrac{1}{4}\|(\mathbf{b} - \mathbf{a}) \times (\mathbf{c} - \mathbf{a})\|^2$$

$$= \tfrac{1}{4}\|\mathbf{b} \times \mathbf{c} - \mathbf{a} \times \mathbf{c} - \mathbf{b} \times \mathbf{a} + \mathbf{a} \times \mathbf{a}\|^2$$

$$= \tfrac{1}{4}\|\mathbf{b} \times \mathbf{c} + \mathbf{c} \times \mathbf{a} + \mathbf{a} \times \mathbf{b}\|^2.$$

But, since **a**, **b**, and **c** are mutually perpendicular, **b** × **c** is a scalar multiple of **a**, **c** × **a** is a scalar multiple of **b**, and **a** × **b** is a sclar multiple of **c**. In particular, the three vectors **b** × **c**, **c** × **a**, and **a** × **b** are mutually perpendicular. It follows that

$$D^2 = \tfrac{1}{4}(\mathbf{b} \times \mathbf{c} + \mathbf{c} \times \mathbf{a} + \mathbf{a} \times \mathbf{b}) \cdot (\mathbf{b} \times \mathbf{c} + \mathbf{c} \times \mathbf{a} + \mathbf{a} \times \mathbf{b})$$

$$= \tfrac{1}{4}\|\mathbf{b} \times \mathbf{c}\|^2 + \tfrac{1}{4}\|\mathbf{c} \times \mathbf{a}\|^2 + \tfrac{1}{4}\|\mathbf{a} \times \mathbf{b}\|^2$$

$$= A^2 + B^2 + C^2. \quad \blacksquare$$

REMARK 3   It is important to understand that although the cross product is a powerful tool in vector calculations, its use is limited to $\mathbb{R}^3$. The cross product **a** × **b** is defined *only* when **a** and **b** are vectors in $\mathbb{R}^3$.

EXERCISES

1. Compute the following cross products.
   (a) $(2, -1, 1) \times (1, 2, 1)$ (b) $(-3, 4, 2) \times (4, 3, -2)$
   (c) $(1, 2, 1) \times (-1, -2, -1)$ (d) $(3, 2, 1) \times (-1, 2, 3)$
   (e) $(1, -1, 1) \times (-1, 1, 1)$ (f) $(2, 2, 1) \times (-1, 2, 2)$

2. Find a vector perpendicular to both of the given vectors.
   (a) $(1, -1, 1)$ and $(2, 3, 2)$ (b) $(2, -1, 1)$ and $(-1, 1, 2)$
   (c) $(3, 0, 2)$ and $(0, 3, 1)$ (d) $(4, -1, 4)$ and $(2, -1, 3)$
   (e) $(1, 2, 4)$ and $(4, 2, 1)$

3. Find an equation for the plane in $\mathbb{R}^3$ through the points
   (a) $(0, 0, 0)$, $(1, 1, 2)$, and $(2, -1, 1)$
   (b) $(1, 1, 1)$, $(1, -1, 1)$, and $(1, 2, 2)$
   (c) $(1, 2, 1)$, $(-1, 5, 3)$, and $(2, 0, 1)$
   (d) $(1, 0, 0)$, $(0, 1, 0)$, and $(0, 0, 1)$
   (e) $(1, 1, 1)$, $(-1, -1, -1)$, and $(2, 3, 0)$
   (f) $(3, 2, 3)$, $(-4, 1, 2)$, and $(-1, 3, 2)$

4. Each of the following pairs of lines intersect. For each pair of lines, find the point of intersection and find a vector equation for the line through that point that is perpendicular to both of the given lines.

   (a) $\mathbf{x} = (-1, 5, 5) + t(-2, 2, 2)$, and $\mathbf{x} = (1, 0, -1) + t(2, 1, 2)$

   (b) $\mathbf{x} = (3, 3, 2) + t(3, -1, 2)$, and $\mathbf{x} = (1, 1, 9) + t(5, 1, -5)$

   (c) $\mathbf{x} = (-1, 4, -3) + t(2, -1, 2)$, and $\mathbf{x} = (-1, 9, 1) + t(1, 2, 3)$

   (d) $\mathbf{x} = (-1, 3, 3) + t(-2, 1, 2)$, and $\mathbf{x} = (4, -1, -3) + t(-1, 2, 2)$

5. Find the area of the parallelogram in $\mathbb{R}^3$ spanned by the vectors

   (a) $(2, -1, 1)$ and $(1, 1, 1)$ (b) $(-1, 3, 1)$ and $(2, 2, 2)$

   (c) $(0, 1, 0)$ and $(0, 1, 1)$ (d) $(1, 2, -1)$ and $(4, 1, 3)$

   (e) $(-1, 2, 2)$ and $(3, 0, 1)$ (f) $(2, 5, 1)$ and $(1, 1, 4)$

6. Find the area of the triangle in $\mathbb{R}^3$ whose vertices are

   (a) $(1, 0, 0)$, $(0, 1, 0)$, $(0, 0, 1)$

   (b) $(0, 1, -1)$, $(5, 4, 3)$, $(-1, 1, -1)$

   (c) $(2, 2, -1)$, $(3, 1, 0)$, $(-1, 1, 2)$

   (d) $(0, 0, 0)$, $(1, 2, 0)$, $(2, 1, 0)$

   (e) $(1, 3, 0)$, $(-3, 1, 0)$, $(1, 1, 1)$

7. Find the area of the parallelogram in $\mathbb{R}^3$ whose vertices are

   (a) $(1, 2, 1)$, $(3, 1, 2)$, $(2, 3, 3)$, $(4, 2, 4)$

   (b) $(3, 5, 7)$, $(4, 6, 8)$, $(3, 4, 5)$, $(4, 5, 6)$

   (c) $(2, -1, 3)$, $(4, 6, -2)$, $(5, 7, 3)$, $(7, 14, -2)$

   (d) $(0, 0, 0)$, $(1, 1, 0)$, $(1, 0, 0)$, $(2, 1, 0)$

   (e) $(0, 0, 0)$, $(1, 2, 0)$, $(2, 1, 0)$, $(3, 3, 0)$

8. Find the volume of the parallelepiped in $\mathbb{R}^3$ spanned by the three vectors

   (a) $(-3, 1, 4)$, $(3, 2, -1)$, $(1, 0, 1)$

   (b) $(2, 1, 1)$, $(3, 1, 1)$, $(1, 3, 2)$

   (c) $(5, 4, 3)$, $(2, 1, 0)$, $(0, 1, -1)$

   (d) $(-1, 1, 2)$, $(0, 1, 1)$, $(0, 1, -1)$

   (e) $(2, 1, 0)$, $(1, 2, 0)$, $(0, 0, 1)$

9. Find the volume of the tetrahedron in $\mathbb{R}^3$ whose vertices are

   (a) $(0, 0, 0)$, $(1, 0, 0)$, $(0, 1, 0)$, $(0, 0, 1)$

   (b) $(0, 0, 0)$, $(1, 0, 0)$, $(0, 2, 0)$, $(1, 1, 3)$

   (c) $(1, 1, 0)$, $(-1, 2, 0)$, $(1, -3, 0)$, $(0, 0, 6)$

   (d) $(1, 3, -4)$, $(-2, 1, 1)$, $(1, -1, 0)$, $(3, -2, -1)$

10. Let (a_1, a_2, a_3) and (b_1, b_2, b_3) be nonzero vectors in $\mathbb{R}^3$ that are not multiples of one another. Verify that the general solution of the linear system

$$a_1x_1 + a_2x_2 + a_3x_3 = 0$$

$$b_1x_1 + b_2x_2 + b_3x_3 = 0$$

is $\mathbf{x} = c(a_2b_3 - a_3b_2, a_3b_1 - a_1b_3, a_1b_2 - a_2b_1)$, where c is an arbitrary real number. (You will need to consider several special cases. Start by assuming that $a_1 \neq 0$ and that $a_1b_2 - a_2b_1 \neq 0$.)

11. Show, by direct calculation, that $(\mathbf{a} \times \mathbf{b}) \cdot \mathbf{a} = 0$ and $(\mathbf{a} \times \mathbf{b}) \cdot \mathbf{b} = 0$ whenever $\mathbf{a}$ and $\mathbf{b}$ are vectors in $\mathbb{R}^3$.

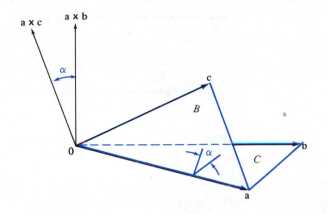

FIGURE 1.68
The angle between two planes is equal to the angle between vectors perpendicular to those planes.

12. Prove parts (a), (c), (d) and (e) of Theorem 1.

13. Prove the 3-dimensional law of cosines: Let A, B, C, and D be the areas of the four faces of a tetrahedron. Then

$$D^2 = A^2 + B^2 + C^2 - 2BC \cos \alpha - 2AC \cos \beta - 2AB \cos \gamma$$

where α is the angle between the faces with areas B and C, β is the angle between the faces with areas A and C, and γ is the angle between the faces with areas A and B. [*Hint:* Imitate the proof of Theorem 3. Use the fact that $\|\mathbf{v}\|^2 = \mathbf{v} \cdot \mathbf{v}$ to evaluate $\|\mathbf{b} \times \mathbf{c} - \mathbf{a} \times \mathbf{c} - \mathbf{b} \times \mathbf{a}\|^2$. Note that the angle between two faces is the same as the angle between vectors perpendicular to these faces (see Figure 1.68).]

1.8
Application: Linear Programming

In the previous sections of this chapter we studied linear equations, or *linear equalities*. In this section we study *linear inequalities*. We shall see how to graph systems of linear inequalities in $\mathbb{R}^1$ and $\mathbb{R}^2$, and we shall discuss some of the elementary ideas of *linear programming*, a branch of mathematics that has many applications in economics, business, and industry.

A ***linear inequality*** in $\mathbb{R}^n$ is an inequality of one of the following four types:

$$a_1 x_1 + \cdots + a_n x_n < b,$$

$$a_1 x_1 + \cdots + a_n x_n > b,$$

$$a_1 x_1 + \cdots + a_n x_n \leq b,$$

or

$$a_1 x_1 + \cdots + a_n x_n \geq b,$$

where $a_1, \ldots, a_n$, and b are real numbers with not all of the a_i's equal to zero. A ***solution*** of a linear inequality in $\mathbb{R}^n$ is any vector $\mathbf{x} = (x_1, \ldots, x_n)$ in $\mathbb{R}^n$ that satisfies the inequality.

A linear inequality in $\mathbb{R}^1$ must be of one of the following types:

$$a_1 x_1 < b, \qquad a_1 x_1 > b, \qquad a_1 x_1 \leq b, \qquad \text{or} \qquad a_1 x_1 \geq b,$$

(a) $x_1 \geq b/a_1$

b/a_1

(b) $x_1 \leq b/a_1$

b/a_1

(c) $x_1 > b/a_1$

b/a_1

(d) $x_1 < b/a_1$

b/a_1

FIGURE 1.69
The solution set of a linear inequality in $\mathbb{R}^1$ is a half-line.

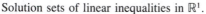

0   3

(a) $2x_1 \leq 6$

-3 0

(b) $-2x_1 < 6$

FIGURE 1.70
Solution sets of linear inequalities in $\mathbb{R}^1$.

where a_1 and b are real numbers, and $a_1 \neq 0$. Any linear inequality in $\mathbb{R}^1$ can be solved by dividing both sides by the coefficient a_1 to obtain one of the four inequalities:

$$x_1 < b/a_1, \qquad x_1 > b/a_1, \qquad x_1 \leq b/a_1, \qquad \text{or} \qquad x_1 \geq b/a_1.$$

(Remember that *if $a_1 < 0$, then the direction of the inequality changes when you divide by a_1!*) Each of these inequalities describes a *half-line* (see Figure 1.69). *If the inequality is strict ($>$ or $<$) then the endpoint of the half-line is not included in the solution set. If the inequality is not strict ($\geq$ or $\leq$) then the endpoint is included in the solution set.*

EXAMPLE 1

In $\mathbb{R}^1$, the solution set of the inequality $2x_1 \leq 6$ is the set of all real numbers x_1 with $x_1 \leq 3$ (see Figure 1.70a). The solution set of the inequality $-2x_1 < 6$ is the set of all real numbers x_1 with $x_1 > -3$ (see Figure 1.70b). ■

Now let us consider linear inequalities in $\mathbb{R}^2$. We know that the equation $a_1x_1 + a_2x_2 = b$ describes a line in $\mathbb{R}^2$ whenever $(a_1, a_2) \neq (0, 0)$. This line divides the plane into two **half-planes**. One of these half-planes is the solution set of the inequality $a_1x_1 + a_2x_2 < b$. The other is the solution set of the inequality $a_1x_1 + a_2x_2 > b$. These half-planes are called *open* half planes: a half-plane is **open** if it does not contain its boundary line (see Figure 1.71a).

FIGURE 1.71
Half-planes in $\mathbb{R}^2$: (a) an open half-plane, (b) a closed half-plane.

(a)

(b)

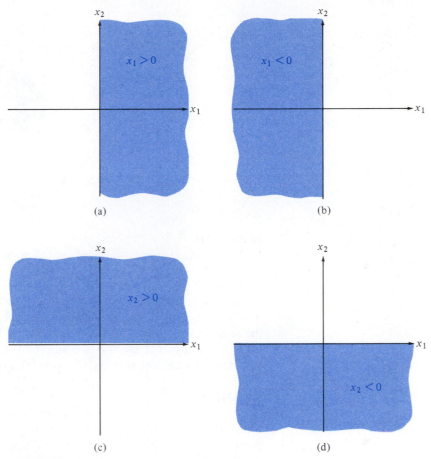

FIGURE 1.72
The open (a) right, (b) left, (c) upper, and (d) lower half-planes in $\mathbb{R}^2$.

On the other hand, the solution sets of the linear inequalities $a_1x_1 + a_2x_2 \leq b$ and $a_1x_1 + a_2x_2 \geq b$ are *closed* half-planes: a half-plane is **closed** if it does contain its boundary line (see Figure 1.71b).

EXAMPLE 2

In $\mathbb{R}^2$, the inequality $x_1 > 0$ describes the **open right half-plane** (see Figure 1.72a). Similarly, the inequalities $x_1 < 0$, $x_2 > 0$, and $x_2 < 0$ describe, respectively, the **open left, upper,** and **lower half-planes** (see Figure 1.72b, c, and d). The half-planes $x_1 \geq 0$, $x_1 \leq 0$, $x_2 \geq 0$, and $x_2 \leq 0$ are the **closed right, left, upper,** and **lower half-planes.** ■

EXAMPLE 3

Let us sketch the half-plane $x_1 + x_2 \leq 1$. First, we sketch the boundary line $x_1 + x_2 = 1$ (see Figure 1.73a). We know that the inequality $x_1 + x_2 \leq 1$ describes the line $x_1 + x_2 = 1$ together with all of the points on one side of it. To see which side of the line is described by the inequality, we need only check

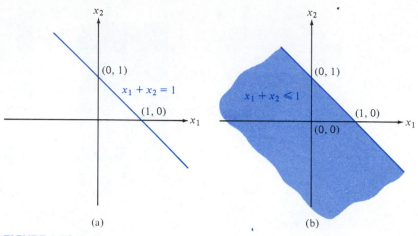

FIGURE 1.73
(a) The line $x_1 + x_2 = 1$. (b) The closed half-plane $x_1 + x_2 \leq 1$.

a point that is not on the line to see whether or not it is in the solution set. Let us try $(x_1, x_2) = (0, 0)$. Since $0 + 0 < 1$, we see that $(0, 0)$ is a solution of the inequality $x_1 + x_2 \leq 1$. We can conclude, then, that the solution set of the linear inequality $x_1 + x_2 \leq 1$ is the closed half-plane that contains $(0, 0)$ and is bounded by the line $x_1 + x_2 = 1$ (see Figure 1.73b). ∎

EXAMPLE 4

Let us sketch the solution set of the linear inequality $-2x_1 + 3x_2 > 0$. This solution set is an open half-plane bounded by the line $-2x_1 + 3x_2 = 0$. The point $(1, 0)$ is not on the boundary line. Since $-2(1) + 3(0) = -2 < 0$, we see that $(1, 0)$ does not satisfy the given inequality. Hence the solution set is the open half-plane not containing $(1, 0)$, bounded by the line $-2x_1 + 3x_2 = 0$ (see Figure 1.74). ∎

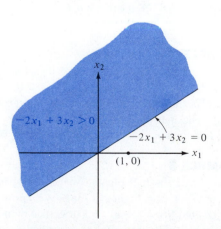

FIGURE 1.74
The open half-plane $-2x_1 + 3x_2 > 0$.

In $\mathbb{R}^n$, when $n \geq 3$, the solution set of a linear inequality is called a *half-space*. The half-space is *open* if the inequality is strict, and it is *closed* if the inequality is not strict.

EXAMPLE 5

In $\mathbb{R}^3$, the solution set of the inequality $x_3 > 0$ is often called the open *upper half-space*. Similarly, the half-space $x_3 \leq 0$ is the closed *lower half-space* in $\mathbb{R}^3$. ∎

Now let us turn our attention to *systems* of linear inequalities.

EXAMPLE 6

Let us sketch the solution set in $\mathbb{R}^2$ of the system of linear inequalities

$$x_1 > 0$$

$$x_2 > 0$$

$$x_1 + x_2 \leq 1.$$

FIGURE 1.75
The first quadrant is the solution set of the pair of inequalities $x_1 > 0$, $x_2 > 0$.

The solution set of the inequality $x_1 > 0$ is the open right half-plane. The solution set of the inequality $x_2 > 0$ is the open upper half-plane. A point (x_1, x_2) satisfies both of these inequalities if and only if it lies on both of these half-planes; that is, if and only if it lies in the first quadrant (see Figure 1.75). The solution set of the third inequality, $x_1 + x_2 \leq 1$, is the closed half-plane sketched in Figure 1.73b. A point (x_1, x_2) satisfies all three inequalities if and only if it is in this half-plane *and* in the first quadrant. Hence the solution set of the system is a triangular region (see Figure 1.76). Notice that part, but not all, of the boundary of the triangle is contained in the solution set. ∎

The technique illustrated in Example 6 may be summarized as follows.

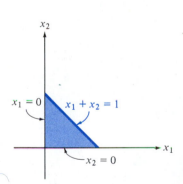

FIGURE 1.76
The solution set of the system of linear inequalities $x_1 > 0$, $x_2 > 0$, $x_1 + x_2 \leq 1$.

To sketch the solution set of a system of linear inequalities in $\mathbb{R}^2$, proceed as follows:

(1) First sketch S_1, the solution set of the first inequality. The set S_1 is a half-plane.

(2) Next sketch S_2, the intersection of S_1 with the half-plane defined by the second inequality. The set S_2 is the set of points that simultaneously satisfy the first two inequalities.

(3) Then sketch S_3, the intersection of S_2 with the half-plane defined by the third inequality. The set S_3 is the set of points that simultaneously satisfies the first three inequalities.

(4) Continue in this way until you have dealt with all of the given inequalities. The resulting set is the set of points that satisfies all of the given inequalities simultaneously.

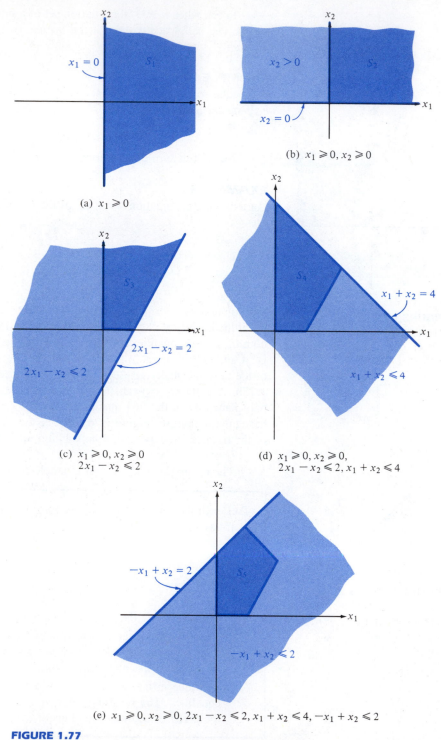

(a) $x_1 \geqslant 0$

(b) $x_1 \geqslant 0, x_2 \geqslant 0$

(c) $x_1 \geqslant 0, x_2 \geqslant 0$
$2x_1 - x_2 \leqslant 2$

(d) $x_1 \geqslant 0, x_2 \geqslant 0,$
$2x_1 - x_2 \leqslant 2, x_1 + x_2 \leqslant 4$

(e) $x_1 \geqslant 0, x_2 \geqslant 0, 2x_1 - x_2 \leqslant 2, x_1 + x_2 \leqslant 4, -x_1 + x_2 \leqslant 2$

FIGURE 1.77
Sketching the solution set of a system of linear inequalities.

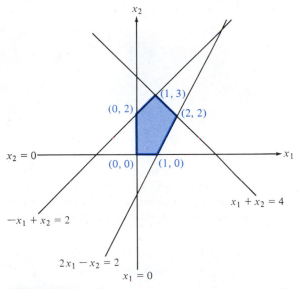

FIGURE 1.78
The boundary lines and vertices of the polygon $x_1 \geq 0$,
$x_2 \geq 0$, $2x_1 - x_2 \leq 2$, $x_1 + x_2 \leq 4$, $-x_1 + x_2 \leq 2$.

EXAMPLE 7

In Figure 1.77 is sketched the solution set of the system of linear inequalities

$$x_1 \geq 0$$

$$x_2 \geq 0$$

$$2x_1 - x_2 \leq 2$$

$$x_1 + x_2 \leq 4$$

$$-x_1 + x_2 \leq 2.$$

Notice that the solution set is a polygon bounded by segments of the lines
$x_1 = 0$, $x_2 = 0$, $2x_1 - x_2 = 2$, $x_1 + x_2 = 4$, and $-x_1 + x_2 = 2$ (see Figure
1.78). We can, if we wish, find the vertices of this polygon by solving pairs of
simultaneous equations. For example, one vertex is the point where the lines
$2x_1 - x_2 = 2$ and $x_1 + x_2 = 4$ intersect. Solving this pair of equations we
find the vertex $(x_1, x_2) = (2, 2)$. ■

The solution set of a system of linear inequalities in $\mathbb{R}^2$ is always a
polygonal set; that is, it is a subset of $\mathbb{R}^2$ bounded by straight line segments.
But not every polygonal set can be described as the solution set of a system of
linear inequalities, because these solution sets are necessarily *convex*. A subset
S of $\mathbb{R}^2$ is ***convex*** if every line segment that joins two points of S lies entirely
in S (see Figure 1.79). The reason that solution sets of systems of linear inequalities in $\mathbb{R}^2$ are necessarily convex is that they are intersections of half-

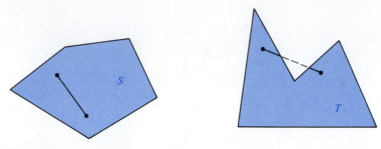

FIGURE 1.79

The set S is convex because every line segment that joins two points of S lies entirely in S. The set T is not convex because there are line segments that join pairs of points of T but that do not lie entirely in T. The set T cannot be described as the solution set of a system of linear inequalities.

planes. Half-planes are obviously convex, and an intersection of convex sets is always convex (see Exercise 20).

Solution sets of systems of linear inequalities in $\mathbb{R}^2$ are convex polygonal sets. Some of these sets are *bounded* and some are *unbounded*. A subset of $\mathbb{R}^2$ is **bounded** if it is contained inside a (possibly very large) circle (see Figure 1.80a). A subset of $\mathbb{R}^2$ is **unbounded** if it is not bounded (see Figure 1.80b).

Much of what has been said here about solution sets of linear inequalities in $\mathbb{R}^2$ carries over directly to solution sets of linear inequalities in $\mathbb{R}^3$. But rather than pursue these ideas here, let us see how linear inequalities arise in applications.

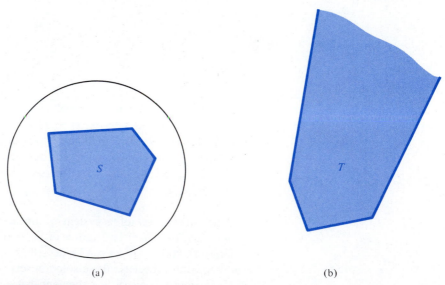

(a)   (b)

FIGURE 1.80

Polygonal sets in $\mathbb{R}^2$: (a) S is bounded because there is a circle that contains it. (b) T is unbounded because there is no circle that contains it.

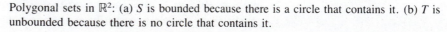

EXAMPLE 8

A tailor has 18 yards of material from which to make two kinds of shirts, short-sleeved shirts and long-sleeved shirts. Each short-sleeved shirt requires $1\frac{1}{2}$ yards of material. Each long-sleeved shirt requires 2 yards of material. Let x_1 be the number of short-sleeved shirts to be made and let x_2 be the number of long-sleeved shirts to be made. Not all values of (x_1, x_2) are possible. First, the tailor cannot make a negative number of shirts of either style, so $x_1 \geq 0$ and $x_2 \geq 0$. Furthermore, no more than 18 yards of material can be used. The amount of material required to make x_1 short-sleeved shirts and x_2 long-sleeved shirts is $\frac{3}{2}x_1 + 2x_2$, so $\frac{3}{2}x_1 + 2x_2 \leq 18$. We can conclude that (x_1, x_2) must be a solution of the system of linear inequalities

$$x_1 \geq 0$$

$$x_2 \geq 0$$

$$\tfrac{3}{2}x_1 + 2x_2 \leq 18.$$

These inequalities are called *constraints* on the values of x_1 and x_2. There may be additional constraints. For example, the tailor may have only 30 hours available that can be devoted to making the shirts. If it takes three hours to make each shirt, then the total number of hours required to make x_1 short-sleeved shirts and x_2 long-sleeved shirts is $3x_1 + 3x_2$, so there is the additional constraint $3x_1 + 3x_2 \leq 30$. Hence (x_1, x_2) must satisfy the system of four linear inequalities

$$x_1 \geq 0$$

$$x_2 \geq 0$$

$$\tfrac{3}{2}x_1 + 2x_2 \leq 18$$

$$3x_1 + 3x_2 \leq 30. \quad \blacksquare$$

The tailor in Example 8 may ask, "Which of the many possible values of (x_1, x_2) will earn the greatest possible profit?" If there is a profit of \$14 on every short-sleeved shirt, and a profit of \$16 on every long-sleeved shirt, then the total profit (in dollars) from making x_1 short-sleeved shirts and x_2 long-sleeved shirts is $P = 14x_1 + 16x_2$. So the tailor wants to know which of the possible values of (x_1, x_2) makes P largest. This is an example of a *linear programming* problem.

A *linear programming problem* is a problem of the following type: *given a system of linear inequalities in* $\mathbb{R}^n$ *and a function P of the form P =* $c_1x_1 + c_2x_2 + \cdots + c_nx_n$, *where* $c_1, c_2, \ldots, c_n$ *are real numbers, find the solutions* $(x_1, \ldots, x_n)$ *(if there are any) of the system of inequalities that make P largest (or smallest).* The function P is called the **objective function** of the linear programming problem. The linear inequalities are the **constraints.** The solution set of the system of linear inequalities is the **constraint set.**

Some linear programming problems are fairly easy to solve geometrically.

EXAMPLE 9

Let us find the largest and smallest values of the objective function $P = 2x_1 + x_2$ subject to the constraints

$$x_1 \geq 0$$

$$x_2 \geq 0$$

$$2x_1 - x_2 \leq 2$$

$$x_1 + x_2 \leq 4$$

$$-x_1 + x_2 \leq 2.$$

First we sketch the constraint set. This has already been done for this set of constraints in Example 7 (Figure 1.77). Now, for each real number c, the set of points (x_1, x_2) where P takes on the value c is the line $2x_1 + x_2 = c$. This line has slope -2 and x_2-intercept c. If we vary c, we obtain a family of parallel lines (see Figure 1.81). If these lines are superimposed on the constraint set (Figures 1.77e and 1.78) we obtain Figure 1.82. From this figure it can be seen that, on the constraint set, the function P takes values between 0 and 6. Furthermore, the maximum value 6 is attained only at the point $(2, 2)$, and the minimum value 0 is attained only at the point $(0, 0)$. ■

The method used in Example 9 can be used to solve any linear programming problem in $\mathbb{R}^2$ when the constraint set is bounded and closed. A polygonal set is ***closed*** if it contains all of its boundary line segments. The constraint set will be closed whenever none of the inequalities that appear in the constraints is strict. If the constraint set is bounded and closed, then it is a closed convex

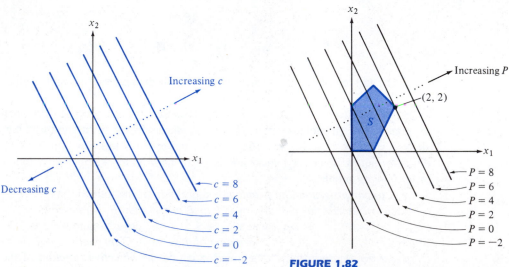

FIGURE 1.81

Taking different values for c in $2x_1 + x_2 = c$ yields a family of parallel lines.

FIGURE 1.82

The maximum value of the objective function $P = 2x_1 + x_2$ on the constraint set S is 6, and this value is achieved only at the point $(2, 2)$.

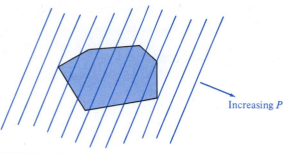

FIGURE 1.84
The objective function is constant along each line of the family. The largest and smallest values of P occur at vertices of the constraint set.

FIGURE 1.83
A closed convex polygon.

polygon (see Figure 1.83). If $P = c_1x_1 + c_2x_2$ is any objective function, then for each real number c, the set of points where $P = c$ ($c_1x_1 + c_2x_2 = c$) is a line in $\mathbb{R}^2$. If we vary c, we obtain a family of parallel lines (Figure 1.84). It is clear from the figure that the largest and smallest values of P must occur at *vertices* of the constraint set. [If the lines $c_1x_1 + c_2x_2 = c$ are parallel to one of the boundary line segments, then the maximum (or minimum) will be attained at each point of that segment. This will be the case when P is largest (or smallest) at *two* vertices (see Exercise 19)]. This means that to find the largest and smallest values of P, all that needs to be done is to compare the values of P at the vertices of the constraint set.

This method can be summarized as follows:

The Vertex Method

To solve a linear programming problem in $\mathbb{R}^2$ when the constraint set is bounded and closed,

1. Sketch the constraint set.
2. Find all the vertices of the constraint set.
3. Evaluate the objective function at each vertex.
4. Identify the largest and smallest values obtained in Step 3.

EXAMPLE 10

Let us use the vertex method to solve the linear programming problem of Example 9. The first step is to sketch the constraint set as was done in Figure 1.77. The next step is to find the vertices of the constraint set as in Figure 1.78. Then we must evaluate the objective function $P = 2x_1 + x_2$ at each of the vertices. The results can be tabulated as follows:

VERTEX	(0, 0)	(1, 0)	(2, 2)	(1, 3)	(0, 2)
VALUE OF P	0	2	6	5	2

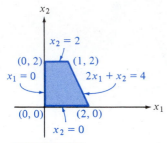

FIGURE 1.85
The constraint set $x_1 \geq 0$, $x_2 \geq 0$, $x_2 \leq 2$, $2x_1 + x_4 \leq 4$.

We see that 6 is the largest value of P and that this value occurs at $(2, 2)$. Also, 0 is the smallest value of P and this value occurs at $(0, 0)$. ■

EXAMPLE 11

Let us find the largest and the smallest values of $P = 4x_1 + 2x_2$ subject to the constraints

$$x_1 \geq 0$$

$$x_2 \geq 0$$

$$x_2 \leq 2$$

$$2x_1 + x_2 \leq 4.$$

The constraint set is sketched in Figure 1.85. The vertices and the values of P at the vertices are:

VERTEX	$(0, 0)$	$(0, 2)$	$(1, 2)$	$(2, 0)$
VALUE OF P	0	4	8	8

It can be seen that the smallest value of P at a vertex, and therefore on the constraint set, is 0, and that value is attained at $(0, 0)$. The largest value is 8. It is attained at two vertices, $(1, 2)$ and $(2, 0)$. Hence the largest value of P on the constraint set is attained at *every* point of the boundary line segment that connects the vertices $(1, 2)$ and $(2, 0)$. (Sketch the lines $4x_1 + 2x_2 = c$ for various values of c to see geometrically why this is true.) ■

EXAMPLE 12

Let us return to Example 8. Recall that the tailor plans to make x_1 short-sleeved shirts and x_2 long-sleeved shirts from 18 yards of material. Each short-sleeved shirt requires $1\frac{1}{2}$ yards of material, and each long-sleeved shirt requires 2 yards of material. The constraints are therefore

$$x_1 \geq 0$$

$$x_2 \geq 0$$

$$\tfrac{3}{2}x_1 + 2x_2 \leq 18.$$

The constraint set is the triangle in Figure 1.86. If the tailor can make a profit of \$14 on each short-sleeved shirt and \$16 on each long-sleeved shirt, how many shirts of each type should be made to maximize the profit?

The objective function here is $P = 14x_1 + 16x_2$. The vertices of the constraint set and the values of P on those vertices are

VERTEX	$(0, 0)$	$(0, 9)$	$(12, 0)$
VALUE OF P	0	144	168

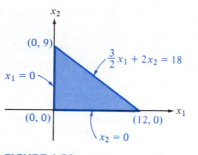

FIGURE 1.86
The constraint set $x_1 \geq 0$, $x_2 \geq 0$, $\frac{3}{2}x_1 + 2x_2 \leq 18$.

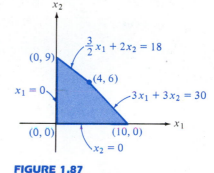

FIGURE 1.87
The constraint set $x_1 \geq 0$, $x_2 \geq 0$, $\frac{3}{2}x_1 + 2x_2 \leq 18$, $3x_1 + 3x_2 \leq 30$.

The profit P will be maximized if 12 long-sleeved shirts and no short-sleeved shirts are made. The profit will be $168.

But suppose now that the tailor has only 30 hours that can be devoted to the task of making these shirts. If three hours are needed to make a shirt of either style, then there is the additional constraint

$$3x_1 + 3x_2 \leq 30.$$

This changes the problem significantly. The new constraint set is the quadrilateral in Figure 1.87. The vertices and the values of P on those vertices are

VERTEX	(0, 0)	(0, 9)	(4, 6)	(10, 0)
VALUE OF P	0	144	152	140

With this additional constraint, the profit P will be maximized if 4 short-sleeved shirts and 6 long-sleeved shirts are made, for a profit of $152. ∎

REMARK The vertex method works fairly well for solving linear programming problems in $\mathbb{R}^2$. Although this method does generalize to a method for solving linear programming problems in $\mathbb{R}^n$, the generalization is not efficient. A much more efficient method, the *simplex method*, has been developed for solving these problems. You can read about the simplex method in *Finite Mathematics* by Althoen and Bumcrot (Norton, 1978).

EXERCISES

1. Sketch the following half-lines in $\mathbb{R}^1$.
   (a) $x_1 > 3$ (b) $x_1 < 7$
   (c) $x_1 \leq 2$ (d) $x_1 \geq -1$
   (e) $2x_1 < 8$ (f) $3x_1 \leq \frac{1}{3}$
   (g) $-x_1 < 5$ (h) $-3x_1 > -9$
2. Sketch the following half-planes in $\mathbb{R}^2$.
   (a) $x_1 \geq 3$ (b) $x_2 < 5$

(c) $x_2 > -1$ (d) $x_1 - x_2 \leq 0$

(e) $x_1 - x_2 \leq 1$ (f) $2x_1 + x_2 < 5$

(g) $3x_1 - 4x_2 \geq -2$ (h) $-5x_1 + x_2 > 3$

3. Sketch the solution sets in $\mathbb{R}^1$ of the following systems of linear inequalities.

(a) $x_1 > 1, x_1 < 3$ (b) $x_1 \geq -2, x_1 < 0$

(c) $x_1 \leq 1, x_1 \geq 0$ (d) $x_1 < 0, x_1 < 1$

(e) $x_1 \leq 1, x_1 \geq 1$ (f) $x_1 < 2, x_1 < 3, x_1 > 0$

4. Explain why the solution set of a system of linear inequalities in $\mathbb{R}^1$ is always either a half-line, an interval, a single point, or the empty set.

5. Sketch the solution sets in $\mathbb{R}^2$ of the following systems of inequalities.

(a) $x_1 > 0, x_2 > 0, x_1 < 3, x_2 < 5$

(b) $x_1 \geq 0, x_2 \geq 0, -x_1 + x_2 \leq 0, x_2 \leq 1, x_1 \leq 2$

(c) $x_1 > 0, x_2 > 0, x_2 < 1, x_1 - x_2 \geq 0$

(d) $x_1 + x_2 \geq 1, x_1 + x_2 \leq 3, x_1 - x_2 \leq 1, x_1 - x_2 \geq -1$

(e) $x_1 \geq 0, x_2 \geq 0, x_1 + x_2 \leq 4, x_1 - x_2 \leq 1, x_1 - x_2 \geq -1$

(f) $x_1 \geq 0, x_2 \geq 0, 2x_1 - x_2 \leq 2, 2x_2 - x_1 \leq 2$

6. Sketch the solution sets in $\mathbb{R}^2$ of the following systems of inequalities.

(a) $x_1 \geq 0, x_2 \geq 0, \frac{1}{2}x_1 - x_2 \leq \frac{1}{2}, x_1 - x_2 \geq -2, x_1 + x_2 \leq 4$

(b) $x_1 + x_2 \leq 2, x_1 + x_2 \geq -2, x_2 - x_1 \geq -1, x_2 - x_1 \leq 1$

(c) $x_1 + x_2 \geq 1, x_2 \geq 0, x_1 \leq 2, 2x_2 - x_1 \leq 2$

(d) $x_1 \geq 0, x_2 \geq 0, x_1 - 5x_2 \leq 0$

(e) $x_1 \geq 0, 2x_2 - x_1 \leq 4, 2x_2 - x_1 \geq 0$

(f) $x_1 \geq 0, x_2 \geq 0, 2x_2 - 3x_1 \leq 1, x_2 \leq 2, x_1 + x_2 \leq 4, x_2 - x_1 \geq -2$

(g) $x_1 \geq 0, x_2 \geq 0, x_1 + 2x_2 \leq 4, 7x_1 + 6x_2 \leq 16, x_1 - 3x_2 \leq 1$

(h) $x_1 \leq 0, x_2 \geq 0, x_1 + x_2 \leq 1, x_2 \geq -1$

7. Which of the following subsets of $\mathbb{R}^2$ are convex?

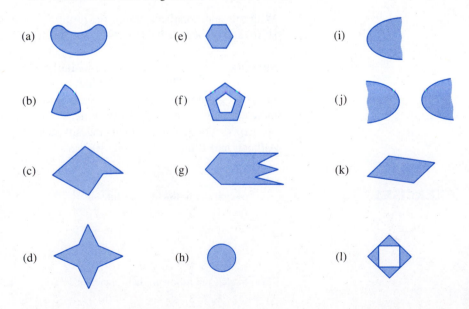

(a)

(b)

(c)

(d)

(e)

(f)

(g)

(h)

(i)

(j)

(k)

(l)

8. Which of the following subsets of $\mathbb{R}^2$ are bounded?
   (a) $x_2 > 10$
   (b) $x_1 > 0, x_2 > 0$
   (c) $x_1 \geq 0, x_2 \geq 0, x_1 + x_2 \leq 2$
   (d) $x_1 \geq 0, x_2 \geq 0, x_1 - x_2 \leq 1$
   (e) $x_1 + x_2 \leq 1, x_1 + x_2 \geq -1, x_1 - x_2 \leq 1, x_1 - x_2 \geq -1$

9. Find the largest and the smallest values of $P = 2x_1 - x_2$ subject to the given constraints. In each case, also find the values of x_1 and x_2 for which these largest and smallest values of P are attained.
   (a) $x_1 \geq 0, x_2 \geq 0, x_1 + x_2 \leq 10$
   (b) $x_1 \geq 0, x_2 \geq 0, x_1 + x_2 \leq 10, x_2 \leq 8$
   (c) $x_1 \geq 0, x_2 \geq 0, x_1 + x_2 \leq 10, x_2 \leq 8, x_1 - x_2 \leq 6$

10. Find the largest and smallest values of $P = -3x_1 + 4x_2$ subject to the given constraints. In each case, also find the values of x_1 and x_2 for which these largest and smallest values of P are attained.
    (a) $x_1 \geq 0, x_1 \leq 2, x_2 \geq 0, x_2 \leq 2$
    (b) $x_1 + x_2 \geq -1, x_1 + x_2 \leq 1, x_1 - x_2 \geq -1, x_1 - x_2 \leq 1$
    (c) $x_1 \geq -4, x_1 + x_2 \leq 2, -x_1 + x_2 \leq 2$
    (d) $x_2 \geq -4, x_1 + x_2 \leq 2, -x_1 + x_2 \leq 2, x_2 \leq 1$

11. Find the largest value of P subject to the constraints $x_1 \geq 0, x_2 \geq 0, x_1 + x_2 \leq 4, x_1 - x_2 \leq 1, x_1 - x_2 \geq -1$, where
    (a) $P = x_2$ (b) $P = x_1$
    (c) $P = x_1 + x_2$ (d) $P = 2x_1 + x_2$
    (e) $P = x_1 + 2x_2$ (f) $P = 2x_1 - x_2$
    (g) $P = x_1 - 2x_2$ (h) $P = x_1 - x_2$

12. A nut store sells two mixtures of peanuts and cashews, a regular mixture and a deluxe mixture. The regular mixture consists of 75% peanuts and 25% cashews (by weight). The deluxe mixture consists of equal weights of each. Suppose that 50 pounds of peanuts and 30 pounds of cashews are available for mixing, that the profit per pound sold of the regular mixture is 74¢, and that the profit per pound sold of the deluxe mixture is 50¢. How many pounds of each mixture should be made in order to maximize profit, assuming all will be sold?

13. In Exercise 12 suppose that the profit per pound for the regular mixture is 75¢ and the profit per pound for the deluxe mixture is 49¢. How many pounds of each should be made to maximize profit?

14. In Exercise 12 suppose that the profit per pound for the regular mixture is 75¢ and the profit per pound for the deluxe mixture is 50¢. How many pounds of each should be made to maximize profit?

15. In Exercise 12 suppose that the profit per pound for the regular mixture is 40¢ and the profit per pound for the deluxe mixture is 80¢. How many pounds of each should be made to maximize profit?

16. Foods A and B each contain nutrients X, Y, and Z. Each unit of food A contains 1 unit of X, 7 units of Y, and 2 units of Z. Each unit of food B contains 1 unit of X, 3 units of Y, and 3 units of Z. Recommended daily allowances of these nutrients have been established as follows: 5 units of X, 21 units of Y, and 12 units of Z. The unit cost of food A is 35¢ and the unit cost of food B is 22¢. How many units

of each food must be consumed daily in order to at least meet recommended daily allowances of each nutrient at lowest cost?

17. In Exercise 16 suppose that the unit cost of food A is 25¢, and the unit cost of food B is 41¢. How many units of each food must be consumed daily to at least meet recommended daily allowances at lowest cost?

18. In Exercise 16 suppose that the unit cost of each food is 30¢. How many units of each must be consumed daily to at least meet recommended daily allowances at lowest cost?

19. (a) Show that if $P = c_1 x_1 + c_2 x_2$ assumes the same value at two distinct points (a_1, a_2) and (b_1, b_2) in $\mathbb{R}^2$, then it assumes that value at all points on the line joining (a_1, a_2) and (b_1, b_2). [*Hint:* $\mathbf{x}$ is on the line joining $\mathbf{a}$ and $\mathbf{b}$ if and only if $\mathbf{x} = t\mathbf{a} + (1 - t)\mathbf{b}$ for some t.]

    (b) Show that if $P = c_1 x_1 + c_2 x_2$ assumes the same value at three noncollinear points (p_1, p_2), (q_1, q_2), and (r_1, r_2), then c_1 and c_2 must be zero and hence P assumes the same value at every point of $\mathbb{R}^2$.

20. Show that every intersection of convex sets in $\mathbb{R}^2$ is convex.

REVIEW EXERCISES

1. True or False? Explain your answer.

    (a) Every linear system of four equations in five unknowns has infinitely many solutions.

    (b) If two systems of linear equations have matrices that row reduce to the same row echelon matrix, then they have the same solution set.

    (c) A system of three linear equations in three unknowns has a *unique* solution if and only if its matrix row reduces to a row echelon matrix of the form

    $$\begin{pmatrix} 1 & 0 & 0 & a \\ 0 & 1 & 0 & b \\ 0 & 0 & 1 & c \end{pmatrix}.$$

    (d) If a system of m linear equations in 4 unknowns has a unique solution, then m must be greater than or equal to 4.

    (e) If a system of m linear equations in 4 unknowns has no solutions, then m must be greater than or equal to 5.

2. Identify, as a point, a line, a plane, or the empty set, the solution set of the linear system whose matrix is

    (a) $\begin{pmatrix} 0 & 1 & 1 \\ 0 & 0 & 0 \end{pmatrix}$ (b) $\begin{pmatrix} 1 & 0 & 1 \\ 0 & 1 & -1 \end{pmatrix}$ (c) $\begin{pmatrix} 1 & 1 & 0 & 2 \\ 0 & 0 & 1 & -1 \end{pmatrix}$

    (d) $\begin{pmatrix} 1 & 0 & 0 & 0 \\ 0 & 1 & 0 & 0 \\ 0 & 0 & 0 & 1 \end{pmatrix}$ (e) $\begin{pmatrix} 1 & -1 & 2 & 1 \\ 0 & 0 & 0 & 0 \\ 0 & 0 & 0 & 0 \end{pmatrix}$

3. Solve the linear system

$$2x_1 + 5x_2 - 3x_3 + x_4 = 1$$

$$3x_1 - 5x_2 + x_3 - 2x_4 = 0$$

$$-x_1 + x_2 - 3x_3 + 3x_4 = 7$$

4. Find an equation that describes

   (a) the line in $\mathbb{R}^3$ that passes through $(1, -1, 1)$ and is perpendicular to the plane $2x_1 - 3x_2 + x_3 = 7$.

   (b) the plane in $\mathbb{R}^3$ that passes through $(0, 0, 0)$ and is perpendicular to the line $\mathbf{x} = (-1, 1, 2) + t(1, -2, -3)$.

   (c) the plane in $\mathbb{R}^3$ that passes through the three points $(-1, 1, 2)$, $(1, 0, 3)$, and $(3, 4, -1)$.

   (d) the line in $\mathbb{R}^4$ that passes through the two points $(-1, 2, 3, 0)$ and $(1, 1, -1, 2)$.

   (e) the plane in $\mathbb{R}^3$ that contains the two lines $\mathbf{x} = (1, 1, 1) + t(-2, 1, 3)$ and $\mathbf{x} = (1, 1, 1) + t(3, -1, 2)$.

5. Let P be the plane $3x_1 - 3x_2 + 4x_3 = 5$.

   (a) Find a vector perpendicular to P.

   (b) Find the distance from $(0, 0, 0)$ to P.

   (c) Find the distance from $(-1, 1, 2)$ to P.

   (d) Find an equation that describes the plane that passes through $(1, 0, 0)$ and is parallel to P.

   (e) Find an equation that describes the line of intersection of P and the plane $x_1 + x_2 + x_3 = 1$.

6. Let $\mathbf{a} = (-3, 5, 4)$ and $\mathbf{b} = (3, 0, -4)$. Find

   (a) $2\mathbf{a} + 5\mathbf{b}$ (b) $\mathbf{a} \cdot \mathbf{b}$ (c) $\mathbf{a} \times \mathbf{b}$ (d) $\|\mathbf{a}\|$

   (e) the angle between $\mathbf{a}$ and $\mathbf{b}$

   (f) the distance from $\mathbf{a}$ to $\mathbf{b}$

   (g) the area of the parallelogram spanned by $\mathbf{a}$ and $\mathbf{b}$

   (h) the midpoint of the line segment that joins $\mathbf{a}$ to $\mathbf{b}$

   (i) the component of $\mathbf{b}$ along $\mathbf{a}$

   (j) the component of $\mathbf{b}$ perpendicular to $\mathbf{a}$.

7. Find the point (or points) of intersection, if any, of

   (a) the lines $2x_1 + 3x_2 = 1$ and $3x_1 - 4x_2 = 7$ in $\mathbb{R}^2$

   (b) the planes $x_1 + x_2 + x_3 = 1$, $x_1 - x_2 - x_3 = 0$, and $x_1 + x_2 = 1$ in $\mathbb{R}^3$

   (c) the lines $\mathbf{x} = (1, 1, -1) + t(-1, 0, 3)$ and $\mathbf{x} = (3, 2, 5) + t(1, 1, 1)$ in $\mathbb{R}^3$

   (d) the line $\mathbf{x} = (5, 5, 2) + t(-1, 1, 1)$ and the plane $x_1 + 2x_2 - x_3 = 7$ in $\mathbb{R}^3$

   (e) the planes $x_1 + x_2 - 2x_3 = 1$ and $x_1 - 2x_2 + x_3 = 1$ in $\mathbb{R}^3$

8. Find an equation that describes the line in $\mathbb{R}^3$ that passes through the point of intersection of the lines $\mathbf{x} = (3, 0, -2) + t(1, -1, 1)$ and $\mathbf{x} = (0, 0, -5) + t(1, 2, 1)$ and is perpendicular to both of these lines.

9. (a) The points in $\mathbb{R}^2$ that are equidistant from the points $(1, 3)$ and $(-1, 2)$ form a line. Find an equation that describes that line.

   (b) The points in $\mathbb{R}^3$ that are equidistant from the points $(0, 1, 2)$ and $(4, -1, 3)$ form a plane. Find an equation for that plane.

10. Show that if $\mathbf{a}$ and $\mathbf{b}$ are distinct points in $\mathbb{R}^n$, then the set of all points in $\mathbb{R}^n$ that are equidistant from $\mathbf{a}$ and $\mathbf{b}$ is the hyperplane that is perpendicular to $\mathbf{b} - \mathbf{a}$ and passes through the midpoint of the line segment from $\mathbf{a}$ to $\mathbf{b}$.

11. Let ℓ be the line in $\mathbb{R}^3$ with equation $\mathbf{x} = \mathbf{a} + t\mathbf{d}$ and let $\mathbf{p}$ be a point in $\mathbb{R}^3$. Show that the distance from $\mathbf{p}$ to ℓ is given by the formula

$$D = \frac{\|(\mathbf{a} - \mathbf{p}) \times \mathbf{d}\|}{\|\mathbf{d}\|}$$

12. Sketch the solution sets in $\mathbb{R}^2$ of the following systems of inequalities and decide which of these sets are closed and which are bounded.

(a) $x_1 \geq 0, x_2 \geq 0, x_1 \leq 3, x_1 + x_2 \leq 4$

(b) $x_1 + x_2 \geq 1, x_1 < 1, x_2 < 1.$

(c) $2x_1 \leq 5, 2x_1 + x_2 \leq 5, x_2 \geq -1$

(d) $x_2 < 2, -x_1 + x_2 < 1, x_1 > 0, x_1 + x_2 > -1, x_2 > -2$

13. Find the largest and the smallest values of P, and the points where they occur, in the constraint set $x_1 \geq 0, x_2 \geq 0, x_1 + x_2 \leq 4, x_2 \leq 2, x_1 - 2x_2 \leq 1$ where

(a) $P = x_1 - 2x_2$ (b) $P = x_1 - x_2$

(c) $P = x_1 + 3x_2$ (d) $P = 3x_1 + x_2$

Linear Maps and Matrix Algebra

In Chapter 1 we saw that matrices are useful in solving systems of linear equations. In this chapter we shall use matrices to define functions with interesting geometric properties. We shall add and multiply matrices. And we shall see that, using matrix multiplication, we can describe systems of linear equations in an especially neat and pleasing way. This description will lead to a new method for solving linear systems and to new insights about their solution sets.

2.1
Linear Maps from the Plane to the Plane

In this section we shall show how to use a 2×2 matrix to define a function from $\mathbb{R}^2$ to $\mathbb{R}^2$.

Recall that a function f from a set S to a set T is a rule that assigns to each element x in the set S some element $f(x)$ in the set T. Thus, for example, the function f defined by $f(x) = x^5$ assigns to each real number x its fifth power. The function g defined by the formula $g(x_1, x_2) = x_1 + x_2$ assigns to each pair of real numbers their sum.

We write $f:S \rightarrow T$ whenever f is a function from S to T. The notation $f:S \rightarrow T$ is read "f maps S to T" or "f, mapping S to T."

The word "map" is used here because a function may be viewed as a way of "mapping" the set S on the set T. This is illustrated in Figure 2.1. On the right is a map of a part of New York City containing Kennedy Airport. Let S be the set of points in the city and let T be the set of points on the map. The map defines, and is defined by, the function f that sends a point in the city to the corresponding point on the map. Hence it is reasonable to call the function f a *mapping*, or simply a *map*, for short.

Let us see how we can use a 2×2 matrix to define a map from $\mathbb{R}^2$ to $\mathbb{R}^2$.

Consider the matrix

$$A = \begin{pmatrix} 2 & 3 \\ -1 & 5 \end{pmatrix}.$$

This matrix is the coefficient matrix of the linear system

$$2x_1 + 3x_2 = b_1$$

$$-x_1 + 5x_2 = b_2.$$

This pair of equations can be viewed in two ways. On the one hand, if b_1 and b_2 are known, then this pair of equations can be solved for the unknowns x_1 and x_2. That is what was done in Chapter 1. On the other hand, if x_1 and x_2 are

FIGURE 2.1
A map is a function.

known, then this pair of equations can be used to determine the numbers b_1 and b_2. In other words, we can use this pair of equations to define a function $f:\mathbb{R}^2 \to \mathbb{R}^2$. This function is defined by the formula $f(x_1, x_2) = (b_1, b_2)$, or

$$f(x_1, x_2) = (2x_1 + 3x_2, -x_1 + 5x_2).$$

This function is called the *linear map associated with the matrix A*.

More generally, given any 2×2 matrix

$$A = \begin{pmatrix} a & b \\ c & d \end{pmatrix}$$

we define the **linear map associated with** A to be the function $f:\mathbb{R}^2 \to \mathbb{R}^2$ defined by the formula

$$f(x_1, x_2) = (ax_1 + bx_2, cx_1 + dx_2).$$

This function is called *linear* because each entry in $f(\mathbf{x}) = f(x_1, x_2)$ is defined by a linear equation in the entries of $\mathbf{x}$. Notice that $f(\mathbf{x})$ *is the vector whose entries are the dot products of the row vectors of A with* $\mathbf{x}$. The matrix A is called the **matrix of the linear map** f.

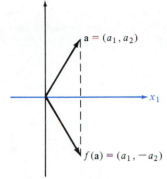

FIGURE 2.2
Reflection in the x_1-axis.

EXAMPLE 1
Let
$$A = \begin{pmatrix} 1 & 0 \\ 0 & -1 \end{pmatrix}.$$

Then the linear map $f:\mathbb{R}^2 \to \mathbb{R}^2$ associated with A is given by the formula
$$f(x_1, x_2) = (1x_1 + 0x_2, 0x_1 - 1x_2)$$

or
$$f(x_1, x_2) = (x_1, -x_2).$$

This map leaves the first entry of each vector alone and changes the sign of the second entry. Geometrically, f reflects each vector in $\mathbb{R}^2$ in the x_1-axis (see Figure 2.2). ∎

EXAMPLE 2
Let
$$A = \begin{pmatrix} 0 & 1 \\ 1 & 0 \end{pmatrix}.$$

The associated linear map $f:\mathbb{R}^2 \to \mathbb{R}^2$ is defined by the formula
$$f(x_1, x_2) = (0x_1 + 1x_2, 1x_1 + 0x_2)$$

or
$$f(x_1, x_2) = (x_2, x_1).$$

This linear map interchanges the entries of each vector in $\mathbb{R}^2$. Geometrically, f reflects each vector in $\mathbb{R}^2$ in the line $x_1 = x_2$ (see Figure 2.3). [This is easy to check. Try it! (See Exercise 8.)] ∎

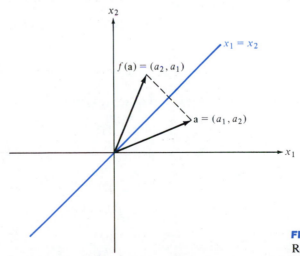

FIGURE 2.3
Reflection in the line $x_1 = x_2$.

EXAMPLE 3

It is easy to find the matrix of a linear map f whenever an explicit formula for $f(x_1, x_2)$ is given. Simply read off the coefficients of x_1 and x_2 in the formula and arrange them in a 2×2 matrix. Thus, if

$$f(x_1, x_2) = (ax_1 + bx_2, cx_1 + dx_2),$$

then the matrix of f is

$$\begin{pmatrix} a & b \\ c & d \end{pmatrix}.$$

In particular, if $f(x_1, x_2) = (2x_1 - 3x_2, x_1)$, then the matrix of f is

$$\begin{pmatrix} 2 & -3 \\ 1 & 0 \end{pmatrix}. \quad \blacksquare$$

The next theorem shows that linear maps behave well with respect to vector addition and scalar multiplication.

Theorem 1 *Let* $f: \mathbb{R}^2 \to \mathbb{R}^2$ *be a linear map, let* $\mathbf{x}$ *and* $\mathbf{y}$ *be vectors in* $\mathbb{R}^2$, *and let* r *and* s *be real numbers. Then*

(i) $f(\mathbf{x} + \mathbf{y}) = f(\mathbf{x}) + f(\mathbf{y})$,

(ii) $f(r\mathbf{x}) = rf(\mathbf{x})$

and

(iii) $f(r\mathbf{x} + s\mathbf{y}) = rf(\mathbf{x}) + sf(\mathbf{y})$.

Proof   Let $f: \mathbb{R}^2 \to \mathbb{R}^2$ be a linear map and let

$$A = \begin{pmatrix} a & b \\ c & d \end{pmatrix}$$

be the matrix of f. Let $\mathbf{x} = (x_1, x_2)$ and $\mathbf{y} = (y_1, y_2)$. Then

$$
\begin{aligned}
f(\mathbf{x} + \mathbf{y}) &= f(x_1 + y_1, x_2 + y_2) \\
&= (a(x_1 + y_1) + b(x_2 + y_2), c(x_1 + y_1) + d(x_2 + y_2)) \\
&= (ax_1 + ay_1 + bx_2 + by_2, cx_1 + cy_1 + dx_2 + dy_2) \\
&= (ax_1 + bx_2, cx_1 + dx_2) + (ay_1 + by_2, cy_1 + dy_2) \\
&= f(x_1, x_2) + f(y_1, y_2) \\
&= f(\mathbf{x}) + f(\mathbf{y}).
\end{aligned}
$$

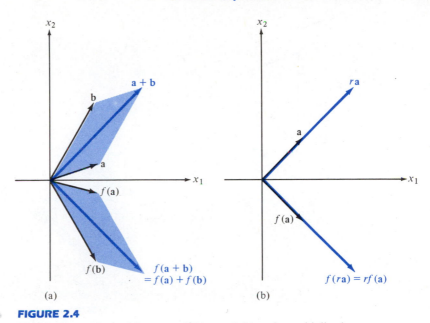

(a) (b)

FIGURE 2.4

Linear maps preserve (a) vector addition and (b) scalar multiplication.

That proves (i). Similarly,

$$f(r\mathbf{x}) = f(rx_1, rx_2)$$

$$= (arx_1 + brx_2, crx_1 + drx_2)$$

$$= r(ax_1 + bx_2, cx_1 + dx_2)$$

$$= rf(x_1, x_2)$$

$$= rf(\mathbf{x}),$$

proving (ii). Finally, (iii) follows from (i) and (ii):

$$f(r\mathbf{x} + s\mathbf{y}) = f(r\mathbf{x}) + f(s\mathbf{y}) = rf(\mathbf{x}) + sf(\mathbf{y}). \quad \blacksquare$$

Often this theorem is stated as follows: *linear maps preserve vector addition and scalar multiplication*.

Theorem 1 is pictured graphically in Figure 2.4.

Given two sets S and T, a function $f: S \to T$, and an element $x \in S$, the element $f(x)$ in T is called the **image of x under f**. This terminology is used because of its connection with optics. If a slide is projected through a lens onto a screen, then each point x on the slide is projected to a point $f(x)$ on the screen (see Figure 2.5). Thus we have a function f that assigns to each point x on the slide a corresponding point $f(x)$, its *image*, on the screen.

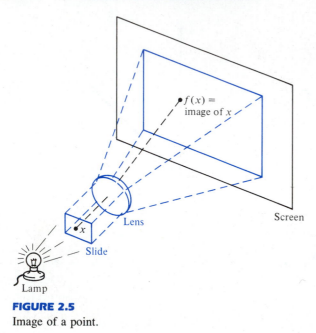

FIGURE 2.5
Image of a point.

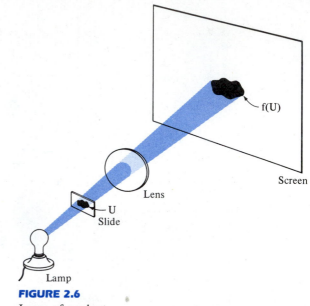

FIGURE 2.6
Image of a subset.

Given any subset U of S and any function $f:S \rightarrow T$, the ***image of U under*** f is the subset

$$f(U) = \{y \in T \,|\, y = f(x) \text{ for some } x \in U\}$$

consisting of the images under f of all the elements in U (see Figure 2.6).

Some functions are complicated in that the images of simple sets are quite complex. A poor lens, for example, may lead to a very distorted representation of the picture being projected. But linear maps are quite good in this respect. For example, the image under a linear map of a straight line is always either a straight line or a single point.

Theorem 2 *Let* $f:\mathbb{R}^2 \rightarrow \mathbb{R}^2$ *be a linear map and let* **a** *and* **d** *be vectors in* $\mathbb{R}^2$, *with* $\mathbf{d} \neq \mathbf{0}$. *Then the image under* f *of the line through* **a** *in the direction of* **d** *is either*

    (i)   *a line, the line through* $f(\mathbf{a})$ *in the direction of* $f(\mathbf{d})$ *(if* $f(\mathbf{d}) \neq \mathbf{0}$*)*

or

    (ii)   *a point, the point* $f(\mathbf{a})$ *(if* $f(\mathbf{d}) = \mathbf{0}$*)*.

Proof The parametric representation of the line through **a** in the direction of **d** is $\mathbf{x} = \mathbf{a} + t\mathbf{d}$. By Theorem 1,

$$f(\mathbf{a} + t\mathbf{d}) = f(\mathbf{a}) + f(t\mathbf{d}) = f(\mathbf{a}) + tf(\mathbf{d}).$$

Hence, as t runs through $\mathbb{R}$, $\mathbf{a} + t\mathbf{d}$ traces out the line through **a** in the direction of **d**, and $f(\mathbf{a} + t\mathbf{d}) = f(\mathbf{a}) + tf(\mathbf{d})$ traces out the line through $f(\mathbf{a})$ in the

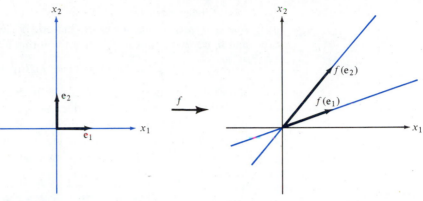

FIGURE 2.7
Coordinate axes and their images under a linear map.

direction of $f(\mathbf{d})$, provided that $f(\mathbf{d}) \neq \mathbf{0}$. Thus, if $f(\mathbf{d}) \neq \mathbf{0}$, then the image under f of the line $\mathbf{x} = \mathbf{a} + t\mathbf{d}$ is the line $\mathbf{x} = f(\mathbf{a}) + tf(\mathbf{d})$.

If, on the other hand, $f(\mathbf{d}) = \mathbf{0}$, then $f(\mathbf{a} + t\mathbf{d}) = f(\mathbf{a})$ for all $t \in \mathbb{R}$ so, under f, the line $\mathbf{x} = \mathbf{a} + t\mathbf{d}$ collapses to the point $f(\mathbf{a})$.   ■

Figure 2.7 shows the coordinate axes $\mathbf{x} = t\mathbf{e}_1$ and $\mathbf{x} = t\mathbf{e}_2$ and their images $\mathbf{x} = tf(\mathbf{e}_1)$ and $\mathbf{x} = tf(\mathbf{e}_2)$ under a linear map f. Figure 2.8 shows a horizontal line $\mathbf{x} = \mathbf{a} + t\mathbf{e}_1$, a vertical line $\mathbf{x} = \mathbf{a} + t\mathbf{e}_2$, and their images $\mathbf{x} = f(\mathbf{a}) + tf(\mathbf{e}_1)$ and $\mathbf{x} = f(\mathbf{a}) + tf(\mathbf{e}_2)$ under the same linear map. Notice that the images of horizontal lines are all parallel to one another (they are all in the direction of $f(\mathbf{e}_1)$) and that the images of vertical lines are all parallel to one another (they are all in the direction of $f(\mathbf{e}_2)$).

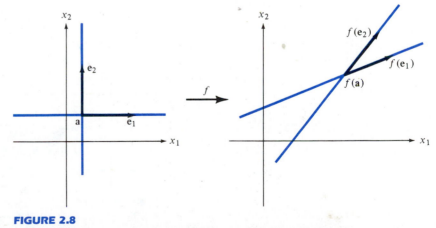

FIGURE 2.8
A horizontal line, a vertical line, and their images under a linear map.

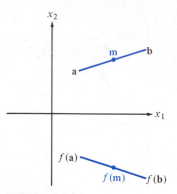

FIGURE 2.9
Linear maps send line segments
to line segments and midpoints
to midpoints.

Linear maps from $\mathbb{R}^2$ to $\mathbb{R}^2$ not only send lines to lines, they also send *line segments* to *line segments*. Indeed, if $f:\mathbb{R}^2 \to \mathbb{R}^2$ is a linear map and if **a** and **b** are two points in $\mathbb{R}^2$, then the formula

$$f((1 - t)\mathbf{a} + t\mathbf{b}) = (1 - t)f(\mathbf{a}) + tf(\mathbf{b}), \qquad 0 \le t \le 1,$$

shows that the image under f of the line segment from **a** to **b** is just the line segment from $f(\mathbf{a})$ to $f(\mathbf{b})$, provided of course that $f(\mathbf{a}) \ne f(\mathbf{b})$. (What happens if $f(\mathbf{a}) = f(\mathbf{b})$?) Moreover, the midpoint of the line segment from **a** to **b** is sent by f to the midpoint of the line segment from $f(\mathbf{a})$ to $f(\mathbf{b})$ (see Figure 2.9).

Linear maps $f:\mathbb{R}^2 \to \mathbb{R}^2$ also send *parallelograms* to *parallelograms*. Indeed, if P is the parallelogram

$$P = \{\mathbf{a} + s\mathbf{b} + t\mathbf{c}|0 \le s \le 1, 0 \le t \le 1\},$$

then the image of P under f is the parallelogram

$$f(P) = \{f(\mathbf{a}) + sf(\mathbf{b}) + tf(\mathbf{c})|0 \le s \le 1, 0 \le t \le 1\}$$

(see Figure 2.10).

EXAMPLE 4

Let us sketch some of the coordinate lines and their images under the linear map $f(x_1, x_2) = (2x_1 + x_2, x_1 + 2x_2)$.

The vertical line $x_1 = c$ is the line through $\mathbf{a} = (c, 0)$ in the direction of $\mathbf{e}_2 = (0, 1)$. Its image under f, therefore, is the line through $f(\mathbf{a}) = (2c, c)$ in the direction of $f(\mathbf{e}_2) = (1, 2)$. These lines are sketched for $c = -2$, $c = -1$, $c = 0$, $c = 1$, $c = 2$, and $c = 3$ in Figure 2.11.

Similarly, the horizontal line $x_2 = d$ is the line through $\mathbf{a} = (0, d)$ in the direction of $\mathbf{e}_1 = (1, 0)$. Its image under f is the line through $f(\mathbf{a}) = (d, 2d)$

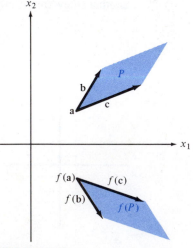

FIGURE 2.10
Linear maps send parallelograms to parallelograms.

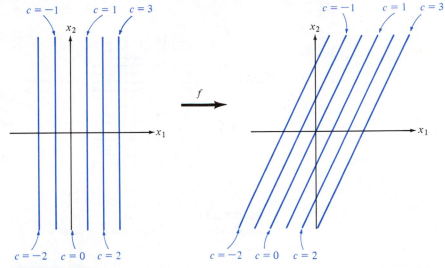

FIGURE 2.11
The effect of a linear map on the vertical lines $x_1 = c$.

in the direction of $f(\mathbf{e}_1) = (2, 1)$. These lines are sketched for $d = -2$, $d = -1$, $d = 0$, $d = 1$, $d = 2$, and $d = 3$ in Figure 2.12.

In Figure 2.13 we have sketched, in one diagram, all of the above lines. In this figure we have also shaded the square

$$\{(x_1, x_2)|1 \le x_1 \le 2, 1 \le x_2 \le 2\}$$

and its image under f. Can you identify the square

$$\{(x_1, x_2)|2 \le x_1 \le 3, 1 \le x_2 \le 2\}$$

and its image under f? ■

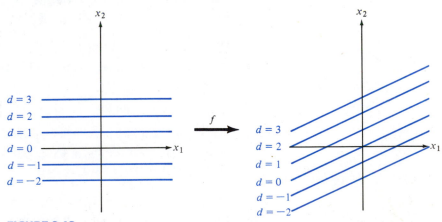

FIGURE 2.12
The effect of a linear map on the horizontal lines $x_2 = d$.

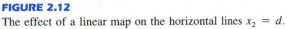

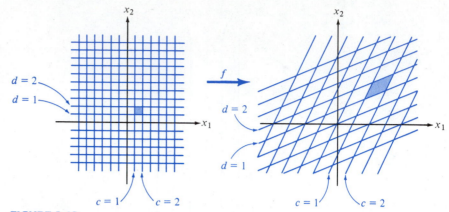

FIGURE 2.13
The effect of a linear map on a coordinate grid.

EXERCISES

1. Let $f:\mathbb{R}^2 \to \mathbb{R}^2$ be the linear map associated with the matrix

$$A = \begin{pmatrix} -1 & 3 \\ 2 & -2 \end{pmatrix}.$$

Find

(a) $f(1, 3)$ (b) $f(3, 1)$ (c) $f(-1, 2)$

(d) $f(\mathbf{e}_1)$ (e) $f(\mathbf{e}_2)$

2. Let $f:\mathbb{R}^2 \to \mathbb{R}^2$ be the linear map associated with the matrix

$$A = \begin{pmatrix} 1 & 1 \\ 0 & 1 \end{pmatrix}.$$

Find

(a) $f(7, -5)$ (b) $f(3, 3)$ (c) $f(-1, 1)$

(d) $f(\mathbf{e}_1)$ (e) $f(\mathbf{e}_2)$

3. Find a formula for $f(x_1, x_2)$, where $f:\mathbb{R}^2 \to \mathbb{R}^2$ is the linear map associated with the matrix

(a) $\begin{pmatrix} 0 & 1 \\ 0 & 0 \end{pmatrix}$ (b) $\begin{pmatrix} 0 & 0 \\ 1 & 0 \end{pmatrix}$ (c) $\begin{pmatrix} 2 & 0 \\ 0 & 3 \end{pmatrix}$ (d) $\begin{pmatrix} 0 & 2 \\ 3 & 0 \end{pmatrix}$

4. Find a formula for $f(x_1, x_2)$, where $f:\mathbb{R}^2 \to \mathbb{R}^2$ is the linear map associated with the matrix

(a) $\begin{pmatrix} -1 & 1 \\ 1 & 1 \end{pmatrix}$ (b) $\begin{pmatrix} 1 & 1 \\ -1 & 1 \end{pmatrix}$ (c) $\begin{pmatrix} 1 & \frac{1}{2} \\ 0 & 1 \end{pmatrix}$ (d) $\begin{pmatrix} \frac{2}{3} & \frac{1}{2} \\ -\frac{1}{2} & 1 \end{pmatrix}$

5. Find the matrix of each of the following linear maps.

(a) $f(x_1, x_2) = (3x_1 - 2x_2, -4x_1 + 5x_2)$

(b) $f(x_1, x_2) = (\frac{1}{2}x_1 + x_2, x_1 + \frac{1}{2}x_2)$

(c) $f(x_1, x_2) = (\sqrt{2}x_1 - x_2, x_2)$

(d) $f(x_1, x_2) = (x_1 - \sqrt{3}x_2, -\sqrt{3}x_1 + 3x_2)$

6. Find the matrix of each of the following linear maps.

(a) $f(x_1, x_2) = (2x_1 + 3x_2, 3x_1 - x_2)$

(b) $f(x_1, x_2) = (2x_2, -2x_1)$

(c) $f(x_1, x_2) = (ax_1 + bx_2, bx_1 + cx_2)$

(d) $f(x_1, x_2) = (ax_2, -ax_1)$

7.   For each of the following matrices A, describe geometrically what f does to vectors in $\mathbb{R}^2$, where f is the linear map associated with A.

(a) $\begin{pmatrix} -1 & 0 \\ 0 & 1 \end{pmatrix}$ (b) $\begin{pmatrix} -1 & 0 \\ 0 & -1 \end{pmatrix}$ (c) $\begin{pmatrix} 2 & 0 \\ 0 & 2 \end{pmatrix}$ (d) $\begin{pmatrix} 1 & 0 \\ 0 & 0 \end{pmatrix}$

(e) $\begin{pmatrix} 0 & -1 \\ 1 & 0 \end{pmatrix}$ (f) $\begin{pmatrix} 0 & 1 \\ -1 & 0 \end{pmatrix}$ (g) $\begin{pmatrix} \frac{1}{2} & 0 \\ 0 & \frac{1}{2} \end{pmatrix}$ (h) $\begin{pmatrix} 0 & 0 \\ 0 & 1 \end{pmatrix}$

8.   Let $(a, b) \in \mathbb{R}^2$ with $a \neq b$. Show that the line segment from (a, b) to (b, a) is perpendicular to the line $x_1 = x_2$. Also show that the points (a, b) and (b, a) are equidistant from the line $x_1 = x_2$. Conclude that the linear map $f(x_1, x_2) = (x_2, x_1)$ really does reflect each vector in the line $x_1 = x_2$.

9.   Let $f : \mathbb{R}^2 \to \mathbb{R}^2$ be defined by $f(x_1, x_2) = (2x_1 - x_2, x_1 + x_2)$. Sketch in $\mathbb{R}^2$ the following vectors and their images under f.

(a) $(1, 0)$ (b) $(0, 1)$ (c) $(1, 1)$

(d) $(-1, 1)$ (e) $(2, 3)$ (f) $(-3, 4)$

(g) $(1, -1)$ (h) $(2, 0)$ (i) $(0, 3)$

10.   Repeat Exercise 9 for the linear map $f : \mathbb{R}^2 \to \mathbb{R}^2$ associated with the matrix

$$A = \begin{pmatrix} 1 & 0 \\ 1 & 1 \end{pmatrix}.$$

11.   Let $f : \mathbb{R}^2 \to \mathbb{R}^2$ be defined by $f(x_1, x_2) = (x_1 - x_2, 2x_1 + x_2)$. Sketch the image under f of each of the following lines.

(a) $x_1 = 0$ (b) $x_2 = 0$ (c) $x_1 = 1$

(d) $x_2 = 1$ (e) $x_1 = 2$ (f) $x_2 = 2$

12.   Repeat Exercise 11 for the linear map $f : \mathbb{R}^2 \to \mathbb{R}^2$ defined by $f(x_1, x_2) = (-x_1 - 2x_2, 2x_1 + x_2)$.

13.   Let $f : \mathbb{R}^2 \to \mathbb{R}^2$ be defined by $f(x_1, x_2) = (x_1 - x_2, 2x_1 + x_2)$, as in Exercise 11. Sketch (shade in) each of the following squares and their images under f.

(a) $\{(x_1, x_2) | 0 \leq x_1 \leq 1, 0 \leq x_2 \leq 1\}$

(b) $\{(x_1, x_2) | 1 \leq x_1 \leq 2, 0 \leq x_2 \leq 1\}$

(c) $\{(x_1, x_2) | 0 \leq x_1 \leq 1, 1 \leq x_2 \leq 2\}$

(d) $\{(x_1, x_2) | 1 \leq x_1 \leq 2, 3 \leq x_2 \leq 4\}$

(e) $\{(x_1, x_2) | -1 \leq x_1 \leq 1, -1 \leq x_2 \leq 1\}$

14.   Repeat Exercise 13 for each of the following linear maps.

(i) $f(x_1, x_2) = (-x_1 - 2x_2, 2x_1 + x_2)$

(ii) $f(x_1, x_2) = (-x_1 - 2x_2, -2x_1 - x_2)$

(iii) $f(x_1, x_2) = (-x_2, x_1)$

(iv) $f(x_1, x_2) = (2x_1, x_2)$

(v) $f(x_1, x_2) = (x_1 + x_2, x_1 + x_2)$

(vi) $f(\mathbf{x}) = 2\mathbf{x}$

(vii) $f(\mathbf{x}) = -\mathbf{x}$

15.   If $f : \mathbb{R}^2 \to \mathbb{R}^2$ is a map that is not linear, then the images under f of straight lines are not necessarily straight lines. Sketch the image under f of the line $\mathbf{x} = t\mathbf{e}_1$ where $f(x_1, x_2) = (x_1 + x_2, x_1^2)$.

16. Which of the following maps are linear?

(a) $f(x_1, x_2) = (x_1 + x_2, 0)$

(b) $f(x_1, x_2) = (x_1 x_2, 0)$

(c) $f(x_1, x_2) = (x_1^2, x_2^2)$

(d) $f(x_1, x_2) = (2x_1, 3x_2)$

(e) $f(x_1, x_2) = (x_1 + x_2, (x_1 + x_2)^2)$

17. Let $f: \mathbb{R}^2 \to \mathbb{R}^2$ be a linear map and let U be a subset of $\mathbb{R}^2$. Explain why, if U is a convex subset of $\mathbb{R}^2$, then the image of U under f must also be convex.

2.2
Expansions, Contractions, Reflections, Rotations, and Shears

In the previous section we learned how to use a 2×2 matrix to define a linear map from $\mathbb{R}^2$ to $\mathbb{R}^2$. In this section we shall study several examples of linear maps, with special attention devoted to the geometric properties of these maps.

EXAMPLE 1
Consider the linear map $f: \mathbb{R}^2 \to \mathbb{R}^2$ whose matrix is

$$\begin{pmatrix} c & 0 \\ 0 & c \end{pmatrix}$$

where c is some fixed real number. For each $\mathbf{x} = (x_1, x_2) \in \mathbb{R}^2$, the vector $f(\mathbf{x})$ is given by the formula

$$f(x_1, x_2) = (cx_1, cx_2)$$

or

$$f(\mathbf{x}) = c\mathbf{x}.$$

The geometric characteristics of this linear map depend on the size of c, and on whether c is positive, negative, or zero.

If $c > 1$, then f is an **expansion:** f magnifies each vector by a factor of c (see Figure 2.14a).

If $c = 1$, then f is the **identity map:** f sends each vector to itself.

If $0 < c < 1$, then f is a **contraction:** f shrinks each vector by a factor of c (see Figure 2.14b).

If $c = 0$ then f is the **zero map:** f sends each vector to $\mathbf{0}$.

If $c = -1$, then f is **reflection** in $\mathbf{0}$: f sends each vector to its negative (see Figure 2.14c).

Whenever $c < 0$ and $c \neq -1$, f can be viewed as an expansion or a contraction (sending $\mathbf{x}$ to $|c|\mathbf{x}$) followed by reflection in $\mathbf{0}$ (sending $|c|\mathbf{x}$ to $-|c|\mathbf{x} = c\mathbf{x}$).

It is always helpful, when studying a linear map f from $\mathbb{R}^2$ to $\mathbb{R}^2$, to sketch some coordinate lines and their images under f. In Figure 2.15 we have sketched several of these lines and we have shaded the **unit square**

$$\{(x_1, x_2) \mid 0 \leq x_1 \leq 1, 0 \leq x_2 \leq 1\}$$

and its image under f, where f is the expansion $f(\mathbf{x}) = 2\mathbf{x}$. ■

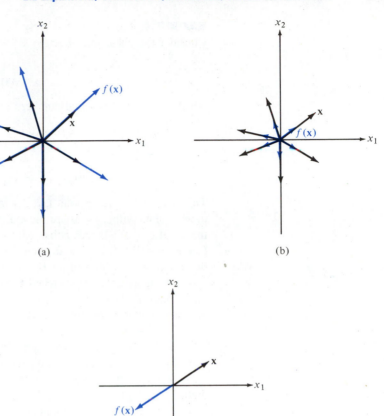

(a) (b)

(c)

FIGURE 2.14

The linear map $f(\mathbf{x}) = c\mathbf{x}$ is (a) an *expansion* if $c > 1$, (b) a *contraction* if $0 < c < 1$, and (c) a *reflection* if $c = -1$.

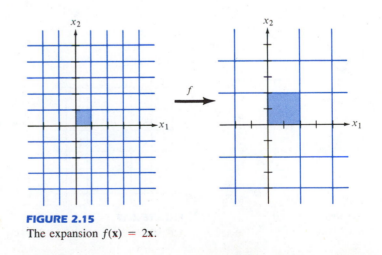

FIGURE 2.15

The expansion $f(\mathbf{x}) = 2\mathbf{x}$.

EXAMPLE 2

Consider the linear map $f:\mathbb{R}^2 \to \mathbb{R}^2$ whose matrix is

$$\begin{pmatrix} 1 & c \\ 0 & 1 \end{pmatrix}$$

where c is some fixed real number. This map can be described by the formula

$$f(x_1, x_2) = (x_1 + cx_2, x_2)$$

or

$$f(x_1, x_2) = (x_1, x_2) + cx_2(1, 0).$$

This map moves each point (x_1, x_2) in $\mathbb{R}^2$ horizontally a distance proportional to x_2. Points with $x_2 > 0$ are moved in one direction (to the right if $c > 0$, to the left if $c < 0$) whereas points with $x_2 < 0$ are moved in the opposite direction (see Figure 2.16). If $c = 0$, then f is the identity map. Notice that points on the x_1-axis are not moved by f. This linear map is called a ***horizontal shear.***

In Figure 2.17 are sketched some coordinate lines and their images under the horizontal shear $f(x_1, x_2) = (x_1 + \frac{1}{2}x_2, x_2)$. As usual, the unit square and its image under f are shaded. ■

Sometimes maps from $\mathbb{R}^2$ to $\mathbb{R}^2$ are defined geometrically rather than algebraically. For example, the map that reflects each vector in the line $x_1 = x_2$ is a linear map (see Example 2 of the previous section). The next theorem will give a criterion for determining when a geometrically defined map $f:\mathbb{R}^2 \to \mathbb{R}^2$ is a linear map.

Recall Theorem 1 of the previous section. That theorem stated that linear maps from $\mathbb{R}^2$ to $\mathbb{R}^2$ preserve vector addition and scalar multiplication. More precisely, if $f:\mathbb{R}^2 \to \mathbb{R}^2$ is a linear map, then

$$f(\mathbf{x} + \mathbf{y}) = f(\mathbf{x}) + f(\mathbf{y})$$

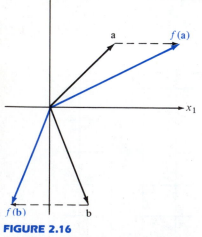

FIGURE 2.16
A horizontal shear $f(x_1, x_2) = (x_1 + cx_2, x_2)$ with $c > 0$.

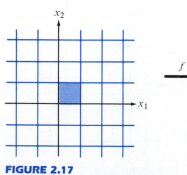
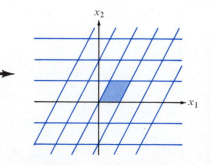

FIGURE 2.17
The horizontal shear $f(x_1, x_2) = (x_1 + \frac{1}{2}x_2, x_2)$.

and

$$f(r\mathbf{x}) = rf(\mathbf{x})$$

for each $\mathbf{x}$ and $\mathbf{y}$ in $\mathbb{R}^2$ and for each real number r. It turns out that every map $f:\mathbb{R}^2 \to \mathbb{R}^2$ that has these two properties is a linear map.

Theorem 1 *Let $f:\mathbb{R}^2 \to \mathbb{R}^2$ be any function that preserves both vector addition and scalar multiplication. Then f is a linear map.*

Proof Let $f:\mathbb{R}^2 \to \mathbb{R}^2$ be any function that preserves both vector addition and scalar multiplication. We shall find a formula for $f(x_1, x_2)$ which will show that f is a linear map. Recalling that $\mathbf{e}_1 = (1, 0)$ and $\mathbf{e}_2 = (0, 1)$, notice that

$$f(x_1, x_2) = f(x_1\mathbf{e}_1 + x_2\mathbf{e}_2)$$

$$= f(x_1\mathbf{e}_1) + f(x_2\mathbf{e}_2)$$

$$= x_1 f(\mathbf{e}_1) + x_2 f(\mathbf{e}_2)$$

$$= x_1(a, b) + x_2(c, d)$$

where $(a, b) = f(\mathbf{e}_1)$ and $(c, d) = f(\mathbf{e}_2)$. Carrying out the indicated vector operations, we see that

$$f(x_1, x_2) = (ax_1 + cx_2, bx_1 + dx_2).$$

Thus f is a linear map, the linear map associated with the matrix

$$\begin{pmatrix} a & c \\ b & d \end{pmatrix}. \quad \blacksquare$$

The proof of Theorem 1 shows more than just that every function $f:\mathbb{R}^2 \to \mathbb{R}^2$ that preserves vector addition and scalar multiplication is a linear map. The proof also shows how to find the matrix of f. The matrix of f is the matrix whose first-column entries are the entries in the vector $f(\mathbf{e}_1)$ and whose second-column entries are the entries in the vector $f(\mathbf{e}_2)$.

In general, given any matrix

$$A = \begin{pmatrix} a_{11} & a_{12} & \cdots & a_{1n} \\ a_{21} & a_{22} & \cdots & a_{2n} \\ & & \cdots & \\ a_{m1} & a_{m2} & \cdots & a_{mn} \end{pmatrix},$$

the vectors

$$(a_{11}, a_{21}, \ldots, a_{m1}), (a_{12}, a_{22}, \ldots, a_{m2}), \ldots, (a_{1n}, a_{2n}, \ldots, a_{mn}),$$

whose entries appear in the columns of A, are called the ***column vectors*** of A. We can, then, deduce from the proof of Theorem 1 of the following additional fact.

Theorem 2 *Let $f:\mathbb{R}^2 \to \mathbb{R}^2$ be any linear map. Then the matrix of f is the matrix whose column vectors are the vectors $f(\mathbf{e}_1)$ and $f(\mathbf{e}_2)$.*

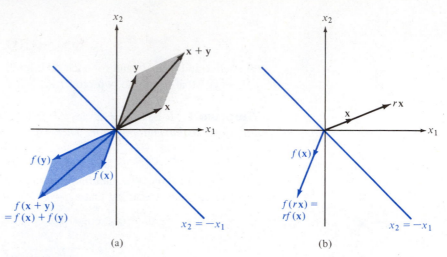

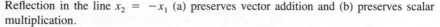

(a)   (b)

FIGURE 2.18
Reflection in the line $x_2 = -x_1$ (a) preserves vector addition and (b) preserves scalar multiplication.

The importance of these theorems lies in the fact that many maps $f:\mathbb{R}^2 \to \mathbb{R}^2$ can be seen to preserve vector addition and scalar multiplication. We can conclude, using the theorems, that each such map is linear and we can even write down its matrix.

EXAMPLE 3

Let $f:\mathbb{R}^2 \to \mathbb{R}^2$ be the map that reflects each vector in $\mathbb{R}^2$ in the line $x_2 = -x_1$. It is geometrically clear (see Figure 2.18) that this map preserves vector addition and scalar multiplication. (For example, we see that f reflects the entire vector addition parallelogram for $\mathbf{x}$ and $\mathbf{y}$ in the line $x_2 = -x_1$.) Hence f is a linear map. To find its matrix we need only compute $f(\mathbf{e}_1)$ and $f(\mathbf{e}_2)$. We find that (see Figure 2.19)

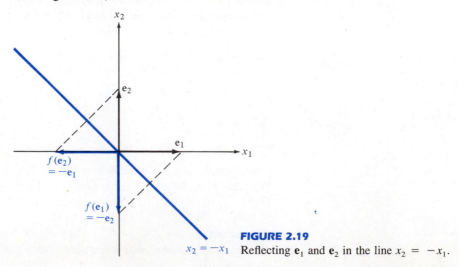

FIGURE 2.19
Reflecting $\mathbf{e}_1$ and $\mathbf{e}_2$ in the line $x_2 = -x_1$.

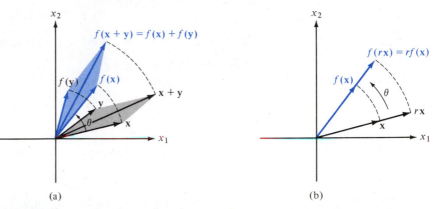

(a)   (b)

FIGURE 2.20

Counterclockwise rotation through the angle θ (a) preserves vector addition and (b) preserves scalar multiplication.

$$f(\mathbf{e}_1) = -\mathbf{e}_2 = (0, -1)$$

$$f(\mathbf{e}_2) = -\mathbf{e}_1 = (-1, 0).$$

The matrix of f is, therefore, the matrix

$$\begin{pmatrix} 0 & -1 \\ -1 & 0 \end{pmatrix}.$$

Finally, we can, if we wish, write down an algebraic formula for f, namely,

$$f(x_1, x_2) = (-x_2, -x_1). \quad \blacksquare$$

EXAMPLE 4

Let $f : \mathbb{R}^2 \to \mathbb{R}^2$ be the map that rotates each vector in $\mathbb{R}^2$ counterclockwise through some fixed angle θ. Once again it is clear geometrically that this map preserves vector addition and scalar multiplication (see Figure 2.20). Hence it, too, is a linear map. To find its matrix we compute $f(\mathbf{e}_1)$ and $f(\mathbf{e}_2)$. We find that (see Figure 2.21)

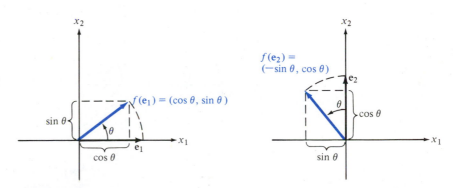

FIGURE 2.21

Rotating $\mathbf{e}_1$ and $\mathbf{e}_2$ counterclockwise through the angle θ.

$$f(\mathbf{e}_1) = (\cos\theta, \sin\theta)$$

$$f(\mathbf{e}_2) = (-\sin\theta, \cos\theta).$$

The matrix of f is, therefore,

$$\begin{pmatrix} \cos\theta & -\sin\theta \\ \sin\theta & \cos\theta \end{pmatrix}.$$

It follows that the formula for f is

$$f(x_1, x_2) = ((\cos\theta)x_1 - (\sin\theta)x_2, (\sin\theta)x_1 + (\cos\theta)x_2)$$

or

$$f(x_1, x_2) = (x_1\cos\theta - x_2\sin\theta, x_1\sin\theta + x_2\cos\theta). \quad \blacksquare$$

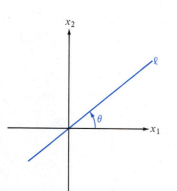

FIGURE 2.22

θ is the inclination angle of the line ℓ.

EXAMPLE 5

Let ℓ be any line in $\mathbb{R}^2$ passing through $\mathbf{0}$. We shall use Theorem 2 to derive a formula for the function f that reflects each point in the line ℓ.

First, observe that f is a linear map because it preserves vector addition and scalar multiplication. This can easily be verified geometrically. So we need only compute the matrix of f.

Let θ ($0 \le \theta < \pi$) be the angle that ℓ makes with the positive x_1-axis (see Figure 2.22). θ is called the *inclination angle* of ℓ. We shall find formulas for the vectors $f(\mathbf{e}_1)$ and $f(\mathbf{e}_2)$ in terms of θ.

The formula for $f(\mathbf{e}_1)$ is quite easy to find (see Figure 2.23a). The angle between the line ℓ and the vector $f(\mathbf{e}_1)$ is equal to the angle between $\mathbf{e}_1$ and ℓ. Both of these angles are θ. Hence $f(\mathbf{e}_1)$ is the unit vector inclined at the angle 2θ with the positive x_1-axis. It is easy to read off the entries in this vector. We see that

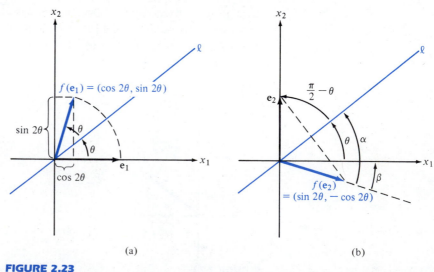

(a)

(b)

FIGURE 2.23

Reflecting $\mathbf{e}_1$ and $\mathbf{e}_2$ in ℓ.

$$f(\mathbf{e}_1) = (\cos 2\theta, \sin 2\theta).$$

The formula for $f(\mathbf{e}_2)$ is only slightly more difficult to find (see Figure 2.23b). The angle from ℓ to $\mathbf{e}_2$ is $(\pi/2) - \theta$. Hence the angle α from $f(\mathbf{e}_2)$ to ℓ is also $(\pi/2) - \theta$:

$$\alpha = \frac{\pi}{2} - \theta.$$

It follows that the angle β from $f(\mathbf{e}_2)$ to the positive x_1-axis is

$$\beta = \alpha - \theta = \frac{\pi}{2} - 2\theta.$$

Thus $f(\mathbf{e}_2)$ is the unit vector inclined at the angle $-\beta = 2\theta - (\pi/2)$ with the positive x_1-axis. Therefore

$$f(\mathbf{e}_2) = \left(\cos\left(2\theta - \frac{\pi}{2}\right), \sin\left(2\theta - \frac{\pi}{2}\right)\right)$$

or

$$f(\mathbf{e}_2) = (\sin 2\theta, -\cos 2\theta).$$

We can now write the matrix for f. It is the matrix

$$\begin{pmatrix} \cos 2\theta & \sin 2\theta \\ \sin 2\theta & -\cos 2\theta \end{pmatrix}$$

whose column vectors are $f(\mathbf{e}_1)$ and $f(\mathbf{e}_2)$. The formula for f, which we set out to derive, is therefore

$$f(x_1, x_2) = ((\cos 2\theta)x_1 + (\sin 2\theta)x_2, (\sin 2\theta)x_1 - (\cos 2\theta)x_2).$$

In other words, *the function $f:\mathbb{R}^2 \to \mathbb{R}^2$ that reflects each vector in the line ℓ through $\mathbf{0}$ inclined at the angle θ has the formula*

$$f(x_1, x_2) = (x_1 \cos 2\theta + x_2 \sin 2\theta, x_1 \sin 2\theta - x_2 \cos 2\theta).$$

Notice that if $\theta = \pi/4$ then ℓ is the line $x_2 = x_1$. The formula for f then becomes

$$f(x_1, x_2) = \left(x_1 \cos \frac{\pi}{2} + x_2 \sin \frac{\pi}{2}, x_1 \sin \frac{\pi}{2} - x_2 \cos \frac{\pi}{2}\right)$$

$$= (x_2, x_1),$$

in agreement with Example 2 of the previous section.

Similarly, you can check that when $\theta = 0$ then

$$f(x_1, x_2) = (x_1, -x_2)$$

and when $\theta = 3\pi/4$ then

$$f(x_1, x_2) = (-x_2, -x_1),$$

in agreement with Example 1 of the previous section and Example 3 of this section. ■

EXERCISES

1. Sketch a few coordinate lines and their images under f for each of the following linear maps f. Also, in each drawing, shade the unit square and its image under f.

   (a) The expansion $f(\mathbf{x}) = 3\mathbf{x}$

   (b) The contraction $f(\mathbf{x}) = \frac{1}{2}\mathbf{x}$

   (c) The reflection $f(\mathbf{x}) = -\mathbf{x}$

   (d) The shear $f(x_1, x_2) = (x_1 + 2x_2, x_2)$

   (e) The shear $f(x_1, x_2) = (x_1 - x_2, x_2)$

2. Let $f : \mathbb{R}^2 \to \mathbb{R}^2$ be the linear map whose matrix is

$$\begin{pmatrix} 1 & 0 \\ \frac{1}{2} & 1 \end{pmatrix}.$$

   (a) Describe geometrically what f does to the point (x_1, x_2) in each of the three cases: $x_1 > 0$, $x_1 < 0$, and $x_1 = 0$.

   (b) Sketch a few coordinate lines and their images under f.

   (c) Sketch the unit square and its image under f.

   REMARK   A linear map $f : \mathbb{R}^2 \to \mathbb{R}^2$ whose matrix is of the form

$$\begin{pmatrix} 1 & 0 \\ c & 1 \end{pmatrix}$$

   is called a *vertical shear*.

3. For each of the following linear maps f, calculate $f(\mathbf{e}_1)$ and $f(\mathbf{e}_2)$, and use these to find the matrix of f.

   (a) $f(x_1, x_2) = (3x_1 + 4x_2, -2x_1 + x_2)$

   (b) $f(x_1, x_2) = (x_1 - x_2, x_1 + x_2)$

   (c) $f(x_1, x_2) = (-x_1 - \frac{1}{2}x_2, x_1 + \frac{1}{2}x_2)$

   (d) $f(x_1, x_2) = (x_2, x_1 + x_2)$

   (e) $f(x_1, x_2) = (x_1, x_1 + x_2)$

4. Proceed as in Example 3 to calculate the matrix of each of the following linear maps from $\mathbb{R}^2$ to $\mathbb{R}^2$:

   (a) Reflection in the x_1-axis

   (b) Reflection in the x_2-axis

   (c) Reflection in the line $x_1 = x_2$.

5. Let $f : \mathbb{R}^2 \to \mathbb{R}^2$ be reflection in ℓ, where ℓ is the line through $\mathbf{0}$ inclined at the angle $\pi/8$ with the positive x_1-axis. Calculate $f(\mathbf{e}_1)$, $f(\mathbf{e}_2)$, and the matrix of f.

6. Let $f : \mathbb{R}^2 \to \mathbb{R}^2$ be the linear map that rotates each vector in $\mathbb{R}^2$ counterclockwise through the angle θ. Find the matrix of f when

   (a) $\theta = \pi/2$ (b) $\theta = \pi$ (c) $\theta = \pi/4$

   (d) $\theta = 3\pi/4$ (e) $\theta = -\pi/2$

7. Let $f : \mathbb{R}^2 \to \mathbb{R}^2$ be the map that assigns to each vector $\mathbf{x}$ in $\mathbb{R}^2$ its component along $\mathbf{e}_1$.

   (a) Verify that f is a linear map.

   (b) Find the matrix of f.

8. Let $f : \mathbb{R}^2 \to \mathbb{R}^2$ be the map that assigns to each vector $\mathbf{x}$ in $\mathbb{R}^2$ its component along $\mathbf{e}_2$.

   (a) Verify that f is a linear map.

   (b) Find the matrix of f.

9. Let $\mathbf{u} = (u_1, u_2)$ be a unit vector and let $f:\mathbb{R}^2 \to \mathbb{R}^2$ be the map that assigns to each vector $\mathbf{x}$ in $\mathbb{R}^2$ its component along $\mathbf{u}$.

   (a) Verify that f is a linear map.

   (b) Find the matrix of f.

10. For each of the following linear maps f, sketch a few coordinate lines and their images under f. In each drawing, shade the unit square and its image under f.

    (a) Rotation counterclockwise through the angle $\pi/4$

    (b) Rotation counterclockwise through the angle $\pi/8$

    (c) Reflection in the line $x_2 = -x_1$

    (d) Reflection in the line $x_2 = 2x_1$

    (e) $f(x_1, x_2) = (x_1, -x_1 + x_2)$

    (f) $f(\mathbf{x})$ is the component of $\mathbf{x}$ along $\mathbf{e}_1$

11. Let $f:\mathbb{R}^2 \to \mathbb{R}^2$ be a linear map, let

$$A = \begin{pmatrix} a & b \\ c & d \end{pmatrix}$$

be its matrix, and let S be the unit square

$$\{x_1\mathbf{e}_1 + x_2\mathbf{e}_2 \mid 0 \le x_1 \le 1,\, 0 \le x_2 \le 1\}.$$

    (a) Show that the area of the parallelogram $f(S)$ is equal to $|ad - bc| = |\det(A)|$.

    (b) Show that if f is a rotation (Example 4), then the area of $f(S)$ is equal to 1.

    (c) Show that if f is reflection in a line ℓ through $\mathbf{0}$ (Example 5), then the area of $f(S)$ is equal to 1.

    (d) Show that if f is a horizontal shear (Example 2), then the area of $f(S)$ is equal to 1.

    (e) What is the area of $f(S)$ if f is the expansion $f(\mathbf{x}) = c\mathbf{x}$ $(c > 1)$?

2.3
Linear Maps from $\mathbb{R}^n$ to $\mathbb{R}^m$

In the first two sections of this chapter we studied linear maps from $\mathbb{R}^2$ to $\mathbb{R}^2$. In this section we shall study linear maps from $\mathbb{R}^n$ to $\mathbb{R}^m$, where n and m are arbitrary positive integers. These maps are described using $m \times n$ matrices. As in the case when $m = n = 2$ (linear maps from the plane to the plane), these maps have many nice geometric properties.

Let A be an $m \times n$ matrix. We can write A as

$$A = \begin{pmatrix} \mathbf{a}_1 \\ \cdots \\ \mathbf{a}_m \end{pmatrix}$$

where $\mathbf{a}_1, \ldots, \mathbf{a}_m \in \mathbb{R}^n$ are the row vectors of A. Then we can use A to define a function $f:\mathbb{R}^n \to \mathbb{R}^m$ by the formula

$$f(\mathbf{x}) = (\mathbf{a}_1 \cdot \mathbf{x}, \ldots, \mathbf{a}_m \cdot \mathbf{x}).$$

Thus $f(\mathbf{x})$ is the vector whose entries are the dot products of the row vectors of A with $\mathbf{x}$. This map f is called the ***linear map associated with the matrix*** A, and the matrix A is called ***the matrix of the linear map*** f.

EXAMPLE 1
Let

$$A = \begin{pmatrix} 3 & -1 & 2 \\ 1 & 5 & -4 \end{pmatrix}.$$

The linear map associated with this 2×3 matrix A is a map $f:\mathbb{R}^3 \to \mathbb{R}^2$. For each vector $\mathbf{x}$ in $\mathbb{R}^3$ we can find $f(\mathbf{x})$ by computing the dot products $\mathbf{a}_1 \cdot \mathbf{x}$ and $\mathbf{a}_2 \cdot \mathbf{x}$ where $\mathbf{a}_1 = (3, -1, 2)$ and $\mathbf{a}_2 = (1, 5, -4)$. Thus, for example,

$$f(1, 2, 3) = (\mathbf{a}_1 \cdot (1, 2, 3), \mathbf{a}_2 \cdot (1, 2, 3))$$

$$= ((3, -1, 2) \cdot (1, 2, 3), (1, 5, -4) \cdot (1, 2, 3))$$

$$= (7, -1).$$

Similarly,

$$f(-1, -1, 1) = ((3, -1, 2) \cdot (-1, -1, 1), (1, 5, -4) \cdot (-1, -1, 1))$$

$$= (0, -10).$$

In general, for $\mathbf{x} = (x_1, x_2, x_3)$, we find that

$$f(x_1, x_2, x_3) = (3x_1 - x_2 + 2x_3, x_1 + 5x_2 - 4x_3). \quad \blacksquare$$

EXAMPLE 2
Let A be the $n \times n$ matrix

$$A = \begin{pmatrix} c & 0 & \cdots & 0 \\ 0 & c & \cdots & 0 \\ & & \cdots & \\ 0 & 0 & \cdots & c \end{pmatrix}$$

where c is a real number. Then the linear map $f:\mathbb{R}^n \to \mathbb{R}^n$ associated with A is given by the formula $f(\mathbf{x}) = c\mathbf{x}$ for all $\mathbf{x} \in \mathbb{R}^n$.

If $c > 1$, f is **expansion** by the factor c.

If $c = 1$, f is the **identity map** of $\mathbb{R}^n$.

If $0 < c < 1$, f is **contraction** by the factor c.

If $c = -1$, f is **reflection** in $\mathbf{0}$:

$$f(\mathbf{x}) = -\mathbf{x} \text{ for all } \mathbf{x} \in \mathbb{R}^n. \quad \blacksquare$$

EXAMPLE 3
Let

$$A = \begin{pmatrix} \cos\theta & -\sin\theta & 0 \\ \sin\theta & \cos\theta & 0 \\ 0 & 0 & 1 \end{pmatrix},$$

where θ is some fixed real number. Then the linear map associated with A is the map $f:\mathbb{R}^3 \to \mathbb{R}^3$ defined by the formula

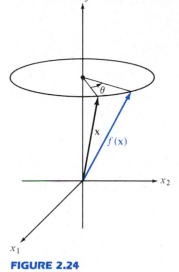

FIGURE 2.24
Rotation about the x_3-axis.

$$f(x_1, x_2, x_3) = ((\cos \theta)x_1 - (\sin \theta)x_2, (\sin \theta)x_1 + (\cos \theta)x_2, x_3)$$

or

$$f(x_1, x_2, x_3) = (x_1 \cos \theta - x_2 \sin \theta, x_1 \sin \theta + x_2 \cos \theta, x_3).$$

This linear map rotates each vector about the x_3-axis through the angle θ (see Figure 2.24). ■

It is easy to write the matrix of a linear map $f:\mathbb{R}^n \rightarrow \mathbb{R}^m$ whenever we are given an explicit formula for $f(x_1, \ldots, x_n)$. Simply read off the coefficients of $x_1, x_2, \ldots, x_n$, and arrange them in an $m \times n$ matrix. Thus if

$$f(x_1, \ldots, x_n) = (a_{11}x_1 + \cdots + a_{1n}x_n, \ldots, a_{m1}x_1 + \cdots + a_{mn}x_n),$$

then the matrix of f is

$$\begin{pmatrix} a_{11} & \cdots & a_{1n} \\ & \cdots & \\ a_{m1} & \cdots & a_{mn} \end{pmatrix}.$$

EXAMPLE 4
Let $f:\mathbb{R}^2 \rightarrow \mathbb{R}^3$ be defined by

$$f(x_1, x_2) = (2x_1 - 3x_2, 3x_1 - x_2, x_1 + x_2).$$

Then f is a linear map. Its matrix is

$$\begin{pmatrix} 2 & -3 \\ 3 & -1 \\ 1 & 1 \end{pmatrix}. \quad ■$$

EXAMPLE 5
Let $f:\mathbb{R}^3 \rightarrow \mathbb{R}^3$ be defined by $f(x_1, x_2, x_3) = (x_1, x_2, -x_3)$. Then f is a linear map. Its matrix is

$$\begin{pmatrix} 1 & 0 & 0 \\ 0 & 1 & 0 \\ 0 & 0 & -1 \end{pmatrix}.$$

Geometrically, this linear map reflects each vector in the plane $x_3 = 0$ (see Figure 2.25). ■

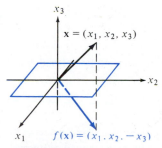

FIGURE 2.25
Reflection in the plane $x_3 = 0$.

EXAMPLE 6
Let $f:\mathbb{R}^3 \rightarrow \mathbb{R}^3$ be defined by $f(\mathbf{x}) = \mathbf{a} \times \mathbf{x}$ where $\mathbf{a} = (a_1; a_2, a_3)$ is some fixed vector in $\mathbb{R}^3$. Then

$$f(x_1, x_2, x_3) = (a_1, a_2, a_3) \times (x_1, x_2, x_3)$$

$$= (a_2x_3 - a_3x_2, -a_1x_3 + a_3x_1, a_1x_2 - a_2x_1)$$

$$= (-a_3x_2 + a_2x_3, a_3x_1 - a_1x_3, -a_2x_1 + a_1x_2).$$

It is evident from this formula that f is a linear map. The matrix of f is

$$\begin{pmatrix} 0 & -a_3 & a_2 \\ a_3 & 0 & -a_1 \\ -a_2 & a_1 & 0 \end{pmatrix}. \quad \blacksquare$$

In Sections 1 and 2 we proved three theorems about linear maps from $\mathbb{R}^2$ to $\mathbb{R}^2$. These theorems are also true for linear maps from $\mathbb{R}^n$ to $\mathbb{R}^m$. The proofs are almost identical to the proofs for $\mathbb{R}^2$ so we will not reproduce them here.

Theorem 1   *Let $f:\mathbb{R}^n \to \mathbb{R}^m$ be a linear map, let $\mathbf{x}$ and $\mathbf{y}$ be vectors in $\mathbb{R}^n$, and let r and s be real numbers. Then*

(i) $f(\mathbf{x} + \mathbf{y}) = f(\mathbf{x}) + f(\mathbf{y})$

(ii) $f(r\mathbf{x}) = rf(\mathbf{x})$

and

(iii) $f(r\mathbf{x} + s\mathbf{y}) = rf(\mathbf{x}) + sf(\mathbf{y}).$

Property (i) of Theorem 1 states that *linear maps $f:\mathbb{R}^n \to \mathbb{R}^m$ preserve vector addition*. Property (ii) states that *linear maps $f:\mathbb{R}^n \to \mathbb{R}^m$ preserve scalar multiplication*. The following theorem states that linear maps are the only maps from $\mathbb{R}^n$ to $\mathbb{R}^m$ with these two properties.

Theorem 2   *Let $f:\mathbb{R}^n \to \mathbb{R}^m$ be any function that preserves both vector addition and scalar multiplication. Then f is a linear map. In fact, f is the linear map associated with the matrix whose column vectors are the vectors $f(\mathbf{e}_1)$, . . . , $f(\mathbf{e}_n)$.*

The third theorem tells us a little about the geometry of linear maps from $\mathbb{R}^n$ to $\mathbb{R}^m$.

Theorem 3   *Let $f:\mathbb{R}^n \to \mathbb{R}^m$ be a linear map and let $\mathbf{a}$ and $\mathbf{d}$ be vectors in $\mathbb{R}^n$, with $\mathbf{d} \neq \mathbf{0}$. Then the image under f of the line through $\mathbf{a}$ in the direction of $\mathbf{d}$ is either*

(i)   *a line—the line through $f(\mathbf{a})$ in the direction of $f(\mathbf{d})$ (if $f(\mathbf{d}) \neq \mathbf{0}$), or*

(ii)   *a point—the point $f(\mathbf{a})$ (if $f(\mathbf{d}) = \mathbf{0}$).*

Theorem 2 can be used to verify that certain geometrically defined maps from $\mathbb{R}^n$ to $\mathbb{R}^m$ are linear maps.

EXAMPLE 7

Let $f:\mathbb{R}^3 \to \mathbb{R}^3$ be the function that reflects each vector in the plane $\mathbf{a} \cdot \mathbf{x} = 0$, where $\mathbf{a}$ is some nonzero vector in $\mathbb{R}^3$. It is geometrically clear that this function preserves both vector addition and scalar multiplication (see Figure 2.26). Hence, by Theorem 2, f is a linear map.

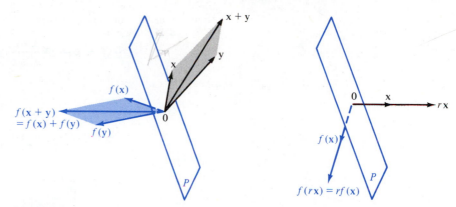

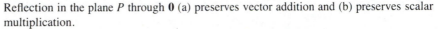

FIGURE 2.26
Reflection in the plane P through $\mathbf{0}$ (a) preserves vector addition and (b) preserves scalar multiplication.

Let us derive a formula for $f(\mathbf{x})$ where $\mathbf{x}$ is any vector in $\mathbb{R}^3$. Recall that the plane $\mathbf{a} \cdot \mathbf{x} = 0$ is the plane that is perpendicular to $\mathbf{a}$ and that passes through $\mathbf{0}$. To obtain the reflection of $\mathbf{x}$ in this plane we must subtract from $\mathbf{x}$ twice the component of $\mathbf{x}$ along $\mathbf{a}$ (see Figure 2.27). Since the component $\mathbf{v}$ of $\mathbf{x}$ along $\mathbf{a}$ is given by the formula

$$\mathbf{v} = \frac{\mathbf{x} \cdot \mathbf{a}}{\mathbf{a} \cdot \mathbf{a}} \, \mathbf{a},$$

we see that *reflection in the plane* $\mathbf{a} \cdot \mathbf{x} = 0$ *is described by the formula*

$$f(\mathbf{x}) = \mathbf{x} - 2 \frac{\mathbf{x} \cdot \mathbf{a}}{\mathbf{a} \cdot \mathbf{a}} \, \mathbf{a}.$$

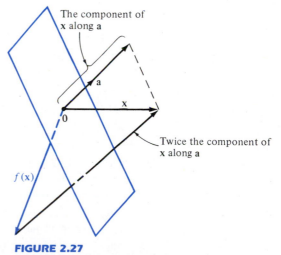

FIGURE 2.27
Reflecting $\mathbf{x}$ in the plane $\mathbf{a} \cdot \mathbf{x} = 0$.

If $\mathbf{a} = \mathbf{e}_3$, then this formula becomes

$$f(\mathbf{x}) = \mathbf{x} - 2\frac{\mathbf{x} \cdot \mathbf{e}_3}{\mathbf{e}_3 \cdot \mathbf{e}_3}\mathbf{e}_3$$

or

$$f(x_1, x_2, x_3) = (x_1, x_2, x_3) - 2x_3(0, 0, 1)$$

$$= (x_1, x_2, -x_3),$$

in agreement with Example 5. ■

EXERCISES

1. Let $f:\mathbb{R}^3 \to \mathbb{R}^2$ be the linear map associated with the matrix

$$\begin{pmatrix} 1 & 0 & 1 \\ 0 & 1 & -1 \end{pmatrix}.$$

Find

(a) $f(1, 0, 0)$ (b) $f(0, 1, 0)$ (c) $f(0, 0, 1)$
(d) $f(-2, 1, 3)$ (e) $f(7, 6, 5)$ (f) $f(x_1, x_2, x_3)$

2. Let $f:\mathbb{R}^2 \to \mathbb{R}^3$ be the linear map associated with the matrix

$$\begin{pmatrix} -1 & 4 \\ 5 & -3 \\ 2 & 0 \end{pmatrix}.$$

Find

(a) $f(1, 0)$ (b) $f(0, 1)$ (c) $f(2, 3)$
(d) $f(-1, 2)$ (e) $f(0, 0)$ (f) $f(x_1, x_2)$

3. Let $f:\mathbb{R}^3 \to \mathbb{R}^3$ be the linear map associated with the matrix

$$\begin{pmatrix} 2 & 1 & 3 \\ -1 & 1 & 2 \\ 0 & 3 & 4 \end{pmatrix}.$$

Find

(a) $f(1, 0, 0)$ (b) $f(0, 1, 0)$ (d) $f(0, 0, 1)$
(d) $f(-1, 1, 1)$ (e) $f(-3, 2, 2)$ (f) $f(x_1, x_2, x_3)$

4. Let

$$A = \begin{pmatrix} 1 & 0 & 0 \\ 0 & \cos\theta & -\sin\theta \\ 0 & \sin\theta & \cos\theta \end{pmatrix}$$

where θ is some fixed real number and let $f:\mathbb{R}^3 \to \mathbb{R}^3$ be the linear map associated with A.

(a) Find a formula for $f(x_1, x_2, x_3)$.

(b) Convince yourself, by drawing an appropriate picture, that f rotates each vector about the x_1-axis through the angle θ.

(c) Find a 3×3 matrix whose associated linear map is rotation about the x_2-axis through the angle θ.

5. Find the matrix of each of the following linear maps.

   (a) $f(x_1, x_2, x_3) = (3x_1 + 2x_2 - x_3, -x_1 + 4x_2 - 5x_3)$

   (b) $f(x_1, x_2) = (2x_1 - 3x_2, 7x_1 - 6x_2, x_1 - x_2)$

   (c) $f(x_1, x_2) = (x_2, x_1, x_1, x_2)$

   (d) $f(x_1, x_2, x_3) = x_1 + x_2 + x_3$

   (e) $f(x_1) = (x_1, 2x_1, 3x_1)$

6. Which of the following maps are linear maps? For each that is, find its matrix.

   (a) $f(x_1, x_2) = (x_1, x_2, x_2)$

   (b) $f(x_1, x_2) = (x_1, x_2^2)$

   (c) $f(x_1, x_2) = x_1 + x_2$

   (d) $f(x_1, x_2) = x_1 x_2$

   (e) $f(x_1, x_2, x_3, x_4) = (x_3, x_4, x_1, x_2)$

7. Find the matrix of the linear map f and find a formula for $f(x_1, x_2, x_3)$, where $f: \mathbb{R}^3 \to \mathbb{R}^3$ is the linear map such that

   (a) $f(1, 0, 0) = (2, 3, 1)$, $f(0, 1, 0) = (-1, 1, 0)$, and $f(0, 0, 1) = (-1, 2, 1)$

   (b) $f(1, 0, 0) = (-1, 1, 1)$, $f(0, 1, 0) = (3, 4, 5)$, and $f(0, 0, 1) = (0, 0, 1)$

   (c) $f(1, 0, 0) = (0, 1, 0)$, $f(0, 1, 0) = (0, 0, 1)$, and $f(0, 0, 1) = (1, 0, 0)$

   (d) $f(1, 0, 0) = (a, b, c)$, $f(0, 1, 0) = (b, c, a)$, and $f(0, 0, 1) = (c, a, b)$

8. Find the matrix of the linear map f and find a formula for $f(x_1, x_2)$ where $f: \mathbb{R}^2 \to \mathbb{R}^4$ is the linear map such that

   (a) $f(1, 0) = (1, 0, 1, 0)$ and $f(0, 1) = (0, 1, 0, 1)$

   (b) $f(1, 0) = (1, 2, 3, 4)$ and $f(0, 1) = (4, 3, 2, 1)$

   (c) $f(1, 0) = f(0, 1) = (1, 1, 1, 1)$

   (d) $f(1, 0) = (a, b, c, d)$ and $f(0, 1) = (0, 0, 0, 0)$

9. Each of the following matrices is associated with a linear map $f: \mathbb{R}^3 \to \mathbb{R}^3$ of the form $f(\mathbf{x}) = \mathbf{a} \times \mathbf{x}$. In each case, find the vector $\mathbf{a}$.

   (a) $\begin{pmatrix} 0 & 1 & 0 \\ -1 & 0 & 0 \\ 0 & 0 & 0 \end{pmatrix}$ (b) $\begin{pmatrix} 0 & 0 & 0 \\ 0 & 0 & 1 \\ 0 & -1 & 0 \end{pmatrix}$

   (c) $\begin{pmatrix} 0 & 0 & 1 \\ 0 & 0 & 0 \\ -1 & 0 & 0 \end{pmatrix}$ (d) $\begin{pmatrix} 0 & a & b \\ -a & 0 & c \\ -b & -c & 0 \end{pmatrix}$

10. Let $f: \mathbb{R}^n \to \mathbb{R}^m$ be a linear map and let $\mathbf{a}, \mathbf{b} \in \mathbb{R}^n$.

    (a) Show that the image under f of the line segment from $\mathbf{a}$ to $\mathbf{b}$ is the line segment from $f(\mathbf{a})$ to $f(\mathbf{b})$, provided that $f(\mathbf{a}) \neq f(\mathbf{b})$.

    (b) What is the image under f of the line segment from $\mathbf{a}$ to $\mathbf{b}$ in the case when $f(\mathbf{a}) = f(\mathbf{b})$?

    (c) Show that if $f(\mathbf{a}) \neq f(\mathbf{b})$ then the midpoint of the line segment from $f(\mathbf{a})$ to $f(\mathbf{b})$ is $f(\mathbf{m})$ where $\mathbf{m}$ is the midpoint of the line segment from $\mathbf{a}$ to $\mathbf{b}$.

11. Let $f: \mathbb{R}^3 \to \mathbb{R}^3$ be a linear map and let P be the parallelepiped

    $$P = \{\mathbf{p} + s\mathbf{a} + t\mathbf{b} + u\mathbf{c} | 0 \leq s \leq 1, 0 \leq t \leq 1, 0 \leq u \leq 1\}.$$

    Show that the image under f of P is the (possibly degenerate) parallelepiped

    $$\{f(\mathbf{p}) + sf(\mathbf{a}) + tf(\mathbf{b}) + uf(\mathbf{c}) | 0 \leq s \leq 1, 0 \leq t \leq 1, 0 \leq u \leq 1\}.$$

12. Find the matrix of $f:\mathbb{R}^3 \to \mathbb{R}^3$ where

    (a) f reflects each vector in the plane $\mathbf{e}_1 \cdot \mathbf{x} = 0$.

    (b) f reflects each vector in the plane $\mathbf{a} \cdot \mathbf{x} = 0$ where $\mathbf{a} = (1, 1, 1)$.

    (c) f reflects each vector in the plane $\mathbf{u} \cdot \mathbf{x} = 0$ where $\mathbf{u} = (u_1, u_2, u_3)$ is a unit vector in $\mathbb{R}^3$.

13. Show that reflection in $\mathbb{R}^3$ in the plane $\mathbf{a} \cdot \mathbf{x} = b$ is *not* a linear map if $b \neq 0$.

14. Let $\mathbf{d}$ be a nonzero vector in $\mathbb{R}^3$ and let $f:\mathbb{R}^3 \to \mathbb{R}^3$ be the function that assigns to each vector in $\mathbb{R}^3$ its component along $\mathbf{d}$. Thus

$$f(\mathbf{x}) = \frac{\mathbf{x} \cdot \mathbf{d}}{\mathbf{d} \cdot \mathbf{d}} \, \mathbf{d}.$$

    (a) Show that f is a linear map.

    (b) Find the matrix of f when $\mathbf{d} = (1, 0, 0)$.

    (c) Find the matrix of f when $\mathbf{d} = (\frac{2}{3}, \frac{2}{3}, \frac{1}{3})$.

    (d) Find the matrix of f when $\mathbf{d} = \mathbf{u}$ where $\mathbf{u} = (u_1, u_2, u_3)$ is a unit vector in $\mathbb{R}^3$.

15. Let $\mathbf{d}$ be a nonzero vector in $\mathbb{R}^3$ and let $f:\mathbb{R}^3 \to \mathbb{R}^3$ be the function that assigns to each vector in $\mathbb{R}^3$ its component perpendicular to $\mathbf{d}$. Thus

$$f(\mathbf{x}) = \mathbf{x} - \frac{\mathbf{x} \cdot \mathbf{d}}{\mathbf{d} \cdot \mathbf{d}} \, \mathbf{d}.$$

    (a) Show that f is a linear map.

    (b) Find the matrix of f when $\mathbf{d} = (0, 0, 1)$.

    (This linear map f is called ***projection onto the plane*** $\mathbf{d} \cdot \mathbf{x} = 0$.)

2.4
The Algebra of Linear Maps

In this section we shall study some ways of combining linear maps to get other linear maps. We shall see that linear maps can be added, subtracted, multiplied by scalars, and composed with one another.

We begin by defining addition, subtraction, and scalar multiplication for linear maps. Suppose $f:\mathbb{R}^n \to \mathbb{R}^m$ and $g:\mathbb{R}^n \to \mathbb{R}^m$ are linear maps. The ***sum*** $f + g$ and the ***difference*** $f - g$ of f and g are the linear maps $f + g:\mathbb{R}^n \to \mathbb{R}^m$ and $f - g:\mathbb{R}^n \to \mathbb{R}^m$ defined by

$$(f + g)(\mathbf{x}) = f(\mathbf{x}) + g(\mathbf{x})$$

and

$$(f - g)(\mathbf{x}) = f(\mathbf{x}) - g(\mathbf{x})$$

for all $\mathbf{x} \in \mathbb{R}^n$. The ***scalar multiple*** cf of the linear map f by the real number c is the linear map $cf:\mathbb{R}^n \to \mathbb{R}^m$ defined by

$$(cf)(\mathbf{x}) = cf(\mathbf{x})$$

for all $\mathbf{x} \in \mathbb{R}^n$.

EXAMPLE 1
Let $a \in \mathbb{R}$, $a > 1$, and let $f:\mathbb{R}^2 \to \mathbb{R}^2$ be expansion by the factor a,

$$f(\mathbf{x}) = a\mathbf{x}$$

for all $\mathbf{x} \in \mathbb{R}^2$. Then $2f:\mathbb{R}^2 \to \mathbb{R}^2$ is expansion by the factor $2a$:

$$(2f)(\mathbf{x}) = (2a)f(\mathbf{x})$$

for all $\mathbf{x} \in \mathbb{R}^2$. Similarly, for any $c > 1$, cf is expansion by the factor ca:

$$cf(\mathbf{x}) = (ca)f(\mathbf{x})$$

for all $\mathbf{x} \in \mathbb{R}^2$. ∎

EXAMPLE 2

Let $f:\mathbb{R}^3 \to \mathbb{R}^3$ be the identity map,

$$f(\mathbf{x}) = \mathbf{x}$$

for all $\mathbf{x} \in \mathbb{R}^3$, and let $g:\mathbb{R}^3 \to \mathbb{R}^3$ be the linear map that assigns to each vector $\mathbf{x} \in \mathbb{R}^3$ its component along $\mathbf{e}_3 = (0, 0, 1)$:

$$f(\mathbf{x}) = \frac{\mathbf{x} \cdot \mathbf{e}_3}{\mathbf{e}_3 \cdot \mathbf{e}_3} \mathbf{e}_3 = x_3\mathbf{e}_3 = (0, 0, x_3),$$

where $\mathbf{x} = (x_1, x_2, x_3)$. Then $f - g$ is the linear map that assigns to each vector $\mathbf{x}$ in $\mathbb{R}^3$ its component perpendicular to $\mathbf{e}_3$:

$$(f - g)(x_1, x_2, x_3) = f(x_1, x_2, x_3) - g(x_1, x_2, x_3)$$

$$= (x_1, x_2, x_3) - (0, 0, x_3)$$

$$= (x_1, x_2, 0).$$

Similarly, $f - 2g$ is the linear map that assigns to each vector $\mathbf{x}$ in $\mathbb{R}^3$ its reflection in the plane $x_3 = 0$:

$$(f - 2g)(x_1, x_2, x_3) = f(x_1, x_2, x_3) - 2g(x_1, x_2, x_3)$$

$$= (x_1, x_2, x_3) - (0, 0, 2x_3)$$

$$= (x_1, x_2, -x_3)$$

(see Examples 5 and 7 of the previous section). ∎

EXAMPLE 3

If $f:\mathbb{R}^3 \to \mathbb{R}^2$ and $g:\mathbb{R}^3 \to \mathbb{R}^2$ are the linear maps defined by

$$f(x_1, x_2, x_3) = (x_1 + 2x_2, -x_1 + 3x_2 + x_3)$$

and

$$g(x_1, x_2, x_3) = (2x_1 + x_2 + 2x_3, x_1 + 4x_2 + 3x_3),$$

then the sum $f + g$ of f and g is the map

$$(f + g)(x_1, x_2, x_3) = f(x_1, x_2, x_3) + g(x_1, x_2, x_3)$$

$$= (3x_1 + 3x_2 + 2x_3, 7x_2 + 4x_3)$$

and the scalar multiple $(-2)f$ of f by -2 is the map

$$(-2f)(x_1, x_2, x_3) = (-2)f(x_1, x_2, x_3)$$

$$= (-2x_1 - 4x_2, 2x_1 - 6x_2 - 2x_3). \quad \blacksquare$$

Now let us verify that the maps $f + g$, $f - g$, and cf are in fact linear maps whenever f and g are linear maps. According to Theorem 2 of Section 2.3 we need only verify that these maps preserve vector addition and scalar multiplication.

These verifications are straightforward. Since

$$
\begin{aligned}
(f + g)(\mathbf{x} + \mathbf{y}) &= f(\mathbf{x} + \mathbf{y}) + g(\mathbf{x} + \mathbf{y}) && \text{(by definition of } f + g) \\
&= f(\mathbf{x}) + f(\mathbf{y}) + g(\mathbf{x}) + g(\mathbf{y}) && \text{(by linearity of } f \text{ and } g) \\
&= f(\mathbf{x}) + g(\mathbf{x}) + f(\mathbf{y}) + g(\mathbf{y}) && \text{(by properties of vector} \\
&&& \text{addition in } \mathbb{R}^m) \\
&= (f + g)(\mathbf{x}) + (f + g)(\mathbf{y}) && \text{(by definition of } f + g),
\end{aligned}
$$

$f + g$ preserves vector addition. Similarly,

$$
\begin{aligned}
(f + g)(r\mathbf{x}) &= f(r\mathbf{x}) + g(r\mathbf{x}) && \text{(by definition of } f + g) \\
&= rf(\mathbf{x}) + rg(\mathbf{x}) && \text{(by linearity of } f \text{ and } g) \\
&= r(f(\mathbf{x}) + g(\mathbf{x})) && \text{(by a property of scalar} \\
&&& \text{multiplication in } \mathbb{R}^m) \\
&= r(f + g)(\mathbf{x}) && \text{(by definition of } f + g),
\end{aligned}
$$

so $f + g$ preserves scalar multiplication. Thus we have shown that $f + g$ is indeed a linear map.

The verification that cf is a linear map is similar and is left as an exercise (Exercise 10). Finally, since $f - g = f + (-1)g$, it follows that $f - g$ is also a linear map.

There is another way to combine linear maps that is of great importance. Given two linear maps, $f: \mathbb{R}^n \to \mathbb{R}^k$ and $g: \mathbb{R}^k \to \mathbb{R}^m$, then, for each $\mathbf{x} \in \mathbb{R}^n$, we can compute $f(\mathbf{x})$, a vector in $\mathbb{R}^k$. Since $f(\mathbf{x}) \in \mathbb{R}^k$, we can also compute $g(f(\mathbf{x}))$, a vector in $\mathbb{R}^m$. The function that assigns to each $\mathbf{x} \in \mathbb{R}^n$ the vector $g(f(\mathbf{x})) \in \mathbb{R}^m$ is called the **composition of** f and g, or f **followed by** g. It is denoted by $g \circ f$. Thus $g \circ f: \mathbb{R}^n \to \mathbb{R}^m$ is defined by the formula

$$g \circ f(\mathbf{x}) = g(f(\mathbf{x}))$$

for all $\mathbf{x} \in \mathbb{R}^n$.

EXAMPLE 4

Let $f: \mathbb{R}^n \to \mathbb{R}^n$ and $g: \mathbb{R}^n \to \mathbb{R}^n$ be defined by $f(\mathbf{x}) = 2\mathbf{x}$ and $g(\mathbf{x}) = 3\mathbf{x}$. Then

$$g \circ f(\mathbf{x}) = g(f(\mathbf{x})) = g(2\mathbf{x}) = 3(2\mathbf{x}) = 6\mathbf{x}.$$

In other words, multiplication by 2 followed by multiplication by 3 is the same as multiplication by 6. ∎

EXAMPLE 5

Let $f:\mathbb{R}^2 \to \mathbb{R}^2$ be the linear map that rotates each vector in $\mathbb{R}^2$ counterclockwise through the angle α and let $g:\mathbb{R}^2 \to \mathbb{R}^2$ be the linear map that rotates each vector counterclockwise through the angle β. Then it is clear geometrically (see Figure 2.28) that $g \circ f$ is the linear map that rotates each vector in $\mathbb{R}^2$ counterclockwise through the angle $\alpha + \beta$. ∎

EXAMPLE 6

Let $f:\mathbb{R}^2 \to \mathbb{R}^3$ and $g:\mathbb{R}^3 \to \mathbb{R}^1$ be defined by

$$f(x_1, x_2) = (2x_1 + 3x_2, -x_1 + x_2, 4x_1 - 5x_2)$$

and

$$g(x_1, x_2, x_3) = x_1 + x_2 + x_3.$$

We can compute $g \circ f(-1, 2)$ as follows. First,

$$f(-1, 2) = (2(-1) + 3(2), -(-1) + 2, 4(-1) - 5(2))$$
$$= (4, 3, -14).$$

Then

$$g \circ f(-1, 2) = g(f(-1, 2))$$
$$= g(4, 3, -14)$$
$$= 4 + 3 + (-14)$$
$$= -7.$$

Similarly, we can compute $g \circ f(x_1, x_2)$ for any $(x_1, x_2) \in \mathbb{R}^2$. Since

$$f(x_1, x_2) = (2x_1 + 3x_2, -x_1 + x_2, 4x_1 - 5x_2),$$

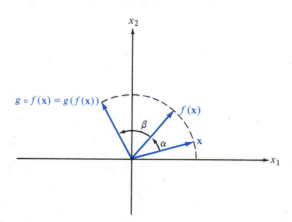

FIGURE 2.28
Rotation through the angle α followed by rotation through the angle β is the same as rotation through the angle $\alpha + \beta$.

we have

$$g \circ f(x_1, x_2) = g(f(x_1, x_2))$$

$$= g(2x_1 + 3x_2, -x_1 + x_2, 4x_1 - 5x_2)$$

$$= (2x_1 + 3x_2) + (-x_1 + x_2) + (4x_1 - 5x_2)$$

$$= 5x_1 - x_2. \quad \blacksquare$$

EXAMPLE 7

Let $f:\mathbb{R}^2 \to \mathbb{R}^2$ and $g:\mathbb{R}^2 \to \mathbb{R}^2$ be defined by

$$f(x_1, x_2) = (x_1 + x_2, x_1 - x_2)$$

$$g(x_1, x_2) = (x_2, x_1).$$

Then

$$g \circ f(x_1, x_2) = g(f(x_1, x_2))$$

$$= g(x_1 + x_2, x_1 - x_2)$$

$$= (x_1 - x_2, x_1 + x_2),$$

since g is the function that assigns to each vector $\mathbf{x} \in \mathbb{R}^2$ the vector obtained by interchanging the entries in $\mathbf{x}$. Similarly,

$$f \circ g(x_1, x_2) = f(g(x_1, x_2))$$

$$= f(x_2, x_1)$$

$$= (x_2 + x_1, x_2 - x_1),$$

since f is the function that assigns to each vector $\mathbf{x}$ in $\mathbb{R}^2$ the vector whose first coordinate is the sum of the entries in $\mathbf{x}$ and whose second coordinate is the difference of the entries in $\mathbf{x}$.

We can, if we wish, rewrite $f \circ g(x_1, x_2)$ as

$$f \circ g(x_1, x_2) = (x_1 + x_2, -x_1 + x_2).$$

Notice that, since

$$g \circ f(x_1, x_2) = (x_1 - x_2, x_1 + x_2),$$

the functions $f \circ g$ and $g \circ f$ are not the same. $\quad \blacksquare$

REMARK If $f:\mathbb{R}^n \to \mathbb{R}^k$ and $g:\mathbb{R}^l \to \mathbb{R}^m$, then $g \circ f$ is defined only when $k = l$. In particular, often you will find that $g \circ f$ is defined ($k = l$) but that $f \circ g$ is not defined ($m \neq n$). Even when both $g \circ f$ and $f \circ g$ are defined, they are usually not the same (see Example 7).

Let us verify now that if $f:\mathbb{R}^n \to \mathbb{R}^k$ and $g:\mathbb{R}^k \to \mathbb{R}^m$ are linear maps, then $g \circ f:\mathbb{R}^n \to \mathbb{R}^m$ is indeed a linear map. We must verify that $g \circ f$ preserves vector addition and scalar multiplication. First

$$g \circ f(\mathbf{x} + \mathbf{y}) = g(f(\mathbf{x} + \mathbf{y})) \qquad \text{(definition of } g \circ f)$$

$$= g(f(\mathbf{x}) + f(\mathbf{y})) \qquad (f \text{ preserves vector addition})$$

$$= g(f(\mathbf{x})) + g(f(\mathbf{y})) \qquad (g \text{ preserves vector addition})$$

$$= g \circ f(\mathbf{x}) + g \circ f(\mathbf{y}) \qquad \text{(definition of } g \circ f),$$

so $g \circ f$ preserves vector addition. Similarly,

$$g \circ f(c\mathbf{x}) = g(f(c\mathbf{x})) \qquad \text{(definition of } g \circ f)$$

$$= g(cf(\mathbf{x})) \qquad (f \text{ preserves scalar multiplication})$$

$$= cg(f(\mathbf{x})) \qquad (g \text{ preserves scalar multiplication})$$

$$= cg \circ f(\mathbf{x}) \qquad \text{(definition of } g \circ f),$$

so $g \circ f$ preserves scalar multiplication. Thus $g \circ f$ is, indeed, a linear map.

The operations of addition and scalar multiplication of linear maps obey the same algebraic rules as do addition and scalar multiplication of vectors. These rules are listed here for future reference. If f, g, and h are linear maps from $\mathbb{R}^n$ to $\mathbb{R}^m$ and c and d are real numbers, then

(A$_1$) $f + g = g + f$

(A$_2$) $(f + g) + h = f + (g + h)$

(A$_3$) $0 + f = f + 0 = f$, where $0\colon\mathbb{R}^n \to \mathbb{R}^m$ is the **zero map** defined by $0(\mathbf{x}) = \mathbf{0}$ for all $\mathbf{x} \in \mathbb{R}^n$

(A$_4$) $f + (-f) = (-f) + f = 0$, where $-f\colon\mathbb{R}^n \to \mathbb{R}^m$ is defined by $(-f)(\mathbf{x}) = -f(\mathbf{x})$ for all $\mathbf{x} \in \mathbb{R}^n$

(S$_1$) $c(f + g) = cf + cg$

(S$_2$) $(c + d)f = cf + df$

(S$_3$) $c(df) = (cd)f$

(S$_4$) $1f = f$.

Similarly, the operation of composition of linear maps obeys some algebraic rules that are worth remembering:

(C$_1$) $h \circ (f + g) = h \circ f + h \circ g$

(C$_2$) $(g + h) \circ f = g \circ f + h \circ f$

(C$_3$) $h \circ (g \circ f) = (h \circ g) \circ f$

(C$_4$) $(cg) \circ f = g \circ (cf) = c(g \circ f)$.

These rules are valid whenever f, g, and h are linear maps for which the indicated compositions make sense and where c is any real number.

These rules are easy to verify, using the appropriate definitions. For example, let us verify (C$_3$). Suppose $f\colon\mathbb{R}^n \to \mathbb{R}^l$, $g\colon\mathbb{R}^l \to \mathbb{R}^k$, and $h\colon\mathbb{R}^k \to \mathbb{R}^m$ are linear maps. In order to verify that

$$h \circ (g \circ f) = (h \circ g) \circ f,$$

we must check that

$$(h \circ (g \circ f))(\mathbf{x}) = ((h \circ g) \circ f)(\mathbf{x})$$

for all $\mathbf{x} \in \mathbb{R}^n$. But, using the definition of composition repeatedly, we find that

$$(h \circ (g \circ f))(\mathbf{x}) = h(g \circ f(\mathbf{x})) = h(g(f(\mathbf{x})))$$

$$= (h \circ g)(f(\mathbf{x})) = ((h \circ g) \circ f)(\mathbf{x}),$$

which proves (C_3).

EXERCISES

1. Find $(f + g)(x_1, x_2)$, $(3f)(x_1, x_2)$, and $(-1)g(x_1, x_2)$ where
   (a) $f(x_1, x_2) = x_1 + x_2$, $g(x_1, x_2) = x_1 - x_2$
   (b) $f(x_1, x_2) = (2x_1 + x_2, -x_1 + 2x_2)$, $g(x_1, x_2) = (-2x_1 - x_2, x_1)$
   (c) $f(x_1, x_2) = (x_1, x_1, x_2)$, $g(x_1, x_2) = (x_1, -x_1, 0)$
   (d) $f(x_1, x_2) = (-x_2, x_1)$, $g(x_1, x_2) = (x_1 + x_2, -x_1 + x_2)$

2. Let $f(x_1, x_2, x_3) = (x_1 + x_2, x_2 + x_3)$ and $g(x_1, x_2, x_3) = (x_1 - x_3, x_2 - x_3)$. Find
   (a) $(2f)(x_1, x_2, x_3)$ (b) $(3g)(x_1, x_2, x_3)$
   (c) $(2f + 3g)(x_1, x_2, x_3)$ (d) $(f - g)(x_1, x_2, x_3)$
   (e) $(2f - g)(x_1, x_2, x_3)$ (f) $(2f - 3g)(x_1, x_2, x_3)$

3. Let $f:\mathbb{R}^3 \to \mathbb{R}^3$ be the identity map, $f(\mathbf{x}) = \mathbf{x}$, and let $g:\mathbb{R}^3 \to \mathbb{R}^3$ be the linear map that assigns to each vector in $\mathbb{R}^3$ its component along $\mathbf{e}_1 = (1, 0, 0)$.
   (a) Find $(f - g)(x_1, x_2, x_3)$ and describe the linear map $f - g$ geometrically.
   (b) Find $(f - 2g)(x_1, x_2, x_3)$ and describe the linear map $f - 2g$ geometrically.

4. Let $p_1:\mathbb{R}^2 \to \mathbb{R}^2$ be the linear map that assigns to each vector in $\mathbb{R}^2$ its component along $\mathbf{e}_1 = (1, 0)$, and let $p_2:\mathbb{R}^2 \to \mathbb{R}^2$ be the linear map that assigns to each vector in $\mathbb{R}^2$ its component along $\mathbf{e}_2 = (0, 1)$. Describe the linear map $p_1 + p_2$.

5. Let $f:\mathbb{R}^3 \to \mathbb{R}^2$ and $g:\mathbb{R}^2 \to \mathbb{R}^2$ be defined by

$$f(x_1, x_2, x_3) = (x_1 + x_2, x_1 - x_3)$$

$$g(x_1, x_2) = (x_1 - x_2, 2x_1)$$

Find
   (a) $g \circ f(1, 0, 0)$ (b) $g \circ f(0, 1, 0)$
   (c) $g \circ f(0, 0, 1)$ (d) $g \circ f(-1, 0, 2)$
   (e) $g \circ f(3, -1, 4)$ (f) $g \circ f(x_1, x_2, x_3)$

6. Find $g \circ f(x_1, x_2)$ where
   (a) $f(x_1, x_2) = (0, x_1, x_2)$, $g(x_1, x_2, x_3) = (x_2, x_3)$
   (b) $f(x_1, x_2) = (3x_1 - 4x_2, -x_1 + 2x_2)$, $g(x_1, x_2) = (x_2, x_1)$
   (c) $f(x_1, x_2) = (x_2, x_1)$, $g(x_1, x_2) = (3x_1, -4x_2, -x_1 + 2x_2)$
   (d) $f(x_1, x_2) = (-x_2, x_1)$, $g(x_1, x_2) = (-x_2, x_1)$

7. Let $f:\mathbb{R}^2 \to \mathbb{R}^2$ be the horizontal shear

$$f(x_1, x_2) = (x_1 + cx_2, x_2)$$

and let $g:\mathbb{R}^2 \to \mathbb{R}^2$ be the horizontal shear

$$g(x_1, x_2) = (x_1 + dx_2, x_2).$$

   (a)   Show that $g \circ f : \mathbb{R}^2 \to \mathbb{R}^2$ is also a horizontal shear.

   (b)   Show that, for this f and g, $g \circ f = f \circ g$.

8.   If $f : \mathbb{R}^2 \to \mathbb{R}^2$ is reflection in the line ℓ, where ℓ is some line through $\mathbf{0}$ in $\mathbb{R}^2$, then what is $f \circ f$?

9.   Let $\mathbf{a}$ be a nonzero vector in $\mathbb{R}^n$ and let $f : \mathbb{R}^n \to \mathbb{R}^n$ be the linear map that assigns to each vector in $\mathbb{R}^n$ its component along $\mathbf{a}$,

$$f(\mathbf{x}) = \frac{\mathbf{a} \cdot \mathbf{x}}{\mathbf{a} \cdot \mathbf{a}} \, \mathbf{a}.$$

Show that $f \circ f = f$.

10.   Verify that if $f : \mathbb{R}^n \to \mathbb{R}^m$ is a linear map, then so is cf.

11.   Verify that addition and scalar multiplication of linear maps have the properties A_1–A_4 and S_1–S_4 listed in this section.

12.   Verify the properties C_1, C_2, and C_4 for the composition of linear maps.

2.5
The Algebra of Matrices

In the previous section we discussed various operations that can be performed on linear maps to get new linear maps. In this section, we will study the corresponding operations for matrices: addition, subtraction, scalar multiplication, and matrix multiplication.

The definitions of addition, subtraction, and scalar multiplication for matrices are exactly as you might expect, based on your experience with addition, subtraction, and scalar multiplication of vectors. Suppose A and B are two matrices of the same size (same number of rows and same number of columns). The **sum** $A + B$ is the matrix obtained by adding corresponding entries in A and B. The difference $A - B$ is the matrix obtained by subtracting the entries in B from the corresponding entries in A. Thus if

$$A = \begin{pmatrix} a_{11} & \cdots & a_{1n} \\ & \cdots & \\ a_{m1} & \cdots & a_{mn} \end{pmatrix} \quad \text{and} \quad B = \begin{pmatrix} b_{11} & \cdots & b_{1n} \\ & \cdots & \\ b_{m1} & \cdots & b_{mn} \end{pmatrix},$$

then

$$A + B = \begin{pmatrix} a_{11} + b_{11} & \cdots & a_{1n} + b_{1n} \\ & \cdots & \\ a_{m1} + b_{m1} & \cdots & a_{mn} + b_{mn} \end{pmatrix}$$

and

$$A - B = \begin{pmatrix} a_{11} - b_{11} & \cdots & a_{1n} - b_{1n} \\ & \cdots & \\ a_{m1} - b_{m1} & \cdots & a_{mn} - b_{mn} \end{pmatrix}.$$

Similarly, if A is any matrix and c is any real number then the **scalar multiple** cA of A by c is the matrix whose entries are obtained by multiplying all of the entries in A by c. Thus,

$$\text{if } A = \begin{pmatrix} a_{11} & \cdots & a_{1n} \\ & \cdots & \\ a_{m1} & \cdots & a_{mn} \end{pmatrix}, \quad \text{then} \quad cA = \begin{pmatrix} ca_{11} & \cdots & ca_{1n} \\ & \cdots & \\ ca_{m1} & \cdots & ca_{mn} \end{pmatrix}.$$

EXAMPLE 1
If

$$A = \begin{pmatrix} 1 & -1 \\ 2 & 3 \\ 0 & 1 \end{pmatrix} \quad \text{and} \quad B = \begin{pmatrix} 2 & 1 \\ 1 & 4 \\ 2 & 3 \end{pmatrix}$$

then

$$A + B = \begin{pmatrix} 1+2 & -1+1 \\ 2+1 & 3+4 \\ 0+2 & 1+3 \end{pmatrix} = \begin{pmatrix} 3 & 0 \\ 3 & 7 \\ 2 & 4 \end{pmatrix},$$

$$A - B = \begin{pmatrix} 1-2 & -1-1 \\ 2-1 & 3-4 \\ 0-2 & 1-3 \end{pmatrix} = \begin{pmatrix} -1 & -2 \\ 1 & -1 \\ -2 & -2 \end{pmatrix},$$

and

$$(-2)A = \begin{pmatrix} (-2)(1) & (-2)(-1) \\ (-2)(2) & (-2)(3) \\ (-2)(0) & (-2)(1) \end{pmatrix} = \begin{pmatrix} -2 & 2 \\ -4 & -6 \\ 0 & -2 \end{pmatrix}. \quad \blacksquare$$

We know that there is one-to-one correspondence between the set of linear maps from $\mathbb{R}^n$ to $\mathbb{R}^m$ and the set of $m \times n$ matrices. Each $m \times n$ matrix A defines a linear map $f:\mathbb{R}^n \to \mathbb{R}^m$, and each linear map $f:\mathbb{R}^n \to \mathbb{R}^m$ corresponds to an $m \times n$ matrix A. This correspondence respects addition, subtraction, and scalar multiplication.

Theorem 1   *Let f and g be linear maps from $\mathbb{R}^n$ to $\mathbb{R}^m$ and let c be a real number. Let A be the matrix of f and let B be the matrix of g. Then*

(i)   *$A + B$ is the matrix of $f + g$,*

(ii)   *$A - B$ is the matrix of $f - g$, and*

(iii)   *cA is the matrix of cf.*

Proof   (i) By Theorem 2 of Section 2.3, the matrix A of f is the matrix whose column vectors are $f(\mathbf{e}_1), \ldots, f(\mathbf{e}_n)$, and the matrix B of g is the matrix whose column vectors are $g(\mathbf{e}_1), \ldots, g(\mathbf{e}_n)$. Therefore, by the definition of matrix addition, the column vectors of $A + B$ are the vectors $f(\mathbf{e}_1) + g(\mathbf{e}_1) = (f + g)(\mathbf{e}_1), \ldots, f(\mathbf{e}_n) + g(\mathbf{e}_n) = (f + g)(\mathbf{e}_n)$. Hence $A + B$ is the matrix of the linear map $f + g$.

(iii) By the definition of scalar multiplication of matrices, the column vectors of cA are the vectors $cf(\mathbf{e}_1) = (cf)(\mathbf{e}_1), \ldots, cf(\mathbf{e}_n) = (cf)(\mathbf{e}_n)$. Therefore cA is the matrix of the linear map cf.

(ii) Since $(-1)B$ is the matrix of $(-1)g$, by (iii), and $A + (-1)B$ is the matrix of $f + (-1)g$, by (i), we see that $A - B = A + (-1)B$ is the matrix of $f - g = f + (-1)g$. $\quad \blacksquare$

According to Theorem 1, addition of matrices corresponds to addition of linear maps, subtraction of matrices corresponds to subtraction of linear maps, and scalar multiplication of matrices corresponds to scalar multiplication of linear maps. It is natural to ask, then, if there is an operation for matrices that corresponds to the operation of composition for linear maps. There is. It is called *matrix multiplication*.

Let A be an $m \times k$ matrix and let B be a $k \times n$ matrix. The ***product*** AB of A and B is the $m \times n$ matrix whose (i, j)-entry (entry in the ith row and jth column) is equal to the dot product of the ith row vector of A and the jth column vector of B, for each i and j ($1 \le i \le m$, $1 \le j \le n$). In other words, the entries in the first row of AB are obtained by computing the dot products of the first row vector of A with the column vectors of B. The entries in the second row of AB are obtained by computing the dot products of the second row vector of A with the column vectors of B. And so on.

EXAMPLE 2

Let $A = \begin{pmatrix} 1 & -1 & 3 \\ 2 & 3 & 4 \end{pmatrix}$ and $B = \begin{pmatrix} 1 & 3 \\ 0 & -1 \\ 1 & 2 \end{pmatrix}$.

Then

$$AB = \begin{pmatrix} 1 & -1 & 3 \\ 2 & 3 & 4 \end{pmatrix} \begin{pmatrix} 1 & 3 \\ 0 & -1 \\ 1 & 2 \end{pmatrix}$$

$$= \begin{pmatrix} (1, -1, 3) \cdot (1, 0, 1) & (1, -1, 3) \cdot (3, -1, 2) \\ (2, 3, 4) \cdot (1, 0, 1) & (2, 3, 4) \cdot (3, -1, 2) \end{pmatrix}$$

$$= \begin{pmatrix} 4 & 10 \\ 6 & 11 \end{pmatrix}.$$

Also,

$$BA = \begin{pmatrix} 1 & 3 \\ 0 & -1 \\ 1 & 2 \end{pmatrix} \begin{pmatrix} 1 & -1 & 3 \\ 2 & 3 & 4 \end{pmatrix}$$

$$= \begin{pmatrix} (1, 3) \cdot (1, 2) & (1, 3) \cdot (-1, 3) & (1, 3) \cdot (3, 4) \\ (0, -1) \cdot (1, 2) & (0, -1) \cdot (-1, 3) & (0, -1) \cdot (3, 4) \\ (1, 2) \cdot (1, 2) & (1, 2) \cdot (-1, 3) & (1, 2) \cdot (3, 4) \end{pmatrix}$$

$$= \begin{pmatrix} 7 & 8 & 15 \\ -2 & -3 & -4 \\ 5 & 5 & 11 \end{pmatrix}.$$

Notice that AB and BA are not of the same size. In particular, $AB \ne BA$. ■

EXAMPLE 3

Let $A = \begin{pmatrix} 1 & -1 \\ 2 & 3 \end{pmatrix}$ and $B = \begin{pmatrix} 1 & 0 \\ -1 & 1 \end{pmatrix}$.

Then

$$AB = \begin{pmatrix} 2 & -1 \\ -1 & 3 \end{pmatrix} \quad \text{and} \quad BA = \begin{pmatrix} 1 & -1 \\ 1 & 4 \end{pmatrix}.$$

Notice that, even though the 2×2 matrices AB and BA are of the same size, they are still not equal. Matrix multiplication, like the composition of linear maps, is not a commutative operation. ■

REMARK   The matrix product AB of two matrices A and B is not always defined. In order for the product AB to be defined, the number of *columns* in A must be the same as the number of *rows* in B. In particular, you will often find that AB is defined but that BA is not (or vice versa). Even when both AB and BA are defined, they are usually not the same. In fact, AB and BA may be of different sizes, as we saw in Example 2.

As mentioned above, multiplication of matrices corresponds to composition of linear maps.

Theorem 2   *Let $f:\mathbb{R}^n \to \mathbb{R}^k$ and $g:\mathbb{R}^k \to \mathbb{R}^m$ be linear maps. Let A be the matrix of g and let B be the matrix of f. Then AB is the matrix of $g \circ f$.*

Proof   Let C be the matrix of $g \circ f$. According to Theorem 2 of Section 2.3, the jth column vector of C is the vector $g \circ f(\mathbf{e}_j)$. Hence the entry c_{ij} in the ith row and jth column of C is equal to the ith entry of the vector $g \circ f(\mathbf{e}_j) = g(f(\mathbf{e}_j))$.

Since g is the linear map associated with the matrix A, the ith entry of the vector $g(f(\mathbf{e}_j))$ is equal to the dot product of the ith row vector of A and the vector $f(\mathbf{e}_j)$. Therefore $c_{ij} = \mathbf{a}_i \cdot f(\mathbf{e}_j)$, where $\mathbf{a}_i$ is the ith row vector of A.

Since the matrix of f is B, the vector $f(\mathbf{e}_j)$ is the jth column vector of B. Thus $c_{ij} = \mathbf{a}_i \cdot \mathbf{b}_j$ where $\mathbf{a}_i$ is the ith row vector of A and $\mathbf{b}_j$ is the jth column vector of B. But $\mathbf{a}_i \cdot \mathbf{b}_j$ is just the (i, j)-entry of AB. Hence $C = AB$, as asserted. ■

EXAMPLE 4

Let $f:\mathbb{R}^3 \to \mathbb{R}^2$ and $g:\mathbb{R}^2 \to \mathbb{R}^4$ be the linear maps defined by

$$f(x_1, x_2, x_3) = (x_1 + x_2, -x_1 + x_3)$$

and

$$g(x_1, x_2) = (x_1 + 2x_2, -x_1 + x_2, x_2, 3x_1 - x_2).$$

We can use Theorem 2 to compute $g \circ f:\mathbb{R}^3 \to \mathbb{R}^4$. The matrices A of g and B of f are

$$A = \begin{pmatrix} 1 & 2 \\ -1 & 1 \\ 0 & 1 \\ 3 & -1 \end{pmatrix} \quad \text{and} \quad B = \begin{pmatrix} 1 & 1 & 0 \\ -1 & 0 & 1 \end{pmatrix}.$$

Hence the matrix C of $g \circ f$ is

$$C = AB = \begin{pmatrix} -1 & 1 & 2 \\ -2 & -1 & 1 \\ -1 & 0 & 1 \\ 4 & 3 & -1 \end{pmatrix}$$

and

$$g \circ f(x_1, x_2, x_3)$$
$$= (-x_1 + x_2 + 2x_3, -2x_1 - x_2 + x_3, -x_1 + x_3, 4x_1 + 3x_2 - x_3). \quad \blacksquare$$

EXAMPLE 5

Using matrix multiplication, we can restate the relationship between a matrix and its associated linear map. Given an $m \times n$ matrix A, the associated linear map is the linear map $f:\mathbb{R}^n \to \mathbb{R}^m$ defined by

$$f(\mathbf{x}) = (\mathbf{a}_1 \bullet \mathbf{x}, \ldots, \mathbf{a}_m \bullet \mathbf{x})$$

where $\mathbf{a}_1, \ldots, \mathbf{a}_m$ are the row vectors of A. If we write the vectors $\mathbf{x}$ and $f(\mathbf{x})$ vertically rather than horizontally, so that

$$\mathbf{x} = \begin{pmatrix} x_1 \\ \vdots \\ x_n \end{pmatrix} \quad \text{and} \quad f(\mathbf{x}) = \begin{pmatrix} \mathbf{a}_1 \bullet \mathbf{x} \\ \vdots \\ \mathbf{a}_m \bullet \mathbf{x} \end{pmatrix},$$

then we see that

$$f(\mathbf{x}) = A\mathbf{x}.$$

In other words, *if vectors are written vertically rather than horizontally, then the linear map associated with the matrix A is just multiplication on the left by A.* $\quad \blacksquare$

EXAMPLE 6

Using matrix multiplication, we can interpret systems of linear equations as matrix equations. Consider the linear system

$$a_{11}x_1 + a_{12}x_2 + \cdots + a_{1n}x_n = b_1$$
$$a_{21}x_1 + a_{22}x_2 + \cdots + a_{2n}x_n = b_2$$
$$\cdots$$
$$a_{m1}x_1 + a_{m2}x_2 + \cdots + a_{mn}x_n = b_m.$$

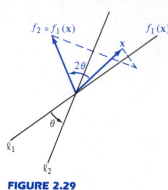

FIGURE 2.29

Reflection in ℓ_1 followed by reflection in ℓ_2 is the same as rotation through the angle 2θ.

You can readily verify that this system of equations is the same as the single matrix equation

$$A\mathbf{x} = \mathbf{b}$$

where

$$A = \begin{pmatrix} a_{11} & a_{12} & \cdots & a_{1n} \\ a_{21} & a_{22} & \cdots & a_{2n} \\ & & \cdots & \\ a_{m1} & a_{m2} & \cdots & a_{mn} \end{pmatrix}, \quad \mathbf{x} = \begin{pmatrix} x_1 \\ x_2 \\ \vdots \\ x_n \end{pmatrix}, \quad \text{and} \quad \mathbf{b} = \begin{pmatrix} b_1 \\ b_2 \\ \vdots \\ b_m \end{pmatrix}$$

The matrix A is called the **coefficient matrix** of the given system.

We shall see, in Section 2.7, that this matrix interpretation of a linear system can sometimes be used to help solve the system. ■

Theorem 2 can be used to prove theorems in geometry. We shall use Theorem 2 to show that, in $\mathbb{R}^2$, any composition of two reflections is a rotation.

Theorem 3   *Let ℓ_1 and ℓ_2 be two lines through $\mathbf{0}$ in $\mathbb{R}^2$. Let $f_1:\mathbb{R}^2 \to \mathbb{R}^2$ be the linear map that reflects each vector in the line ℓ_1 and let $f_2:\mathbb{R}^2 \to \mathbb{R}^2$ be the linear map that reflects each vector in the line ℓ_2. Then $f_2 \circ f_1$ is the linear map that rotates each vector counterclockwise through the angle 2θ, where θ is the angle, measured counterclockwise, from ℓ_1 to ℓ_2 (see Figure 2.29).*

Proof   Recall from Example 5 in Section 2.2 that the matrices A of f_2 and B of f_1 are

$$A = \begin{pmatrix} \cos 2\alpha & \sin 2\alpha \\ \sin 2\alpha & -\cos 2\alpha \end{pmatrix} \quad \text{and} \quad B = \begin{pmatrix} \cos 2\beta & \sin 2\beta \\ \sin 2\beta & -\cos 2\beta \end{pmatrix}$$

where α and β are the inclination angles of the lines ℓ_2 and ℓ_1 (see Figure 2.30). By Theorem 2, the matrix of $f_2 \circ f_1$ is

$$AB = \begin{pmatrix} \cos 2\alpha \cos 2\beta + \sin 2\alpha \sin 2\beta & \cos 2\alpha \sin 2\beta - \sin 2\alpha \cos 2\beta \\ \sin 2\alpha \cos 2\beta - \cos 2\alpha \sin 2\beta & \sin 2\alpha \sin 2\beta + \cos 2\alpha \cos 2\beta \end{pmatrix}$$

$$= \begin{pmatrix} \cos(2\alpha - 2\beta) & -\sin(2\alpha - 2\beta) \\ \sin(2\alpha - 2\beta) & \cos(2\alpha - 2\beta) \end{pmatrix}$$

But Example 4 of Section 2.2 shows that this matrix is the matrix of the linear map that rotates each vector in $\mathbb{R}^2$ counterclockwise through the angle $2\alpha - 2\beta = 2(\alpha - \beta)$. Since $\alpha - \beta = \theta$, $f_2 \circ f_1$ rotates each vector counterclockwise through the angle 2θ, as asserted. ■

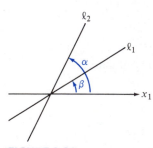

FIGURE 2.30

α and β are the inclination angles of ℓ_1 and ℓ_2.

We conclude this section by listing the fundamental algebraic properties of matrix addition, scalar multiplication, and matrix multiplication.

Properties of Matrix Addition

(A$_1$) $A + B = B + A$

(A$_2$) $(A + B) + C = A + (B + C)$

(A$_3$) $O + A = A + O = A$, where O is the matrix with all entries equal to zero.

(A$_4$) $A + (-A) = (-A) + A = O$ where $-A = (-1)A$.

Properties of Scalar Multiplication

(S$_1$) $c(A + B) = cA + cB$

(S$_2$) $(c + d)A = cA + dA$

(S$_3$) $c(dA) = (cd)A$

(S$_4$) $1A = A$

Properties of Matrix Multiplication

(M$_1$) $A(B + C) = AB + AC$

(M$_2$) $(A + B)C = AC + BC$

(M$_3$) $A(BC) = (AB)C$

(M$_4$) $(cA)B = A(cB) = c(AB)$

These properties hold whenever A, B, and C are matrices of appropriate sizes so that the indicated operations make sense, and c and d are any real numbers. The verification of these properties is straightforward (see Exercises 15 and 16).

EXERCISES

1. Let
$$A = \begin{pmatrix} 1 & -1 \\ 2 & 3 \end{pmatrix} \quad \text{and} \quad B = \begin{pmatrix} -2 & 5 \\ 4 & 3 \end{pmatrix}.$$

    Find
    (a) $A + B$ (b) $2A$ (c) $(-3)B$
    (d) $2A - 3B$ (e) $A - B$ (f) $7A + B$

2. Repeat Exercise 1 with
$$A = \begin{pmatrix} -2 & 0 & 1 \\ 1 & 0 & -1 \\ 2 & 3 & 2 \end{pmatrix} \quad \text{and} \quad B = \begin{pmatrix} -1 & 1 & 0 \\ 0 & 3 & 4 \\ 2 & -1 & 1 \end{pmatrix}.$$

3. Let $A = \begin{pmatrix} -1 & 3 & 2 \\ 2 & -1 & 1 \end{pmatrix}$, $B = \begin{pmatrix} 1 & -1 \\ -1 & 2 \\ 3 & -3 \end{pmatrix}$, $C = (1 \ -1 \ 1)$, and

$$D = \begin{pmatrix} 2 \\ -1 \\ 2 \end{pmatrix}.$$

    Compute
    (a) AB (b) AD (c) BA (d) CB (e) CD (f) DC

4. Check that the six products listed in Exercise 3 are the only pairwise products of A, B, C, D that are defined. (There are 16 possibilities!)

5. Compute the following matrix products:

(a) $\begin{pmatrix} -1 & 1 \\ 3 & 2 \end{pmatrix} \begin{pmatrix} 5 & 7 \\ 9 & -3 \end{pmatrix}$

(b) $\begin{pmatrix} 2 & -3 \\ -3 & 2 \end{pmatrix} \begin{pmatrix} 4 \\ 1 \end{pmatrix}$

(c) $\begin{pmatrix} -1 & 3 & 2 \\ 1 & 4 & -5 \\ 3 & 6 & 9 \end{pmatrix} \begin{pmatrix} -1 \\ 2 \\ 1 \end{pmatrix}$

(d) $\begin{pmatrix} -1 & 1 & 2 \\ 0 & 3 & -2 \\ 1 & 1 & 1 \end{pmatrix} \begin{pmatrix} 3 & 7 & 1 \\ -5 & 0 & 0 \\ 1 & -3 & -1 \end{pmatrix}$

(e) $(-1 \quad 2 \quad 1) \begin{pmatrix} -1 & 3 & 2 \\ 1 & 4 & 5 \\ 3 & 6 & 9 \end{pmatrix}$

(f) $\begin{pmatrix} 1 & 3 & 4 \\ 2 & 5 & 3 \\ 0 & 1 & 4 \end{pmatrix} \begin{pmatrix} 17 & -8 & -11 \\ -8 & 4 & 5 \\ 2 & -1 & -1 \end{pmatrix}$

6. Let $A = \begin{pmatrix} 1 & 1 \\ 0 & 1 \end{pmatrix}$.

   (a) Find $A^2 = AA$, $A^3 = A^2A$, and $A^4 = A^3A$.

   (b) Find $A^n = A^{n-1}A$ for every positive integer n.

   (c) Find a square root of A; that is, find a matrix B such that $B^2 = A$. [*Hint:* $B = $ "$A^{1/2}$."]

7. Compute A^2 and A^3 for

   (a) $A = \begin{pmatrix} 0 & a & b \\ 0 & 0 & c \\ 0 & 0 & 0 \end{pmatrix}$

   (b) $A = \begin{pmatrix} 1 & 0 & a \\ 0 & 1 & 0 \\ 0 & 0 & 1 \end{pmatrix}$

8. Compute A^k for $k = 2, 3, 4$ when

$$A = \begin{pmatrix} 0 & 1 & 0 & 0 \\ 0 & 0 & 1 & 0 \\ 0 & 0 & 0 & 1 \\ 0 & 0 & 0 & 0 \end{pmatrix}.$$

9. Let $f: \mathbb{R}^2 \to \mathbb{R}^2$ and $g: \mathbb{R}^2 \to \mathbb{R}^2$ be the linear maps defined by

$$f(x_1, x_2) = (x_1 + x_2, x_1 - x_2)$$

and

$$g(x_1, x_2) = (x_2, x_1).$$

   (a) Write out the matrices A of g and B of f.

   (b) Compute the matrix product AB and use it to find a formula for $g \circ f(x_1, x_2)$.

   (c) Compute the matrix product BA and use it to find a formula for $f \circ g(x_1, x_2)$.

   (You may find it interesting to compare these computations with the computations in Example 7 of Section 2.4 where the same formulas were derived by the direct approach.)

10. For each of the following, find $g \circ f(x_1, x_2, x_3)$ by computing the matrix product AB, where A is the matrix of g and B is the matrix of f.

   (a) $f(x_1, x_2, x_3) = (x_1 + x_2, x_1 + x_3)$, $g(x_1, x_2) = (2x_1 - x_2, x_1 + 3x_2)$

   (b) $f(x_1, x_2, x_3) = (x_2, x_3, x_1)$, $g(x_1, x_2, x_3) = (x_1 + x_2, x_3)$

   (c) $f(x_1, x_2, x_3) = (5x_1 + 4x_2 + 3x_3, -x_1 + 2x_2 - 2x_3, x_1 + x_2 - x_3)$, $g(x_1, x_2, x_3) = (-x_1 + x_2, 2x_1 + x_2 - x_3, x_1 + 3x_2 - 4x_3)$.

11. For each of the following pairs of linear maps f and g, (i) find the matrices A of g and B of f, (ii) compute the matrix product AB, and (iii) write a formula for $g \circ f(x_1, \ldots, x_n)$.

   (a) $f(x_1, x_2) = (2x_1 - x_2, 3x_1 + 4x_2)$
   $g(x_1, x_2) = (-3x_1 + x_2, x_1 + x_2)$ $(n = 2)$

   (b) $f(x_1, x_2) = (x_1 + x_2, x_1 - x_2, 2x_1)$
   $g(x_1, x_2, x_3) = (x_3, x_2, x_1)$ $(n = 2)$

   (c) $f(x_1, x_2, x_3) = (x_1 + x_2 + x_3, x_1 + x_2, x_1)$
   $g(x_1, x_2, x_3) = (x_1 - x_2, x_2 - x_3, x_3 - x_1)$ $(n = 3)$

   (d) $f(x_1, x_2, x_3) = (x_1 + x_2, x_2 + x_3)$
   $g(x_1, x_2) = x_1 - x_2$ $(n = 3)$

   (e) $f(x) = (x, -x, 2x, 3x)$
   $g(x_1, x_2, x_3, x_4) = x_1 + x_2 + x_3 + x_4$ $(n = 1)$

12. Let $f : \mathbb{R}^2 \to \mathbb{R}^2$ be the linear map that rotates each vector in $\mathbb{R}^2$ counterclockwise through the angle α and let g be the linear map that rotates each vector counterclockwise through the angle β. Verify, by multiplying the matrix of g by the matrix of f, that $g \circ f$ rotates each vector counterclockwise through the angle $\alpha + \beta$.

13. Let f be the linear map that reflects vectors in the line through $\mathbf{0}$ with inclination angle α, and let g be the linear map that rotates vectors counterclockwise through the angle θ. Show that

   (a) $g \circ f$ is the linear map that reflects vectors in the line through $\mathbf{0}$ with inclination angle $\alpha + \frac{1}{2}\theta$.

   (b) $f \circ g$ is the linear map that reflects vectors in the line through $\mathbf{0}$ with inclination angle $\alpha - \frac{1}{2}\theta$.

14. Show, by multiplying matrices, that the composition of two horizontal shears is a horizontal shear.

15. Verify properties A_1–A_4 and S_1–S_4 for addition and scalar multiplication of matrices.

16. Use Theorem 2 and the results of Section 2.4 to show that matrix multiplication has the properties M_1–M_4 listed in this section. [*Hint:* To verify M_1, for example, use the facts that if A is the matrix of f, B is the matrix of g, and C is the matrix of h then $A(B + C)$ is the matrix of $f \circ (g + h)$ and $AB + AC$ is the matrix of $f \circ g + f \circ h$.]

2.6
Inverse Maps

In this section we shall study inverse maps. These are pairs of maps such that each map reverses the effect of the other. That is, when we apply either one of them to a vector and then apply the other to the result, we get back to the vector we started with (see Figure 2.31). Then the two maps are *inverse to each other*, and each is called the *inverse* of the other. We shall learn how to tell if a given *linear* map has an inverse, and, in the next section, we shall learn how to calculate the inverse map when it does exist.

Let S and T be sets and let f and g be maps, with $f : S \to T$ and $g : T \to S$. The maps f and g are **inverse to each other** if

   (i) $g \circ f(x) = x$ for all $x \in S$, and

   (ii) $f \circ g(x) = x$ for all $x \in T$.

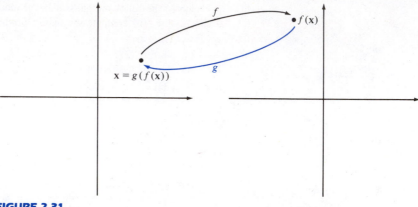

FIGURE 2.31
The maps f and g are inverse to each other.

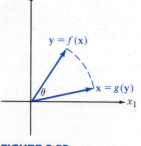

FIGURE 2.32
Counterclockwise rotation through the angle θ and clockwise rotation through the angle θ are inverse to each other.

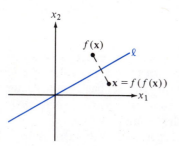

FIGURE 2.33
The linear map that reflects each vector in the line ℓ is its own inverse.

EXAMPLE 1

Let $f:\mathbb{R}^n \to \mathbb{R}^n$ be expansion by the factor 2 and let $g:\mathbb{R}^n \to \mathbb{R}^n$ be contraction by the factor $\frac{1}{2}$. Then

(i) $g \circ f(\mathbf{x}) = g(f(\mathbf{x})) = g(2\mathbf{x}) = \frac{1}{2}(2\mathbf{x}) = \mathbf{x}$ for all $\mathbf{x} \in \mathbb{R}^n$, and

(ii) $f \circ g(\mathbf{x}) = f(g(\mathbf{x})) = f(\frac{1}{2}\mathbf{x}) = 2(\frac{1}{2}\mathbf{x}) = \mathbf{x}$

for all $\mathbf{x} \in \mathbb{R}^n$, so f and g are inverse to each other. ∎

EXAMPLE 2

Let $f:\mathbb{R}^2 \to \mathbb{R}^2$ be counterclockwise rotation through the angle θ and let $g:\mathbb{R}^2 \to \mathbb{R}^2$ be clockwise rotation through the angle θ. Then $g \circ f(\mathbf{x}) = \mathbf{x}$ and $f \circ g(\mathbf{x}) = \mathbf{x}$ for all $\mathbf{x} \in \mathbb{R}^2$ (see Figure 2.32). Hence f and g are inverse to each other. ∎

EXAMPLE 3

Let ℓ be any line through $\mathbf{0}$ in $\mathbb{R}^2$ and let $f:\mathbb{R}^2 \to \mathbb{R}^2$ be the map that reflects each vector in the line ℓ. Then $f \circ f(\mathbf{x}) = \mathbf{x}$ for all $\mathbf{x} \in \mathbb{R}^2$ (see Figure 2.33). Hence f is inverse to itself. ∎

A map $f:S \to T$, where S and T are any two sets, is said to be **_invertible_** if there exists a map $g:T \to S$ such that g and f are inverse to each other. The map g, if it exists, is unique.* It is called the **_inverse_** of f and is often denoted by the symbol f^{-1}. Thus, according to Example 1 above, expansion by the factor 2 in $\mathbb{R}^n$ is invertible. Its inverse is contraction by the factor $\frac{1}{2}$. Similarly, counterclockwise rotation in $\mathbb{R}^2$ through the angle θ is invertible (Example 2). Its inverse is clockwise rotation through the angle θ. Reflection in a line ℓ in $\mathbb{R}^2$ is also invertible (Example 3). It is its own inverse.

---

* There can be only one map $g:T \to S$ that is inverse to f. For if $h:T \to S$ were another, then $h(x) = (g \circ f)(h(x)) = g(f(h(x))) = g(f \circ h(x)) = g(x)$ for all $x \in T$ so, in fact, $h = g$.

A map $f:S \to T$ that is invertible must be ***one-to-one***; that is, $f(a)$ must be different from $f(b)$ whenever $a \neq b$ ($a, b \in S$). To see this, assume that g is inverse to f. Then $a = g(f(a))$ and $b = g(f(b))$, so $a = b$ whenever $f(a) = f(b)$.

A map $f:S \to T$ that is invertible must also be ***onto***; that is, for each $b \in T$ there must be an $a \in S$ such that $f(a) = b$. To see this, suppose that g is inverse to f. Then we may set $a = g(b)$ and find that $f(a) = f(g(b)) = b$.

It turns out, conversely, that every map $f:S \to T$ that is both one-to-one and onto is invertible.

Theorem 1   *A map $f:S \to T$ is invertible if and only if it is both one-to-one and onto.*

Proof   It has already been shown that each invertible map $f:S \to T$ is both one-to-one and onto. So we need only show that each map $f:S \to T$ that is both one-to-one and onto has an inverse. But it is easy to find the inverse of such a map. Let $y \in T$. Since f is onto, there is an $x \in S$ such that $f(x) = y$. Moreover, since f is also one-to-one, there is only one such x. So we define $g:T \to S$ to be the map that assigns to each $y \in T$ the unique $x \in S$ with $f(x) = y$. Then it is clear that $g(f(x)) = x$ and that $f(g(y)) = y$ for all $x \in S$ and $y \in T$, so f is invertible and g is the inverse of f.   ■

If $f:\mathbb{R}^n \to \mathbb{R}^m$ is a *linear* map, there is a simple test that will determine if f is one-to-one: f is one-to-one if and only if the solution set of the vector equation $f(\mathbf{x}) = \mathbf{0}$ consists solely of the zero vector. The solution set of the equation $f(\mathbf{x}) = \mathbf{0}$ is called the ***kernel*** of f.

Theorem 2   *Let $f:\mathbb{R}^n \to \mathbb{R}^m$ be a linear map. Then f is one-to-one if and only if the only vector in the kernel of f is the vector $\mathbf{0}$.*

Proof   Certainly $\mathbf{0}$ is in the kernel of f because $f(\mathbf{0}) = \mathbf{0}$ for every linear map f.

Suppose f is one-to-one. Then $f(\mathbf{a}) \neq f(\mathbf{0})$ whenever $\mathbf{a} \neq \mathbf{0}$ ($\mathbf{a} \in \mathbb{R}^n$). Since $f(\mathbf{0}) = \mathbf{0}$, this means that $f(\mathbf{a}) \neq \mathbf{0}$ whenever $\mathbf{a} \neq \mathbf{0}$. Hence $\mathbf{0}$ is the *only* vector in the kernel of f.

Conversely, suppose $\mathbf{0}$ is the only vector in the kernel of f. Then $f(\mathbf{a} - \mathbf{b}) \neq \mathbf{0}$ whenever $\mathbf{a} \neq \mathbf{b}$ ($\mathbf{a}, \mathbf{b} \in \mathbb{R}^n$). Since $f(\mathbf{a} - \mathbf{b}) = f(\mathbf{a}) - f(\mathbf{b})$ it follows that $f(\mathbf{a}) - f(\mathbf{b}) \neq \mathbf{0}$, that is, $f(\mathbf{a}) \neq f(\mathbf{b})$, whenever $\mathbf{a} \neq \mathbf{b}$. Hence f is one-to-one.   ■

EXAMPLE 4

Let $f:\mathbb{R}^3 \to \mathbb{R}^4$ be the linear map defined by

$f(x_1, x_2, x_3)$

$$= (x_1 + x_2 + x_3, x_1 + x_2 - x_3, x_1 - x_2 + x_3, -x_1 + x_2 + x_3).$$

According to Theorem 2, we can find out if f is one-to-one by finding the kernel of f, that is, by solving the vector equation $f(\mathbf{x}) = \mathbf{0}$. This vector equation is equivalent to the linear system

$$x_1 + x_2 + x_3 = 0$$

$$x_1 + x_2 - x_3 = 0$$

$$x_1 - x_2 + x_3 = 0$$

$$-x_1 + x_2 + x_3 = 0.$$

The matrix of this system is

$$\begin{pmatrix} 1 & 1 & 1 & 0 \\ 1 & 1 & -1 & 0 \\ 1 & -1 & 1 & 0 \\ -1 & 1 & 1 & 0 \end{pmatrix}.$$

Its echelon matrix is

$$\begin{pmatrix} 1 & 0 & 0 & 0 \\ 0 & 1 & 0 & 0 \\ 0 & 0 & 1 & 0 \\ 0 & 0 & 0 & 0 \end{pmatrix}.$$

From the echelon matrix we can see that the only solution of the system is $\mathbf{x} = \mathbf{0}$, so the kernel of f is $\{\mathbf{0}\}$ and f is one-to-one. ■

EXAMPLE 5

Let us see if the linear map f defined in Example 4 is onto. We must find out if, for each $\mathbf{b} \in \mathbb{R}^4$, there is an $\mathbf{a} \in \mathbb{R}^3$ such that $f(\mathbf{a}) = \mathbf{b}$. In other words we need to know if, for every choice of $\mathbf{b} = (b_1, b_2, b_3, b_4) \in \mathbb{R}^4$, there is a solution of the vector equation

$$f(x_1, x_2, x_3) = (b_1, b_2, b_3, b_4).$$

Since

$f(x_1, x_2, x_3)$

$$= (x_1 + x_2 + x_3, x_1 + x_2 - x_3, x_1 - x_2 + x_3, -x_1 + x_2 + x_3),$$

the vector equation $f(x_1, x_2, x_3) = (b_1, b_2, b_3, b_4)$ is equivalent to the linear system

$$x_1 + x_2 + x_3 = b_1$$

$$x_1 + x_2 - x_3 = b_2$$

$$x_1 - x_2 + x_3 = b_3$$

$$-x_1 + x_2 + x_3 = b_4.$$

The matrix of this system reduces as follows:

$$
\begin{pmatrix}
1 & 1 & 1 & b_1 \\
1 & 1 & -1 & b_2 \\
1 & -1 & 1 & b_3 \\
-1 & 1 & 1 & b_4
\end{pmatrix}
\longrightarrow
\begin{pmatrix}
1 & 1 & 1 & b_1 \\
0 & 0 & -2 & b_2 - b_1 \\
0 & -2 & 0 & b_3 - b_1 \\
0 & 2 & 2 & b_1 + b_4
\end{pmatrix}
$$

$$
\longrightarrow
\begin{pmatrix}
1 & 1 & 1 & b_1 \\
0 & -2 & 0 & b_3 - b_1 \\
0 & 0 & -2 & b_2 - b_1 \\
0 & 2 & 2 & b_1 + b_4
\end{pmatrix}
\longrightarrow
\begin{pmatrix}
1 & 1 & 1 & b_1 \\
0 & 1 & 0 & \frac{1}{2}(b_1 - b_3) \\
0 & 0 & 1 & \frac{1}{2}(b_1 - b_2) \\
0 & 2 & 2 & b_1 + b_4
\end{pmatrix}
$$

$$
\longrightarrow
\begin{pmatrix}
1 & 0 & 1 & \frac{1}{2}(b_1 + b_3) \\
0 & 1 & 0 & \frac{1}{2}(b_1 - b_3) \\
0 & 0 & 1 & \frac{1}{2}(b_1 - b_2) \\
0 & 0 & 2 & b_3 + b_4
\end{pmatrix}
\longrightarrow
\begin{pmatrix}
1 & 0 & 0 & \frac{1}{2}(b_2 + b_3) \\
0 & 1 & 0 & \frac{1}{2}(b_1 - b_3) \\
0 & 0 & 1 & \frac{1}{2}(b_1 - b_2) \\
0 & 0 & 0 & -b_1 + b_2 + b_3 + b_4
\end{pmatrix}.
$$

It can be seen from this echelon matrix that there is no solution of the system when $-b_1 + b_2 + b_3 + b_4 \neq 0$. For example, there is no vector **a** in $\mathbb{R}^3$ such that $f(\mathbf{a}) = \mathbf{b}$ when $\mathbf{b} = (1, 0, 0, 0)$, because $-1 + 0 + 0 + 0 \neq 0$. Therefore, f is *not* onto. ■

Suppose now that $f:\mathbb{R}^n \to \mathbb{R}^m$ is any linear map. Then, by Theorem 2, f is one-to-one if and only if the vector **0** is the unique solution of the vector equation $f(\mathbf{x}) = \mathbf{0}$. If the equation $f(\mathbf{x}) = \mathbf{0}$ is written out, the result is the linear system

$$a_{11}x_1 + \cdots + a_{1n}x_n = 0$$

$$\cdots$$

$$a_{m1}x_1 + \cdots + a_{mn}x_n = 0$$

where

$$
A = \begin{pmatrix}
a_{11} & \cdots & a_{1n} \\
 & \cdots & \\
a_{m1} & \cdots & a_{mn}
\end{pmatrix}
$$

is the matrix of f. This system has a unique solution if and only if its matrix,

$$
\begin{pmatrix}
a_{11} & \cdots & a_{1n} & 0 \\
 & \cdots & & \\
a_{m1} & \cdots & a_{mn} & 0
\end{pmatrix},
$$

can be transformed by a sequence of row operations into an echelon matrix with a corner 1 in every column *except* the last. This is the same as saying that the echelon matrix of A has a corner 1 in *every* column. Thus the linear map

$f:\mathbb{R}^n \to \mathbb{R}^m$ is one-to-one if and only if the echelon matrix of A has n corner 1's.

The number of corner 1's in the echelon matrix of A, where A is the matrix of $f:\mathbb{R}^n \to \mathbb{R}^m$, is called the **rank** of the linear map f. This discussion proves the following theorem.

Theorem 3   *A linear map $f:\mathbb{R}^n \to \mathbb{R}^m$ is one-to-one if and only if its rank is n.*

EXAMPLE 6

It can be seen that the linear map $f:\mathbb{R}^3 \to \mathbb{R}^4$ defined in Example 4 is one-to-one by noting that its matrix

$$A = \begin{pmatrix} 1 & 1 & 1 \\ 1 & 1 & -1 \\ 1 & -1 & 1 \\ -1 & 1 & 1 \end{pmatrix} \text{ has echelon matrix } \begin{pmatrix} 1 & 0 & 0 \\ 0 & 1 & 0 \\ 0 & 0 & 1 \\ 0 & 0 & 0 \end{pmatrix}$$

and hence that f has rank 3.

Similarly, the linear map $g:\mathbb{R}^3 \to \mathbb{R}^3$ defined by

$$g(x_1, x_2, x_3) = (x_1 - x_2, x_2 - x_3, -x_1 + x_3)$$

has matrix

$$\begin{pmatrix} 1 & -1 & 0 \\ 0 & 1 & -1 \\ -1 & 0 & 1 \end{pmatrix}$$

with corresponding echelon matrix

$$\begin{pmatrix} 1 & 0 & -1 \\ 0 & 1 & -1 \\ 0 & 0 & 0 \end{pmatrix},$$

so g has rank 2. Since its rank is not 3, g is *not* one-to-one.   ■

We can also tell whether or not a linear map $f:\mathbb{R}^n \to \mathbb{R}^m$ is onto by calculating its rank.

Theorem 4   *A linear map $f:\mathbb{R}^n \to \mathbb{R}^m$ is onto if and only if its rank is m.*

Proof.   Let

$$A = \begin{pmatrix} a_{11} & \cdots & a_{1n} \\ & \cdots & \\ a_{m1} & \cdots & a_{mn} \end{pmatrix}$$

be the matrix of f. The linear map f is onto if and only if the equation $f(\mathbf{x}) = \mathbf{b}$ has a solution for every $\mathbf{b} \in \mathbb{R}^m$. This is the same as the statement that the linear system

$$a_{11}x_1 + \cdots + a_{1n}x_n = b_1$$

$$\cdots$$

$$a_{m1}x_1 + \cdots + a_{mn}x_n = b_m$$

has a solution for every $b_1, \ldots, b_m \in \mathbb{R}$. But this system has a solution for every $b_1, \ldots, b_m$ if and only if its echelon matrix *never* has a corner 1 in the last column. This is the case if and only if the echelon matrix of the coefficient matrix A has a corner 1 in every row (that is, it does not have a row of zeros). Thus $f:\mathbb{R}^n \to \mathbb{R}^m$ is onto if and only if the echelon matrix of A has m corner 1's, that is, if and only if the rank of f is m. ■

EXAMPLE 7
Let $f:\mathbb{R}^3 \to \mathbb{R}^2$ be defined by

$$f(x_1, x_2, x_3) = (x_1 + x_2, x_1 + x_3).$$

The matrix of f is

$$A = \begin{pmatrix} 1 & 1 & 0 \\ 1 & 0 & 1 \end{pmatrix}.$$

Its echelon matrix is

$$\begin{pmatrix} 1 & 0 & 1 \\ 0 & 1 & -1 \end{pmatrix}.$$

Hence f has rank 2 and is therefore onto. This map f is *not* one-to-one. ■

By combining Theorems 3 and 4, we can prove the following important theorem.

Theorem 5   *Suppose $f:\mathbb{R}^n \to \mathbb{R}^m$ is an invertible linear map. Then $m = n$.*

Proof   If f is invertible then f is one-to-one and onto. Since f is one-to-one, the rank of f must be equal to n, by Theorem 3. Since f is onto, the rank of f must be equal to m, by Theorem 4. Hence m must be equal to n. ■

Theorem 5 tells us that we will never find an invertible linear map $f:\mathbb{R}^n \to \mathbb{R}^m$, when $m \neq n$. So let us turn our attention to linear maps from $\mathbb{R}^n$ to $\mathbb{R}^n$. The next theorem tells us how to determine whether or not a given linear map $f:\mathbb{R}^n \to \mathbb{R}^n$ is invertible.

Theorem 6   *Let $f:\mathbb{R}^n \to \mathbb{R}^n$ be a linear map. Then the following statements are equivalent (each implies the others):*

(i)   *f is invertible*

(ii)   *f is one-to-one*

(iii)   *The kernel of f is $\{\mathbf{0}\}$.*

(iv)   *f is onto*

(v)   *f has rank n.*

(vi)   *The matrix of f has the identity matrix as its echelon matrix.*

Proof (ii) and (iii) are equivalent, by Theorem 2. (ii) and (v) are equivalent, by Theorem 3. (v) and (iv) are equivalent by Theorem 4. (v) and (vi) are equivalent by the definition of rank.

Hence (ii), (iii), (iv), (v), and (vi) are equivalent, for linear maps from $\mathbb{R}^n$ to $\mathbb{R}^n$.

(i) implies (ii), by Theorem 1. Hence (i) also implies (iii), (iv), (v) and (vi). On the other hand, each of the statements (ii), (iii), (iv), (v) and (vi) implies the others, and (ii) and (iii) together imply (i), by Theorem 1.

Hence each of the statements (i), (ii), (iii), (iv), (v), and (vi) implies all of the others. ■

EXAMPLE 8

Let $f:\mathbb{R}^3 \to \mathbb{R}^3$ be the linear map defined by

$$f(x_1, x_2, x_3) = (x_2 + x_3, x_1 + x_3, x_1 + x_2).$$

The matrix of f is

$$A = \begin{pmatrix} 0 & 1 & 1 \\ 1 & 0 & 1 \\ 1 & 1 & 0 \end{pmatrix}.$$

The echelon matrix of A is

$$\begin{pmatrix} 1 & 0 & 0 \\ 0 & 1 & 0 \\ 0 & 0 & 1 \end{pmatrix}.$$

Hence f has rank 3. This map is invertible. ■

This section concludes with a theorem that will be needed in the next section. We know from Theorem 6 that if a linear map $f:\mathbb{R}^n \to \mathbb{R}^n$ has rank n then it has an inverse. But is the inverse of f a *linear* map? The next theorem says that it is.

Theorem 7 *Let $f:\mathbb{R}^n \to \mathbb{R}^n$ be an invertible linear map, and let $g:\mathbb{R}^n \to \mathbb{R}^n$ be the inverse of f. Then g is a linear map.*

Proof By Theorem 2 of Section 2.3, we need only check that g preserves vector addition and scalar multiplication. To see this, notice that

$$f(g(\mathbf{x} + \mathbf{y})) = \mathbf{x} + \mathbf{y} = f(g(\mathbf{x})) + f(g(\mathbf{y}))$$
$$= f(g(\mathbf{x}) + g(\mathbf{y}))$$

and that

$$f(g(c\mathbf{x})) = c\mathbf{x} = cf(g(\mathbf{x})) = f(cg(\mathbf{x}))$$

for all $\mathbf{x}$ and $\mathbf{y}$ in $\mathbb{R}^n$ and every real number c. Since, by Theorem 1, f is one-to-one, it follows that

$$g(\mathbf{x} + \mathbf{y}) = g(\mathbf{x}) + g(\mathbf{y})$$

and that

$$g(c\mathbf{x}) = cg(\mathbf{x}),$$

for all $\mathbf{x}$, $\mathbf{y}$, and c. Hence g does preserve vector addition and scalar multiplication, as required. ∎

EXERCISES

1. Describe the inverse of each of the following linear maps.
   - (a) $f(\mathbf{x}) = 3\mathbf{x}$
   - (b) $f(\mathbf{x}) = \frac{1}{5}\mathbf{x}$
   - (c) $f(\mathbf{x}) = -\mathbf{x}$
   - (d) $f(x_1, x_2) = (2x_1, 3x_2)$
   - (e) $f(x_1, x_2) = (x_2, x_1)$
   - (f) $f(x_1, x_2, x_3) = (x_2, x_3, x_1)$

2. Let $f:\mathbb{R}^2 \to \mathbb{R}^2$ be the shear

   $$f(x_1, x_2) = (x_1 + cx_2, x_2),$$

   where c is some real number, and let $g:\mathbb{R}^2 \to \mathbb{R}^2$ be the shear

   $$g(x_1, x_2) = (x_1 - cx_2, x_2).$$

   Verify that f and g are inverse to each other.

3. Find the kernel of each of the following linear maps and use Theorem 2 to decide which of these maps are one-to-one.
   - (a) $f(x_1, x_2) = (2x_1 - x_2, 3x_1 - 2x_2)$
   - (b) $f(x_1, x_2) = (2x_1 - x_2, 4x_1 - 2x_2)$
   - (c) $f(x_1, x_2) = (x_1 + x_2, x_1 - x_2, x_2)$
   - (d) $f(x_1, x_2, x_3) = (x_1 + x_2 + x_3, x_1 + 2x_2 + 3x_3)$
   - (e) $f(x_1, x_2, x_3) = (x_1 + x_2, x_1 + x_3, x_2 - x_3)$

4. Use the method of Example 5 to determine which of the linear maps in Exercise 3 are onto.

5. Find the rank of each of the linear maps in Exercise 3.

6. Find the rank of each of the following linear maps and determine which of these maps are one-to-one and which are onto.
   - (a) $f(x_1, x_2) = (x_1, x_2, 0)$
   - (b) $f(x_1, x_2, x_3) = (x_1, x_2)$
   - (c) $f(x_1, x_2, x_3) = (x_1 + x_2 + x_3, -x_1 + x_2 - x_3, x_2, x_1 + x_3)$
   - (d) $f(x_1, x_2, x_3) = (x_1, x_2, x_3, x_1, x_2, x_3)$
   - (e) $f(x_1, x_2, x_3, x_4) = (x_2, x_3, x_4)$

7. Find the rank of each of the following linear maps and determine which of these maps are invertible.
   - (a) $f(x_1, x_2) = (5x_1 - 4x_2, x_1 + 2x_2)$
   - (b) $f(x_1, x_2) = (-2x_1 + 2x_2, 3x_1 - 3x_2)$
   - (c) $f(x_1, x_2, x_3) = (x_1 + x_2 + x_3, x_1 + 2x_2 + 3x_3, x_1 - x_2 - 3x_3)$
   - (d) $f(x_1, x_2, x_3) = (x_1 + x_2 + x_3, x_1 + 2x_2 + 3x_3, x_1 + 4x_2 + 9x_3)$
   - (e) $f(x_1, x_2, x_3, x_4) = (x_1, x_1 + x_2, x_1 + x_2 + x_3, x_1 + x_2 + x_3 + x_4)$

8. Let $f:\mathbb{R}^n \to \mathbb{R}^n$ and $g:\mathbb{R}^n \to \mathbb{R}^n$ be invertible linear maps, and let f^{-1} and g^{-1} denote the inverses of f and g. Show that $f \circ g$ is invertible and that the inverse of $f \circ g$ is equal to $g^{-1} \circ f^{-1}$.

9. Let $f:S \to T$ be a map from a set S to a set T. A map $g:T \to S$ is called a *left inverse* of f if $g \circ f(x) = x$ for all $x \in S$. Show that if f has a left inverse then f must be one-to-one.

10. Let $f:S \to T$ be a map from a set S to a set T. A map $g:T \to S$ is called a *right inverse* of f if $f \circ g(x) = x$ for all $x \in T$. Show that if f has a right inverse then f is onto.

2.7
Inverse Matrices

In the previous section we studied inverse maps. We learned that a linear map $f:\mathbb{R}^n \to \mathbb{R}^m$ cannot be invertible unless $m = n$, and we learned how to tell if a linear map $f:\mathbb{R}^n \to \mathbb{R}^n$ is invertible. This section will explain how to calculate the inverse of a linear map $f:\mathbb{R}^n \to \mathbb{R}^n$ whenever f is invertible. This is done by transforming the problem for linear maps into a corresponding problem about matrices.

Suppose $f:\mathbb{R}^n \to \mathbb{R}^n$ and $g:\mathbb{R}^n \to \mathbb{R}^n$ are linear maps. Let A be the matrix of f and let B be the matrix of g. Then AB is the matrix of $f \circ g$ and BA is the matrix of $g \circ f$. The matrix of the identity map $i:\mathbb{R}^n \to \mathbb{R}^n$, defined by $i(\mathbf{x}) = \mathbf{x}$ for all $\mathbf{x} \in \mathbb{R}^n$, is the $n \times n$ *identity matrix*

$$I = \begin{pmatrix} 1 & 0 & \cdots & 0 \\ 0 & 1 & \cdots & 0 \\ & & \cdots & \\ 0 & 0 & \cdots & 1 \end{pmatrix}.$$

It follows that

$$f \circ g = i \text{ if and only if } AB = I$$

and that

$$g \circ f = i \text{ if and only if } BA = I.$$

Thus f and g are inverse to each other if and only if their matrices A and B satisfy the equations $AB = I$ and $BA = I$.

An $n \times n$ matrix A is said to be *invertible* if there is an $n \times n$ matrix B such that $AB = BA = I$. The matrix B, if it exists, is called the *inverse* of A, and we write $B = A^{-1}$. Notice that if A is invertible then there is only one matrix B that satisfies $AB = BA = I$. If C were another, so that $AC = CA = I$, then

$$C = CI = C(AB) = (CA)B = IB = B$$

so, in fact, $C = B$.

According to Theorem 7 of the previous section, the inverse of an invertible linear map $f:\mathbb{R}^n \to \mathbb{R}^n$ is always a linear map. Therefore, we have proved the following theorem.

Theorem 1 *Let $f:\mathbb{R}^n \to \mathbb{R}^n$ be a linear map and let A be the matrix of f. Then f is invertible if and only if A is invertible. Moreover, if f is invertible, then the inverse of f is the linear map associated with A^{-1}.*

According to Theorem 1, the inverse of any invertible linear map can be found from the inverse of the matrix associated with that map. Let us now investigate the problem of how to find the inverse of an invertible matrix.

EXAMPLE 1

Let us try to find A^{-1}, where

$$A = \begin{pmatrix} 3 & 6 \\ 2 & 5 \end{pmatrix}.$$

First, we shall find all 2×2 matrices B such that $AB = I$. Let x_1, x_2, x_3, and x_4 be the entries in B, so that

$$B = \begin{pmatrix} x_1 & x_2 \\ x_3 & x_4 \end{pmatrix}.$$

Then the equation $AB = I$ becomes

$$\begin{pmatrix} 3 & 6 \\ 2 & 5 \end{pmatrix} \begin{pmatrix} x_1 & x_2 \\ x_3 & x_4 \end{pmatrix} = \begin{pmatrix} 1 & 0 \\ 0 & 1 \end{pmatrix}.$$

This matrix equation is equivalent to the four scalar equations

$$3x_1 + 6x_3 = 1 \qquad\qquad 3x_2 + 6x_4 = 0$$
$$2x_1 + 5x_3 = 0 \qquad\qquad 2x_2 + 5x_4 = 1$$

These four scalar equations can be solved as follows.

To find (x_1, x_3) we solve the first pair of equations. Since

$$\begin{pmatrix} 3 & 6 & | & 1 \\ 2 & 5 & | & 0 \end{pmatrix} \longrightarrow \begin{pmatrix} 1 & 2 & | & \frac{1}{3} \\ 2 & 5 & | & 0 \end{pmatrix} \longrightarrow \begin{pmatrix} 1 & 2 & | & \frac{1}{3} \\ 0 & 1 & | & -\frac{2}{3} \end{pmatrix} \longrightarrow \begin{pmatrix} 1 & 0 & | & \frac{5}{3} \\ 0 & 1 & | & -\frac{2}{3} \end{pmatrix},$$

we find that $x_1 = \frac{5}{3}$ and $x_2 = -\frac{2}{3}$. (A vertical line has been drawn in each matrix in order to separate the coefficient matrix of the system from the last column, for reasons that will become evident.)

To find (x_2, x_4) we solve the second pair of equations. Since

$$\begin{pmatrix} 3 & 6 & | & 0 \\ 2 & 5 & | & 1 \end{pmatrix} \longrightarrow \begin{pmatrix} 1 & 2 & | & 0 \\ 2 & 5 & | & 1 \end{pmatrix} \longrightarrow \begin{pmatrix} 1 & 2 & | & 0 \\ 0 & 1 & | & 1 \end{pmatrix} \longrightarrow \begin{pmatrix} 1 & 0 & | & -2 \\ 0 & 1 & | & 1 \end{pmatrix}$$

we find that $x_2 = -2$ and $x_4 = 1$.

Hence the only matrix B such that $AB = I$ is

$$B = \begin{pmatrix} \frac{5}{3} & -2 \\ -\frac{2}{3} & 1 \end{pmatrix}.$$

Since there is only one matrix B such that $AB = I$, this matrix must be the inverse of A, if A is invertible. Since

$$BA = \begin{pmatrix} \frac{5}{3} & -2 \\ -\frac{2}{3} & 1 \end{pmatrix} \begin{pmatrix} 3 & 6 \\ 2 & 5 \end{pmatrix} = \begin{pmatrix} 1 & 0 \\ 0 & 1 \end{pmatrix}$$

we see that $BA = I$ as well, so A is invertible and

$$A^{-1} = B = \begin{pmatrix} \frac{5}{3} & -2 \\ -\frac{2}{3} & 1 \end{pmatrix}. \quad \blacksquare$$

The computations in Example 1 can be simplifed somewhat. You may have noticed that the two row reduction processes that were carried out involved the same row operations. In fact, both linear systems to be solved had the same coefficient matrix,

$$A = \begin{pmatrix} 3 & 6 \\ 2 & 5 \end{pmatrix},$$

and the row operations that were used were the ones required to transform this matrix into an echelon matrix. Hence both systems could have been solved at the same time by including both third columns in a 2×4 matrix and reducing as follows:

$$\begin{pmatrix} 3 & 6 & 1 & 0 \\ 2 & 5 & 0 & 1 \end{pmatrix} \longrightarrow \begin{pmatrix} 1 & 2 & \frac{1}{3} & 0 \\ 2 & 5 & 0 & 1 \end{pmatrix} \longrightarrow \begin{pmatrix} 1 & 2 & \frac{1}{3} & 0 \\ 0 & 1 & -\frac{2}{3} & 1 \end{pmatrix}$$

$$\longrightarrow \begin{pmatrix} 1 & 0 & \frac{5}{3} & -2 \\ 0 & 1 & -\frac{2}{3} & 1 \end{pmatrix}.$$

Notice that, in this reduction process, we have used row operations to transform the matrix $(A|I)$ into an echelon matrix. The echelon matrix turned out to be $(I|A^{-1})$. This process works in general.

Theorem 2 *Let A be an $n \times n$ matrix. Then A is invertible if and only if the echelon matrix of A is the $n \times n$ identity matrix I. If A is invertible, then the inverse matrix A^{-1} can be found by row reducing the matrix $(A|I)$:*

$$(A|I) \xrightarrow[\text{operations}]{\text{row}} (I|A^{-1}).$$

The proof of Theorem 2 will be postponed until the end of this section.

EXAMPLE 2
Let

$$A = \begin{pmatrix} 2 & 3 & 1 \\ -1 & 1 & 0 \\ 1 & 0 & 1 \end{pmatrix}.$$

Then row reduction of $(A|I)$ yields

$$\begin{pmatrix} 2 & 3 & 1 & 1 & 0 & 0 \\ -1 & 1 & 0 & 0 & 1 & 0 \\ 1 & 0 & 1 & 0 & 0 & 1 \end{pmatrix} \longrightarrow \begin{pmatrix} 1 & 0 & 1 & 0 & 0 & 1 \\ -1 & 1 & 0 & 0 & 1 & 0 \\ 2 & 3 & 1 & 1 & 0 & 0 \end{pmatrix}$$

$$\longrightarrow \begin{pmatrix} 1 & 0 & 1 & 0 & 0 & 1 \\ 0 & 1 & 1 & 0 & 1 & 1 \\ 0 & 3 & -1 & 1 & 0 & -2 \end{pmatrix} \longrightarrow \begin{pmatrix} 1 & 0 & 1 & 0 & 0 & 1 \\ 0 & 1 & 1 & 0 & 1 & 1 \\ 0 & 0 & -4 & 1 & -3 & -5 \end{pmatrix}$$

$$\longrightarrow \begin{pmatrix} 1 & 0 & 1 & 0 & 0 & 1 \\ 0 & 1 & 1 & 0 & 1 & 1 \\ 0 & 0 & 1 & -\frac{1}{4} & \frac{3}{4} & \frac{5}{4} \end{pmatrix} \longrightarrow \begin{pmatrix} 1 & 0 & 0 & \frac{1}{4} & -\frac{3}{4} & -\frac{1}{4} \\ 0 & 1 & 0 & \frac{1}{4} & \frac{1}{4} & -\frac{1}{4} \\ 0 & 0 & 1 & -\frac{1}{4} & \frac{3}{4} & \frac{5}{4} \end{pmatrix}.$$

Hence A is invertible and

$$A^{-1} = \begin{pmatrix} \frac{1}{4} & -\frac{3}{4} & -\frac{1}{4} \\ \frac{1}{4} & \frac{1}{4} & -\frac{1}{4} \\ -\frac{1}{4} & \frac{3}{4} & \frac{5}{4} \end{pmatrix}. \quad \blacksquare$$

EXAMPLE 3

If

$$A = \begin{pmatrix} 1 & 2 \\ 2 & 4 \end{pmatrix}$$

then row reduction of $(A|I)$ yields

$$\begin{pmatrix} 1 & 2 & | & 1 & 0 \\ 2 & 4 & | & 0 & 1 \end{pmatrix} \longrightarrow \begin{pmatrix} 1 & 2 & | & 1 & 0 \\ 0 & 0 & | & -2 & 1 \end{pmatrix},$$

at which point it is clear that A is not invertible, since the echelon matrix of A is not equal to I. $\blacksquare$

REMARK The number of corner 1's in the echelon matrix of A, where A is any $m \times n$ matrix, is called the **rank** of the matrix A. When A is an $n \times n$ matrix, the statement that the echelon matrix of A is equal to the identity matrix is the same as the statement that the rank of A is equal to n. Therefore, *an $n \times n$ matrix A is invertible if and only if its rank is n.*

Notice that *if $f:\mathbb{R}^n \to \mathbb{R}^m$ is a linear map and A is its matrix, then the rank of f is the same as the rank of A.*

Now that we know how to find the inverse of any invertible matrix, we can also find the inverse of any invertible linear map $f:\mathbb{R}^n \to \mathbb{R}^n$.

EXAMPLE 4

Let us determine if the linear map $f:\mathbb{R}^3 \to \mathbb{R}^3$ defined by

$$f(x_1, x_2, x_3) = (x_1, x_1 + x_2, x_1 + x_2 + x_3)$$

is invertible and, if it is, let us find its inverse.

The matrix of f is

$$A = \begin{pmatrix} 1 & 0 & 0 \\ 1 & 1 & 0 \\ 1 & 1 & 1 \end{pmatrix}.$$

Since

$$(A|I) = \begin{pmatrix} 1 & 0 & 0 & | & 1 & 0 & 0 \\ 1 & 1 & 0 & | & 0 & 1 & 0 \\ 1 & 1 & 1 & | & 0 & 0 & 1 \end{pmatrix}$$

$$\xrightarrow[\text{operations}]{\text{row}} \begin{pmatrix} 1 & 0 & 0 & | & 1 & 0 & 0 \\ 0 & 1 & 0 & | & -1 & 1 & 0 \\ 0 & 0 & 1 & | & 0 & -1 & 1 \end{pmatrix},$$

A is invertible and

$$A^{-1} = \begin{pmatrix} 1 & 0 & 0 \\ -1 & 1 & 0 \\ 0 & -1 & 1 \end{pmatrix}.$$

The inverse *g* of *f* is, therefore,

$$g(x_1, x_2, x_3) = (x_1, -x_1 + x_2, -x_2 + x_3). \quad \blacksquare$$

Matrix inverses can be used to describe solutions of linear systems of *n* equations in *n* unknowns when the coefficient matrix is invertible. Consider the linear system

$$a_{11}x_1 + \cdots + a_{1n}x_n = b_1$$

$$\cdots$$

$$a_{n1}x_1 + \cdots + a_{nn}x_n = b_n.$$

This system of equations is equivalent to the matrix equation

$$A\mathbf{x} = \mathbf{b}$$

where

$$A = \begin{pmatrix} a_{11} & \cdots & a_{1n} \\ & \cdots & \\ a_{n1} & \cdots & a_{nn} \end{pmatrix}, \qquad \mathbf{x} = \begin{pmatrix} x_1 \\ \vdots \\ x_n \end{pmatrix}, \qquad \text{and} \qquad \mathbf{b} = \begin{pmatrix} b_1 \\ \vdots \\ b_n \end{pmatrix}.$$

If *A* is invertible, we can solve the equation $A\mathbf{x} = \mathbf{b}$ for $\mathbf{x}$ by multiplying both sides of the equation by A^{-1}. We get

$$A^{-1}A\mathbf{x} = A^{-1}\mathbf{b}$$

or

$$I\mathbf{x} = A^{-1}\mathbf{b}$$

or

$$\mathbf{x} = A^{-1}\mathbf{b}.$$

Hence, *if A is invertible then the unique solution of the equation $A\mathbf{x} = \mathbf{b}$ is* $\mathbf{x} = A^{-1}\mathbf{b}$.

This formula is sometimes useful when we are required to solve $A\mathbf{x} = \mathbf{b}$ for many different vectors $\mathbf{b}$.

EXAMPLE 5

Let us solve the following linear systems:

(a) $\begin{array}{l} 3x_1 + 5x_2 = 1 \\ 4x_1 + 7x_2 = 2 \end{array}$ (b) $\begin{array}{l} 3x_1 + 5x_2 = -3 \\ 4x_1 + 7x_2 = 8 \end{array}$ (c) $\begin{array}{l} 3x_1 + 5x_2 = \pi \\ 4x_1 + 7x_2 = -\pi. \end{array}$

All these systems are of the form $A\mathbf{x} = \mathbf{b}$ where

$$A = \begin{pmatrix} 3 & 5 \\ 4 & 7 \end{pmatrix}.$$

Since

$$A^{-1} = \begin{pmatrix} 7 & -5 \\ -4 & 3 \end{pmatrix},$$

we can rapidly compute the solutions $\mathbf{x} = A^{-1}\mathbf{b}$:

(a) $\mathbf{x} = \begin{pmatrix} 7 & -5 \\ -4 & 3 \end{pmatrix} \begin{pmatrix} 1 \\ 2 \end{pmatrix} = \begin{pmatrix} -3 \\ 2 \end{pmatrix}$

(b) $\mathbf{x} = \begin{pmatrix} 7 & -5 \\ -4 & 3 \end{pmatrix} \begin{pmatrix} -3 \\ 8 \end{pmatrix} = \begin{pmatrix} -61 \\ 36 \end{pmatrix}$

(c) $\mathbf{x} = \begin{pmatrix} 7 & -5 \\ -4 & 3 \end{pmatrix} \begin{pmatrix} \pi \\ -\pi \end{pmatrix} = \begin{pmatrix} 12\pi \\ -7\pi \end{pmatrix}.$ ■

This section concludes with a proof of Theorem 2.

Proof of Theorem 2 Let A by an $n \times n$ matrix and let $f : \mathbb{R}^n \to \mathbb{R}^n$ be the linear map associated with A. Theorem 1 states that A is invertible if and only if f is invertible. Theorem 6 of Section 2.6 states that f is invertible if and only if f has rank n. But f has rank n if and only if A has rank n, that is, if and only if the echelon matrix of A is equal to I. This proves the first statement of Theorem 2.

Now suppose that A is invertible. Let $B = A^{-1}$. Then $AB = I$. From the definition of matrix multiplication you can see that the matrix equation $AB = I$ is equivalent to the n equations

$$A\mathbf{b}_1 = \mathbf{e}_1, \ldots, A\mathbf{b}_n = \mathbf{e}_n$$

where $\mathbf{b}_1, \ldots, \mathbf{b}_n$ are the column vectors of B and $\mathbf{e}_1, \ldots, \mathbf{e}_n$ are the column vectors of I. Thus the ith column vector $\mathbf{b}_i$ of B is the solution of the equation $A\mathbf{x} = \mathbf{e}_i$. To solve the equation $A\mathbf{x} = \mathbf{e}_i$, reduce the matrix $(A|\mathbf{e}_i)$ to its echelon matrix. The result is the matrix $(I|\mathbf{b}_i)$. Carrying out these n reductions simultaneously yields

$$(A|\mathbf{e}_1, \ldots, \mathbf{e}_n) \xrightarrow[\text{operations}]{\text{row}} (I|\mathbf{b}_1, \ldots, \mathbf{b}_n)$$

or

$$(A|I) \xrightarrow[\text{operations}]{\text{row}} (I|B).$$

Since $B = A^{-1}$, this completes the proof of Theorem 2. ■

1. Determine which of the following matrices are invertible and, for each that is, find its inverse.

    (a) $\begin{pmatrix} 1 & 1 \\ 1 & -1 \end{pmatrix}$ (b) $\begin{pmatrix} 1 & 3 \\ -2 & 2 \end{pmatrix}$

    (c) $\begin{pmatrix} 2 & 3 & -1 \\ 1 & 2 & 3 \\ -1 & -1 & 4 \end{pmatrix}$ (d) $\begin{pmatrix} 1 & -1 & 1 \\ -1 & 2 & -1 \\ 2 & -1 & 1 \end{pmatrix}$

    (e) $\begin{pmatrix} 1 & 1 & 1 \\ 1 & 2 & 3 \\ 1 & 4 & 9 \end{pmatrix}$ (f) $\begin{pmatrix} 2 & 1 & 4 \\ 3 & 2 & 5 \\ 0 & -1 & 1 \end{pmatrix}$

2. Find the inverse of

    (a) $\begin{pmatrix} 1 & 2 & -1 & 3 \\ 0 & 1 & 4 & -2 \\ 0 & 0 & 1 & 5 \\ 0 & 0 & 0 & 1 \end{pmatrix}$ (b) $\begin{pmatrix} 1 & 2 & 3 & 4 \\ 1 & 3 & 2 & 6 \\ -2 & -2 & -6 & 0 \\ 1 & 4 & 1 & 7 \end{pmatrix}$

3. Find the inverse of each of the given matrices.

    (a) $\begin{pmatrix} 1 & a \\ 0 & 1 \end{pmatrix}$ (b) $\begin{pmatrix} a & 0 \\ 0 & b \end{pmatrix}$ where $ab \neq 0$

    (c) $\begin{pmatrix} 1 & 0 & a \\ 0 & 1 & 0 \\ 0 & 0 & 1 \end{pmatrix}$ (d) $\begin{pmatrix} 1 & a & b \\ 0 & 1 & c \\ 0 & 0 & 1 \end{pmatrix}$

    (e) $\begin{pmatrix} 1 & 0 & 0 & a \\ 0 & 1 & 0 & 0 \\ 0 & 0 & 1 & 0 \\ 0 & 0 & 0 & 1 \end{pmatrix}$

4. Use the result of Example 1 to help you find the inverse of the linear map

    $$f(x_1, x_2) = (3x_1 + 6x_2, 2x_1 + 5x_2).$$

5. Use the result of Example 2 to help you find the inverse of the linear map

    $$f(x_1, x_2, x_3) = (2x_1 + 3x_2 + x_3, -x_1 + x_2, x_1 + x_3).$$

6. Determine which of the following linear maps are invertible. For each map that is invertible, find its inverse.

    (a) $f(x_1, x_2) = (x_1 + x_2, x_1)$
    (b) $f(x_1, x_2) = (2x_1 - 3x_2, -x_1 + x_2)$
    (c) $f(x_1, x_2) = (2x_1 + x_2, x_1 + 2x_2)$
    (d) $f(x_1, x_2, x_3) = (x_1 + 2x_2 + 3x_3, 4x_1 + 5x_2 + 6x_3, 7x_1 + 8x_2 + 9x_3)$
    (e) $f(x_1, x_2, x_3) = (x_1 + x_2 - x_3, x_1 - x_2 + x_3, -x_1 + x_2 + x_3)$

7. For each of the linear maps in Exercise 7 of Section 2.6 that is invertible, find the inverse map.

8. Find the inverse of each of the following linear maps.

    (a) $f(x_1, x_2) = (4x_1 + 3x_2, x_1 + x_2)$
    (b) $f(x_1, x_2) = (\sqrt{3}x_1 + x_2, -x_1 + \sqrt{3}x_2)$

(c) $f(x_1, x_2, x_3) = (x_3, -x_2, x_1)$

(d) $f(x_1, x_2, x_3, x_4) = (x_2, x_4, x_1, x_3)$

(e) $f(x_1, x_2, \ldots, x_n) = (x_1, x_1 + x_2, \ldots, x_1 + x_2 + \cdots + x_n)$

9. Find the rank of each of the following matrices.

(a) $\begin{pmatrix} 1 & 1 & 1 \\ 1 & 1 & 1 \\ 1 & 1 & 1 \end{pmatrix}$ (b) $\begin{pmatrix} 0 & 1 & 2 \\ -1 & 0 & 3 \\ -2 & -2 & 0 \end{pmatrix}$

(c) $\begin{pmatrix} 1 & 2 \\ 3 & 4 \\ 5 & 6 \end{pmatrix}$ (d) $\begin{pmatrix} 0 & 1 & 2 \\ 3 & 4 & 5 \end{pmatrix}$

10. Find the rank of each of the following matrices.

(a) $\begin{pmatrix} 1 & 1 & 1 \\ 1 & 0 & 0 \\ 1 & 0 & 0 \end{pmatrix}$ (b) $\begin{pmatrix} 1 & 1 & 1 \\ 1 & 1 & 0 \\ 1 & 0 & 0 \end{pmatrix}$

(c) $\begin{pmatrix} 0 & 0 & 0 & 0 \\ 0 & 0 & 0 & 0 \\ 0 & 0 & 0 & 0 \end{pmatrix}$ (d) $\begin{pmatrix} 1 & -1 & 2 \\ -1 & 2 & 1 \\ 1 & 0 & 5 \\ -1 & 3 & 4 \end{pmatrix}$

11. Find A^{-1} and use it to solve $A\mathbf{x} = \mathbf{b}$ where $A = \begin{pmatrix} 3 & 4 \\ 17 & 23 \end{pmatrix}$ and $\mathbf{b} =$

(a) $\begin{pmatrix} 2 \\ 1 \end{pmatrix}$ (b) $\begin{pmatrix} -3 \\ 5 \end{pmatrix}$ (c) $\begin{pmatrix} \pi \\ 0 \end{pmatrix}$ (d) $\begin{pmatrix} 0 \\ 1 \end{pmatrix}$.

12. Find A^{-1} and use it to solve $A\mathbf{x} = \mathbf{b}$ where

$$A = \begin{pmatrix} -3 & 1 & 1 \\ 2 & 6 & 5 \\ -5 & 7 & 6 \end{pmatrix}$$

and $\mathbf{b} =$

(a) $\begin{pmatrix} 1 \\ -1 \\ 0 \end{pmatrix}$ (b) $\begin{pmatrix} 8 \\ 4 \\ 8 \end{pmatrix}$ (c) $\begin{pmatrix} -2 \\ 3 \\ 7 \end{pmatrix}$ (d) $\begin{pmatrix} 16 \\ -16 \\ 32 \end{pmatrix}$.

13. Let A and B be $n \times n$ matrices. Show that if A and B are both invertible then so is AB, and $(AB)^{-1} = B^{-1}A^{-1}$.

14. Let A be an invertible $n \times n$ matrix. Show that $(A^{-1})^{-1} = A$.

15. Let A be an $m \times n$ matrix. Show that there is an $n \times m$ matrix B such that $AB = I$ if and only if the rank of A is m. (Here, I is the $m \times m$ identity matrix.) Is the matrix B unique?

2.8
Application: Markov Processes

Some of the matrix algebra that we have learned in this chapter can be used to study *Markov processes*. These processes are studied in courses on probability theory, but it is not necessary to know a lot about probability before you can study this section.

Consider an experiment with a finite number of possible outcomes. For example, if we toss a coin, there are two possible outcomes: heads or tails. If we roll a die, there are six possible outcomes: 1, 2, 3, 4, 5, or 6. The *probability* of a given outcome is the number p, $0 \le p \le 1$, that measures the expected frequency that this outcome will occur if the experiment is repeated many times. If the given outcome is expected to occur m times in n experiments, then $p = m/n$. If $p = 0$, the outcome will *never* occur. If $p = 1$, the outcome will *always* occur. If $p = \frac{1}{2}$ then, on the average, the outcome will occur in half of the experiments performed and it will fail to occur in half of the experiments. In the toss of a coin, the probability of tossing a head is $\frac{1}{2}$, because there are two possible outcomes and each will occur, on the average, half of the time. In the roll of a die, the probability of rolling a 1 is $\frac{1}{6}$, because there are six possible outcomes and each of these will occur, on average, one sixth of the time. Notice that the sum of the probabilities of all the possible outcomes of any given experiment is 1.

The situation that we will study in this chapter is illustrated by the following two examples.

EXAMPLE 1

A large midwestern university has discovered, by analyzing contributions to its annual alumni fund campaign, that on the average,

(i)   an alumnus who contributes to the fund in any given year will contribute again the following year with probability $\frac{3}{4}$, and

(ii)   an alumnus who does *not* contribute in any given year will nevertheless contribute the following year with probability $\frac{1}{2}$.

The university is interested in finding the answers to questions such as "What is the probability that an alumnus who contributes in 1984 will also contribute in 1987?" ■

EXAMPLE 2

A jar containing flies is partitioned into three layers as in Figure 2.34. Each partition contains a small opening through which flies may pass from one layer to an adjoining layer, and the top of the jar contains a small opening through which flies can escape. It is observed that, in any given one minute interval,

(i)   a fly will move from one layer to a *higher* layer, or from the top layer out of the jar, with probability $\frac{3}{10}$,

(ii)   a fly will move from any layer other than the bottom one to a *lower* layer with probability $\frac{1}{10}$,

(iii) the probability that a fly will pass through two partitions in any one minute interval is so small as to be negligible, and

(iv)   the probability that a fly that escapes will return to the jar is also negligible.

The keeper of the flies is interested in knowing the probability that a fly which is in the bottom layer will escape from the jar within a five minute interval. ■

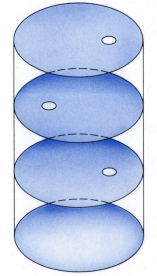

FIGURE 2.34
A three-layered fly jar.

These are examples of *Markov processes*. A ***Markov process*** is a probabilistic model in which an object can reside in any one of a finite number of ***states***. Let us call these states $A_1, A_2, \ldots, A_n$. Objects can move from one state to another. Associated with each pair (A_i, A_j) of states is a ***transition probability*** p_{ij}, the probability that an object which is in the jth state, A_j, at one time t will be in the ith state, A_i, at time $t + 1$. The matrix

$$T = \begin{pmatrix} p_{11} & p_{12} & \cdots & p_{1n} \\ p_{21} & p_{22} & \cdots & p_{2n} \\ & & \cdots & \\ p_{n1} & p_{n2} & \cdots & p_{nn} \end{pmatrix}$$

is called the ***transition matrix*** of the Markov process. The sum of the entries in each column of T is 1, since the entries in each column represent the probabilities of all possible outcomes of an experiment in which an object is initially in one specified state.

EXAMPLE 3

The alumni fund campaign described in Example 1 is an example of a 2-state Markov process. Let us say that an alumnus who contributes to the alumni fund in the year under consideration is in State A_1 and that an alumnus who does not contribute that year is in State A_2. Then the transition probability p_{11} is the probability that an alumnus who is in State A_1 (contributor) one year remains in State A_1 the following year. Thus $p_{11} = \frac{3}{4}$. Similarly, p_{12} is the probability that an alumnus who is in State A_2 (noncontributor) one year moves to State A_1 (contributor) the following year. Thus $p_{12} = \frac{1}{2}$.

The remaining probabilities, p_{21} and p_{22}, can be computed using the fact that the sum of the probabilities of the various outcomes of each experiment is 1. Since an alumnus who contributes in any one year will contribute the following year with probability $\frac{3}{4}$, this alumnus will fail to contribute the following year with probability $1 - \frac{3}{4} = \frac{1}{4}$. Thus $\frac{1}{4}$ is the probability that an alumnus who is in State A_1 one year will be in State A_2 the following year: $p_{21} = \frac{1}{4}$.

Similarly, an alumnus who does not contribute in any one year will nevertheless contribute the following year with probability $\frac{1}{2}$ and hence will fail to contribute the following year with probability $1 - \frac{1}{2} = \frac{1}{2}$. Thus $\frac{1}{2}$ is the probability that an alumnus who is in State A_2 one year will remain in State A_2 the following year: $p_{22} = \frac{1}{2}$. Therefore the transition matrix is

$$T = \begin{pmatrix} \frac{3}{4} & \frac{1}{2} \\ \frac{1}{4} & \frac{1}{2} \end{pmatrix}. \quad \blacksquare$$

EXAMPLE 4

The fly-in-the-jar experiment described in Example 2 is an example of a 4-state Markov process. The four states are

A_1: the fly is in the first (bottom) layer,

A_2: the fly is in the second (middle) layer,

A_3:   the fly is in the third (top) layer, and

A_4:   the fly is outside the jar.

The transition matrix is

$$T = \begin{pmatrix} .7 & .1 & 0 & 0 \\ .3 & .6 & .1 & 0 \\ 0 & .3 & .6 & 0 \\ 0 & 0 & .3 & 1 \end{pmatrix}.$$

The entries in the second column, for example, tell us that in one unit of time, a fly in the second layer will move to the first layer with probability $\frac{1}{10}$ ($p_{12} = .1$), will move to the third layer with probability $\frac{3}{10}$ ($p_{32} = .3$), and will not be able to escape from the jar ($p_{42} = 0$). The remaining entry in the second column ($p_{22} = .6$) is computed from the fact that the sum of the entries in that column must be 1.   ■

The transition matrix T of a Markov process has two important properties:

(1)   Each entry p_{ij} in T is nonnegative.

(2)   The sum of the entries in each column of T is equal to 1.

Property (1) simply states that probabilities are always nonnegative. Property (2) states that, for each j, every object that is in State A_j at time t must be in *some* state A_i at time $t + 1$. This second property was used to fill in the missing probabilities in the transition matrices of Examples 3 and 4.

It turns out that these two properties characterize transition matrices of Markov processes. That is, *if T is any $n \times n$ matrix that satisfies* (1) *and* (2) *then T is the transition matrix of some (perhaps hypothetical) Markov process.*

Suppose now that we have a Markov process with states $A_1, \ldots, A_n$, that we have an object in State A_1 at time t, and that we want to calculate the probability p that this object is in State A_2 at time $t + 2$. To discover how this is done, look at the alumni fund process described in Examples 1 and 3 and examine the following intuitive argument.

In that process $p_{11} = \frac{3}{4}$ and $p_{21} = \frac{1}{4}$. Hence, out of every four alumni in State A_1 at time t three are expected to be in State A_1 at time $t + 1$ and one is expected to be in State A_2. Out of *sixteen* alumni in State A_1 at time t *twelve* ($\frac{3}{4} \times 16$) are expected to be in State A_1 at time $t + 1$ and *four* ($\frac{1}{4} \times 16$) to be in State A_2. Now, of the *twelve* alumni in State A_1 at time $t + 1$ *three* ($\frac{1}{4} \times 12$) are expected to be in State A_2 at time $t + 2$, and of the *four* alumni in State A_2 at time $t + 1$ *two* ($\frac{1}{2} \times 4$) are expected to still be in State A_2 at time $t + 2$. Hence, *out of sixteen alumni in State A_1 at time t five* (3 + 2) *are expected to be in State A_2 at time $t + 2$.* Therefore $p = \frac{5}{16}$. This is the probability that an alumnus who contributes in a given year will fail to contribute two years later.

This calculation shows that the probability that an alumnus moves from State A_1 (contributor) to State A_2 (noncontributor) in two years is given by the formula (work backward through the calculations!)

$$p = \tfrac{5}{16} = (3 + 2)/16 = (\tfrac{1}{4} \times 12 + \tfrac{1}{2} \times 4)/16$$
$$= (\tfrac{1}{4} \times \tfrac{3}{4} \times 16 + \tfrac{1}{2} \times \tfrac{1}{4} \times 16)/16$$
$$= \tfrac{1}{4} \times \tfrac{3}{4} + \tfrac{1}{2} \times \tfrac{1}{4}$$
$$= p_{21}p_{11} + p_{22}p_{21}$$

Notice that this is just the (2, 1)-entry in the matrix T^2.

It can be shown that, if T is the transition matrix of *any* Markov process, then the probability that an object which is in State A_1 at time t will be in State A_2 at time $t + 2$ is equal to the (2, 1)-entry in the matrix T^2. More generally, we have the following theorem.

Theorem 1 *Let T be the transition matrix of an n-state Markov process. Then the probability that an object which is in the jth state at time t is in the ith state at time $t + k$ ($1 \le i \le n$, $1 \le j \le n$, $k = 1, 2, 3, \ldots$) appears as the (i, j)-entry in the matrix T^k.*

EXAMPLE 5

For the Markov process described in Examples 1 and 3 (the alumni fund campaign),

$$T = \begin{pmatrix} \frac{3}{4} & \frac{1}{2} \\ \frac{1}{4} & \frac{1}{2} \end{pmatrix}, \qquad T^2 = \begin{pmatrix} \frac{11}{16} & \frac{5}{8} \\ \frac{5}{16} & \frac{3}{8} \end{pmatrix}, \qquad \text{and} \qquad T^3 = \begin{pmatrix} \frac{43}{64} & \frac{42}{64} \\ \frac{21}{64} & \frac{22}{64} \end{pmatrix}.$$

The probability that an alumnus who contributes to the alumni fund campaign in 1984 will also contribute in 1987 (State $A_1 \to$ State A_1, $k = 3$) is the (1, 1)-entry in T^3, namely $\frac{43}{64} \approx .672$. This answers the question raised in Example 1. ■

EXAMPLE 6

For the Markov process described in Examples 2 and 4 (the fly-in-the-jar)

$$T = \begin{pmatrix} .7 & .1 & 0 & 0 \\ .3 & .6 & .1 & 0 \\ 0 & .3 & .6 & 0 \\ 0 & 0 & .3 & 1 \end{pmatrix} \quad \text{and} \quad T^2 = \begin{pmatrix} .52 & .13 & .01 & 0 \\ .39 & .42 & .12 & 0 \\ .09 & .36 & .39 & 0 \\ 0 & .09 & .48 & 1 \end{pmatrix}.$$

It can be seen from the second column of this matrix that a fly which is initially in the middle layer (State A_2) will be, two minutes later, in the bottom layer (State A_1) with probability .13, in the middle layer (State A_2) with probability .42, in the top layer (State A_3) with probability .36, and will have escaped from the jar (State A_4) with probability .09.

If we compute the matrix T^5 we find that

$$T^5 \approx \begin{pmatrix} .263 & .115 & .027 & 0 \\ .345 & .230 & .088 & 0 \\ .247 & .263 & .148 & 0 \\ .145 & .392 & .737 & 1 \end{pmatrix}.$$

The (4, 1) entry in this matrix, .145, is the probability that a fly, which is initially in the bottom layer (State A_1), is outside the jar (State A_4) five minutes later. This answers the question raised in Example 2.   ∎

The transition matrix and its powers can be used to keep track of population shifts in a Markov process. Suppose, for example, in the alumni fund campaign of Examples 1, 3, and 5, that there were 800 alumni from the class of '82 who contributed to the campaign this year and 1200 who did not. Since $\frac{3}{4}(= p_{11})$ of those who contributed this year and $\frac{1}{2}(= p_{12})$ of those who did not contribute this year will probably contribute next year, the total expected number of contributors next year is

$$\tfrac{3}{4}(800) + \tfrac{1}{2}(1200) = 1200.$$

Similarly, the total expected number of noncontributors next year is

$$\tfrac{1}{4}(800) + \tfrac{1}{2}(1200) = 800.$$

For the year after next, the expected numbers in these two categories are (based on the projected numbers for next year)

$$\tfrac{3}{4}(1200) + \tfrac{1}{2}(800) = 1300 \text{ contributors}$$

and

$$\tfrac{1}{4}(1200) + \tfrac{1}{2}(800) = 700 \text{ noncontributors.}$$

Notice that these calculations can be rewritten in matrix form. The calculations for next year correspond to the matrix equation

$$\begin{pmatrix} \tfrac{3}{4} & \tfrac{1}{2} \\ \tfrac{1}{4} & \tfrac{1}{2} \end{pmatrix} \begin{pmatrix} 800 \\ 1200 \end{pmatrix} = \begin{pmatrix} 1200 \\ 800 \end{pmatrix}$$

and the calculations for the year after next to the equation

$$\begin{pmatrix} \tfrac{3}{4} & \tfrac{1}{2} \\ \tfrac{1}{4} & \tfrac{1}{2} \end{pmatrix} \begin{pmatrix} 1200 \\ 800 \end{pmatrix} = \begin{pmatrix} 1300 \\ 700 \end{pmatrix}$$

or, equivalently, to the equation

$$\begin{pmatrix} \tfrac{3}{4} & \tfrac{1}{2} \\ \tfrac{1}{4} & \tfrac{1}{2} \end{pmatrix} \begin{pmatrix} \tfrac{3}{4} & \tfrac{1}{2} \\ \tfrac{1}{4} & \tfrac{1}{2} \end{pmatrix} \begin{pmatrix} 800 \\ 1200 \end{pmatrix} = \begin{pmatrix} 1300 \\ 700 \end{pmatrix}.$$

Arguments similar to these can be used to prove the following theorem.

Theorem 2   *Let T be the transition matrix of an n-state Markov process. Let x_j $(1 \le j \le n)$ denote the number of objects in the jth state at time t, and let y_i $(1 \le i \le n)$ denote the number of objects expected to be in the ith state at time $t + k$ $(k = 1, 2, 3, \ldots)$. Then*

$$\begin{pmatrix} y_1 \\ \vdots \\ y_n \end{pmatrix} = T^k \begin{pmatrix} x_1 \\ \vdots \\ x_n \end{pmatrix}.$$

EXAMPLE 7

Let us continue the calculations begun above, calculating the expected number of contributors from the Class of '82 to the alumni fund campaign in each of the next six years. Recall that during this year 800 contributed and 1200 did not. The transition matrix is

$$T = \begin{pmatrix} \frac{3}{4} & \frac{1}{2} \\ \frac{1}{4} & \frac{1}{2} \end{pmatrix}$$

so, by Theorem 2, if c_k is the expected number of contributors k years from now and n_k is the expected number of noncontributors that year, then

$$\begin{pmatrix} c_1 \\ n_1 \end{pmatrix} = T\begin{pmatrix} 800 \\ 1200 \end{pmatrix} = \begin{pmatrix} 1200 \\ 800 \end{pmatrix}$$

$$\begin{pmatrix} c_2 \\ n_2 \end{pmatrix} = T^2\begin{pmatrix} 800 \\ 1200 \end{pmatrix} = T\begin{pmatrix} 1200 \\ 800 \end{pmatrix} = \begin{pmatrix} 1300 \\ 700 \end{pmatrix}$$

$$\begin{pmatrix} c_3 \\ n_3 \end{pmatrix} = T^3\begin{pmatrix} 800 \\ 1200 \end{pmatrix} = T\begin{pmatrix} 1300 \\ 700 \end{pmatrix} = \begin{pmatrix} 1325 \\ 675 \end{pmatrix}$$

$$\begin{pmatrix} c_4 \\ n_4 \end{pmatrix} = T^4\begin{pmatrix} 800 \\ 1200 \end{pmatrix} = T\begin{pmatrix} 1325 \\ 675 \end{pmatrix} = \begin{pmatrix} 1331.25 \\ 668.75 \end{pmatrix}$$

$$\begin{pmatrix} c_5 \\ n_5 \end{pmatrix} = T^5\begin{pmatrix} 800 \\ 1200 \end{pmatrix} = T\begin{pmatrix} 1331.25 \\ 668.75 \end{pmatrix} \approx \begin{pmatrix} 1332.8 \\ 667.2 \end{pmatrix}$$

and

$$\begin{pmatrix} c_6 \\ n_6 \end{pmatrix} = T^6\begin{pmatrix} 800 \\ 1200 \end{pmatrix} \approx T\begin{pmatrix} 1332.8 \\ 667.2 \end{pmatrix} \approx \begin{pmatrix} 1333.2 \\ 666.8 \end{pmatrix}.$$

Thus, the expected number of contributors from the Class of '82 to the alumni fund campaign in each of the next six years is approximately 1200, 1300, 1325, 1331, 1333, and 1333. ∎

EXAMPLE 8

Suppose the jar in Examples 2, 4, and 6 initially contains 600 flies, 200 in each layer. Let us find the distribution of the flies after five minutes.

In Example 6, the fifth power of the transition matrix T was calculated, yielding

$$T^5 \approx \begin{pmatrix} .263 & .115 & .027 & 0 \\ .345 & .230 & .088 & 0 \\ .247 & .263 & .148 & 0 \\ .145 & .392 & .737 & 1 \end{pmatrix}$$

Hence

$$T^5 \begin{pmatrix} 200 \\ 200 \\ 200 \\ 0 \end{pmatrix} \approx \begin{pmatrix} 81.0 \\ 132.6 \\ 131.6 \\ 254.8 \end{pmatrix}.$$

Therefore, by Theorem 2, the expected distribution of flies after five minutes is approximately 81 in the bottom layer, 133 in the middle layer, and 132 in the top layer. Approximately 255 flies will have escaped from the jar. ■

You may have noticed an interesting phenomenon that occurred in Example 7. The expected number of contributors to the alumni fund campaign grew each year until it reached 1333 and then it stayed at that level the following year. In fact, it will stay at that level in all future years as well. To see this, simply note that, rounding off to nearest whole numbers,

$$T\begin{pmatrix} 1333 \\ 667 \end{pmatrix} = \begin{pmatrix} \frac{3}{4} & \frac{1}{2} \\ \frac{1}{4} & \frac{1}{2} \end{pmatrix}\begin{pmatrix} 1333 \\ 667 \end{pmatrix} \approx \begin{pmatrix} 1333 \\ 667 \end{pmatrix}.$$

Hence the ratio of the number of contributors to the number of noncontributors is such that the population shift each year from contributors to noncontributors is balanced exactly by the shift from noncontributors to contributors.

This phenomenon occurs in most Markov processes. Consider an n-state Markov process with transition matrix T, and a population distributed among the various states of the process. The vector

$$\mathbf{x} = \begin{pmatrix} x_1 \\ x_2 \\ \vdots \\ x_n \end{pmatrix},$$

where x_i is the number of residents in the ith state ($1 \leq i \leq n$), is called the **state vector** of the population. The population is said to be **stationary** if its state vector $\mathbf{x}$ satisfies the equation

$$T\mathbf{x} = \mathbf{x}.$$

Thus, a population is stationary if and only if, in each unit time interval, there are no net shifts in the distribution of the population.

Theorem 3   *Every Markov process has stationary populations.*

Proof   A population is stationary if its state vector $\mathbf{x}$ satisfies the equation

$$T\mathbf{x} = \mathbf{x}$$

where T is the transition matrix of the Markov process. This equation can be rewritten as

$$(T - I)\mathbf{x} = \mathbf{0}$$

where I is the $n \times n$ identity matrix, n being the number of states in the process. We must show that this homogeneous linear system has nonzero solutions. But this follows from the fact that the sum of the entries in each column of T is equal to 1. (This is Property 2 of a Markov process.) Since the sum of the entries in each column of I is also 1, we can conclude that the sum of the entries in each column of $T - I$ is equal to zero. This fact can be expressed as the matrix equation

$$(1 \quad 1 \quad \cdots \quad 1)(T - I) = (0 \quad 0 \quad \cdots \quad 0).$$

Since the matrix $T - I$ satisfies this equation, it cannot be invertible; if it were then

$$(1 \quad 1 \quad \cdots \quad 1) = (0 \quad 0 \quad \cdots \quad 0)(T - I)^{-1}$$

which is clearly impossible. So $T - I$ is not invertible, hence its rank is less than n, and hence the homogeneous linear system

$$(T - I)\mathbf{x} = \mathbf{0}$$

must have nonzero solutions. And, as we have seen, these nonzero solutions are the stationary populations. ∎

REMARK Although ideally the state vector of a population would have only nonnegative integer entries, the proof of Theorem 2 guarantees only that there is a nonzero vector $\mathbf{x}$ with *real* (possibly negative) entries such that $T\mathbf{x} = \mathbf{x}$. It can be shown, however, that every Markov process does have stationary populations whose state vectors have nonnegative real entries (see Exercise 22). In problems involving real populations, each noninteger entry can be rounded off to the nearest integer in order to interpret the results.

We shall insist that all state vectors have only nonnegative entries. Given a state vector $\mathbf{x}$ we shall denote by $|\mathbf{x}|$ the sum of the entries in $\mathbf{x}$. Thus $|\mathbf{x}|$ represents the total number of objects in the population.

EXAMPLE 9

For the Markov process of Examples 1, 3, 5, and 7 (the alumni fund campaign), the transition matrix is

$$T = \begin{pmatrix} \frac{3}{4} & \frac{1}{2} \\ \frac{1}{4} & \frac{1}{2} \end{pmatrix}$$

so the linear system $(T - I)\mathbf{x} = \mathbf{0}$ is

$$\begin{pmatrix} -\frac{1}{4} & \frac{1}{2} \\ \frac{1}{4} & -\frac{1}{2} \end{pmatrix} \begin{pmatrix} x_1 \\ x_2 \end{pmatrix} = \begin{pmatrix} 0 \\ 0 \end{pmatrix}$$

with general solution

$$\mathbf{x} = c \begin{pmatrix} 2 \\ 1 \end{pmatrix}, \qquad c \in \mathbb{R}.$$

Hence the stationary populations are those populations that have $2c$ contributors and c noncontributors, for some c.

If the total number, N, of alumni in the population is known, then we can evaluate c using the equation $|\mathbf{x}| = N$. We find that

$$2c + c = N$$

or

$$c = N/3.$$

In particular, if $N = 2000$, then $c = 666\frac{2}{3}$ and $2c = 1333\frac{1}{3}$, so a population consisting of 1333 (rounding off) contributors and 667 noncontributors will be stationary, in agreement with Example 7. ∎

EXAMPLE 10

For the Markov process of Examples 2, 4, 6, and 8 (flies in a jar), the transition matrix is

$$T = \begin{pmatrix} .7 & .1 & 0 & 0 \\ .3 & .6 & .1 & 0 \\ 0 & .3 & .6 & 0 \\ 0 & 0 & .3 & 1 \end{pmatrix}$$

so the linear system $(T - I)\mathbf{x} = \mathbf{0}$ is

$$\begin{pmatrix} -.3 & .1 & 0 & 0 \\ .3 & -.4 & .1 & 0 \\ 0 & .3 & -.4 & 0 \\ 0 & 0 & .3 & 0 \end{pmatrix} \begin{pmatrix} x_1 \\ x_2 \\ x_3 \\ x_4 \end{pmatrix} = \begin{pmatrix} 0 \\ 0 \\ 0 \\ 0 \end{pmatrix}$$

with general solution

$$\mathbf{x} = c \begin{pmatrix} 0 \\ 0 \\ 0 \\ 1 \end{pmatrix}, \qquad c \in \mathbb{R}.$$

Hence the stationary populations are those populations that have no flies in any of the first three states. In a stationary population, all the flies are outside the jar! ∎

Example 7 indicated that, with time, the projected contributor/non-contributor population of alumni from the Class of '82 shifts closer and closer to the stationary population, which consists of 1333 contributors and 667 non-contributors. This behavior is present in most Markov processes. In particular, it is present whenever the transition matrix, or some power of it, has all entries different from zero.

Theorem 4 *Let T be the transition matrix of a Markov process. Assume that, for some k ($k = 1, 2, 3, \ldots$), the matrix T^k has all entries different from zero. Then, for each $N > 0$, there is a unique stationary population, with state vector* **y**, *such that* $|\mathbf{y}| = N$. *Moreover, for each state vector* **x** *with* $|\mathbf{x}| = N$, *the vectors* $T^k\mathbf{x}$ *approach* **y**, *in the sense that the numbers*

$$\|T^k\mathbf{x} - \mathbf{y}\|$$

*approach 0, as k gets larger and larger.**

---

* The statement that the numbers $\|T^k\mathbf{x} - \mathbf{y}\|$ approach 0 as k gets larger is usually written $\lim_{k \to \infty} \|T^k\mathbf{x} - \mathbf{y}\| = 0$.

This theorem will not be proved here. You can find these matters discussed in more depth in *Finite Mathematical Structures* by Kemeny, Mirkil, Snell, and Thompson (Prentice Hall, 1959).

EXERCISES

1. Suppose we draw a card from a well shuffled standard 52-card deck. What is the probability of drawing

   (a) an ace?   (b) an ace of spades?

   (c) a spade?   (d) a black queen?

2. Suppose an urn contains colored balls. There are 30 red balls, 50 white balls, and 20 green balls. Suppose a ball is drawn from the urn. What is the probability of drawing

   (a) a red ball?   (b) a white ball?   (c) a green ball?

3. Suppose

$$T = \begin{pmatrix} \frac{1}{3} & \frac{3}{4} \\ \frac{2}{3} & \frac{1}{4} \end{pmatrix}$$

   is the transition matrix of a 2-state Markov process with States A_1 and A_2. What is the probability that

   (a) an object which is in State A_2 at time $t = 0$ will be in State A_1 at time $t = 1$?

   (b) an object which is in State A_1 at time $t = 1$ will still be in State A_1 at time $t = 2$?

   (c) an object which is in State A_1 at time $t = 1$ will be in State A_2 at time $t = 2$?

4. Let

$$T = \begin{pmatrix} \frac{1}{4} & 0 & \frac{1}{2} \\ \frac{3}{4} & \frac{1}{2} & \frac{1}{4} \\ 0 & \frac{1}{2} & \frac{1}{4} \end{pmatrix}$$

   be the transition matrix of a 3-state Markov process with states A_1, A_2, and A_3. What is the probability that

   (a) an object which is in State A_2 at time $t = 0$ will be in State A_3 at time $t = 1$?

   (b) an object which is in State A_3 at time $t = 3$ will be in State A_1 at time $t = 4$?

   (c) an object which is in State A_2 at time $t = 4$ will still be in State A_2 at time $t = 5$?

   (d) an object which is in State A_1 at time $t = 17$ will be in State A_3 at time $t = 18$?

5. Which of the following matrices are transition matrices of Markov processes?

   (a) $\begin{pmatrix} \frac{1}{3} & \frac{1}{2} \\ \frac{2}{3} & 0 \end{pmatrix}$ (b) $\begin{pmatrix} \frac{1}{2} & \frac{1}{2} \\ \frac{3}{4} & \frac{1}{4} \end{pmatrix}$

   (c) $\begin{pmatrix} 1 & \frac{1}{3} \\ 0 & \frac{2}{3} \end{pmatrix}$ (d) $\begin{pmatrix} 0 & \frac{1}{2} & \frac{1}{2} \\ \frac{1}{2} & 0 & \frac{1}{2} \\ \frac{1}{2} & \frac{1}{2} & 0 \end{pmatrix}$

   (e) $\begin{pmatrix} \frac{1}{3} & \frac{2}{3} & 0 \\ \frac{1}{3} & -\frac{1}{3} & \frac{1}{3} \\ \frac{1}{3} & \frac{2}{3} & \frac{2}{3} \end{pmatrix}$

6. For each of the following matrices, find all values of a and b such that the given matrix is the transition matrix of a Markov process.

(a) $\begin{pmatrix} a & b \\ \frac{1}{2} & \frac{1}{3} \end{pmatrix}$ (b) $\begin{pmatrix} a & \frac{1}{5} \\ \frac{2}{5} & b \end{pmatrix}$

(c) $\begin{pmatrix} \frac{1}{3} & a \\ \frac{2}{3} & b \end{pmatrix}$ (d) $\begin{pmatrix} \frac{1}{3} & \frac{1}{2} & \frac{1}{3} \\ \frac{2}{3} & \frac{1}{4} & b \\ a & \frac{1}{4} & \frac{1}{6} \end{pmatrix}$

(e) $\begin{pmatrix} .1 & .4 & .6 \\ .2 & b & .2 \\ a & .3 & .2 \end{pmatrix}$

7. For the Markov process of Exercise 3, find the probability that
   (a) an object which is in State A_2 at time $t = 0$ will be in State A_1 at time $t = 2$.
   (b) an object which is in State A_1 at time $t = 0$ will be in State A_1 at time $t = 2$.
   (c) an object which is in State A_1 at time $t = 0$ will be in State A_1 at time $t = 3$.
   (d) an object which is in State A_1 at time $t = 3$ will be in State A_2 at time $t = 7$.

8. Repeat Exercise 7 for the Markov process described in Exercise 4.

9. Suppose that, in the Markov process of Exercise 3, there are 144 objects in State A_1 at time $t = 0$ and 288 objects in State A_2. Find the number of objects that are expected to be in
   (a) State A_1 at time $t = 1$.   (b) State A_1 at time $t = 2$.
   (c) State A_2 at time $t = 2$.   (d) State A_2 at time $t = 3$.
   (e) State A_2 at time $t = 4$.

10. Suppose that, in the Markov process described in Exercise 4, there are 1600 objects in each of the three states at time $t = 1$. How many objects are expected to be in each state when $t = 3$?

11. Find the stationary population for the Markov process of Example 3 for which the total number of objects in the population is 432.

12. Find the stationary population for the Markov process of Example 4 for which the total number of objects in the population is 4800.

13. Find all stationary populations for the Markov process whose transition matrix is

(a) $\begin{pmatrix} \frac{1}{2} & \frac{1}{2} \\ \frac{1}{2} & \frac{1}{2} \end{pmatrix}$ (b) $\begin{pmatrix} \frac{1}{3} & \frac{2}{3} \\ \frac{2}{3} & \frac{1}{3} \end{pmatrix}$

(c) $\begin{pmatrix} 1 & \frac{1}{4} \\ 0 & \frac{3}{4} \end{pmatrix}$ (d) $\begin{pmatrix} .3 & .5 & .1 \\ .4 & .3 & .2 \\ .3 & .2 & .7 \end{pmatrix}$

(e) $\begin{pmatrix} \frac{1}{2} & \frac{1}{2} & 0 \\ \frac{1}{2} & \frac{1}{2} & 0 \\ 0 & 0 & 1 \end{pmatrix}$

14. Show that if the transition matrix of a Markov process is *symmetric* (that is, $p_{ij} = p_{ji}$ for all i and j), then every population which has the same number of objects in each state is stationary.

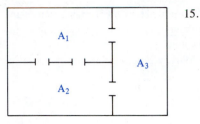

FIGURE 2.35
A mouse cage.

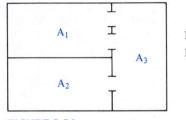

FIGURE 2.36
Another mouse cage.

15. Suppose that a white mouse, put into a cage with many compartments, chooses to pass once each hour from one compartment to another. Suppose also that the choice of which compartment to enter depends only on the number of doors leading to that compartment: the mouse chooses to go through any available door with the same probability as any other. Assume the configuration of the cage is as in Figure 2.35.

   (a) Find the transition matrix of this Markov process, where the mouse is in the ith state if it is in Compartment A_i.

   (b) For each i ($i = 1, 2, 3$), find the probability that the mouse will be in Compartment A_i after three hours, if it is initially in Compartment A_1.

   (c) Suppose 27 mice with this same behavior pattern are put into each compartment. How many mice are expected to be in each compartment after three hours?

   (d) Find the stationary population for a total population of 81 mice.

16. Repeat Exercise 15 for the cage shown in Figure 2.36.

17. A consumer buys a one-pound can of coffee each week, always choosing one of three brands, A, B, or C. Over a period of time it is observed that he purchases the same brand that he purchased the previous week with probability $\frac{1}{2}$, that if he buys Brand A one week then he is equally likely to buy either Brand B or Brand C the following week, and that if he buys either Brand B or Brand C one week then he will buy Brand A the following week with probability $\frac{1}{3}$.

   (a) Find a transition matrix that describes this Markov process.

   (b) Find the probability that, if he buys Brand A this week then he will buy Brand A again two weeks hence.

   (c) Suppose there are 100 consumers like this one, and that they all buy Brand B this week. How many are expected to buy each brand two weeks from now?

   (d) Find the stable population for a total population of 100 consumers.

18. A message consisting of either the word "yes" or the word "no" is relayed from person to person along a chain of people. Suppose the probability that a person relays the same message that he received is .9, and suppose the initial message is "yes". Find the probability that the message is "yes" after n relays, where

   (a) $n = 2$ (b) $n = 3$ (c) $n = 4$

19. Mary has one dollar and her friend has two dollars. They flip a coin. If the coin turns up heads, Mary receives a dollar from her friend. If the coin turns up tails, Mary gives a dollar to her friend. If at any stage either is out of money, no money changes hands.

   (a) Find the transition matrix that describes this process, where State A_i ($0 \le i \le 3$) corresponds to the situation that Mary has i dollars.

   (b) Find the probabilities that Mary has no money after 2 flips, after 3 flips, and after 4 flips.

20. Let

$$T = \begin{pmatrix} 0 & 1 \\ 1 & 0 \end{pmatrix}.$$

   (a) Find all stationary populations for the Markov process whose transition matrix is T.

(b)   Show that if $\mathbf{x}$ is the state vector of any population other than a stationary one, then $T^k\mathbf{x}$ does *not* approach a stationary population as k gets larger and larger. Why doesn't this contradict Theorem 4?

21.   Let T be the transition matrix of an n-state Markov process. Suppose there exists a vector

$$\mathbf{y} = \begin{pmatrix} y_1 \\ \vdots \\ y_n \end{pmatrix}$$

such that, for every $\mathbf{x} \in \mathbb{R}^n$ with nonnegative entries and with $|\mathbf{x}| = 1$, the vectors $T^k\mathbf{x}$ approach $\mathbf{y}$ as k gets larger and larger.

(a)   Show that $y_i \geq 0$ for all i and that $|\mathbf{y}| = 1$.

(b)   Show that $\mathbf{y}$ is the state vector of a stationary population for this Markov process. [*Hint:* Use the fact that $\|T\mathbf{y} - \mathbf{y}\| \leq \|T\mathbf{y} - T^{k+1}\mathbf{y}\| + \|T^{k+1}\mathbf{y} - \mathbf{y}\|$ for all k.]

(c)   Show that the matrices T^k approach the matrix

$$\begin{pmatrix} y_1 & y_1 & & y_1 \\ y_2 & y_2 & & y_2 \\ \vdots & \vdots & \vdots & \vdots \\ y_n & y_n & & y_n \end{pmatrix}$$

(that is, the columns of the matrices T^k all approach $\mathbf{y}$) as k gets larger and larger. [*Hint:* Let $\mathbf{x} = \mathbf{e}_i$.]

22.   Let T be the transition matrix of a 2-state Markov process and let $\mathbf{x} \in \mathbb{R}^2$ be a vector such that

$$T\mathbf{x} = \mathbf{x}.$$

Show that either $\mathbf{x}$ or $-\mathbf{x}$ has both entries nonnegative. [*Hint:* Write the transition matrix as

$$T = \begin{pmatrix} p & q \\ 1 - p & 1 - q \end{pmatrix}.]$$

(This exercise, when combined with Theorem 3, shows that every 2-state Markov process has a stationary population whose state vector has all entries nonnegative. This result can also be proved for n-state Markov processes, but the proof is much more difficult.)

REVIEW EXERCISES

1.   Which of the following maps are linear? For each that is linear, write down its matrix.

(a) $f(x_1, x_2) = (x_1^2 - x_2^2, x_1^2 + x_2^2)$

(b) $f(x_1, x_2) = (2x_1 - 2x_2, 2x_1 + 2x_2)$

(c) $f(x_1, x_2, x_3, x_4, x_5, x_6) = (x_1 + x_2, x_3 + x_4, x_5 + x_6)$

(d) $f(x_1, x_2, x_3) = (x_1 + x_2 + x_3, x_1x_2x_3)$

(e) $f(x_1, x_2) = (x_1, x_1, x_1, x_1)$

2. Find the matrix of f, and find a formula for $f(x_1, x_2, x_3)$, where $f: \mathbb{R}^3 \to \mathbb{R}^2$ is the linear map such that
   (a) $f(1, 0, 0) = (1, 0)$, $f(0, 1, 0) = (1, 0)$, and $f(0, 0, 1) = (0, 1)$
   (b) $f(1, 0, 0) = (-1, 1)$, $f(0, 1, 0) = (-2, 3)$, and $f(0, 0, 1) = (2, -3)$
   (c) $f(1, 0, 0) = f(0, 1, 0) = f(0, 0, 1) = (1, 0)$
   (d) $f(1, 0, 0) = (a, b)$, $f(0, 1, 0) = (b, a)$, and $f(0, 0, 1) = (0, 0)$

3. Identify f as an expansion, a contraction, a rotation, a reflection, or a shear, where f is the linear map associated with the matrix

   (a) $\begin{pmatrix} 1 & 5 \\ 0 & 1 \end{pmatrix}$ (b) $\begin{pmatrix} 3 & 0 \\ 0 & 3 \end{pmatrix}$ (c) $\begin{pmatrix} 0 & 1 \\ 1 & 0 \end{pmatrix}$

   (d) $\begin{pmatrix} 0 & -1 \\ 1 & 0 \end{pmatrix}$ (e) $\begin{pmatrix} \frac{1}{2} & 0 \\ 0 & \frac{1}{2} \end{pmatrix}$ (f) $\begin{pmatrix} \dfrac{\sqrt{3}}{2} & -\dfrac{1}{2} \\ \dfrac{1}{2} & \dfrac{\sqrt{3}}{2} \end{pmatrix}$

4. Sketch a few coordinate lines and their images under the given linear map f. Also shade the unit square $\{(x_1, x_2) | 0 \le x_1 \le 1, 0 \le x_2 \le 1\}$ and its image under f.
   (a) $f(x_1, x_2) = (2x_1 - x_2, x_1 - 2x_2)$
   (b) $f(x_1, x_2) = (\frac{3}{5}x_1 + \frac{4}{5}x_2, -\frac{4}{5}x_1 + \frac{3}{5}x_2)$
   (c) $f(x_1, x_2) = (x_1 - x_2, -x_1 + x_2)$

5. For each of the following pairs of linear maps, f and g, (i) find the matrix A of f and the matrix B of g, (ii) calculate the matrix product AB, and (iii) use the matrix AB to find a formula for the linear map $f \circ g$.
   (a) $f(x_1, x_2) = (3x_1 + 2x_2, -x_1 + 4x_2)$, $g(x_1, x_2) = (4x_1 - 5x_2, 2x_1 - 3x_2)$
   (b) $f(x_1, x_2) = (x_1, x_1 + x_2, x_1 - x_2)$,
       $g(x_1, x_2, x_3) = (x_1 - x_2, x_2 - x_3)$
   (c) $f(x_1, x_2) = (x_1, x_2, x_2, x_1)$, $g(x_1, x_2) = (x_2, x_1)$
   (d) $f(x_1, x_2, x_3) = (x_1 - x_2, x_2 - x_3, x_3 - x_1)$, $g(x_1, x_2) = (x_1, x_2, x_1 + x_2)$
   (e) $f(x_1, x_2, x_3) = x_1 + x_2 + x_3$, $g(x_1, x_2) = (3x_1 + 4x_2, -x_1 + x_2, x_1)$

6. Let

$$A = \begin{pmatrix} 1 & -1 & 0 \\ 3 & 4 & -1 \\ -9 & 3 & 1 \end{pmatrix} \quad \text{and} \quad B = \begin{pmatrix} 5 & 2 & -1 \\ 4 & 1 & 1 \\ 3 & 0 & -1 \end{pmatrix}$$

   Find
   (a) $A + B$ (b) $A - B$ (c) AB
   (d) A^{-1} (e) B^{-1} (f) $(AB)^{-1}$

7. For each of the following linear maps, f, (i) find the matrix A of f, (ii) find the inverse matrix A^{-1}, and (iii) use the matrix A^{-1} to find a formula for the inverse g of f.
   (a) $f(x_1, x_2) = (7x_1 + 5x_2, 10x_1 + 7x_2)$
   (b) $f(x_1, x_2) = (4x_1 - 3x_2, 6x_1 - 5x_2)$
   (c) $f(x_1, x_2, x_3) = (x_1 - x_2, 3x_1 + 4x_2 - x_3, -9x_1 + 3x_2 + x_3)$
   (d) $f(x_1, x_2, x_3) = (5x_1 + 4x_2 + 3x_3, 2x_1 + x_2, -x_1 + x_2 - x_3)$

8. Find the rank of each of the following matrices.

(a) $\begin{pmatrix} 1 & 1 \\ 1 & 1 \\ 1 & 1 \end{pmatrix}$ (b) $\begin{pmatrix} 1 & 2 & 3 \\ 4 & 5 & 6 \\ 7 & 8 & 9 \end{pmatrix}$ (c) $\begin{pmatrix} 3 & 1 & 4 \\ 1 & 5 & 9 \\ 2 & 6 & 5 \end{pmatrix}$

(d) $\begin{pmatrix} 1 & 1 & 1 \\ 1 & 1 & 0 \\ 1 & 0 & 0 \end{pmatrix}$ (e) $\begin{pmatrix} 1 & 3 & 5 & 7 \\ 2 & 4 & 6 & 8 \\ 1 & 1 & 1 & 1 \end{pmatrix}$

9. True or false?

   (a) If $f:\mathbb{R}^n \to \mathbb{R}^m$ is a linear map then $f(\mathbf{0}) = \mathbf{0}$.

   (b) If A and B are $n \times n$ matrices then $(A - B)(A + B) = A^2 - B^2$.

   (c) If A and B are invertible $n \times n$ matrices then $(AB)^{-1} = A^{-1}B^{-1}$.

   (d) If A is an $n \times n$ matrix and A has rank n then A is invertible.

   (e) If A is the matrix of the linear map f and B is the matrix of the linear map g then BA is the matrix of $g \circ f$.

   (f) If A is an $n \times n$ matrix and A has rank n, then the linear system $A\mathbf{x} = \mathbf{b}$ has a unique solution for each $\mathbf{b} \in \mathbb{R}^n$.

   (g) If T is the transition matrix of an n-state Markov process, then there is a unique state vector $\mathbf{x} \in \mathbb{R}^n$ such that $T\mathbf{x} = \mathbf{x}$ and $|\mathbf{x}| = 1$.

10. Find the kernel of each of the following linear maps.

    (a) $f(x_1, x_2) = 4x_1 + 3x_2$

    (b) $f(x_1, x_2) = (2x_1 + x_2, 2x_1 + 3x_2)$

    (c) $f(x_1, x_2, x_3) = (x_1 - x_2 + 2x_3, 3x_1 - x_2 - x_3)$

    (d) $f(x_1, x_2, x_3) = (x_1 + 3x_2 - 4x_3, 2x_1 + x_2 + 2x_3, 8x_1 + 9x_2 - 2x_3)$

    (e) $f(x_1, x_2, x_3, x_4) = (x_1 - x_2 - x_3 - x_4, x_1 - 2x_2 - 2x_3 - 2x_4)$

11. Determine which of the linear maps in Exercise 10 are one-to-one and which are onto.

12. Let

$$T = \begin{pmatrix} p & 1 - q \\ 1 - p & q \end{pmatrix}$$

be the transition matrix of a 2-state Markov process and let N be a positive real number. Show that, unless $p = q = 1$, there is a unique state vector $\mathbf{y}$ such that $T\mathbf{y} = \mathbf{y}$ and $|\mathbf{y}| = N$. Do the vectors $T^k\mathbf{x}$ necessarily approach $\mathbf{y}$ as k gets larger and larger for each state vector $\mathbf{x}$ with $|\mathbf{x}| = N$?

13. Let

$$T = \begin{pmatrix} .3 & .1 & .2 \\ .3 & .7 & .5 \\ .4 & .2 & .3 \end{pmatrix}.$$

   (a) Show that T is the transition matrix of a Markov process.

   (b) What is the probability that an object that is in the second state at time t will be in the third state at time $t + 1$?

   (c) What is the probability that an object that is in the second state at time t will be in the third state at time $t + 2$?

   (d) If a population has, at time $t = 0$, 100 residents in the first state and none in the others, find the expected distribution of the population at time $t = 2$.

   (e) Find the state vector $\mathbf{y}$ of the stationary population for this Markov process for which $|\mathbf{y}| = 100$.

3

Determinants

This chapter is devoted to a study of determinants. You used determinants in section 1.7 to calculate the cross product of two vectors. We shall see in this chapter how to use determinants to find the inverse of a matrix, if it exists, and how to solve certain systems of linear equations. We shall also see how to use determinants to calculate areas and volumes.

3.1
Elementary Properties of Determinants

In this section we shall define the determinant of an $n \times n$ matrix and describe a method for evaluating these determinants. We begin by recalling the definition of the determinant of a 2×2 matrix and examining some of its properties.

The **determinant** of the 2×2 matrix

$$A = \begin{pmatrix} a & b \\ c & d \end{pmatrix},$$

where a, b, c, and d are real numbers, is the real number defined by the formula

$$\det A = ad - bc.$$

Notice that det is a function that assigns a real number to each 2×2 matrix. This function has the following four important properties, the first three of which describe the effect that row operations have on the determinant of a 2×2 matrix.

(i)   If the matrix B is obtained from the matrix A by multiplying a row of A by a real number r, then

$$\det B = r \det A.$$

(ii)   If B is obtained from A by interchanging the rows of A, then

$$\det B = -\det A.$$

(iii)   If B is obtained from A by adding a scalar multiple of one row of A to the other row of A, then

$$\det B = \det A.$$

(iv) $\det \begin{pmatrix} 1 & 0 \\ 0 & 1 \end{pmatrix} = 1.$

These four properties are easy to check. For example, if

$$A = \begin{pmatrix} a & b \\ c & d \end{pmatrix} \quad \text{and} \quad B = \begin{pmatrix} ra & rb \\ c & d \end{pmatrix},$$

then

$$\det B = (ra)(d) - (rb)(c) = r(ad - bc) = r \det A,$$

as asserted in (i). If

$$A = \begin{pmatrix} a & b \\ c & d \end{pmatrix} \quad \text{and} \quad B = \begin{pmatrix} a + rc & b + rd \\ c & d \end{pmatrix}$$

then

$$\det B = (a + rc)(d) - (b + rd)c = ad - bc = \det A,$$

as asserted in (iii).

These properties are important because they characterize the determinant of a 2×2 matrix; that is, the determinant function is the *only* function that assigns a real number to each 2×2 matrix and that has these four properties.

Theorem 1   *The determinant function, det, is the only function from the set of 2×2 matrices into $\mathbb{R}$ that has properties* (i), (ii), (iii), *and* (iv).

Proof   Suppose that f is any function that assigns to each 2×2 matrix A a real number, $f(A)$, and suppose f has properties (i), (ii), (iii), and (iv). That is, suppose

(i) $f(B) = rf(A)$ whenever B is obtained from A by multiplying a row of A by the real number r,

(ii) $f(B) = -f(A)$, whenever B is obtained from A by interchanging the rows of A,

(iii) $f(B) = f(A)$, whenever B is obtained from A by adding a scalar multiple of one row of A to the other row of A, and

(iv) $f\begin{pmatrix} 1 & 0 \\ 0 & 1 \end{pmatrix} = 1.$

We shall show that $f(A) = \det A$ for every 2×2 matrix A.

In the computations that follow, we will indicate above each equality which of these four properties of f we are using.

Let $A = \begin{pmatrix} a & b \\ c & d \end{pmatrix}$.

If $a \neq 0$, then

$$f(A) = f\begin{pmatrix} a & b \\ c & d \end{pmatrix} \overset{(i)}{=} af\begin{pmatrix} 1 & b/a \\ c & d \end{pmatrix} \overset{(iii)}{=} af\begin{pmatrix} 1 & b/a \\ 0 & d - cb/a \end{pmatrix}$$

$$\overset{(i)}{=} a(d - bc/a)f\begin{pmatrix} 1 & b/a \\ 0 & 1 \end{pmatrix} \overset{(iii)}{=} (ad - bc)f\begin{pmatrix} 1 & 0 \\ 0 & 1 \end{pmatrix}$$

$$\overset{(iv)}{=} ad - bc = \det A.$$

If $a = 0$, then

$$f(A) = f\begin{pmatrix} 0 & b \\ c & d \end{pmatrix} \overset{(i)}{=} bf\begin{pmatrix} 0 & 1 \\ c & d \end{pmatrix} \overset{(iii)}{=} bf\begin{pmatrix} 0 & 1 \\ c & 0 \end{pmatrix}$$

$$\overset{(i)}{=} bcf\begin{pmatrix} 0 & 1 \\ 1 & 0 \end{pmatrix} \overset{(ii)}{=} -bcf\begin{pmatrix} 1 & 0 \\ 0 & 1 \end{pmatrix} \overset{(iv)}{=} -bc = \det A.$$

So in both cases, $f(A) = \det A$, as asserted. ■

The importance of Theorem 1 is that it tells us some properties that we might expect the determinant of an $n \times n$ matrix to have, namely, properties analogous to (i)–(iv). It turns out, for $n \times n$ matrices as well as for 2×2 matrices, that there is one and only one function with these properties.

Theorem 2 *There is one and only one function,* det, *that assigns a real number to each $n \times n$ matrix and that has the following four properties:*

  (i) *$\det B = c \det A$ whenever B is obtained from A by multiplying a row of A by the real number c,*

  (ii) *$\det B = -\det A$ whenever B is obtained from A by interchanging two rows,*

  (iii) *$\det B = \det A$ whenever B is obtained from A by adding a scalar multiple of one row to another, and*

  (iv) *$\det I = 1$, where*

$$I = \begin{pmatrix} 1 & 0 & \cdots & 0 \\ 0 & 1 & \cdots & 0 \\ & & \cdots & \\ 0 & 0 & \cdots & 1 \end{pmatrix}.$$

The function, det, whose existence and uniqueness is asserted in Theorem 2, is called the $n \times n$ ***determinant function***. For each $n \times n$ matrix A, the real number $\det A$ is called the ***determinant*** of A.

Properties (i), (ii), and (iii) of Theorem 2 tell what happens to the determinant of a matrix when row operations are applied to that matrix. These properties serve as a basis for computing determinants using row operations, as you will soon see.

We shall postpone our discussion of the proof of Theorem 2 until the next section.

The following theorem describes two additional useful properties of the determinant function:

Theorem 3 *Let A be an $n \times n$ matrix.*

  (i) *If A has two rows that are equal, then $\det A = 0$.*

  (ii) *If A has a row of zeros, then $\det A = 0$.*

Proof (i) Suppose A has two rows that are equal. Let B be the matrix obtained from A by interchanging the equal rows. By property (ii) of Theorem 2, $\det B = -\det A$. However, since the interchanged rows are equal, we see that $B = A$, and consequently that $\det B = \det A$. It follows that $\det A = \det B = -\det A$, and therefore that $\det A = 0$.

(ii) Suppose now that A has a row of zeros. Choose any other row of A and add it to the row of zeros to obtain a matrix B. By property (iii) of Theorem 2, $\det B = \det A$. But the matrix B has two equal rows, so $\det B = 0$, by (i). Therefore, $\det A = 0$. ∎

EXAMPLE 1

The determinant of the matrix

$$\begin{pmatrix} 1 & -1 & 2 & 3 \\ 1 & 5 & 4 & 3 \\ -7 & 6 & 5 & 1 \\ 1 & -1 & 2 & 3 \end{pmatrix}$$

is zero because its first and fourth rows are equal. Also, the determinant of the matrix

$$\begin{pmatrix} 3 & 2 & -8 & 7 & 4 \\ -1 & 5 & 6 & 3 & 2 \\ 1 & 5 & 0 & 2 & 1 \\ 0 & 0 & 0 & 0 & 0 \\ 1 & 4 & 3 & -9 & 10 \end{pmatrix}$$

is zero because it has a row of zeros. ∎

There are some $n \times n$ matrices whose determinants are particularly easy to compute. For example, an $n \times n$ matrix

$$A = \begin{pmatrix} a_{11} & a_{12} & \cdots & a_{1n} \\ a_{21} & a_{22} & \cdots & a_{2n} \\ & & \cdots & \\ a_{n1} & a_{n2} & \cdots & a_{nn} \end{pmatrix}$$

is a ***diagonal matrix*** if $a_{ij} = 0$ whenever $i \neq j$. In other words, A is diagonal if it looks like this:

$$A = \begin{pmatrix} a_{11} & 0 & \cdots & 0 \\ 0 & a_{22} & \cdots & 0 \\ & & \cdots & \\ 0 & 0 & & a_{nn} \end{pmatrix}.$$

The entries $a_{11}, a_{22}, \ldots, a_{nn}$ are called the ***diagonal entries*** of A.

Theorem 4   *The determinant of a diagonal matrix is equal to the product of its diagonal entries.*

Proof Let

$$A = \begin{pmatrix} a_{11} & 0 & \cdots & 0 \\ 0 & a_{22} & \cdots & 0 \\ & & \cdots & \\ 0 & 0 & \cdots & a_{nn} \end{pmatrix}.$$

Then, using property (i) of Theorem 2 repeatedly, we see that

$$\det A = \det \begin{pmatrix} a_{11} & 0 & \cdots & 0 \\ 0 & a_{22} & \cdots & 0 \\ & & \cdots & \\ 0 & 0 & \cdots & a_{nn} \end{pmatrix} = a_{11} \det \begin{pmatrix} 1 & 0 & \cdots & 0 \\ 0 & a_{22} & \cdots & 0 \\ & & \cdots & \\ 0 & 0 & \cdots & a_{nn} \end{pmatrix}$$

$$= a_{11}a_{22} \det \begin{pmatrix} 1 & 0 & \cdots & 0 \\ 0 & 1 & \cdots & 0 \\ & & \cdots & \\ 0 & 0 & \cdots & a_{nn} \end{pmatrix}$$

$$= \cdots = a_{11}a_{22} \cdots a_{nn} \det \begin{pmatrix} 1 & 0 & \cdots & 0 \\ 0 & 1 & \cdots & 0 \\ & & \cdots & \\ 0 & 0 & \cdots & 1 \end{pmatrix}$$

$$= a_{11}a_{22} \cdots a_{nn} \det I$$

$$= a_{11}a_{22} \cdots a_{nn}. \quad \blacksquare$$

Another type of matrix whose determinant is easy to compute is an *upper triangular matrix*. A matrix

$$A = \begin{pmatrix} a_{11} & a_{12} & \cdots & a_{1n} \\ a_{21} & a_{22} & \cdots & a_{2n} \\ & & \cdots & \\ a_{n1} & a_{n2} & \cdots & a_{nn} \end{pmatrix}$$

is **upper triangular** if $a_{ij} = 0$ whenever $i > j$. In other words, A is upper triangular if it looks like this:

$$A = \begin{pmatrix} a_{11} & a_{12} & \cdots & a_{1n} \\ 0 & a_{22} & \cdots & a_{2n} \\ & & \cdots & \\ 0 & 0 & \cdots & a_{nn} \end{pmatrix}.$$

All of the entries *below* the diagonal entries in an upper triangular matrix are equal to zero.

Theorem 5   *The determinant of an upper triangular matrix is equal to the product of its diagonal entries.*

Proof   We shall write out the proof for 3×3 matrices. You will be able to see from this how the proof goes for $n \times n$ matrices when $n \neq 3$.
    Suppose

$$A = \begin{pmatrix} a_{11} & a_{12} & a_{13} \\ 0 & a_{22} & a_{23} \\ 0 & 0 & a_{33} \end{pmatrix}.$$

Then, using properties (i) and (iii) of Theorem 2 repeatedly we see that

$$\det A = \det \begin{pmatrix} a_{11} & a_{12} & a_{13} \\ 0 & a_{22} & a_{23} \\ 0 & 0 & a_{33} \end{pmatrix} \overset{(i)}{=} a_{33} \det \begin{pmatrix} a_{11} & a_{12} & a_{13} \\ 0 & a_{22} & a_{23} \\ 0 & 0 & 1 \end{pmatrix}$$

$$\overset{(iii)}{=} a_{33} \det \begin{pmatrix} a_{11} & a_{12} & 0 \\ 0 & a_{22} & 0 \\ 0 & 0 & 1 \end{pmatrix} \overset{(i)}{=} a_{22}a_{33} \det \begin{pmatrix} a_{11} & a_{12} & 0 \\ 0 & 1 & 0 \\ 0 & 0 & 1 \end{pmatrix}$$

$$\overset{(iii)}{=} a_{22}a_{33} \det \begin{pmatrix} a_{11} & 0 & 0 \\ 0 & 1 & 0 \\ 0 & 0 & 1 \end{pmatrix} \overset{(i)}{=} a_{11}a_{22}a_{33} \det \begin{pmatrix} 1 & 0 & 0 \\ 0 & 1 & 0 \\ 0 & 0 & 1 \end{pmatrix}$$

$$= a_{11}a_{22}a_{33}. \quad \blacksquare$$

EXAMPLE 2
The matrix

$$A = \begin{pmatrix} 3 & -6 & 6 \\ 0 & 2 & 4 \\ 0 & 0 & -1 \end{pmatrix}$$

is upper triangular. Its determinant is

$$\det A = (3)(2)(-1) = -6. \quad \blacksquare$$

Properties (i), (ii), and (iii) of Theorem 2 tell us how the determinant changes when we apply row operations to a matrix. Since we already know how to use row operations to transform any matrix into an upper triangular matrix (an echelon matrix is upper triangular!) we can use these properties, together with Theorem 5, to compute the determinant of any square matrix. Notice however, in the examples below, that we do not need to transform a matrix all the way to its echelon matrix in order to find the determinant. We need only transform it into an upper triangular matrix.

EXAMPLE 3
Using property (iii) of Theorem 2 repeatedly, we find that

$$\det\begin{pmatrix} 1 & 2 & 3 \\ 4 & 5 & 6 \\ 7 & 8 & 9 \end{pmatrix} = \det\begin{pmatrix} 1 & 2 & 3 \\ 0 & -3 & -6 \\ 0 & -6 & -12 \end{pmatrix} = \det\begin{pmatrix} 1 & 2 & 3 \\ 0 & -3 & -6 \\ 0 & 0 & 0 \end{pmatrix}$$

$$= (1)(-3)(0) = 0. \quad \blacksquare$$

EXAMPLE 4

Using properties (i), (ii), and (iii) of Theorem 2, we find that

$$\det\begin{pmatrix} 1 & 3 & -2 \\ 2 & 0 & 3 \\ 1 & -1 & 2 \end{pmatrix} = \det\begin{pmatrix} 1 & 3 & -2 \\ 0 & -6 & 7 \\ 0 & -4 & 4 \end{pmatrix} = -\det\begin{pmatrix} 1 & 3 & -2 \\ 0 & -4 & 4 \\ 0 & -6 & 7 \end{pmatrix}$$

$$= (-1)(-4)\det\begin{pmatrix} 1 & 3 & -2 \\ 0 & 1 & -1 \\ 0 & -6 & 7 \end{pmatrix} = 4\det\begin{pmatrix} 1 & 3 & -2 \\ 0 & 1 & -1 \\ 0 & 0 & 1 \end{pmatrix}$$

$$= (4)(1)(1)(1) = 4. \quad \blacksquare$$

EXAMPLE 5

In the same way, we find that

$$\det\begin{pmatrix} 1 & 2 & 3 & 4 \\ 1 & 3 & 5 & 7 \\ 2 & 3 & 5 & 7 \\ 1 & 4 & 8 & 16 \end{pmatrix} = \det\begin{pmatrix} 1 & 2 & 3 & 4 \\ 0 & 1 & 2 & 3 \\ 0 & -1 & -1 & -1 \\ 0 & 2 & 5 & 12 \end{pmatrix}$$

$$= \det\begin{pmatrix} 1 & 2 & 3 & 4 \\ 0 & 1 & 2 & 3 \\ 0 & 0 & 1 & 2 \\ 0 & 0 & 1 & 6 \end{pmatrix} = \det\begin{pmatrix} 1 & 2 & 3 & 4 \\ 0 & 1 & 2 & 3 \\ 0 & 0 & 1 & 2 \\ 0 & 0 & 0 & 4 \end{pmatrix}$$

$$= (1)(1)(1)(4) = 4. \quad \blacksquare$$

Here is a summary of the method for computing the determinant of an $n \times n$ matrix.

To evaluate det A, where A is an $n \times n$ matrix:

(1) Use row operations to transform A into an upper triangular matrix, keeping track of the effect that each row operation has on the determinant. Then,

(2) use the fact that the determinant of an upper triangular matrix is equal to the product of its diagonal entries.

You may, if you wish, use this method to evaluate determinants of 2×2 matrices, although you will find it faster in most cases simply to use the formula

$$\det\begin{pmatrix} a & b \\ c & d \end{pmatrix} = ad - bc.$$

EXAMPLE 6

The determinant of the 2×2 matrix $A = \begin{pmatrix} 3 & -1 \\ 1 & 4 \end{pmatrix}$ can be computed in two ways:

$$\det A = \det\begin{pmatrix} 3 & -1 \\ 1 & 4 \end{pmatrix} = (3)(4) - (-1)(1) = 13$$

or

$$\det A = \det\begin{pmatrix} 3 & -1 \\ 1 & 4 \end{pmatrix} = -\det\begin{pmatrix} 1 & 4 \\ 3 & -1 \end{pmatrix} = -\det\begin{pmatrix} 1 & 4 \\ 0 & -13 \end{pmatrix}$$
$$= -(-13) = 13. \quad \blacksquare$$

If you have studied determinants before, you may remember a formula for finding the determinant of a 3×3 matrix that is analogous to the formula for the determinant of a 2×2 matrix. We will discuss this formula, along with other methods for evaluating determinants, in later sections of this chapter. However, the method presented in this section is usually the simplest method for computing an $n \times n$ determinant, whenever $n > 2$.

EXERCISES

1. Evaluate:

    (a) $\det\begin{pmatrix} 1 & 3 \\ 4 & 2 \end{pmatrix}$ (b) $\det\begin{pmatrix} -1 & -1 \\ 4 & -1 \end{pmatrix}$ (c) $\det\begin{pmatrix} 0 & -3 \\ 2 & 1 \end{pmatrix}$

    (d) $\det\begin{pmatrix} 10 & 11 \\ 12 & 10 \end{pmatrix}$ (e) $\det\begin{pmatrix} -1 & 3 \\ 3 & -1 \end{pmatrix}$ (f) $\det\begin{pmatrix} 5 & -3 \\ 2 & -1 \end{pmatrix}$

2. Evaluate:

    (a) $\det\begin{pmatrix} \cos\theta & -\sin\theta \\ \sin\theta & \cos\theta \end{pmatrix}$ (b) $\det\begin{pmatrix} 1 & 1 \\ a & b \end{pmatrix}$

    (c) $\det\begin{pmatrix} 1 & a \\ 0 & 1 \end{pmatrix}$

3. Find the determinant of each of the following matrices.

    (a) $\begin{pmatrix} 3 & 5 \\ 0 & -9 \end{pmatrix}$ (b) $\begin{pmatrix} 3 & 0 & 0 \\ 0 & -2 & 0 \\ 0 & 0 & 21 \end{pmatrix}$

    (c) $\begin{pmatrix} -5 & 3 & 2 \\ 0 & 1 & 15 \\ 0 & 0 & 100 \end{pmatrix}$ (d) $\begin{pmatrix} -5 & 0 & 1 \\ 0 & 2 & 10 \\ 0 & 0 & -1 \end{pmatrix}$

(e) $\begin{pmatrix} 1 & -3 & 15 & 20 \\ 0 & 2 & 25 & 80 \\ 0 & 0 & -1 & 40 \\ 0 & 0 & 0 & 3 \end{pmatrix}$

4. Find the determinant of each of the following matrices:

(a) $\begin{pmatrix} 2 & 0 & 0 \\ 1 & -1 & 5 \\ 2 & 3 & -1 \end{pmatrix}$
(b) $\begin{pmatrix} 2 & 1 & 5 \\ 1 & 0 & 3 \\ -1 & 2 & 0 \end{pmatrix}$
(c) $\begin{pmatrix} 0 & -1 & 1 \\ 1 & 0 & 3 \\ 2 & -1 & 0 \end{pmatrix}$

(d) $\begin{pmatrix} 0 & 1 & 0 \\ 0 & 0 & 1 \\ 1 & 0 & 0 \end{pmatrix}$
(e) $\begin{pmatrix} 7 & -1 & 5 \\ 3 & 4 & -5 \\ 2 & 3 & 0 \end{pmatrix}$
(f) $\begin{pmatrix} 2 & 1 & 3 \\ -1 & 3 & -5 \\ 0 & 2 & -2 \end{pmatrix}$

5. Find the determinant of each of the following matrices.

(a) $\begin{pmatrix} 1 & a & b \\ 0 & 1 & c \\ 0 & 0 & 1 \end{pmatrix}$
(b) $\begin{pmatrix} 1 & 1 & 1 \\ a & b & c \\ a^2 & b^2 & c^2 \end{pmatrix}$

6. Evaluate:

(a) $\det \begin{vmatrix} 1 & -1 & 1 & -1 \\ 1 & 2 & 4 & 8 \\ 1 & -2 & 4 & -8 \\ 1 & 1 & 1 & 1 \end{vmatrix}$
(b) $\det \begin{vmatrix} 1 & 2 & 3 & 4 \\ 0 & 1 & 2 & 3 \\ 0 & 0 & 1 & 2 \\ 0 & 0 & 0 & 1 \end{vmatrix}$

(c) $\det \begin{vmatrix} 0 & 1 & -2 & 3 \\ -1 & 0 & 1 & 2 \\ 2 & -1 & 0 & 1 \\ -3 & -2 & -1 & 0 \end{vmatrix}$
(d) $\det \begin{vmatrix} 3 & 1 & 2 & 0 \\ -2 & -1 & 5 & -2 \\ 1 & -3 & 1 & 1 \\ 4 & 1 & 2 & -3 \end{vmatrix}$

(e) $\begin{pmatrix} 1 & 1 & 1 & 1 \\ 1 & 2 & 4 & 8 \\ 1 & 3 & 9 & 27 \\ 1 & 4 & 16 & 64 \end{pmatrix}$

7. Explain why, if A has one row that is a scalar multiple of another, then $\det A = 0$.

8. Explain why it is obvious that each of the following matrices has determinant equal to zero.

(a) $\begin{pmatrix} 1 & -1 & 2 \\ 3 & 0 & 1 \\ -1 & 1 & -2 \end{pmatrix}$
(b) $\begin{pmatrix} 0 & 1 & 2 & 3 \\ 0 & -2 & 5 & 4 \\ 0 & 0 & 1 & 2 \\ 0 & 0 & 0 & -1 \end{pmatrix}$

(c) $\begin{pmatrix} 1 & -1 & 0 & 2 \\ 0 & 0 & 0 & 0 \\ -1 & 2 & 1 & 3 \\ -2 & 3 & -1 & 1 \end{pmatrix}$
(d) $\begin{pmatrix} 1 & 2 & -1 & 3 \\ 2 & 4 & -2 & 6 \\ -1 & 1 & 0 & 2 \\ 3 & -1 & 3 & 2 \end{pmatrix}$

9. Find the determinant of each of the following matrices.

(a) $\begin{pmatrix} 0 & 1 \\ 1 & 0 \end{pmatrix}$

(b) $\begin{pmatrix} 0 & 0 & 1 \\ 0 & 1 & 0 \\ 1 & 0 & 0 \end{pmatrix}$

(c) $\begin{pmatrix} 0 & 0 & 0 & 1 \\ 0 & 0 & 1 & 0 \\ 0 & 1 & 0 & 0 \\ 1 & 0 & 0 & 0 \end{pmatrix}$

(d) $\begin{pmatrix} 0 & 0 & 0 & 0 & 1 \\ 0 & 0 & 0 & 1 & 0 \\ 0 & 0 & 1 & 0 & 0 \\ 0 & 1 & 0 & 0 & 0 \\ 1 & 0 & 0 & 0 & 0 \end{pmatrix}$

10. Show that if

$$A = \begin{pmatrix} 0 & \cdots & 0 & 1 \\ 0 & \cdots & 1 & 0 \\ & \cdots & & \\ & \cdots & & \\ 1 & \cdots & 0 & 0 \end{pmatrix}$$

($a_{ij} = 1$ if $i + j = n + 1$ and $a_{ij} = 0$ otherwise), then $\det A = (-1)^{n(n-1)/2}$ (see Exercise 9).

11. For each real number x, let $f(x)$ be defined by the formula

$$f(x) = \det \begin{pmatrix} 1 & x & x^2 \\ 1 & 1 & 1 \\ 1 & 2 & 4 \end{pmatrix}.$$

Without evaluating the determinant, find two values of x for which $f(x) = 0$.

12. Write out the proof of Theorem 5 for $n \times n$ matrices.

13. An $n \times n$ matrix

$$A = \begin{pmatrix} a_{11} & \cdots & a_{1n} \\ & \cdots & \\ a_{n1} & \cdots & a_{nn} \end{pmatrix}$$

is *lower triangular* if all entries *above* the diagonal entries are equal to zero; that is, if $a_{ij} = 0$ whenever $i < j$. The determinant of a lower triangular matrix is equal to the product of its diagonal entries. Prove this statement for 3×3 matrices.

14. Use Exercise 13 to evaluate

(a) $\det \begin{pmatrix} 1 & 0 & 0 \\ 7 & -2 & 0 \\ 3 & -4 & 3 \end{pmatrix}$

(b) $\det \begin{pmatrix} -3 & 0 & 0 \\ 14 & -1 & 0 \\ 2 & 7 & 4 \end{pmatrix}$

(c) $\det \begin{pmatrix} 3 & 0 & 0 \\ 13 & -10 & 0 \\ 4 & 4 & 2 \end{pmatrix}$

(d) $\det \begin{pmatrix} -1 & 0 & 0 \\ 100 & -1 & 0 \\ 1000 & 200 & -1 \end{pmatrix}$

3.2
Evaluating Determinants by Minors

In the previous section you learned how to evaluate the determinant of an $n \times n$ matrix using row operations. In this section, we discuss another method for evaluating determinants. We shall derive a formula that expresses the determinant of any $n \times n$ matrix in terms of the determinants of certain $(n - 1) \times (n - 1)$ matrices. Since we already know how to find the determinant of any 2×2 matrix, this formula will enable us to find the determinant of any 3×3 matrix. Once we can do that, we can use the formula to find the determinant of any 4×4 matrix, and so on.

The $(n - 1) \times (n - 1)$ matrices that will appear in the formula for the determinant of an $n \times n$ matrix A are the *minors* of A. For each i and j, where $1 \le i \le n$, $1 \le j \le n$, the (i, j)-*minor*, A_{ij}, of A is the $(n - 1) \times (n - 1)$ matrix obtained by deleting the ith row and the jth column of A. For example, the minors of the 3×3 matrix

$$A = \begin{pmatrix} 1 & 2 & 3 \\ 4 & 5 & 6 \\ 7 & 8 & 9 \end{pmatrix}$$

are

$$A_{11} = \begin{pmatrix} 1 & 2 & 3 \\ 4 & 5 & 6 \\ 7 & 8 & 9 \end{pmatrix} = \begin{pmatrix} 5 & 6 \\ 8 & 9 \end{pmatrix},$$

$$A_{12} = \begin{pmatrix} 1 & 2 & 3 \\ 4 & 5 & 6 \\ 7 & 8 & 9 \end{pmatrix} = \begin{pmatrix} 4 & 6 \\ 7 & 9 \end{pmatrix},$$

$$A_{13} = \begin{pmatrix} 1 & 2 & 3 \\ 4 & 5 & 6 \\ 7 & 8 & 9 \end{pmatrix} = \begin{pmatrix} 4 & 5 \\ 7 & 8 \end{pmatrix},$$

$$A_{21} = \begin{pmatrix} 1 & 2 & 3 \\ 4 & 5 & 6 \\ 7 & 8 & 9 \end{pmatrix} = \begin{pmatrix} 2 & 3 \\ 8 & 9 \end{pmatrix},$$

and so on.

In general, if

$$A = \begin{pmatrix} a_{11} & a_{12} & \cdots & a_{1n} \\ & & \cdots & \\ a_{n1} & a_{n2} & \cdots & a_{nn} \end{pmatrix},$$

then the (i, j)-minor, A_{ij}, of A is

$$A_{ij} = \begin{pmatrix} a_{11} & \cdots & a_{1j} & \cdots & a_{1n} \\ & & \cdots & & \\ a_{i1} & \cdots & a_{ij} & \cdots & a_{in} \\ & & \cdots & & \\ a_{n1} & \cdots & a_{nj} & \cdots & a_{nn} \end{pmatrix}.$$

Theorem 1 *Let*

$$A = \begin{pmatrix} a_{11} & a_{12} & \cdots & a_{1n} \\ a_{21} & a_{22} & \cdots & a_{2n} \\ & & \cdots & \\ a_{n1} & a_{n2} & \cdots & a_{nn} \end{pmatrix}$$

be any $n \times n$ *matrix. Then*

$$\det A = a_{11} \det A_{11} - a_{21} \det A_{21} + \cdots + (-1)^n a_{n1} \det A_{n1}.$$

The formula in Theorem 1 is called **expansion of** det A **by minors along the first column.** It looks formidable but it is really quite easy to use. This formula can be displayed more graphically as follows:

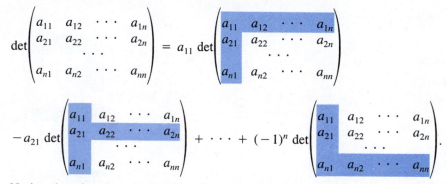

Notice that the signs in front of the terms in this sum alternate between $+$ and $-$.

We will postpone the discussion of the proof of Theorem 1 until the end of this section. Let us now work out some examples.

EXAMPLE 1

Let

$$A = \begin{pmatrix} 2 & 3 & -3 \\ 1 & -2 & 1 \\ 4 & -1 & -1 \end{pmatrix}.$$

Then, expanding by minors along the first column, we find that

$$\det A = a_{11} \det A_{11} - a_{21} \det A_{21} + a_{31} \det A_{31}$$

$$= 2 \det \begin{pmatrix} 2 & 3 & -3 \\ 1 & -2 & 1 \\ 4 & -1 & -1 \end{pmatrix} - 1 \det \begin{pmatrix} 2 & 3 & -3 \\ 1 & -2 & 1 \\ 4 & -1 & -1 \end{pmatrix}$$

$$+ 4 \det \begin{pmatrix} 2 & 3 & -3 \\ 1 & -2 & 1 \\ 4 & -1 & -1 \end{pmatrix}$$

$$= 2 \det\begin{pmatrix} -2 & 1 \\ -1 & -1 \end{pmatrix} - 1 \det\begin{pmatrix} 3 & -3 \\ -1 & -1 \end{pmatrix} + 4 \det\begin{pmatrix} 3 & -3 \\ -2 & 1 \end{pmatrix}$$

$$= 2(3) - 1(-6) + 4(-3)$$

$$= 0.$$

You may want to check that you get the same answer using the method you learned in Section 3.1. ∎

EXAMPLE 2

Let

$$A = \begin{pmatrix} 2 & 0 & 0 & 1 \\ 0 & 2 & 1 & 0 \\ 0 & 1 & 2 & 0 \\ 1 & 0 & 0 & 2 \end{pmatrix}.$$

Then, expanding by minors along the first column, we find that

$$\det A = 2 \det\begin{pmatrix} 2 & 1 & 0 \\ 1 & 2 & 0 \\ 0 & 0 & 2 \end{pmatrix} - 0 \det\begin{pmatrix} 0 & 0 & 1 \\ 1 & 2 & 0 \\ 0 & 0 & 2 \end{pmatrix}$$

$$+ 0 \det\begin{pmatrix} 0 & 0 & 1 \\ 2 & 1 & 0 \\ 0 & 0 & 2 \end{pmatrix} - 1 \det\begin{pmatrix} 0 & 0 & 1 \\ 2 & 1 & 0 \\ 1 & 2 & 0 \end{pmatrix}$$

$$= 2 \det\begin{pmatrix} 2 & 1 & 0 \\ 1 & 2 & 0 \\ 0 & 0 & 2 \end{pmatrix} - \det\begin{pmatrix} 0 & 0 & 1 \\ 2 & 1 & 0 \\ 1 & 2 & 0 \end{pmatrix}.$$

We can evaluate these last two determinants also by expanding by minors along the first column to obtain

$$\det A = 2[2 \det\begin{pmatrix} 2 & 0 \\ 0 & 2 \end{pmatrix} - 1 \det\begin{pmatrix} 1 & 0 \\ 0 & 2 \end{pmatrix} + 0 \det\begin{pmatrix} 1 & 0 \\ 2 & 0 \end{pmatrix}]$$

$$- [0 \det\begin{pmatrix} 1 & 0 \\ 2 & 0 \end{pmatrix} - 2 \det\begin{pmatrix} 0 & 1 \\ 2 & 0 \end{pmatrix} + 1 \det\begin{pmatrix} 0 & 1 \\ 1 & 0 \end{pmatrix}]$$

$$= 2(8 - 2 + 0) - (0 + 4 - 1)$$

$$= 9. ∎$$

EXAMPLE 3

Let us work out a general formula for the determinant of any 3×3 matrix. If

$$A = \begin{pmatrix} a_{11} & a_{12} & a_{13} \\ a_{21} & a_{22} & a_{23} \\ a_{31} & a_{32} & a_{33} \end{pmatrix},$$

then, expanding by minors along the first column, we get

$$\det A = a_{11} \det\begin{pmatrix} a_{22} & a_{23} \\ a_{32} & a_{33} \end{pmatrix} - a_{21} \det\begin{pmatrix} a_{12} & a_{13} \\ a_{32} & a_{33} \end{pmatrix}$$

$$+ a_{31} \det\begin{pmatrix} a_{12} & a_{13} \\ a_{22} & a_{23} \end{pmatrix}$$

$$= a_{11}(a_{22}a_{33} - a_{23}a_{32}) - a_{21}(a_{12}a_{33} - a_{13}a_{32}) + a_{31}(a_{12}a_{23} - a_{13}a_{22})$$

or

$$\det A = a_{11}a_{22}a_{33} - a_{11}a_{23}a_{32} - a_{21}a_{12}a_{33} + a_{21}a_{13}a_{32}$$

$$+ a_{31}a_{12}a_{23} - a_{31}a_{13}a_{22}. \quad \blacksquare$$

You may already be familiar with the formula derived in Example 3 for the determinant of a 3×3 matrix. It is often remembered by writing the matrix A with its first two columns reproduced to yield the 3×5 matrix

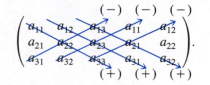

If six arrows are superimposed on this matrix, as shown, then the determinant of A is equal to the sum of the products of the entries hit by the arrows going downward, minus the sum of the products of the entries hit by the arrows going upward. This procedure yields

$$\det A = a_{11}a_{22}a_{33} + a_{12}a_{23}a_{31} + a_{13}a_{21}a_{32}$$

$$- (a_{31}a_{22}a_{13} + a_{32}a_{23}a_{11} + a_{33}a_{21}a_{12}),$$

in agreement with Example 3.

A similar mnemonic device works for the determinant of a 2×2 matrix:

$$\det\begin{pmatrix} a_{11} & a_{12} \\ a_{21} & a_{22} \end{pmatrix} = a_{11}a_{22} - a_{21}a_{12}.$$

This procedure does not, however, generalize to $n \times n$ matrices for $n > 3$ (see Exercise 8).

EXAMPLE 4

Let us use the mnemonic device just described to evaluate the determinant of the 3×3 matrix

$$\begin{pmatrix} -1 & 3 & 5 \\ 2 & -3 & 4 \\ -2 & 1 & 2 \end{pmatrix}.$$

From the matrix

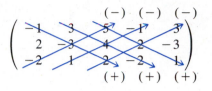

we see that

$$\det\begin{pmatrix} -1 & 3 & 5 \\ 2 & -3 & 4 \\ -2 & 1 & 2 \end{pmatrix} = (-1)(-3)(2) + (3)(4)(-2) + (5)(2)(1)$$

$$- (-2)(-3)(5) - (1)(4)(-1) - (2)(2)(3)$$

$$= 6 - 24 + 10 - 30 + 4 - 12$$

$$= -46. \quad \blacksquare$$

Expansion by minors along the first column is the most efficient way to evaluate the determinant of a 2×2 matrix. It is simply the formula

$$\det\begin{pmatrix} a_{11} & a_{12} \\ a_{21} & a_{22} \end{pmatrix} = a_{11}a_{22} - a_{21}a_{12}.$$

However, for $n \times n$ matrices with $n > 2$, it is usually more efficient to use the row operation method described in Section 3.1. An even more efficient technique combines the two methods, as illustrated in the following example.

EXAMPLE 5
Let

$$A = \begin{pmatrix} 1 & -3 & 2 & 5 \\ -1 & 2 & 1 & 3 \\ 2 & -1 & 0 & 1 \\ 1 & 3 & 4 & 2 \end{pmatrix}.$$

Using row operations, we see that

$$\det A = \det\begin{pmatrix} 1 & -3 & 2 & 5 \\ 0 & -1 & 3 & 8 \\ 0 & 5 & -4 & -9 \\ 0 & 6 & 2 & -3 \end{pmatrix}.$$

Expanding by minors along the first column we get

$$\det A = 1 \det\begin{pmatrix} -1 & 3 & 8 \\ 5 & -4 & -9 \\ 6 & 2 & -3 \end{pmatrix}.$$

Using row operations once again we see that

$$\det A = -\det\begin{pmatrix} 1 & -3 & -8 \\ 5 & -4 & -9 \\ 6 & 2 & -3 \end{pmatrix} = -\det\begin{pmatrix} 1 & -3 & -8 \\ 0 & 11 & 31 \\ 0 & 20 & 45 \end{pmatrix}.$$

Finally, expanding by minors along the first column yields

$$\det A = -1 \det\begin{pmatrix} 11 & 31 \\ 20 & 45 \end{pmatrix} = -[(11)(45) - (20)(31)] = 125. \quad \blacksquare$$

The formula of Theorem 1 describes the expansion of det A by minors along the first column. It is also possible to compute det A by expanding by minors along any column.

Theorem 2 (*Expansion of* det A *by minors along the jth column.*) *Let A be any $n \times n$ matrix and let j be any integer between 1 and n. Then*

$$\det A = (-1)^{1+j}a_{1j} \det A_{1j} + (-1)^{2+j}a_{2j} \det A_{2j}$$

$$+ \cdots + (-1)^{n+j}a_{nj} \det A_{nj}$$

where a_{ij} is the (i, j)-entry of A and A_{ij} is the (i, j)-minor of A ($1 \leq i \leq n$, $1 \leq j \leq n$).

We will discuss the proof of Theorem 2 at the same time that we discuss the proof of Theorem 1.

Notice that, in the formula for expansion by minors along the jth column, the signs in front of the terms alternate between $+$ and $-$, just as they do in expanding along the first column. In writing out the expansion along a given column, you need only determine the sign in front of the first term (it is $+$ for odd numbered columns and $-$ for even numbered columns) and then alternate the signs.

EXAMPLE 6

Let us evaluate det A, where A is the matrix of Example 1, by expanding along the second column. We get

$$\det A = \det\begin{pmatrix} 2 & 3 & -3 \\ 1 & -2 & 1 \\ 4 & -1 & -1 \end{pmatrix}$$

$$= (-1)^{1+2}a_{12} \det A_{12} + (-1)^{2+2}a_{22} \det A_{22} + (-1)^{3+2}a_{32} \det A_{32}$$

$$= -(3) \det\begin{pmatrix} 2 & 3 & -3 \\ 1 & -2 & 1 \\ 4 & -1 & -1 \end{pmatrix} + (-2) \det\begin{pmatrix} 2 & 3 & -3 \\ 1 & -2 & 1 \\ 4 & -1 & -1 \end{pmatrix}$$

$$-(-1) \det\begin{pmatrix} 2 & 3 & -3 \\ 1 & -2 & 1 \\ 4 & -1 & -1 \end{pmatrix}$$

$$= -3 \det\begin{pmatrix} 1 & 1 \\ 4 & -1 \end{pmatrix} - 2 \det\begin{pmatrix} 2 & -3 \\ 4 & -1 \end{pmatrix} + \det\begin{pmatrix} 2 & -3 \\ 1 & 1 \end{pmatrix}$$

$$= -3(-5) - 2(10) + (5)$$

$$= 0,$$

in agreement with Example 1.   ■

There is also a formula for expansion by minors along a *row* of a matrix rather than along a *column*. We state the result here for easy reference, but we shall not discuss its proof until the next section.

Theorem 3   (*Expansion of* det *A by minors along the ith row.*) *Let A be any n × n matrix and let i be any integer between 1 and n. Then*

$$\det A = (-1)^{i+1} a_{i1} \det A_{i1} + (-1)^{i+2} a_{i2} \det A_{i2}$$

$$+ \cdots + (-1)^{i+n} a_{in} \det A_{in}$$

where a_{ij} is the (i, j)-entry of A and A_{ij} is the (i, j)-minor of A (1 ≤ i ≤ n, 1 ≤ j ≤ n).

We conclude this section with a discussion of the proof of Theorem 1. We will not give every detail of the proof but we will try to give you an idea of how the proof works. For clarity, we will restrict the discussion to 3 × 3 matrices.

We want to verify the formula

$$\det\begin{pmatrix} a_{11} & a_{12} & a_{13} \\ a_{21} & a_{22} & a_{23} \\ a_{31} & a_{32} & a_{33} \end{pmatrix} = a_{11} \det\begin{pmatrix} a_{22} & a_{23} \\ a_{32} & a_{33} \end{pmatrix} - a_{21} \det\begin{pmatrix} a_{12} & a_{13} \\ a_{32} & a_{33} \end{pmatrix}$$

$$+ a_{31} \det\begin{pmatrix} a_{12} & a_{13} \\ a_{22} & a_{23} \end{pmatrix}$$

for the determinant of a 3 × 3 matrix. We do this by defining a function f by

$$f\begin{pmatrix} a_{11} & a_{12} & a_{13} \\ a_{21} & a_{22} & a_{23} \\ a_{31} & a_{32} & a_{33} \end{pmatrix} = a_{11} \det\begin{pmatrix} a_{22} & a_{23} \\ a_{32} & a_{33} \end{pmatrix} - a_{21} \det\begin{pmatrix} a_{12} & a_{13} \\ a_{32} & a_{33} \end{pmatrix}$$

$$+ a_{31} \det\begin{pmatrix} a_{12} & a_{13} \\ a_{22} & a_{23} \end{pmatrix}.$$

This function assigns to each 3 × 3 matrix A a real number $f(A)$. We want to show that $f(A) = \det A$.

According to Theorem 2 of Section 3.1, it suffices to check that this function f has the properties (i), (ii), (iii), and (iv) of the determinant function. Since there is only one function, det, that assigns to each 3 × 3 matrix a real number and that has these properties, it will follow that $f(A) = \det(A)$ for every 3×3 matrix A.

Let us check property (i). Suppose

$$A = \begin{pmatrix} a_{11} & a_{12} & a_{13} \\ a_{21} & a_{22} & a_{23} \\ a_{31} & a_{32} & a_{33} \end{pmatrix} \quad \text{and} \quad B = \begin{pmatrix} ca_{11} & ca_{12} & ca_{13} \\ a_{21} & a_{22} & a_{23} \\ a_{31} & a_{32} & a_{33} \end{pmatrix}.$$

Thus B is obtained from A by multiplying the first row of A by c. Then,

$$f(B) = ca_{11} \det\begin{pmatrix} a_{22} & a_{23} \\ a_{32} & a_{33} \end{pmatrix} - a_{21} \det\begin{pmatrix} ca_{12} & ca_{13} \\ a_{32} & a_{33} \end{pmatrix} + a_{31} \det\begin{pmatrix} ca_{12} & ca_{13} \\ a_{22} & a_{23} \end{pmatrix}$$

$$= ca_{11} \det\begin{pmatrix} a_{22} & a_{23} \\ a_{32} & a_{33} \end{pmatrix} - ca_{21} \det\begin{pmatrix} a_{12} & a_{13} \\ a_{32} & a_{33} \end{pmatrix} + ca_{31} \det\begin{pmatrix} a_{12} & a_{13} \\ a_{22} & a_{23} \end{pmatrix}$$

$$= c[a_{11} \det\begin{pmatrix} a_{22} & a_{23} \\ a_{32} & a_{33} \end{pmatrix} - a_{21} \det\begin{pmatrix} a_{12} & a_{13} \\ a_{32} & a_{33} \end{pmatrix} + a_{31} \det\begin{pmatrix} a_{12} & a_{13} \\ a_{22} & a_{23} \end{pmatrix}]$$

$$= cf(A).$$

In the same way we can show that $f(B) = cf(A)$ whenever B is obtained from A by multiplying the second or the third row of A by c. This establishes property (i) for f.

Now let us check property (ii). Suppose

$$A = \begin{pmatrix} a_{11} & a_{12} & a_{13} \\ a_{21} & a_{22} & a_{23} \\ a_{31} & a_{32} & a_{33} \end{pmatrix} \quad \text{and} \quad B = \begin{pmatrix} a_{21} & a_{22} & a_{23} \\ a_{11} & a_{12} & a_{13} \\ a_{31} & a_{32} & a_{33} \end{pmatrix}.$$

Thus B is obtained from A by interchanging the first two rows. Then

$$f(B) = a_{21} \det\begin{pmatrix} a_{12} & a_{13} \\ a_{32} & a_{33} \end{pmatrix} - a_{11} \det\begin{pmatrix} a_{22} & a_{23} \\ a_{32} & a_{33} \end{pmatrix} + a_{31} \det\begin{pmatrix} a_{22} & a_{23} \\ a_{12} & a_{13} \end{pmatrix}$$

$$= -a_{11} \det\begin{pmatrix} a_{22} & a_{23} \\ a_{32} & a_{33} \end{pmatrix} + a_{21} \det\begin{pmatrix} a_{12} & a_{13} \\ a_{32} & a_{33} \end{pmatrix} - a_{31} \det\begin{pmatrix} a_{12} & a_{13} \\ a_{22} & a_{23} \end{pmatrix}$$

$$= -f(A).$$

In the same way we can show that $f(B) = -f(A)$ whenever B is obtained from A by interchanging any other pair of rows. This establishes property (ii) for f.

Property (iii) is a bit harder to verify. Suppose

$$A = \begin{pmatrix} a_{11} & a_{12} & a_{13} \\ a_{21} & a_{22} & a_{23} \\ a_{31} & a_{32} & a_{33} \end{pmatrix} \text{ and } B = \begin{pmatrix} a_{11} + ca_{21} & a_{12} + ca_{22} & a_{13} + ca_{23} \\ a_{21} & a_{22} & a_{23} \\ a_{31} & a_{32} & a_{33} \end{pmatrix}.$$

Thus B is obtained from A by adding c times the second row of A to the first. Then

$$f(B) = (a_{11} + ca_{21}) \det\begin{pmatrix} a_{22} & a_{23} \\ a_{32} & a_{33} \end{pmatrix} - a_{21} \det\begin{pmatrix} a_{12} + ca_{22} & a_{13} + ca_{23} \\ a_{32} & a_{33} \end{pmatrix}$$

$$+ a_{31} \det\begin{pmatrix} a_{12} + ca_{22} & a_{13} + ca_{23} \\ a_{22} & a_{23} \end{pmatrix}.$$

Now

$$\det\begin{pmatrix} a_{11} + ca_{22} & a_{13} + ca_{23} \\ a_{32} & a_{33} \end{pmatrix} = (a_{12} + ca_{22})a_{33} - a_{32}(a_{13} + ca_{23})$$

$$= (a_{12}a_{33} - a_{32}a_{13}) + c(a_{22}a_{33} - a_{32}a_{23})$$

$$= \det\begin{pmatrix} a_{12} & a_{13} \\ a_{32} & a_{33} \end{pmatrix} + c \det\begin{pmatrix} a_{22} & a_{23} \\ a_{32} & a_{33} \end{pmatrix},$$

and

$$\det\begin{pmatrix} a_{12} + ca_{22} & a_{13} + ca_{23} \\ a_{22} & a_{23} \end{pmatrix} = \det\begin{pmatrix} a_{12} & a_{13} \\ a_{22} & a_{23} \end{pmatrix}$$

because adding a scalar multiple of one row to another in a 2×2 matrix does not change its determinant. Hence

$$f(B) = a_{11} \det\begin{pmatrix} a_{22} & a_{23} \\ a_{32} & a_{33} \end{pmatrix} + ca_{21} \det\begin{pmatrix} a_{22} & a_{23} \\ a_{32} & a_{33} \end{pmatrix} - a_{21} \det\begin{pmatrix} a_{12} & a_{13} \\ a_{32} & a_{33} \end{pmatrix}$$

$$- a_{21}c \det\begin{pmatrix} a_{22} & a_{23} \\ a_{32} & a_{33} \end{pmatrix} + a_{31} \det\begin{pmatrix} a_{12} & a_{13} \\ a_{22} & a_{23} \end{pmatrix}$$

$$= a_{11} \det\begin{pmatrix} a_{22} & a_{23} \\ a_{32} & a_{33} \end{pmatrix} - a_{21} \det\begin{pmatrix} a_{12} & a_{13} \\ a_{32} & a_{33} \end{pmatrix} + a_{31} \det\begin{pmatrix} a_{12} & a_{13} \\ a_{22} & a_{23} \end{pmatrix}$$

$$= f(A).$$

In the same way we can show that $f(B) = f(A)$ whenever B is obtained from A by adding a scalar multiple of any row to any other row. This establishes property (iii) for f.

Finally,

$$f\begin{pmatrix} 1 & 0 & 0 \\ 0 & 1 & 0 \\ 0 & 0 & 1 \end{pmatrix} = 1 \det\begin{pmatrix} 1 & 0 \\ 0 & 1 \end{pmatrix} - 0 \det\begin{pmatrix} 0 & 0 \\ 0 & 1 \end{pmatrix} + 0 \det\begin{pmatrix} 0 & 0 \\ 1 & 0 \end{pmatrix}$$

$$= \det\begin{pmatrix} 1 & 0 \\ 0 & 1 \end{pmatrix}$$

$$= 1.$$

This establishes property (iv) for f.

We have verified that the function f has the four properties that characterize the determinant function for 3×3 matrices. We can conclude, therefore, that $f(A) = \det A$ for all 3×3 matrices A, as asserted.

The proof of Theorem 1 for $n \times n$ matrices is similar. The only difficult step is proving property (iii). In that step we need the fact that the determinant function is a linear function in each row. That fact can be established simultaneously with the expansion by minors formula. The expansion by minors formula for $n = 3$ can be used to prove the linearity for $n = 3$. The linearity for $n = 3$ can then be used to prove the expansion by minors formula for $n = 4$. Then the expansion by minors formula for $n = 4$ can be used to prove the linearity for $n = 4$, and so on.

The proof of Theorem 2 is similar to the proof of Theorem 1. Simply define the function f using the formula for expansion by minors of the jth column and verify that this function f also has the properties (i), (ii), (iii), and (iv) that characterize the determinant function.

The arguments in the proof of Theorem 1 can also be used to prove Theorem 2 of Section 3.1. That theorem asserts that there is one and only one function that assigns a real number to each $n \times n$ matrix and that has the four properties required of the determinant. The arguments just given show that, for 3×3 matrices, the function defined by the expansion by minors formula does have the required four properties. From the generalization of these arguments to the case $n = 4$ it follows that the function defined on 4×4 matrices by the expansion by minors formula has the required four properties, and so on. That there is only one such function is a consequence of the fact that these four properties are sufficient to allow us to calculate, using row operations, only one number that can be equal to det A, for each $n \times n$ matrix A.

EXERCISES

1. Find the determinant of each of the following matrices using expansion by minors along the first column.

   (a) $\begin{pmatrix} 2 & -1 & 3 \\ 4 & 1 & 2 \\ -2 & 3 & 4 \end{pmatrix}$ (b) $\begin{pmatrix} -1 & 1 & 2 \\ 5 & 0 & 3 \\ 1 & 4 & 10 \end{pmatrix}$

   (c) $\begin{pmatrix} 3 & 0 & -3 \\ -1 & 2 & 4 \\ 8 & 1 & 5 \end{pmatrix}$ (d) $\begin{pmatrix} 1 & 8 & 1 \\ 2 & 1 & 2 \\ 3 & -4 & -1 \end{pmatrix}$

   (e) $\begin{pmatrix} 1 & -1 & 0 & 1 \\ 0 & 1 & 2 & 4 \\ -1 & 3 & 3 & 2 \\ 1 & -1 & 2 & 1 \end{pmatrix}$ (f) $\begin{pmatrix} 3 & 1 & 4 & 1 \\ 5 & 9 & 2 & 7 \\ 2 & 7 & 1 & 8 \\ 2 & 8 & 1 & 8 \end{pmatrix}$

2. Find the determinant of each of the matrices in Exercise 1 using expansion by minors along the *second* column.

3. Find the determinant of each of the matrices in Exercise 1 using expansion by minors along the *last* column.

4. Evaluate the determinant of each of the 3×3 matrices in Exercise 1 using the mnemonic device described in the text preceding Example 4.

5. Make up four 3×3 matrices and evaluate their determinants using two different methods to see that both methods yield the same answer.

6. Find the determinant of each of the following matrices.

(a) $\begin{pmatrix} a & b & 1 \\ c & 1 & 0 \\ 1 & 0 & 0 \end{pmatrix}$
   (b) $\begin{pmatrix} \lambda & 2 & -2 \\ 3 & \lambda - 1 & -3 \\ 1 & -1 & \lambda - 3 \end{pmatrix}$

(c) $\begin{vmatrix} a_{11} & a_{12} & 0 & 0 \\ a_{21} & a_{22} & 0 & 0 \\ 0 & 0 & b_{11} & b_{12} \\ 0 & 0 & b_{21} & b_{22} \end{vmatrix}$

7. Find the determinants of

(a) $\begin{pmatrix} 0 & -a & -b \\ a & 0 & -c \\ b & c & 0 \end{pmatrix}$
   (b) $\begin{pmatrix} 0 & a & b \\ a & 0 & c \\ b & c & 0 \end{pmatrix}$

(c) $\begin{vmatrix} 0 & -a & -b & -c \\ a & 0 & -d & -e \\ b & d & 0 & -f \\ c & e & f & 0 \end{vmatrix}$

8. Let

$$A = \begin{pmatrix} 2 & 0 & 0 & 1 \\ 0 & 2 & 1 & 0 \\ 0 & 1 & 2 & 0 \\ 1 & 0 & 0 & 2 \end{pmatrix}.$$

Compute the sum of the products of the entries hit by the arrows going downward minus the sum of the products of the entries hit by the arrows going upward in the 4×7 matrix

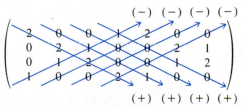

and observe that your result is *not* equal to det A.

9. Show that if $\mathbf{a} = (a_1, a_2)$ and $\mathbf{b} = (b_1, b_2)$ are distinct points in $\mathbb{R}^2$, then

$$\det \begin{pmatrix} 1 & x_1 & x_2 \\ 1 & a_1 & a_2 \\ 1 & b_1 & b_2 \end{pmatrix} = 0$$

is an equation for the line in $\mathbb{R}^2$ through $\mathbf{a}$ and $\mathbf{b}$. [*Hint:* Use expansion by minors along the first row to show that this equation does describe a line. Why is this equation satisfied when $\mathbf{x} = \mathbf{a}$ and when $\mathbf{x} = \mathbf{b}$?]

10. Show that if $\mathbf{a} = (a_1, a_2, a_3)$, $\mathbf{b} = (b_1, b_2, b_3)$, and $\mathbf{c} = (c_1, c_2, c_3)$ are noncollinear points in $\mathbb{R}^3$, then

$$\det \begin{pmatrix} 1 & x_1 & x_2 & x_3 \\ 1 & a_1 & a_2 & a_3 \\ 1 & b_1 & b_2 & b_3 \\ 1 & c_1 & c_2 & c_3 \end{pmatrix} = 0$$

is an equation for the plane in $\mathbb{R}^3$ through $\mathbf{a}$, $\mathbf{b}$, and $\mathbf{c}$.

11. Let $\mathbf{a} = (a_1, a_2, a_3)$, $\mathbf{b} = (b_1, b_2, b_3)$, and $\mathbf{c} = (c_1, c_2, c_3)$ be vectors in $\mathbb{R}^3$. Show that

$$\det \begin{pmatrix} a_1 & a_2 & a_3 \\ b_1 & b_2 & b_3 \\ c_1 & c_2 & c_3 \end{pmatrix} = \mathbf{a} \cdot (\mathbf{b} \times \mathbf{c}).$$

[*Hint:* Expand by minors along the first row.]

3.3
Permutations and Determinants

In this section we shall discuss another formula for the determinant. This formula expresses the determinant of any matrix directly in terms of its entries.

The formula that we will discuss uses permutations. A **permutation** of the set $\{1, 2, \ldots, n\}$ is a one-to-one function σ that maps this set onto itself. There are two permutations of the set $\{1, 2\}$, the functions σ_1 and σ_2 defined by

$$\begin{cases} \sigma_1(1) = 1 \\ \sigma_1(2) = 2 \end{cases} \quad \text{and} \quad \begin{cases} \sigma_2(1) = 2 \\ \sigma_2(2) = 1. \end{cases}$$

There are six permutations of the set $\{1, 2, 3\}$. We shall denote the set of all permutations of $\{1, 2, \ldots, n\}$ by S_n.

Each permutation $\sigma \in S_n$ determines a vector $(\sigma(1), \sigma(2), \ldots, \sigma(n)) \in \mathbb{R}^n$. This vector is called the **vector of** σ. It is simply an arrangement of the numbers $1, 2, \ldots, n$ into a specific order. The vectors of the permutations σ_1 and σ_2 in S_2 are

$$(\sigma_1(1), \sigma_1(2)) = (1, 2)$$

and

$$(\sigma_2(1), \sigma_2(2)) = (2, 1).$$

Notice that each permutation is completely determined by its vector. Thus, for example, if σ is the permutation of $\{1, 2, 3, 4\}$ with vector $(2, 3, 4, 1)$, then $\sigma(1) = 2$, $\sigma(2) = 3$, $\sigma(3) = 4$, and $\sigma(4) = 1$.

Since a permutation σ determines, and is determined by, its vector $(\sigma(1), \sigma(2), \ldots, \sigma(n))$, it is often written

$$\sigma = (\sigma(1), \sigma(2), \ldots, \sigma(n)).$$

Thus, for example, we can list the six permutations of $\{1, 2, 3\}$ as

$$\sigma_1 = (1, 2, 3), \quad \sigma_2 = (2, 3, 1), \quad \sigma_3 = (3, 1, 2),$$

$$\sigma_4 = (2, 1, 3), \quad \sigma_5 = (1, 3, 2), \quad \sigma_6 = (3, 2, 1).$$

We can associate an $n \times n$ *permutation matrix* with each permutation in S_n. Let $\mathbf{e}_1, \mathbf{e}_2, \ldots, \mathbf{e}_n$ be the row vectors in the $n \times n$ identity matrix. Thus

$$\mathbf{e}_1 = (1, 0, 0, \ldots, 0),$$

$$\mathbf{e}_2 = (0, 1, 0, \ldots, 0),$$

$$\ldots$$

$$\mathbf{e}_n = (0, 0, 0, \ldots, 1).$$

The *matrix of the permutation* $\sigma \in S_n$ is the matrix M_σ whose row vectors are $\mathbf{e}_{\sigma(1)}, \mathbf{e}_{\sigma(2)}, \ldots, \mathbf{e}_{\sigma(n)}$, in that order.

EXAMPLE 1

The matrices of the two permutations in S_2 are

$$M_{(1,2)} = \begin{pmatrix} \mathbf{e}_1 \\ \mathbf{e}_2 \end{pmatrix} = \begin{pmatrix} 1 & 0 \\ 0 & 1 \end{pmatrix} \quad \text{and} \quad M_{(2,1)} = \begin{pmatrix} \mathbf{e}_2 \\ \mathbf{e}_1 \end{pmatrix} = \begin{pmatrix} 0 & 1 \\ 1 & 0 \end{pmatrix}.$$

The matrices of the six permutations in S_3 are

$$M_{(1,2,3)} = \begin{pmatrix} 1 & 0 & 0 \\ 0 & 1 & 0 \\ 0 & 0 & 1 \end{pmatrix} \quad M_{(2,1,3)} = \begin{pmatrix} 0 & 1 & 0 \\ 1 & 0 & 0 \\ 0 & 0 & 1 \end{pmatrix}$$

$$M_{(2,3,1)} = \begin{pmatrix} 0 & 1 & 0 \\ 0 & 0 & 1 \\ 1 & 0 & 0 \end{pmatrix} \quad M_{(1,3,2)} = \begin{pmatrix} 1 & 0 & 0 \\ 0 & 0 & 1 \\ 0 & 1 & 0 \end{pmatrix}$$

$$M_{(3,1,2)} = \begin{pmatrix} 0 & 0 & 1 \\ 1 & 0 & 0 \\ 0 & 1 & 0 \end{pmatrix} \quad M_{(3,2,1)} = \begin{pmatrix} 0 & 0 & 1 \\ 0 & 1 & 0 \\ 1 & 0 & 0 \end{pmatrix}. \quad \blacksquare$$

The determinant of a permutation matrix is always either $+1$ or -1. This is because the identity matrix can be obtained from any permutation matrix by a sequence of row interchanges. Since each row interchange changes the sign of the determinant, we see that

$$\det M_\sigma = (-1)^k \det I = (-1)^k,$$

where k is the number of row interchanges required to transform M_σ into I. The number $\det M_\sigma (= \pm 1)$ is called the *sign* of the permutation σ. We write

$$\text{sign } \sigma = \det M_\sigma.$$

EXAMPLE 2

The signs of the permutations in S_2 are

$$\text{sign } (1, 2) = \det M_{(1,2)} = \det \begin{pmatrix} 1 & 0 \\ 0 & 1 \end{pmatrix} = +1$$

and

$$\text{sign}(2,1) = \det M_{(2,1)} = \det\begin{pmatrix} 0 & 1 \\ 1 & 0 \end{pmatrix} = -1.$$

The signs of the permutations in S_3 are

$$\text{sign}(1, 2, 3) = \det\begin{pmatrix} 1 & 0 & 0 \\ 0 & 1 & 0 \\ 0 & 0 & 1 \end{pmatrix} = +1,$$

$$\text{sign}(2, 3, 1) = \det\begin{pmatrix} 0 & 1 & 0 \\ 0 & 0 & 1 \\ 1 & 0 & 0 \end{pmatrix} = +1,$$

$$\text{sign}(3, 1, 2) = \det\begin{pmatrix} 0 & 0 & 1 \\ 1 & 0 & 0 \\ 0 & 1 & 0 \end{pmatrix} = +1,$$

$$\text{sign}(2, 1, 3) = \det\begin{pmatrix} 0 & 1 & 0 \\ 1 & 0 & 0 \\ 0 & 0 & 1 \end{pmatrix} = -1,$$

$$\text{sign}(1, 3, 2) = \det\begin{pmatrix} 1 & 0 & 0 \\ 0 & 0 & 1 \\ 0 & 1 & 0 \end{pmatrix} = -1,$$

and

$$\text{sign}(3, 2, 1) = \det\begin{pmatrix} 0 & 0 & 1 \\ 0 & 1 & 0 \\ 1 & 0 & 0 \end{pmatrix} = -1. \quad \blacksquare$$

We mentioned above that the sign of a permutation σ is equal to $(-1)^k$ where k is the number of row interchanges required to transform the matrix M_σ into the identity matrix I. This is the same as the number of entry interchanges required to transform the vector $(\sigma(1), \sigma(2), \ldots, \sigma(n))$ into $(1, 2, \ldots, n)$.

EXAMPLE 3

Let us find the sign of the permutation $\sigma = (2, 5, 3, 1, 4)$ in S_5. We can transform σ into $(1, 2, 3, 4, 5)$ by the following interchanges:

$$(2, 5, 3, 1, 4) \rightarrow (1, 5, 3, 2, 4) \rightarrow (1, 2, 3, 5, 4) \rightarrow (1, 2, 3, 4, 5)$$

We used three interchanges so

$$\text{sign}\ \sigma = (-1)^3 = -1. \quad \blacksquare$$

The number, k, of interchanges required to transform the vector $(\sigma(1), \sigma(2), \ldots, \sigma(n))$ into $(1, 2, \ldots, n)$ is not unique. If you carry out the interchanges in two different ways you may get two different values of k. But, although k may vary, the number $(-1)^k = \text{sign } \sigma$ cannot vary. If the sign of σ is $+1$ then an even number of interchanges will always be required to transform $(\sigma(1), \sigma(2), \ldots, \sigma(n))$ into $(1, 2, \ldots, n)$. If the sign of σ is -1 then an odd number of interchanges will always be required. For this reason, we say that σ is an ***even permutation*** if $\text{sign } \sigma = +1$ and that σ is an ***odd permutation*** if $\text{sign } \sigma = -1$.

Our goal in this section is to find a formula that expresses that determinant of any $n \times n$ matrix directly in terms of its entries. For 3×3 matrices, we already have such a formula. In Example 3 of Section 3.2 we showed that if

$$A = \begin{pmatrix} a_{11} & a_{12} & a_{13} \\ a_{21} & a_{22} & a_{23} \\ a_{31} & a_{32} & a_{33} \end{pmatrix}$$

then

$$\det A = a_{11}a_{22}a_{33} + a_{12}a_{23}a_{31} + a_{13}a_{21}a_{32}$$

$$- a_{12}a_{21}a_{33} - a_{11}a_{23}a_{32} - a_{13}a_{22}a_{31}.$$

Notice that this formula expresses $\det A$ as a sum of six terms. Each term is of the form

$$(\pm 1)a_{1\sigma(1)}a_{2\sigma(2)}a_{3\sigma(3)}$$

where σ is a permutation of $\{1, 2, 3\}$. The factor ± 1 is just the sign of the permutation σ. Moreover, the six terms correspond to the six permutations in S_3. So $\det A$ is equal to the sum of all products of the form

$$(\text{sign } \sigma)a_{1\sigma(1)}a_{2\sigma(2)}a_{3\sigma(3)}$$

where σ runs through all the permutations in S_3.

This formula generalizes to $n \times n$ matrices as follows.

Theorem 1 *Let*

$$A = \begin{pmatrix} a_{11} & a_{12} & \cdots & a_{1n} \\ a_{21} & a_{22} & \cdots & a_{2n} \\ & & \cdots & \\ a_{n1} & a_{n2} & \cdots & a_{nn} \end{pmatrix}.$$

Then $\det A$ *is equal to the sum of all products of the form*

$$(\text{sign } \sigma)a_{1\sigma(1)}a_{2\sigma(2)} \cdots a_{n\sigma(n)}$$

where σ *runs through all the permutations in* S_n.

EXAMPLE 4

Applying the formula of Theorem 1 to the 2×2 matrix

$$A = \begin{pmatrix} a_{11} & a_{12} \\ a_{21} & a_{22} \end{pmatrix}$$

yields the familiar formula

$$\det A = \text{sign}(1, 2)a_{11}a_{22} + \text{sign}(2, 1)a_{12}a_{21}$$

$$= a_{11}a_{22} - a_{12}a_{21}. \quad \blacksquare$$

Theorem 1 can be proved by checking that the function f that assigns to each $n \times n$ matrix A the sum of all the products $(\text{sign } \sigma)a_{1\sigma(1)} \cdots a_{n\sigma(n)}$ $(\sigma \in S_n)$ has the four properties that characterize the determinant function. We will not carry out the details here, but you will be asked to check some of these properties in the exercises.

Given any $m \times n$ matrix

$$A = \begin{pmatrix} a_{11} & a_{12} & \cdots & a_{1n} \\ a_{21} & a_{22} & \cdots & a_{2n} \\ & & \cdots & \\ a_{m1} & a_{m2} & \cdots & a_{mn} \end{pmatrix},$$

the $n \times m$ matrix

$$A^t = \begin{pmatrix} a_{11} & a_{21} & \cdots & a_{m1} \\ a_{12} & a_{22} & \cdots & a_{m2} \\ & & \cdots & \\ a_{1n} & a_{2n} & \cdots & a_{mn} \end{pmatrix}$$

is called the **transpose** of A. Notice that the row vectors of A^t are the column vectors of A and that the column vectors of A^t are the row vectors of A. For example,

$$\text{if } A = \begin{pmatrix} 3 & 1 & 4 \\ 1 & 5 & 9 \end{pmatrix}, \quad \text{then} \quad A^t = \begin{pmatrix} 3 & 1 \\ 1 & 5 \\ 4 & 9 \end{pmatrix}.$$

If A is a square matrix, then A^t is also a square matrix. We can use Theorem 1 to prove that A and A^t have the same determinant.

Theorem 2   *Let A be a square matrix. Then*

$$\det A^t = \det A.$$

Proof   According to Theorem 1, $\det A$ is the sum of the products

$$(\text{sign } \sigma)a_{1\sigma(1)}a_{2\sigma(2)} \cdots a_{n\sigma(n)}$$

$(\sigma \in S_n)$. Notice that apart from the sign these products are all the products that can be formed by choosing one entry from each row of A in such a way that all the entries chosen come from different columns. But that is the same as choosing one entry from each column of A in a way such that all the entries chosen come from different rows. Thus, the set of products

$$\{a_{1\sigma(1)}a_{2\sigma(2)} \cdots a_{n\sigma(n)} | \sigma \in S_n\}$$

that appear in det A is the same as the set of products

$$\{a_{\tau(1)1}a_{\tau(2)2} \cdots a_{\tau(n)n} | \tau \in S_n\}$$

that appear in det A^t. Moreover, it can be shown (see Exercise 14) that if

$$a_{1\sigma(1)}a_{2\sigma(2)} \cdots a_{n\sigma(n)} = a_{\tau(1)1}a_{\tau(2)2} \cdots a_{\tau(n)n},$$

then sign σ = sign τ. Hence the sets

$$\{(\text{sign } \sigma)a_{1\sigma(1)}a_{2\sigma(2)} \cdots a_{n\sigma(n)} | \sigma \in S_n\}$$

and

$$\{(\text{sign } \tau)a_{\tau(1)1}a_{\tau(2)2} \cdots a_{\tau(n)n} | \tau \in S_n\}$$

are also the same. The sum of the numbers in the first set is det A. The sum of the numbers in the second set is det A^t. Hence det A = det A^t. ∎

We can use Theorem 2 to prove the formula stated in the previous section for the expansion of det A by minors along the ith row. Let a_{ij} denote the (i, j)-entry of A and let A_{ij} denote the (i, j)-minor of A. Then the (i, j)-entry of A^t is a_{ji} and the (i, j)-minor of A^t is $(A_{ji})^t$. Using the formula for the expansion of det A^t by minors along the ith *column*, we obtain the formula for the expansion of det A along the ith *row*, as follows:

$$\det A = \det A^t$$

$$= (-1)^{1+i}a_{i1} \det(A_{i1})^t + (-1)^{2+i}a_{i2} \det(A_{i2})^t$$

$$+ \cdots + (-1)^{n+i}a_{in} \det(A_{in})^t$$

or

$$\det A = (-1)^{i+1}a_{i1} \det A_{i1} + (-1)^{i+2}a_{i2} \det A_{i2}$$

$$+ \cdots + (-1)^{i+n}a_{in} \det A_{in}.$$

EXAMPLE 5
Let us evaluate the determinant of the 5×5 matrix

$$A = \begin{pmatrix} -1 & 2 & 3 & -1 & 4 \\ 0 & 0 & 2 & 0 & 0 \\ 1 & 3 & -2 & 1 & 2 \\ 0 & 2 & -1 & 3 & 1 \\ -1 & 0 & 2 & 0 & -4 \end{pmatrix}.$$

Since the second row of A contains four zeros, it is convenient to expand by minors along that row. We get

$$\det A = -2 \det \begin{pmatrix} -1 & 2 & -1 & 4 \\ 1 & 3 & 1 & 2 \\ 0 & 2 & 3 & 1 \\ -1 & 0 & 0 & -4 \end{pmatrix}.$$

This 4 × 4 matrix has two zeros in the fourth row, so we expand by minors along that row to get

$$\det A = -2\left[-(-1)\det\begin{pmatrix} 2 & -1 & 4 \\ 3 & 1 & 2 \\ 2 & 3 & 1 \end{pmatrix} + (-4)\det\begin{pmatrix} -1 & 2 & -1 \\ 1 & 3 & 1 \\ 0 & 2 & 3 \end{pmatrix}\right].$$

Finally, we evaluate the two 3 × 3 determinants using row operations and expansion by minors along appropriate columns:

$$\det A = -2\left[\det\begin{pmatrix} 5 & 0 & 6 \\ 3 & 1 & 2 \\ -7 & 0 & -5 \end{pmatrix} - 4\det\begin{pmatrix} 0 & 5 & 0 \\ 1 & 3 & 1 \\ 0 & 2 & 3 \end{pmatrix}\right]$$

$$= -2\left[\det\begin{pmatrix} 5 & 6 \\ -7 & -5 \end{pmatrix} - 4(-1)\det\begin{pmatrix} 5 & 0 \\ 2 & 3 \end{pmatrix}\right]$$

$$= -2(17 + 60)$$

$$= -154. \quad\blacksquare$$

EXERCISES

1. List the 24 permutations of the set $\{1, 2, 3, 4\}$.

2. Write out the matrix of each permutation.
   (a) $(1, 3, 4, 2)$ (b) $(3, 4, 1, 2)$
   (c) $(2, 1, 4, 3)$ (d) $(1, 4, 3, 2)$
   (e) $(3, 1, 2, 4)$ (f) $(4, 1, 2, 3)$

3. Find the sign of each of the permutations in Exercise 2.

4. Find the sign of each permutation.
   (a) $(2, 1)$ (b) $(3, 2, 1)$
   (c) $(4, 3, 2, 1)$ (d) $(5, 4, 3, 2, 1)$
   (e) $(n, n - 1, n - 2, \ldots, 1)$

5. Each of the following products is of the form

$$a_{1\sigma(1)}a_{2\sigma(2)}a_{3\sigma(3)}a_{4\sigma(4)}a_{5\sigma(5)}$$

   for some $\sigma \in S_5$. For each, find the permutation σ and calculate its sign.
   (a) $a_{13}a_{25}a_{34}a_{42}a_{51}$ (b) $a_{12}a_{23}a_{34}a_{45}a_{51}$
   (c) $a_{15}a_{24}a_{31}a_{42}a_{53}$ (d) $a_{14}a_{23}a_{32}a_{41}a_{55}$

6. (a) How many permutations are there in S_5?
   (b) How many terms are there in the formula of Theorem 1 when $n = 5$?

7. Write out the transpose of each of the following matrices.

   (a) $\begin{pmatrix} -1 & 1 & 0 \\ 2 & 3 & -1 \end{pmatrix}$ (b) $\begin{pmatrix} 1 & 3 \\ -1 & 0 \\ 1 & 1 \end{pmatrix}$ (c) $\begin{pmatrix} 2 & 4 \\ 3 & 1 \end{pmatrix}$

   (d) $\begin{pmatrix} 1 & -1 & 1 \\ 2 & 3 & 5 \\ 0 & -1 & 2 \end{pmatrix}$ (e) $\begin{pmatrix} 0 & 2 \\ 2 & 3 \end{pmatrix}$ (f) $\begin{pmatrix} 0 & 1 & 2 \\ -1 & 0 & 3 \\ -2 & -3 & 0 \end{pmatrix}$

8. A matrix A is **symmetric** if $A = A^t$. Which of the following matrices are symmetric?

(a) $\begin{pmatrix} 1 & 2 & -1 \\ 2 & 0 & 3 \\ -1 & 3 & 2 \end{pmatrix}$ (b) $\begin{pmatrix} 2 & 1 \\ -1 & 2 \end{pmatrix}$ (c) $\begin{pmatrix} 2 & 0 & 0 \\ 0 & 1 & 0 \\ 0 & 0 & 3 \end{pmatrix}$

9. A matrix A is *skew-symmetric* if $A = -A^t$. Which of the following matrices are skew-symmetric?

(a) $\begin{pmatrix} 1 & -1 & 2 \\ 1 & 1 & 3 \\ -2 & -3 & 1 \end{pmatrix}$ (b) $\begin{pmatrix} 0 & 1 \\ -1 & 0 \end{pmatrix}$ (c) $\begin{pmatrix} 0 & a & b \\ -a & 0 & c \\ -b & -c & 0 \end{pmatrix}$

10. Explain why the determinant of a skew-symmetric $n \times n$ matrix must be equal to zero whenever n is odd.

11. Evaluate the determinant of the given matrix using expansion by minors along a row of your choice.

(a) $\begin{pmatrix} 2 & 3 & -2 \\ 5 & 1 & 4 \\ 0 & 0 & -3 \end{pmatrix}$ (b) $\begin{pmatrix} 0 & 0 & 2 \\ -1 & 1 & 3 \\ 5 & 6 & -4 \end{pmatrix}$

(c) $\begin{pmatrix} a & 0 & b \\ 0 & 1 & 0 \\ c & 0 & d \end{pmatrix}$ (d) $\begin{pmatrix} 1 & x & x^2 \\ 2 & 3 & 1 \\ -1 & -1 & 1 \end{pmatrix}$

12. Evaluate

$$\det \begin{pmatrix} 2 & 0 & 0 & -3 & 0 \\ 3 & 4 & -2 & 1 & 5 \\ -3 & -1 & 0 & 1 & 3 \\ 1 & 1 & 1 & 1 & 1 \\ -1 & 1 & -1 & 1 & -1 \end{pmatrix}.$$

13. Let f be the function that assigns to each $n \times n$ matrix

$$A = \begin{pmatrix} a_{11} & \cdots & a_{1n} \\ & \cdots & \\ a_{n1} & \cdots & a_{nn} \end{pmatrix}$$

the sum of all products of the form

$$(\text{sign } \sigma) a_{1\sigma(1)} a_{2\sigma(2)} \cdots a_{n\sigma(n)}$$

where σ runs through all the permutations in S_n.

(a) Show that if B is obtained from A by multiplying a row of A by the real number c, then $f(B) = cf(A)$.

(b) Show that if B is obtained from A by interchanging the first row with the second, then $f(B) = -f(A)$.

(c) Show that $f(I) = 1$.

14. Let

$$A = \begin{pmatrix} a_{11} & a_{12} & a_{13} \\ a_{21} & a_{22} & a_{23} \\ a_{31} & a_{32} & a_{33} \end{pmatrix}.$$

For each $\sigma \in S_3$, define $\tau \in S_3$ to be the permutation such that

$$a_{1\sigma(1)} a_{2\sigma(2)} a_{3\sigma(3)} = a_{\tau(1)1} a_{\tau(2)2} a_{\tau(3)3}.$$

(a) List the six permutations in S_3 and next to each list the corresponding permutation τ.

(b) Verify that in each case τ is the inverse of σ.

(c) Verify that in each case sign τ = sign σ.

REMARK Exercise 14 provides, for 3×3 matrices, the missing step in our proof that det A^t = det A. This exercise also suggests a way to complete the proof for $n \times n$ matrices:

(i) Verify that, for each $\sigma \in S_n$,

$$a_{1\sigma(1)}a_{2\sigma(2)} \cdots a_{n\sigma(n)} = a_{\tau(1)1}a_{\tau(2)2} \cdots a_{\tau(n)n}$$

where $\tau \in S_n$ is the inverse of σ, and

(ii) verify that sign τ = sign σ whenever τ is the inverse of σ.

Statement (i) is not difficult to verify. You will be asked to verify statement (ii) in Exercise 9 of Section 3.5.

15. For each permutation $\sigma \in S_n$ we can define a linear map $f_\sigma : \mathbb{R}^n \to \mathbb{R}^n$ by

$$f_\sigma(x_1, x_2, \ldots, x_n) = (x_{\sigma(1)}, x_{\sigma(2)}, \ldots, x_{\sigma(n)}).$$

(a) Show that the matrix of f_σ is the same as the permutation matrix M_σ.

(b) Show that if $\sigma \in S_n$ and $\tau \in S_n$ then $f_\sigma \circ f_\tau = f_{\sigma\tau}$.

(c) Conclude that if $\sigma \in S_n$ and $\tau \in S_n$, then $M_{\sigma\tau} = M_\sigma M_\tau$.

16. Let A be an $m \times n$ matrix and let B be an $n \times k$ matrix. Show that

    (a) $(A^t)^t = A$ (b) $(AB)^t = B^t A^t$

3.4
Matrix Inverses and Cramer's Rule

In this section we shall relate determinants to invertible matrices. We shall see that a square matrix is invertible if and only if its determinant is not zero. Then we shall derive a formula for the inverse of an invertible matrix in terms of determinants. Finally, we shall discuss Cramer's Rule, which can be used to solve any system of n linear equations in n unknowns when the coefficient matrix is invertible.

Theorem 1 *Let A be a square matrix. Then A is invertible if and only if* det $A \neq 0$.

Proof Let A be an $n \times n$ matrix. Let B be the echelon matrix of A. Then A can be transformed into B by a sequence of row operations. Each row operation either leaves the determinant unchanged or multiplies the determinant by a nonzero real number. Hence det $A \neq 0$ if and only if det $B \neq 0$.

The matrix B is upper triangular. Its determinant is nonzero if and only if all of its diagonal entries are nonzero. Since B is an echelon matrix, this means that B contains n corner 1's. But the number of corner 1's in B is equal to the rank of A. Therefore, det $B \neq 0$ if and only if A has rank n.

We can conclude, then, that det $A \neq 0$ if and only if A has rank n. But, by Theorem 2 of Section 2.7, A has rank n if and only if A is invertible. Hence det $A \neq 0$ if and only if A is invertible. ∎

According to Theorem 1, the determinant of an invertible matrix A is nonzero. Using this fact, we can find a formula for A^{-1}.

Recall the formula for expanding the determinant of an $n \times n$ matrix A by minors along the ith row:

$$\det A = (-1)^{i+1}a_{i1} \det A_{i1} + (-1)^{i+2}a_{i2} \det A_{i2}$$

$$+ \cdots + (-1)^{i+n}a_{in} \det A_{in}$$

where a_{ij} is the (i, j)-entry of A and A_{ij} is the (i, j)-minor of A. If we let

$$c_{ij} = (-1)^{i+j} \det A_{ij}$$

then this formula becomes

$$\det A = a_{i1}c_{i1} + a_{i2}c_{i2} + \cdots + a_{in}c_{in}.$$

The right-hand side of this formula is the dot product of the ith row vector of A and the ith row vector of the matrix

$$C = \begin{pmatrix} c_{11} & c_{12} & \cdots & c_{1n} \\ c_{21} & c_{22} & \cdots & c_{2n} \\ & & \cdots & \\ c_{n1} & c_{n2} & \cdots & c_{nn} \end{pmatrix}.$$

This is the same as the dot product of the ith row vector of A and the ith *column* vector of C^t, which is the same as the (i, i)-entry in the matrix AC^t. In other words, the diagonal entries in the product matrix AC^t are all equal to $\det A$. We shall soon see that the off-diagonal entries in AC^t are all equal to zero, and hence that

$$AC^t = (\det A)I,$$

where I is the $n \times n$ identity matrix. If A is invertible, then $\det A \neq 0$ so we can divide both sides of this equation by $\det A$ to get

$$\frac{1}{\det A} AC^t = I$$

or

$$A\left(\frac{1}{\det A} C^t\right) = I.$$

It follows that the matrix

$$\frac{1}{\det A} C^t$$

is the inverse of the matrix A.

The number $c_{ij} = (-1)^{i+j} \det A_{ij}$ is called the (i, j)-***cofactor*** of the matrix A. The matrix C whose entries are the cofactors of A is the ***cofactor matrix*** of A.

This discussion leads us to the following theorem.

Theorem 2 *Let A be an invertible matrix. Then*

$$A^{-1} = \frac{1}{\det A} C^t$$

where C is the cofactor matrix of A.

Proof In light of the preceding discussion, we only need to verify that the off-diagonal entries in the matrix product AC^t are all equal to zero. In other words, we must check that

$$a_{i1}c_{j1} + a_{i2}c_{j2} + \cdots + a_{in}c_{jn} = 0$$

whenever $i \neq j$, where a_{ij} is the (i, j)-entry of A and c_{ij} is the (i, j)-cofactor of A $(1 \leq i \leq n, 1 \leq j \leq n)$. To see this, let $\mathbf{b} = (b_1, \ldots, b_n)$ be any vector in $\mathbb{R}^n$ and let B be the matrix obtained from A by replacing the jth row vector of A by $\mathbf{b}$. Thus

$$B = \begin{pmatrix} a_{11} & \cdots & a_{1n} \\ & \cdots & \\ b_1 & \cdots & b_n \\ & \cdots & \\ a_{n1} & \cdots & a_{nn} \end{pmatrix} \leftarrow j\text{th row}$$

Notice that the (j, k)-cofactor of B is equal to the (j, k)-cofactor c_{jk} of A, for all k $(1 \leq k \leq n)$, since the entries in the jth row do not appear in the jth row minors and hence do not appear in the jth row cofactors. Hence, evaluating the determinant of B by minors along the jth row yields

(∗) $$\det B = b_1 c_{j1} + b_2 c_{j2} + \cdots + b_n c_{jn},$$

where the c_{jk} $(1 \leq k \leq n)$ are cofactors of A.
      If we take $\mathbf{b}$ to the jth row vector of A, then $B = A$ and (∗) becomes

$$\det A = a_{j1} c_{j1} + a_{j2} c_{j2} + \cdots + a_{jn} c_{jn},$$

which is the expansion formula for $\det A$ by minors along the jth row.
      But if we take $\mathbf{b}$ to be the ith row vector of A, where $i \neq j$, then B has two equal rows (the ith and the jth). Hence $\det B = 0$ and (∗) becomes

$$0 = a_{i1} c_{j1} + a_{i2} c_{j2} + \cdots + a_{in} c_{jn},$$

which is the formula that we needed to check. ∎

EXAMPLE 1
Let

$$A = \begin{pmatrix} 2 & 3 & 1 \\ -1 & 1 & 0 \\ 1 & 0 & 1 \end{pmatrix}.$$

The cofactors of A are

$$c_{11} = (-1)^{1+1} \det A_{11} = \det\begin{pmatrix} 1 & 0 \\ 0 & 1 \end{pmatrix} = 1,$$

$$c_{12} = (-1)^{1+2} \det A_{12} = -\det\begin{pmatrix} -1 & 0 \\ 1 & 1 \end{pmatrix} = 1,$$

$$c_{13} = (-1)^{1+3} \det A_{13} = \det\begin{pmatrix} -1 & 1 \\ 1 & 0 \end{pmatrix} = -1,$$

$$c_{21} = (-1)^{2+1} \det A_{21} = -\det\begin{pmatrix} 3 & 1 \\ 0 & 1 \end{pmatrix} = -3,$$

and so on. The cofactor matrix is

$$C = \begin{pmatrix} 1 & 1 & -1 \\ -3 & 1 & 3 \\ -1 & -1 & 5 \end{pmatrix}.$$

Since $AC^t = (\det A)I$, we need only compute one of the diagonal entries in the matrix AC^t in order to find $\det A$. Taking the dot product of the first row vector of A with the first column vector of C^t, we find that

$$\det A = (2, 3, 1) \bullet (1, 1, -1) = 4.$$

Hence

$$A^{-1} = \frac{1}{\det A} C^t = \tfrac{1}{4}C^t = \begin{pmatrix} \frac{1}{4} & -\frac{3}{4} & -\frac{1}{4} \\ \frac{1}{4} & \frac{1}{4} & -\frac{1}{4} \\ -\frac{1}{4} & \frac{3}{4} & \frac{5}{4} \end{pmatrix}. \quad \blacksquare$$

EXAMPLE 2

We can use Theorem 2 to find a formula for the inverse of any invertible 2×2 matrix

$$A = \begin{pmatrix} a & b \\ c & d \end{pmatrix}.$$

The cofactor matrix of A is

$$C = \begin{pmatrix} d & -c \\ -b & a \end{pmatrix},$$

so

$$A^{-1} = \frac{1}{\det A} C^t = \frac{1}{ad - bc} \begin{pmatrix} d & -b \\ -c & a \end{pmatrix}. \quad \blacksquare$$

We can use Theorem 2 to prove *Cramer's Rule*, which can be used to solve any system of n linear equations in n unknowns whenever the coefficient matrix is invertible.

Theorem 3 (Cramer's Rule). *Let A be an invertible $n \times n$ matrix and let $\mathbf{b} \in \mathbb{R}^n$. Then the unique solution of the linear system $A\mathbf{x} = \mathbf{b}$ is*

$$\mathbf{x} = \frac{1}{\det A} \, (\det A_1, \det A_2, \ldots, \det A_n)$$

where, for each i $(1 \leq i \leq n)$, A_i is the matrix obtained from A by replacing the ith column vector by $\mathbf{b}$.

Before proving Theorem 3, let us work two examples.

EXAMPLE 3

Let us use Cramer's Rule to solve the linear system

$$3x_1 - 2x_2 = 1$$

$$x_1 - x_2 = 2.$$

In matrix form, this system is $A\mathbf{x} = \mathbf{b}$ where

$$A = \begin{pmatrix} 3 & -2 \\ 1 & -1 \end{pmatrix} \quad \text{and} \quad \mathbf{b} = \begin{pmatrix} 1 \\ 2 \end{pmatrix}.$$

Hence

$$A_1 = \begin{pmatrix} 1 & -2 \\ 2 & -1 \end{pmatrix}, \quad A_2 = \begin{pmatrix} 3 & 1 \\ 1 & 2 \end{pmatrix},$$

and the unique solution of $A\mathbf{x} = \mathbf{b}$ is

$$\mathbf{x} = \frac{1}{\det A} \, (\det A_1, \det A_2)$$

$$= \frac{1}{-1} \, (3, 5)$$

$$= (-3, -5). \quad \blacksquare$$

EXAMPLE 4

Let us use Cramer's Rule to solve the linear system

$$x_1 - x_2 + x_3 = 4$$

$$x_1 + x_2 - x_3 = -2$$

$$2x_1 - x_1 - x_3 = 1.$$

This system can be written as $A\mathbf{x} = \mathbf{b}$ where

$$A = \begin{pmatrix} 1 & -1 & 1 \\ 1 & 1 & -1 \\ 2 & -1 & -1 \end{pmatrix} \quad \text{and} \quad \mathbf{b} = \begin{pmatrix} 4 \\ -2 \\ 1 \end{pmatrix}.$$

Hence

$$A_1 = \begin{pmatrix} 4 & -1 & 1 \\ -2 & 1 & -1 \\ 1 & -1 & -1 \end{pmatrix}, \qquad A_2 = \begin{pmatrix} 1 & 4 & 1 \\ 1 & -2 & -1 \\ 2 & 1 & -1 \end{pmatrix},$$

$$A_3 = \begin{pmatrix} 1 & -1 & 4 \\ 1 & 1 & -2 \\ 2 & -1 & 1 \end{pmatrix},$$

and the unique solution is

$$\mathbf{x} = \frac{1}{\det A} (\det A_1, \det A_2, \det A_3)$$

$$= \frac{1}{-4} (-4, 4, -8)$$

$$= (1, -1, 2). \quad \blacksquare$$

REMARK Although Cramer's Rule provides a quick method for solving systems of 2 linear equations in 2 unknowns, and it is sometimes useful for solving systems of 3 linear equations in 3 unknowns, the method of Gaussian elimination is much faster for solving systems of n linear equations in n unknowns when n is larger than 3.

Proof of Theorem 3 Since A is invertible, we know that the unique solution of $A\mathbf{x} = \mathbf{b}$ is $\mathbf{x} = A^{-1}\mathbf{b}$. By Theorem 2,

$$A^{-1} = \frac{1}{\det A} C^t$$

where C is the cofactor matrix of A. Hence

$$\mathbf{x} = \frac{1}{\det A} C^t \mathbf{b}$$

or

$$x_j = \frac{1}{\det A} (c_{1j}b_1 + c_{2j}b_2 + \cdots + c_{nj}b_n)$$

where x_j is the jth entry of $\mathbf{x}$, c_{ij} is the (i, j)-cofactor of A, and b_i is the ith entry of $\mathbf{b}$ ($1 \le i \le n$, $1 \le j \le n$). The quantity in parentheses is just the expansion by minors along the jth column of

$$\det \begin{pmatrix} a_{11} & \cdots & b_1 & \cdots & a_{1n} \\ & & \cdots & & \\ a_{n1} & \cdots & b_n & \cdots & a_{nn} \end{pmatrix} = \det A_j. \quad \swarrow \text{ } j\text{th column}$$

Therefore, for $1 \le j \le n$,

$$x_j = \frac{1}{\det A} (\det A_j)$$

and

$$\mathbf{x} = \frac{1}{\det A} (\det A_1, \det A_2, \ldots, \det A_n),$$

as asserted. ■

EXERCISES

1. For each of the given matrices A, compute $\det A$ and decide if A is invertible.

(a) $\begin{pmatrix} 1 & 2 \\ -2 & -4 \end{pmatrix}$ (b) $\begin{pmatrix} 3 & 4 \\ -4 & -3 \end{pmatrix}$

(c) $\begin{pmatrix} 1 & 3 \\ 0 & 1 \end{pmatrix}$ (d) $\begin{pmatrix} 1 & -1 & 2 \\ 0 & 1 & -3 \\ 0 & 0 & 1 \end{pmatrix}$

2. Find the cofactor matrix of

(a) $\begin{pmatrix} -1 & 1 \\ 0 & 2 \end{pmatrix}$ (b) $\begin{pmatrix} 4 & 6 \\ 6 & 9 \end{pmatrix}$

(c) $\begin{pmatrix} -1 & 0 & 1 \\ 2 & 3 & -1 \\ 4 & 1 & 2 \end{pmatrix}$ (d) $\begin{pmatrix} 2 & 3 & -1 \\ 0 & 1 & 1 \\ -1 & 0 & 2 \end{pmatrix}$

(e) $\begin{pmatrix} 2 & -1 & 4 \\ 3 & 0 & -6 \\ 5 & 1 & 3 \end{pmatrix}$ (f) $\begin{pmatrix} 1 & a & b \\ 0 & 1 & c \\ 0 & 0 & 1 \end{pmatrix}$

3. Compute the product of each of the matrices in Exercise 2 with the transpose of its cofactor matrix.

4. Find the inverse of each of the matrices in Exercise 2 that is invertible.

5. Use cofactors to find the inverse of each of the following matrices.

(a) $\begin{pmatrix} -1 & 1 \\ 3 & 4 \end{pmatrix}$ (b) $\begin{pmatrix} 1 & 0 \\ a & 1 \end{pmatrix}$

(c) $\begin{pmatrix} 1 & -1 & 2 \\ 5 & 4 & -1 \\ -3 & 0 & 7 \end{pmatrix}$ (d) $\begin{pmatrix} 1 & 0 & 0 \\ a & 1 & 0 \\ b & c & 1 \end{pmatrix}$

(e) $\begin{pmatrix} \cos\theta & -\sin\theta \\ \sin\theta & \cos\theta \end{pmatrix}$ (f) $\begin{pmatrix} 1 & 0 & 0 \\ 0 & a & b \\ 0 & c & d \end{pmatrix}$, $ad - bc \neq 0$

6. Use Cramer's Rule to solve each of the following systems of linear equations.

(a) $3x_1 + 4x_2 = 1$
$2x_1 + 5x_2 = -1$

(b) $7x_1 + 3x_2 = 5$
$-4x_1 + 2x_2 = 1$

(c) $2x_1 + x_2 + 5x_3 = 0$
$x_1 + 3x_3 = 1$
$-x_1 + 2x_2 = -1$

(d) $2x_1 + 3x_2 - 5x_3 = 2$
$3x_1 - x_2 + 2x_3 = -1$
$5x_1 + 4x_2 - 6x_3 = 3$

7. Let A be an upper triangular $n \times n$ matrix.

(a)   Show that the cofactor matrix of A is lower triangular.

(b)   Show that, if A is invertible, then A^{-1} is upper triangular.

8.   Let A be a square matrix with integer entries. Show that, if $\det A = 1$, then

(a) A^{-1} has integer entries.

(b)   the solution of the system $A\mathbf{x} = \mathbf{b}$ has integer entries, whenever the vector $\mathbf{b}$ has integer entries.

3.5
The Determinant of a Product

We shall show in this section that if A and B are square matrices of the same size, then $\det(AB) = (\det A)(\det B)$. We shall do this using *elementary matrices*.

An $n \times n$ matrix E is an *elementary matrix* if it can be obtained from the $n \times n$ identity matrix I using a single row operation.

EXAMPLE 1
The following 2×2 matrices are elementary matrices:

(i) $\begin{pmatrix} c & 0 \\ 0 & 1 \end{pmatrix}$ and $\begin{pmatrix} 1 & 0 \\ 0 & c \end{pmatrix}$, where $c \neq 0$. These matrices are obtained by multiplying a row of the 2×2 identity matrix I by the nonzero real number c.

(ii) $\begin{pmatrix} 0 & 1 \\ 1 & 0 \end{pmatrix}$. This matrix is obtained by interchanging the rows of I.

(iii) $\begin{pmatrix} 1 & 0 \\ c & 1 \end{pmatrix}$ and $\begin{pmatrix} 1 & c \\ 0 & 1 \end{pmatrix}$, where c is any real number. These matrices are obtained by adding a multiple of one row of I to another.

Since we have used every type of row operation here, every 2×2 elementary matrix is one of these three types. ■

EXAMPLE 2
The matrix

$$\begin{pmatrix} 1 & 0 & -2 \\ 0 & 1 & 0 \\ 0 & 0 & 1 \end{pmatrix}$$

is an elementary matrix since it can be obtained from the 3×3 identity matrix by adding -2 times the 3rd row to the first. On the other hand, the matrix

$$\begin{pmatrix} 0 & 0 & 1 \\ 1 & 0 & 0 \\ 0 & 1 & 0 \end{pmatrix}$$

is *not* an elementary matrix since there is no single row operation that will transform the 3×3 identity matrix into this matrix. ■

Let us see what happens when we multiply an elementary matrix by some other matrix.

EXAMPLE 3

Suppose

$$A = \begin{pmatrix} a_{11} & a_{12} & a_{13} \\ a_{21} & a_{22} & a_{23} \end{pmatrix}.$$

(i) If $E = \begin{pmatrix} c & 0 \\ 0 & 1 \end{pmatrix}$, then $EA = \begin{pmatrix} ca_{11} & ca_{12} & ca_{13} \\ a_{21} & a_{22} & a_{23} \end{pmatrix}$; EA is obtained from A by multiplying the first row of A by c.

If $E = \begin{pmatrix} 1 & 0 \\ 0 & c \end{pmatrix}$, then $EA = \begin{pmatrix} a_{11} & a_{12} & a_{13} \\ ca_{21} & ca_{22} & ca_{23} \end{pmatrix}$; EA is obtained from A by multiplying the second row of A by c.

(ii) If $E = \begin{pmatrix} 0 & 1 \\ 1 & 0 \end{pmatrix}$, then $EA = \begin{pmatrix} a_{21} & a_{22} & a_{23} \\ a_{11} & a_{12} & a_{13} \end{pmatrix}$; EA is obtained from A by interchanging the rows of A.

(iii) If $E = \begin{pmatrix} 1 & 0 \\ c & 1 \end{pmatrix}$, then $EA = \begin{pmatrix} a_{11} & a_{12} & a_{13} \\ ca_{11} + a_{21} & ca_{12} + a_{22} & ca_{13} + a_{23} \end{pmatrix}$; EA is obtained from A by adding c times the first row to the second.

(iv) If $E = \begin{pmatrix} 1 & c \\ 0 & 1 \end{pmatrix}$, then $EA = \begin{pmatrix} a_{11} + ca_{21} & a_{12} + ca_{22} & a_{13} + ca_{23} \\ a_{21} & a_{22} & a_{23} \end{pmatrix}$; EA is obtained from A by adding c times the second row to the first.

In each case, the matrix EA is obtained from the matrix A by a row operation, the same row operation that transforms the identity matrix I to E. ∎

You can readily convince yourself that the phenomenon illustrated in Example 3 holds in general. That is the content of the following theorem.

Theorem 1 *Let E be an $n \times n$ elementary matrix. Then multiplying an $n \times k$ matrix A on the left by E has the same effect as applying a row operation to the matrix A. The row operation that transforms A into EA is the same as the row operation that transforms I into E.*

The next two theorems are consequences of Theorem 1.

Theorem 2 *Let A be any matrix and let B be its echelon matrix. Then there is a sequence $E_1, E_2, \ldots, E_k$ of elementary matrices such that*

$$E_k \cdots E_2 E_1 A = B.$$

Proof We know that B can be obtained from A by a sequence of row operations. Let $E_1, E_2, \ldots, E_k$ be the corresponding elementary matrices. Then, by Theorem 1, $E_1 A$ is the matrix obtained from A by applying the first row operation to A, $E_2 E_1 A$ is the matrix obtained from A by applying the first row operation to A and then the second row operation to the result, and so on. After applying the entire sequence of row operations we get

$$E_k \cdots E_2 E_1 A = B. \quad ∎$$

Theorem 3 *Let A be an invertible matrix. Then A can be factored into a product of elementary matrices.*

Proof We know from Theorem 2 of Section 2.7 that if A is an invertible $n \times n$ matrix, then the echelon matrix of A is the $n \times n$ identity matrix I. Hence, by Theorem 2,

$$E_k \cdots E_2 E_1 A = I$$

where $E_1, E_2, \ldots, E_k$ are the elementary matrices corresponding to the row operations that transform A to I.

Each elementary matrix is invertible. The inverse of an elementary matrix E is the elementary matrix corresponding to the row operation that reverses the row operation corresponding to E. Hence the product $E_k \cdots E_2 E_1$ is also invertible. Its inverse is

$$(E_k \cdots E_2 E_1)^{-1} = E_1^{-1} E_2^{-1} \cdots E_k^{-1}.$$

Multiplying both sides of the equation $E_k \cdots E_2 E_1 A = I$ on the left by $(E_k \cdots E_2 E_1)^{-1}$, we get

$$A = (E_k \cdots E_2 E_1)^{-1} I = E_1^{-1} E_2^{-1} \cdots E_k^{-1},$$

so A is a product of elementary matrices, as asserted. ■

The following example illustrates Theorem 3 for a particular invertible matrix A.

EXAMPLE 4

Let us factor the invertible matrix

$$A = \begin{pmatrix} 1 & 2 \\ 2 & 1 \end{pmatrix}$$

into a product of elementary matrices.

We can use row operations to transform A into the identity matrix as follows:

$$\begin{pmatrix} 1 & 2 \\ 2 & 1 \end{pmatrix} \longrightarrow \begin{pmatrix} 1 & 2 \\ 0 & -3 \end{pmatrix} \longrightarrow \begin{pmatrix} 1 & 2 \\ 0 & 1 \end{pmatrix} \longrightarrow \begin{pmatrix} 1 & 0 \\ 0 & 1 \end{pmatrix}.$$

The elementary matrices corresponding to these row operations are

$$E_1 = \begin{pmatrix} 1 & 0 \\ -2 & 1 \end{pmatrix}, E_2 = \begin{pmatrix} 1 & 0 \\ 0 & -\frac{1}{3} \end{pmatrix}, \text{ and } E_3 = \begin{pmatrix} 1 & -2 \\ 0 & 1 \end{pmatrix}.$$

Hence $E_3 E_2 E_1 A = I$, and

$$A = E_1^{-1} E_2^{-1} E_3^{-1} = \begin{pmatrix} 1 & 0 \\ 2 & 1 \end{pmatrix} \begin{pmatrix} 1 & 0 \\ 0 & -3 \end{pmatrix} \begin{pmatrix} 1 & 2 \\ 0 & 1 \end{pmatrix}. ■$$

We are now ready to prove the main result of this section.

Theorem 4 *Let A and B be $n \times n$ matrices. Then*

$$\det(AB) = (\det A)(\det B).$$

Proof   First notice that if E is any $n \times n$ elementary matrix, then

$$\det(EB) = (\det E)(\det B).$$

Let us check each of the three cases.

(i)   Suppose E is obtained from I by multiplying a row of I by c. Then, by Theorem 1, EB is obtained from B by multiplying a row of B by c. Hence

$$\det(EB) = c \det B \qquad \text{and} \qquad \det E = c \det I = c,$$

so

$$\det(EB) = (\det E)(\det B).$$

(ii)   Suppose E is obtained from I by interchanging two rows. Then, by Theorem 1, EB is obtained from B by interchanging two rows. Hence

$$\det EB = -\det B \qquad \text{and} \qquad \det E = -\det I = -1,$$

so

$$\det (EB) = (\det E)(\det B).$$

(iii)   Suppose E is obtained from I by adding a scalar multiple of one row to another. Then, by Theorem 1, EB is obtained from B by adding a scalar multiple of one row to another. Hence

$$\det EB = \det B \qquad \text{and} \qquad \det E = \det I = 1,$$

so

$$\det(EB) = (\det E)(\det B).$$

Thus we have shown that $\det(AB) = (\det A)(\det B)$ whenever A is an elementary matrix.

It follows that $\det(AB) = (\det A)(\det B)$ whenever A is a product of elementary matrices. If $A = E_1 E_2$ where E_1 and E_2 are elementary matrices, then

$$\det(AB) = \det(E_1 E_2 B) = (\det E_1)\,\det(E_2 B) = (\det E_1)(\det E_2)(\det B)$$

$$= (\det E_1 E_2)(\det B) = (\det A)(\det B).$$

The proof is similar when $A = E_1 E_2 \cdots E_k$ for $k > 2$.

By Theorem 3, every invertible matrix is a product of elementary matrices. Therefore, we have shown that

$$\det(AB) = (\det A)(\det B)$$

whenever A is invertible.

If A is not invertible then you can check (see Exercise 8) that $\det(AB) = 0$ and that $\det A = 0$, so

$$\det(AB) = (\det A)(\det B)$$

in this case as well.   ■

EXERCISES

1. Determine which of the following matrices are elementary matrices. For each that is, describe the row operation that transforms I into the given matrix.

(a) $\begin{pmatrix} 1 & 3 \\ 0 & 1 \end{pmatrix}$ (b) $\begin{pmatrix} -1 & 0 \\ 0 & -1 \end{pmatrix}$

(c) $\begin{pmatrix} 1 & -1 & 0 \\ 0 & 1 & 0 \\ 0 & 0 & 1 \end{pmatrix}$ (d) $\begin{pmatrix} 0 & 0 & 1 \\ 0 & 1 & 0 \\ 1 & 0 & 0 \end{pmatrix}$

(e) $\begin{pmatrix} 1 & 0 & 0 & 0 \\ 0 & 1 & 0 & -4 \\ 0 & 0 & 1 & 0 \\ 0 & 0 & 0 & 1 \end{pmatrix}$

2. Find the inverse of each of the following elementary matrices.

(a) $\begin{pmatrix} 1 & 0 \\ 0 & 2 \end{pmatrix}$ (b) $\begin{pmatrix} 0 & 1 \\ 1 & 0 \end{pmatrix}$

(c) $\begin{pmatrix} 1 & 0 \\ 2 & 1 \end{pmatrix}$ (d) $\begin{pmatrix} 1 & 0 & -2 \\ 0 & 1 & 0 \\ 0 & 0 & 1 \end{pmatrix}$

(e) $\begin{pmatrix} 1 & 0 & 0 & 0 \\ 0 & 1 & 0 & 0 \\ 5 & 0 & 1 & 0 \\ 0 & 0 & 0 & 1 \end{pmatrix}$

3. Find a sequence of elementary matrices $E_1, E_2, \ldots, E_k$ such that $E_k \cdots E_2 E_1 A$ is an echelon matrix, where $A =$

(a) $\begin{pmatrix} -1 & 0 & 2 \\ 2 & 1 & 3 \end{pmatrix}$ (b) $\begin{pmatrix} 1 & -1 & 3 \\ -1 & 1 & 3 \end{pmatrix}$

(c) $\begin{pmatrix} 0 & 1 & 0 & 0 \\ 1 & 2 & -1 & 3 \end{pmatrix}$ (d) $\begin{pmatrix} -1 & 1 \\ 2 & 3 \\ 1 & -2 \end{pmatrix}$

(e) $\begin{pmatrix} 1 & 2 & -1 \\ 3 & -1 & 2 \\ -1 & 5 & -4 \end{pmatrix}$

4. Factor each of the following invertible matrices into a product of elementary matrices.

(a) $\begin{pmatrix} 1 & 3 \\ 3 & 1 \end{pmatrix}$ (b) $\begin{pmatrix} 3 & -1 \\ 2 & 5 \end{pmatrix}$ (c) $\begin{pmatrix} 2 & 4 \\ 7 & 6 \end{pmatrix}$

(d) $\begin{pmatrix} 1 & 0 & 0 \\ 2 & 1 & 0 \\ 0 & -1 & 1 \end{pmatrix}$ (e) $\begin{pmatrix} 0 & 1 & 1 \\ 1 & 0 & 1 \\ 1 & 1 & 0 \end{pmatrix}$

5. Use Theorem 4 to show that if A is an invertible matrix, then

$$\det(A^{-1}) = \frac{1}{\det A}.$$

6. Let

$$A = \begin{pmatrix} a_{11} & a_{12} \\ a_{21} & a_{22} \\ a_{31} & a_{32} \end{pmatrix}.$$

For each of the following elementary matrices E, calculate the matrix AE and describe its relationship to the matrix A.

(a) $E = \begin{pmatrix} c & 0 \\ 0 & 1 \end{pmatrix}$ (b) $E = \begin{pmatrix} 1 & 0 \\ 0 & c \end{pmatrix}$ (c) $E = \begin{pmatrix} 0 & 1 \\ 1 & 0 \end{pmatrix}$

(d) $E = \begin{pmatrix} 1 & 0 \\ c & 1 \end{pmatrix}$ (e) $E = \begin{pmatrix} 1 & c \\ 0 & 1 \end{pmatrix}$

7. If A is an $m \times n$ matrix and E is an $n \times n$ elementary matrix, then how is the matrix AE related to the matrix A? [*Hint:* Do Exercise 6 before you attempt this exercise.]

8. Let A be an $n \times n$ matrix that is not invertible and let B be any $n \times n$ matrix. Explain why each of the following statements is true.

(a) There is a sequence $E_1, E_2, \ldots, E_k$ of elementary matrices such that the matrix $E_k \cdots E_2 E_1 A$ contains a row of zeros.

(b) The matrix $E_k \cdots E_2 E_1 AB$ also has a row of zeros.

(c) $\det(E_k \cdots E_2 E_1 A) = 0$ and $\det(E_k \cdots E_2 E_1 AB) = 0$.

(d) $\det A = 0$ and $\det(AB) = 0$.

(e) $\det(AB) = (\det A)(\det B)$.

[In explaining why (d) and (e) are true, do not use Theorem 4. This exercise is designed to supply the last step in the proof of that theorem. You may, if you wish, use the fact that $\det(AB) = (\det A)(\det B)$ whenever A is invertible.]

9. Let σ and τ be permutations in S_n.

(a) Show that sign $(\sigma \circ \tau) = (\text{sign } \sigma)(\text{sign } \tau)$.

(b) Show that if τ is the inverse of σ, then sign $\tau = \text{sign } \sigma$.

3.6

Application: Areas and Volumes

We know that there is a linear map $f: \mathbb{R}^n \to \mathbb{R}^n$ associated with each $n \times n$ matrix A. This linear map multiplies each vector $\mathbf{x} \in \mathbb{R}^n$ on the left by the matrix A:

$$\text{if } \mathbf{x} = \begin{pmatrix} x_1 \\ \vdots \\ x_n \end{pmatrix} \text{ then } f(\mathbf{x}) = A\mathbf{x}.$$

In this section we shall study the relationship between the determinant of A and the geometry of the linear map f. We will see that, when $n = 2$, $\det A$ measures magnification of area under f and, when $n = 3$, $\det A$ measures magnification of volume under f.

Let us begin with the case $n = 2$. Suppose $f: \mathbb{R}^2 \to \mathbb{R}^2$ is any linear map. Recall from Section 2.1 that f maps parallel lines to parallel lines and that f maps parallelograms to parallelograms. In fact, if the plane $\mathbb{R}^2$ is divided into

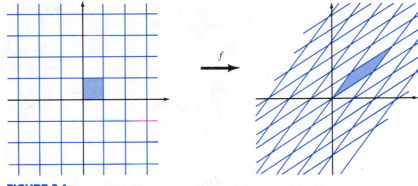

FIGURE 3.1
The squares defined by a coordinate grid are mapped by f to parallelograms.

coordinate squares by means of equally spaced coordinate lines, then these squares are all mapped by f to parallelograms (see Figure 3.1). The coordinate squares are all congruent to each other so they all have the same area. Also, the parallelograms are all congruent to each other so they too all have the same area. Hence the ratio

$$r = \frac{\text{Area of } f(S)}{\text{Area of } S},$$

where S is one of these coordinate squares, is a number that does not depend on the particular coordinate square used to compute it.

EXAMPLE 1
Let $f:\mathbb{R}^2 \to \mathbb{R}^2$ be the horizontal shear $f(\mathbf{x}) = A\mathbf{x}$ where

$$A = \begin{pmatrix} 1 & c \\ 0 & 1 \end{pmatrix}.$$

Then equally spaced coordinate lines are mapped by f as in Figure 3.2 (in the figure, $c = \frac{1}{2}$). If a is the width of each of the coordinate squares, then each

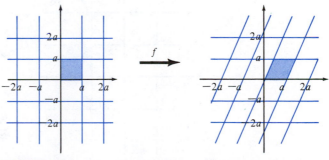

FIGURE 3.2
Under the shear $f(x_1, x_2) = (x_1 + cx_2, x_2)$, coordinate squares with area a^2 are mapped to parallelograms with area a^2.

coordinate square S has area a^2. Each parallelogram $f(S)$ also has area a^2 (its base and its height are both equal to a). Hence the ratio

$$r = \frac{\text{Area of } f(S)}{\text{Area of } S}$$

is equal to 1, for every coordinate square S.

You can also check that $r = 1$ when $f:\mathbb{R}^2 \to \mathbb{R}^2$ is the vertical shear $f(x) = A\mathbf{x}$ where

$$A = \begin{pmatrix} 1 & 0 \\ c & 1 \end{pmatrix}.$$

Sketch the appropriate figure. ■

EXAMPLE 2

Let $f:\mathbb{R}^2 \to \mathbb{R}^2$ be reflection in the line $x_2 = x_1$: $f(\mathbf{x}) = A\mathbf{x}$ where

$$A = \begin{pmatrix} 0 & 1 \\ 1 & 0 \end{pmatrix}.$$

Then equally spaced coordinate lines are mapped by f as in Figure 3.3. Each coordinate square has area a^2 and is mapped by f to a coordinate square with area a^2. Hence the ratio

$$r = \frac{\text{Area of } f(S)}{\text{Area of } S}$$

is equal to 1, for every coordinate square S. ■

EXAMPLE 3

Let c be a positive real number and let $f:\mathbb{R}^2 \to \mathbb{R}^2$ be the linear map $f(x_1, x_2) = (cx_1, x_2)$ with matrix

$$A = \begin{pmatrix} c & 0 \\ 0 & 1 \end{pmatrix}.$$

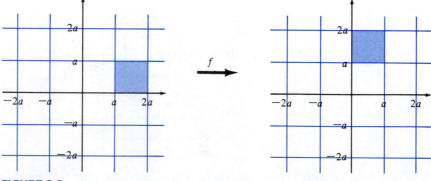

FIGURE 3.3
Under the reflection $f(x_1, x_2) = (x_2, x_1)$, coordinate squares with area a^2 are mapped to coordinate squares with area a^2.

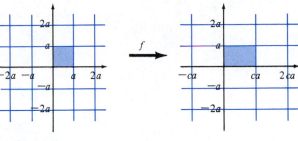

FIGURE 3.4
Under the linear map $f(x_1, x_2) = (cx_1, x_2)$, $c > 0$, coordinate
squares with area a^2 are mapped to rectangles with area ca^2.

Then equally spaced coordinate lines are mapped by f as in Figure 3.4 (in the
figure, $c = \frac{3}{2}$). Each coordinate square has area a^2 and is mapped by f to a
rectangle with area ca^2. Hence the ratio

$$r = \frac{\text{Area of } f(S)}{\text{Area of } S}$$

is equal to c, for every coordinate square S.

If c is a negative real number, then the linear map $f(x_1, x_2) = (cx_1, x_2)$
with matrix

$$A = \begin{pmatrix} c & 0 \\ 0 & 1 \end{pmatrix}$$

maps equally spaced coordinate lines as in Figure 3.5 (in the figure, $c = -\frac{1}{2}$).
Each coordinate square is mapped by f to a rectangle with area $(-c)a^2$
$(c < 0)$. Hence the ratio

$$r = \frac{\text{Area of } f(S)}{\text{Area of } S}$$

is equal to $-c$ for every coordinate square S.

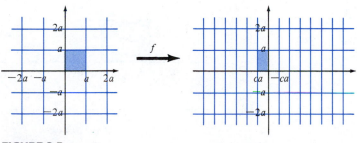

FIGURE 3.5
Under the linear map $f(x_1, x_2) = (cx_1, x_2)$, $c < 0$, coordinate squares
with area a^2 are mapped to rectangles with area $|c|a^2$.

Since $|c| = c$ when $c > 0$ and $|c| = -c$ when $c < 0$, we can conclude that if $f(x_1, x_2) = (cx_1, x_2)$ for any $c \neq 0$, then

$$r = \frac{\text{Area of } f(S)}{\text{Area of } S} = |c|.$$

You can check that $r = 0$ if $c = 0$, so $r = |c|$ in this case as well. You can also check that $r = |c|$ when $f(x_1, cx_2)$, for some $c \in \mathbb{R}$. ∎

Notice that, in each of the above examples, the ratio

$$r = \frac{\text{Area of } f(S)}{\text{Area of } S}$$

is not only independent of the coordinate square S used to compute it, but it is also independent of the particular coordinate grid used. That is, if we use a different spacing between the coordinate lines, we will still get the same ratio of areas.

It is easy to see that this is always true. Let S_a ($a > 0$) be the square with vertices $(0, 0)$, $(a, 0)$, (a, a), and $(0, a)$ (see Figure 3.6). This is a coordinate square of the coordinate grid that has spacing a between coordinate lines. Then S_a is the square spanned by the vectors $a\mathbf{e}_1$ and $a\mathbf{e}_2$. If $f:\mathbb{R}^2 \to \mathbb{R}^2$ is any linear map, then $f(S_a)$ is the parallelogram spanned by $f(a\mathbf{e}_1) = af(\mathbf{e}_1)$ and $f(a\mathbf{e}_2) = af(\mathbf{e}_2)$. The area of S_a is equal to a^2 times the area of the square S_1 spanned by $\mathbf{e}_1$ and $\mathbf{e}_2$. The area of $f(S_a)$ is equal to a^2 times the area of $f(S_1)$ (Figure 3.6). Hence

$$r = \frac{\text{Area of } f(S_a)}{\text{Area of } S_a} = \frac{a^2(\text{Area of } f(S_1))}{a^2(\text{Area of } S_1)} = \frac{\text{Area of } f(S_1)}{\text{Area of } S_1}.$$

This shows that the ratio r does not depend on the choice of a.

Now suppose R is any region in $\mathbb{R}^2$ with finite area. Then, for any coordinate grid, the sum of the areas of the coordinate squares that lie inside R

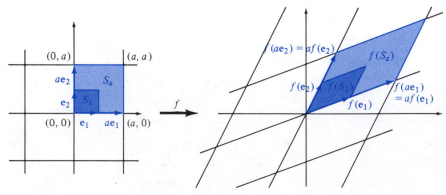

FIGURE 3.6
The area of the square S_a is a^2 times the area of the square S_1. The area of the parallelogram $f(S_a)$ is a^2 times the area of the parallelogram $f(S_1)$.

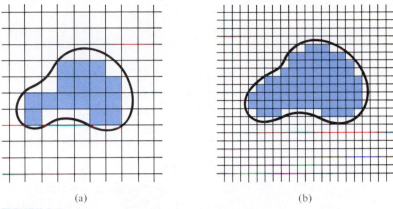

(a)   (b)

FIGURE 3.7

(a) The area of R can be approximated by the sum of the areas of those squares of a coordinate grid that lie inside R. (b) If the spacing between the coordinate lines is reduced by a factor of 2, the approximation improves.

approximates the area of R (see Figure 3.7a). If we use a grid with half as much spacing between the coordinate lines, then the sum of the areas of the new coordinate squares that lie inside R gives a better approximation to the area of R (see Figure 3.7b). If we use an infinite sequence of grids, with the spacing between coordinate lines at each stage half the spacing of the previous grid, the coordinate squares that are inside R will fill up more and more of R and the sum of their areas will approach the area of R.

If $f:\mathbb{R}^2 \to \mathbb{R}^2$ is a linear map, then f will map coordinate squares inside R to parallelograms inside $f(R)$. If we consider an infinite sequence of grids with the spacing constantly decreasing by a factor of 2, these parallelograms will fill up more and more of $f(R)$ and the sum of their areas will approach the area of $f(R)$ (see Figure 3.8). Since f maps each coordinate square S to a

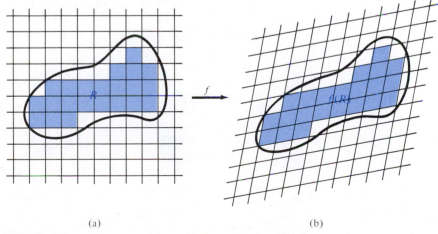

(a)   (b)

FIGURE 3.8

(a) The blue coordinate squares nearly fill R. (b) Their images nearly fill $f(R)$.

parallelogram whose area is r times the area of S, it follows that f maps R to a region whose area is r times the area of R.

This discussion leads to the following theorem.

Theorem 1 *Let $f{:}\mathbb{R}^2 \to \mathbb{R}^2$ be a linear map. Then there is a real number $r \geq 0$ such that f multiplies areas by r; that is,*

$$\text{Area of } f(R) = r(\text{Area of } R)$$

for every region R in $\mathbb{R}^2$ that has finite area.

EXAMPLE 4

From Example 1 we see that horizontal and vertical shears multiply areas by 1. From Example 2 we see that the reflection $f(x_1, x_2) = (x_2, x_1)$ also multiplies areas by 1. Example 3 shows that the linear maps $f(x_1, x_2) = (cx_1, x_2)$ and $f(x_1, x_2) = (x_1, cx_2)$ multiply areas by $|c|$. ■

Since the area of the unit square

$$S_1 = \{(x_1, x_2) | 0 \leq x_1 \leq 1, 0 \leq x_2 \leq 1\}$$

is equal to 1, we can compute the area multiplication factor r of any linear map $f{:}\mathbb{R}^2 \to \mathbb{R}^2$ by computing the area of $f(S_1)$:

$$r = \frac{\text{Area of } f(S_1)}{\text{Area of } S_1} = \text{Area of } f(S_1).$$

EXAMPLE 5

Let $f{:}\mathbb{R}^2 \to \mathbb{R}^2$ be the linear map $f(x_1, x_2) = (ax_1, bx_2)$ where $a > 0$ and $b > 0$. The unit square S_1 is spanned by $\mathbf{e}_1$ and $\mathbf{e}_2$. Its image under f is the parallelogram spanned by $f(\mathbf{e}_1) = (a, 0)$ and $f(\mathbf{e}_2) = (0, b)$. This parallelogram is a rectangle (see Figure 3.9) with area

$$\text{Area of } f(S_1) = ab.$$

Hence the linear map $f(x_1, x_2) = (ax_1, bx_2)$ $(a > 0, b > 0)$ multiplies areas by ab.

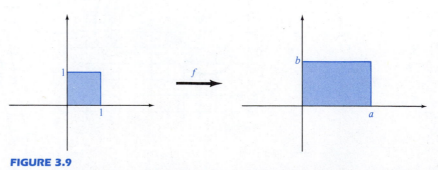

FIGURE 3.9
The linear map $f(x_1, x_2) = (ax_1, bx_2)$, $a > 0$, $b > 0$, maps the unit square S_1 to a rectangle with base a and height b. This linear map multiplies areas by ab.

You can check for yourself that if a or b (or both) is less than zero, then $f(x_1, x_2) = (ax_1, bx_2)$ multiplies areas by $|ab|$. What happens if $a = 0$ or $b = 0$? ∎

Let us arrange the results of Examples 4 and 5 in a table (see Table 3.1). You will notice that in every case f multiplies areas by the absolute value of the determinant of A. This is true in general.

Theorem 2 *Let $f:\mathbb{R}^2 \to \mathbb{R}^2$ be a linear map and let A be the matrix of f. Then f multiplies areas by the absolute value of the determinant of A.*

Proof Suppose first that f is invertible. Then A is invertible. Hence, by Theorem 3 of Section 3.5, A can be factored into a product of elementary matrices:

$$A = E_1 E_2 \cdots E_k,$$

where each E_i is an elementary matrix. Hence f is a composition of linear maps

$$f = f_1 \circ f_2 \circ \cdots \circ f_k,$$

where, for each i $(1 \leq i \leq k)$, $f_i:\mathbb{R}^2 \to \mathbb{R}^2$ is the linear map with matrix E_i. We know, from Table 3.1, that f_i multiplies areas by $|\det E_i|$. (Every 2×2 elementary matrix appears in this table.) Hence f multiplies areas by the factor

$$r = |\det E_1| \, |\det E_2| \cdots |\det E_k|$$

$$= |(\det E_1)(\det E_2) \cdots (\det E_k)|$$

$$= |\det(E_1 E_2 \cdots E_k)|$$

$$= |\det A|$$

as asserted.

TABLE 3.1

LINEAR MAP f	MATRIX A OF f	DETERMINAINT OF A	AREA MULTIPLICATION FACTOR OF f
$f(x_1, x_2) = (x_1 + cx_2, x_2)$ or $f(x_1, x_2) = (x_1, cx_1 + x_2)$	$\begin{pmatrix} 1 & c \\ 0 & 1 \end{pmatrix}$ or $\begin{pmatrix} 1 & 0 \\ c & 1 \end{pmatrix}$	1	1
$f(x_1, x_2) = (x_2, x_1)$	$\begin{pmatrix} 0 & 1 \\ 1 & 0 \end{pmatrix}$	-1	1
$f(x_1, x_2) = (cx_1, x_2)$ or $f(x_1, x_2) = (x_1, cx_2)$	$\begin{pmatrix} c & 0 \\ 0 & 1 \end{pmatrix}$ or $\begin{pmatrix} 1 & 0 \\ 0 & c \end{pmatrix}$	c	$|c|$
$f(x_1, x_2) = (ax_1, bx_2)$	$\begin{pmatrix} a & 0 \\ 0 & b \end{pmatrix}$	ab	$|ab|$

If f is not invertible, then A is not invertible, so $\det A = 0$. You can check that, in this case, the area multiplication factor of f is also zero (see Exercise 13). Hence $r = |\det A|$ in this case as well. ■

EXAMPLE 6

Let $f:\mathbb{R}^2 \to \mathbb{R}^2$ be the linear map $f(x_1, x_2) = (3x_1 - 2x_2, 5x_1 - 4x_2)$. Then the matrix of f is

$$A = \begin{pmatrix} 3 & -2 \\ 5 & -4 \end{pmatrix}.$$

The determinant of A is -2. Hence f multiplies areas by the factor

$$r = |\det A| = |-2| = 2. \quad ■$$

EXAMPLE 7

Let $f(x_1, x_2) = (ax_1, bx_2)$ where $a > 0$ and $b > 0$. Then the matrix of f is

$$A = \begin{pmatrix} a & 0 \\ 0 & b \end{pmatrix}.$$

Therefore f multiplies areas by the factor

$$r = |\det A| = ab,$$

in agreement with Example 5. It follows that, if R is the region bounded by the unit circle $x_1^2 + x_2^2 = 1$ then

$$\text{Area of } f(R) = ab(\text{Area of } R) = \pi ab.$$

The region $f(R)$ is bounded by an ellipse. To see this, let $(x_1, x_2) \in R$. Then $x_1^2 + x_2^2 \leq 1$. If we let $(y_1, y_2) = f(x_1, x_2) = (ax_1, bx_2)$, then $y_1 = ax_1$ and $y_2 = bx_2$, hence

$$x_1^2 + x_2^2 = \left(\frac{y_1}{a}\right)^2 + \left(\frac{y_2}{b}\right)^2 = \frac{y_1^2}{a^2} + \frac{y_2^2}{b^2}.$$

Therefore

$$f(R) = \{(y_1, y_2) | \frac{y_1^2}{a^2} + \frac{y_2^2}{b^2} \leq 1\} = \{(x_1, x_2) | \frac{x_1^2}{a^2} + \frac{x_2^2}{b^2} \leq 1\}.$$

$f(R)$ is bounded by the ellipse $x_1^2/a^2 + x_2^2/b^2 = 1$ (see Figure 3.10).

This proves, using only the fact that the area of the unit circle is π, that the area enclosed by the ellipse $x_1^2/a^2 + x_2^2/b^2 = 1$ ($a > 0$, $b > 0$) is equal to πab. ■

For linear maps from $\mathbb{R}^3$ to $\mathbb{R}^3$, we can use arguments nearly identical to those used above to establish the following theorem.

Theorem 3   *Let $f:\mathbb{R}^3 \to \mathbb{R}^3$ be any linear map and let A be the matrix of f. Then f multiplies volumes by $|\det A|$; that is, if R is any region in $\mathbb{R}^3$ that has finite volume, then*

$$\text{Volume of } f(R) = |\det A| \ (\text{Volume of } R).$$

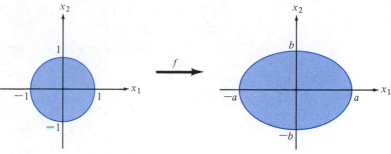

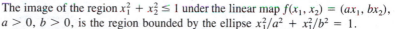

FIGURE 3.10
The image of the region $x_1^2 + x_2^2 \leq 1$ under the linear map $f(x_1, x_2) = (ax_1, bx_2)$, $a > 0$, $b > 0$, is the region bounded by the ellipse $x_1^2/a^2 + x_1^2/b^2 = 1$.

Proof Since the proof is almost the same as for linear maps from $\mathbb{R}^2$ to $\mathbb{R}^2$, we will merely review the arguments.

It is easy to check that any linear map from $\mathbb{R}^3$ to $\mathbb{R}^3$ whose matrix is an elementary matrix E multiplies volumes by $|\det E|$ (see, for example, Figure 3.11). It follows that if $f(\mathbf{x}) = A\mathbf{x}$ where A is a product of elementary matrices, then f multiplies volumes by $|\det A|$. Since every invertible matrix is a product of elementary matrices, we see that f multiplies volumes by $|\det A|$, where A is the matrix of f, whenever f is invertible. This establishes the theorem in the case where f is invertible.

If f is not invertible then you can check that the volume of $f(R)$ is equal to zero for every region R in $\mathbb{R}^3$ that has finite volume (see Exercise 14). Since $\det A = 0$ whenever A is not invertible, this completes the proof. ■

EXAMPLE 8
Let $f:\mathbb{R}^3 \to \mathbb{R}^3$ be the linear map $f(x_1, x_2, x_3) = (ax_1, bx_2, cx_3)$, where $a > 0$, $b > 0$, and $c > 0$. Then the matrix of f is

$$A = \begin{pmatrix} a & 0 & 0 \\ 0 & b & 0 \\ 0 & 0 & c \end{pmatrix}.$$

Therefore, f multiplies volumes by the factor

$$r = |\det A| = abc.$$

Thus, if R is any region in $\mathbb{R}^3$ with finite volume, then

$$\text{Volume of } f(R) = abc(\text{Volume of } R).$$

If we take R to be the region bounded by the unit sphere

$$x_1^2 + x_2^2 + x_3^2 = 1,$$

then $f(R)$ is bounded by the ellipsoid

$$\frac{x_1^2}{a^2} + \frac{x_2^2}{b^2} + \frac{x_3^2}{c^2} = 1$$

(see Figure 3.12). It follows that the volume V of this ellipsoid is given by the formula

$$V = abc(\text{Volume of the sphere of radius } 1)$$
$$= abc(\tfrac{4}{3}\pi)$$

or

$$V = \tfrac{4}{3}\pi abc. \quad \blacksquare$$

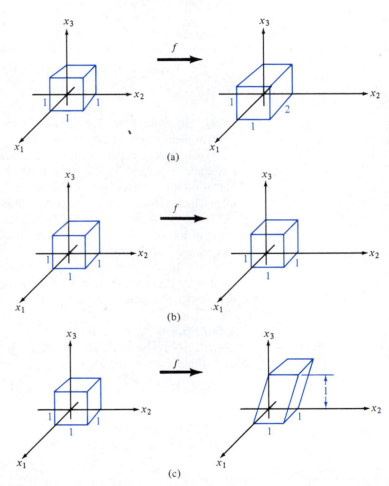

(a)

(b)

(c)

FIGURE 3.11

If $f:\mathbb{R}^3 \to \mathbb{R}^3$ is the linear map whose matrix is E, then f multiplies volumes by $|\det E|$.

(a) $E = \begin{pmatrix} 2 & 0 & 0 \\ 0 & 1 & 0 \\ 0 & 0 & 1 \end{pmatrix}$ (b) $E = \begin{pmatrix} 0 & 1 & 0 \\ 1 & 0 & 0 \\ 0 & 0 & 1 \end{pmatrix}$

(c) $E = \begin{pmatrix} 1 & 0 & -1 \\ 0 & 1 & 0 \\ 0 & 0 & 1 \end{pmatrix}$

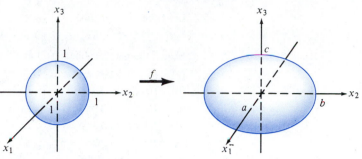

FIGURE 3.12

The image of the sphere $x_1^2 + x_2^2 + x_3^2 = 1$ under the linear map $f(x_1, x_2, x_3) = (ax_1, bx_2, cx_3)$, $a > 0$, $b > 0$, $c > 0$, is the ellipsoid $x_1^2/a^2 + x_2^2/b^2 + x_3^2/c^2 = 1$.

EXERCISES

1. For each of the following linear maps f, sketch the coordinate grid with coordinate lines separated by unit distance and sketch the image of this grid under f. Shade in the unit square $\{(x_1, x_2)|0 \le x_1 \le 1, 0 \le x_2 \le 1\}$ and its image under f. Finally, determine the area multiplication factor of f.

   (a) $f(x_1, x_2) = (x_1 - x_2, x_2)$ (b) $f(x_1, x_2) = (x_1, 3x_1 + x_2)$
   (c) $f(x_1, x_2) = (3x_1, x_2)$ (d) $f(x_1, x_2) = (x_1, -x_2)$
   (e) $f(x_1, x_2) = (3x_1, -x_2)$

2. Repeat Exercise 1 for each of the following linear maps.

   (a) $f(x_1, x_2) = (2x_1, 3x_2)$
   (b) $f(x_1, x_2) = (-2x_1, 3x_2)$
   (c) $f(x_1, x_2) = (x_1 + 2x_2, 2x_1 + x_2)$
   (d) $f(x_1, x_2) = (x_2, -x_1 + x_2)$
   (e) $f(x_1, x_2) = (x_1 + 2x_2, -x_1 - 2x_2)$

3. Repeat Exercise 1 for the rotation

   $$f(x_1, x_2) = (x_1 \cos \theta - x_2 \sin \theta, x_1 \sin \theta + x_2 \cos \theta).$$

4. Repeat Exercise 1 for the reflection

   $$f(x_1, x_2) = (x_1 \cos 2\theta + x_2 \sin 2\theta, x_1 \sin 2\theta - x_2 \cos 2\theta).$$

5. Let S be the unit square

   $$S = \{(x_1, x_2)|0 \le x_1 \le 1, 0 \le x_2 \le 1\}.$$

   Sketch the parallelogram $f(S)$ and find its area when

   (a) $f(x_1, x_2) = (5x_1, 3x_2)$ (b) $f(x_1, x_2) = (-5x_1, 3x_2)$
   (c) $f(x_1, x_2) = (5x_1, -3x_2)$ (d) $f(x_1, x_2) = (-5x_1, -3x_2)$

6. Repeat Exercise 5 for

   (a) $f(x_1, x_2) = (-x_2, x_1)$
   (b) $f(x_1, x_2) = (2x_1 - 4x_2, -3x_1 + 6x_2)$
   (c) $f(x_1, x_2) = (2x_1 - 4x_2, 3x_1 + 6x_2)$
   (d) $f(x_1, x_2) = (x_1 + 3x_2, 3x_1 + x_2)$
   (e) $f(x_1, x_2) = (x_1, 0)$

7. Let C be the unit cube
$$C = \{(x_1, x_2, x_3) | 0 \le x_1 \le 1, 0 \le x_2 \le 1, 0 \le x_3 \le 1\}.$$
Sketch the parallelepiped $f(C)$ and find its volume when
(a) $f(x_1, x_2, x_3) = (3x_1, x_2, x_3)$
(b) $f(x_1, x_2, x_3) = (x_1, -3x_2, x_3)$
(c) $f(x_1, x_2, x_3) = (3x_1, 3x_2, x_3)$
(d) $f(x_1, x_2, x_3) = (3x_1, 3x_2, 3x_3)$
(e) $f(x_1, x_2, x_3) = (-3x_1, -3x_2, -3x_3)$

8. Repeat Exercise 7 for
(a) $f(x_1, x_2, x_3) = (x_1 + 2x_2, x_2, x_3)$
(b) $f(x_1, x_2, x_3) = (x_1, 2x_1 + x_2, x_3)$
(c) $f(x_1\ x_2, x_3) = (x_1 + 2x_3, x_2, x_3)$
(d) $f(x_1, x_2, x_3) = (2x_1, 3x_2, 4x_3)$
(e) $f(x_1, x_2, x_3) = (2x_1, 3x_2, -4x_3)$

9. Let $\mathbf{u}$ and $\mathbf{v}$ be vectors in $\mathbb{R}^2$. Use Theorem 2 to show that the area of the parallelogram spanned by $\mathbf{u}$ and $\mathbf{v}$ is equal to the absolute value of the determinant of A, where A is the matrix whose column vectors are $\mathbf{u}$ and $\mathbf{v}$. [Since det $A =$ det A', this area is also equal to the absolute value of the determinant $\det \begin{pmatrix} \mathbf{u} \\ \mathbf{v} \end{pmatrix}$.]

10. Let $\mathbf{u} = (u_1, u_2)$, $\mathbf{v} = (v_1, v_2)$, and $\mathbf{w} = (w_1, w_2)$ be points in $\mathbb{R}^2$. Show that the area of the triangle with vertices $\mathbf{u}$, $\mathbf{v}$, and $\mathbf{w}$ is equal to
$$\tfrac{1}{2}\left|\det\begin{pmatrix} 1 & u_1 & u_2 \\ 1 & v_1 & v_2 \\ 1 & w_1 & w_2 \end{pmatrix}\right|.$$

[*Hint:* This area is the same as the area of the triangle spanned by $\mathbf{v} - \mathbf{u}$ and $\mathbf{w} - \mathbf{u}$. Use row operations and the parenthetical remark in Exercise 9.]

11. Let $\mathbf{u}$, $\mathbf{v}$, and $\mathbf{w}$ be vectors in $\mathbb{R}^3$. Use Theorem 3 to show that the volume of the parallelepiped spanned by $\mathbf{u}$, $\mathbf{v}$, and $\mathbf{w}$ is equal to the absolute value of the determinant of A, where A is the matrix whose column vectors are $\mathbf{u}$, $\mathbf{v}$, and $\mathbf{w}$. [Since det $A =$ det A', this volume is also equal to the absolute value of the determinant
$$\det\begin{pmatrix} \mathbf{u} \\ \mathbf{v} \\ \mathbf{w} \end{pmatrix}.]$$

12. Let $\mathbf{u} = (u_1, u_2, u_3)$, $\mathbf{v} = (v_1, v_2, v_3)$, $\mathbf{w} = (w_1, w_2, w_3)$, and $\mathbf{x} = (x_1, x_2, x_3)$ be points in $\mathbb{R}^3$. Show that the volume of the tetrahedron with vertices $\mathbf{u}$, $\mathbf{v}$, $\mathbf{w}$, and $\mathbf{x}$ is equal to
$$\tfrac{1}{6}\left|\det\begin{pmatrix} 1 & u_1 & u_2 & u_3 \\ 1 & v_1 & v_2 & v_3 \\ 1 & w_1 & w_2 & w_3 \\ 1 & x_1 & x_2 & x_3 \end{pmatrix}\right|.$$

[See Exercise 11 and the hint to Exercise 10.]

13. (a) Show that if B is a 2×2 echelon matrix that is not the identity matrix and $g:\mathbb{R}^2 \to \mathbb{R}^2$ is the linear map associated with B, then
$$g(x_1, x_2) = (ax_1 + bx_2, 0)$$
for some real numbers a and b.

(b) Without using Theorem 2, show that g multiplies areas by zero.

(c) Conclude that if $f:\mathbb{R}^2 \to \mathbb{R}^2$ is any linear map that is not invertible, then f multiplies areas by zero. [*Hint:* Factor the matrix A of f into a product $E_1 \cdots E_k B$, where $E_1, \ldots, E_k$ are elementary matrices and B is an echelon matrix. Then $f = f_1 \circ \cdots \circ f_k \circ g$ where f_i $(1 \le i \le k)$ is the linear map with matrix E_i and g is the linear map with matrix B.]

14. Without using Theorem 3, show that if $f:\mathbb{R}^3 \to \mathbb{R}^3$ is a linear map that is not invertible, then f multiplies volumes by zero. [*Hint:* Imitate the procedure outlined in Exercise 13. Start by showing that each 3×3 echelon matrix that is not the identity matrix is the matrix of a linear map g of the form

$$g(x_1, x_2, x_3) = (a_{11}x_1 + a_{12}x_2 + a_{13}x_3, a_{22}x_2 + a_{23}x_3, 0).]$$

REVIEW EXERCISES

1. Evaluate

(a) $\det \begin{pmatrix} 2 & 3 & -1 & 4 \\ 6 & -1 & 2 & 0 \\ 3 & 5 & 7 & -2 \\ 1 & 1 & 3 & 4 \end{pmatrix}$ (b) $\det \begin{pmatrix} 2 & 7 & 1 & 8 \\ 2 & 8 & 1 & 8 \\ 2 & 8 & 4 & 5 \\ 9 & 0 & 4 & 5 \end{pmatrix}$

2. Evaluate

(a) $\det \begin{pmatrix} 1 & 2 & 1 \\ x & 3 & 3 \\ x^2 & -1 & 2 \end{pmatrix}$ (b) $\det \begin{pmatrix} x-1 & 1 & 3 \\ 2 & x-3 & 1 \\ -2 & 4 & x+1 \end{pmatrix}$

3. True or false? In each part, A is an $n \times n$ matrix.

(a) If A has one row that is a scalar multiple of another, then $\det A = 0$.

(b) If A has two columns that are equal, then $\det A = 0$.

(c) If $\det A = 0$, then the echelon matrix of A has a row of zeros.

(d) If $A^3 = 0$ then $\det A = 0$.

(e) If $A^3 = I$ then $\det A = 1$.

(f) If $A^3 = A$ then either $\det A = 0$ or $\det A = 1$.

(g) If A is upper triangular and has determinant equal to zero, then at least one of the diagonal entries in A must be zero.

(h) If B is obtained from A by adding a scalar multiple of one column to another then $\det B = \det A$.

(i) $\det 2A = 2 \det A$.

4. Let A be a 3×3 matrix with $\det A = 2$. Find

(a) $\det A^5$ (b) $\det 5A$ (c) $\det (-A)$

(d) $\det A^{-1}$ (e) $\det A^t$

5. Find the sign of each of the following permutations.

(a) $(3, 1, 2)$ (b) $(4, 1, 3, 2)$

(c) $(2, 1, 4, 3, 6, 5)$ (d) $(1, 2, 3, 4, 5)$

6. Find the cofactor matrix of A and use it to find A^{-1}, where

$$A = \begin{pmatrix} 2 & -1 & 5 \\ 0 & 3 & -2 \\ 4 & 3 & 1 \end{pmatrix}.$$

7. Use Cramer's rule to solve

(a) $3x_1 + 5x_2 = 9$ (b) $7x_1 - 6x_2 = 1$
     $5x_1 + 8x_2 = 7$ $3x_1 + x_2 = 0$

(c) $2x_1 + 3x_2 + 4x_3 = 5$
     $3x_1 + 2x_2 + x_3 = -5$
   $-x_1 + 2x_2 + 2x_3 = 0$

8. Evaluate the following determinants by inspection.

(a) $\det \begin{pmatrix} 1 & -3 & 1 & 7 \\ 0 & 2 & 5 & 6 \\ 0 & 0 & 4 & -1 \\ 0 & 0 & 0 & 2 \end{pmatrix}$ (b) $\det \begin{pmatrix} 2 & 3 & 6 & 4 \\ -1 & 1 & 2 & 3 \\ 0 & 4 & 3 & -2 \\ -1 & 1 & 2 & 3 \end{pmatrix}$

(c) $\det \begin{pmatrix} 0 & 1 & 0 & 0 \\ 0 & 0 & 1 & 0 \\ 0 & 0 & 0 & 1 \\ a & b & c & d \end{pmatrix}$ (d) $\det \begin{pmatrix} 0 & 0 & 0 & d \\ 0 & 0 & c & 0 \\ 0 & b & 0 & 0 \\ a & 0 & 0 & 0 \end{pmatrix}$

9. For each of the following linear maps f, sketch the coordinate grid with coordinate lines separated by unit distance, and sketch the images of this grid under f. Shade in the unit square $\{(x_1, x_2) | 0 \le x_1 \le 1, 0 \le x_2 \le 1\}$ and its image under f. Finally, find the area multiplication factor of f.

(a) $f(x_1, x_2) = (x_1, \frac{1}{3}x_1 + x_2)$

(b) $f(x_1, x_2) = (\frac{3}{5}x_1 + \frac{4}{5}x_2, -\frac{4}{5}x_1 + \frac{3}{5}x_2)$

(c) $f(x_1, x_2) = (x_1 + 2x_2, 2x_1 + 2x_2)$

10. Compute the volume multiplication factor of the linear map

(a) $f(x_1, x_2, x_3) = (x_1, x_2, x_1 + x_2 + x_3)$

(b) $f(x_1, x_2, x_3) = (2x_3, 3x_2, 5x_1)$

(c) $f(x_1, x_2, x_3) = (2x_1 + x_2 - x_3, x_1 - 2x_2 + 2x_3, -x_1 - x_2 + 3x_3)$

11. Let σ be a permutation of $(1, 2, \ldots, n)$, let M_σ be the matrix of σ, and let A be any matrix with n rows. Describe the matrix $M_\sigma A$.

4 Subspaces of $\mathbb{R}^n$

In this chapter we shall study subspaces of $\mathbb{R}^n$. These are subsets of $\mathbb{R}^n$ that are closed under vector addition and scalar multiplication. Subspaces arise frequently in linear algebra, as solution sets of homogeneous linear systems, for example. We shall see how to associate with each of these spaces a number, called the *dimension* of the subspace. Using subspaces, we will be able to obtain a better understanding of linear maps. We will also gain insight into an important statistical method, the method of least squares, for fitting lines and curves to experimental data.

4.1 Subspaces

In this section we shall study special subsets of $\mathbb{R}^n$ called *subspaces*. Let us begin by looking at an example of a subspace of $\mathbb{R}^3$.

Let P be any plane passing through $\mathbf{0}$ in $\mathbb{R}^3$. If $\mathbf{v}$ and $\mathbf{w}$ are any two vectors in P, then the addition parallelogram for $\mathbf{v}$ and $\mathbf{w}$ lies in the plane P and so, in particular, the sum $\mathbf{v} + \mathbf{w}$ of $\mathbf{v}$ and $\mathbf{w}$ is in P (see Figure 4.1a). We say, therefore, that P is *closed under vector addition*.

Similarly, if $\mathbf{v}$ is any vector in P and c is any real number, then the vector $c\mathbf{v}$ is also in P (see Figure 4.1b). We say, therefore, that P is *closed under scalar multiplication*.

A subset S of $\mathbb{R}^n$ is said to be **closed under vector addition** if $\mathbf{v} + \mathbf{w} \in S$ whenever $\mathbf{v} \in S$ and $\mathbf{w} \in S$. The subset S is **closed under scalar multiplication** if $c\mathbf{v} \in S$ whenever $c \in \mathbb{R}$ and $\mathbf{v} \in S$. A subset S of $\mathbb{R}^n$ is called a **subspace** of $\mathbb{R}^n$ if

(i) S contains the vector $\mathbf{0}$,

(ii) S is closed under vector addition,

and

(iii) S is closed under scalar multiplication.

EXAMPLE 1

Let P be any plane through $\mathbf{0}$ in $\mathbb{R}^3$. We showed above, using a geometric argument, that P is a subspace of $\mathbb{R}^3$. We can also verify that P is a subspace of $\mathbb{R}^3$ using an algebraic argument.

Since P passes through $\mathbf{0}$, P is described by an equation of the form

$$\mathbf{b} \cdot \mathbf{x} = 0$$

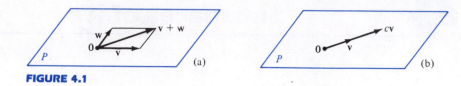

FIGURE 4.1

A plane P passing through $\mathbf{0}$ in $\mathbb{R}^3$. (a) P is closed under vector addition. (b) P is closed under scalar multiplication.

where $\mathbf{b}$ is a nonzero vector perpendicular to P. Certainly,

(i) $\mathbf{0} \in P$, because $\mathbf{b} \cdot \mathbf{0} = 0$. Also,

(ii) if $\mathbf{v} \in P$ and $\mathbf{w} \in P$, then $\mathbf{b} \cdot \mathbf{v} = 0$ and $\mathbf{b} \cdot \mathbf{w} = 0$, hence

$$\mathbf{b} \cdot (\mathbf{v} + \mathbf{w}) = \mathbf{b} \cdot \mathbf{v} + \mathbf{b} \cdot \mathbf{w} = 0 + 0 = 0,$$

and so $\mathbf{v} + \mathbf{w} \in P$. Thus P is closed under vector addition. Finally,

(iii) if $\mathbf{v} \in P$ and $c \in \mathbb{R}$, then $\mathbf{b} \cdot \mathbf{v} = 0$, hence

$$\mathbf{b} \cdot (c\mathbf{v}) = c(\mathbf{b} \cdot \mathbf{v}) = c0 = 0,$$

so $c\mathbf{v} \in P$. Thus P is closed scalar multiplication.

We have verified that P has the three properties required of a subspace. We can conclude, then, that P is a subspace of $\mathbb{R}^3$. ■

REMARK   The same argument shows that each hyperplane through $\mathbf{0}$ in $\mathbb{R}^n$ is a subspace of $\mathbb{R}^n$.

EXAMPLE 2

Let ℓ be any line in $\mathbb{R}^n$ that passes through $\mathbf{0}$ (see Figure 4.2). Take $\mathbf{d}$ to be any nonzero vector along ℓ. Then we know that $\mathbf{x} \in \ell$ if and only if $\mathbf{x} = t\mathbf{d}$ for some $t \in \mathbb{R}$. We shall show that ℓ is a subspace of $\mathbb{R}^n$. To do this, we must verify conditions (i), (ii), and (iii).

(i) $\mathbf{0} \in \ell$. Simply take $t = 0$.

(ii) ℓ is closed under vector addition: if $\mathbf{v} = t_1\mathbf{b}$ and $\mathbf{w} = t_2\mathbf{b}$ are any two points on ℓ, then

$$\mathbf{v} + \mathbf{w} = t_1\mathbf{b} + t_2\mathbf{b} = (t_1 + t_2)\mathbf{b},$$

so $\mathbf{v} + \mathbf{w} \in \ell$.

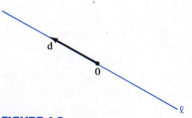

FIGURE 4.2

Each line ℓ in $\mathbb{R}^n$ that passes through $\mathbf{0}$ is a subspace of $\mathbb{R}^n$.

(iii) ℓ is closed under scalar multiplication: if $\mathbf{v} = t\mathbf{b}$ is any point on ℓ and c is any real number, then

$$c\mathbf{v} = c(t\mathbf{b}) = (ct)\mathbf{b},$$

so $c\mathbf{v} \in \ell$.

Thus ℓ is a subspace of $\mathbb{R}^n$. ■

REMARK Notice that, if ℓ is a line in $\mathbb{R}^n$ that does *not* pass through $\mathbf{0}$, then ℓ is *not* a subspace of $\mathbb{R}^n$, because condition (i) is not satisfied. Similarly, if P is a plane in $\mathbb{R}^3$ that does not pass through $\mathbf{0}$, then P is not a subspace of $\mathbb{R}^3$.

EXAMPLE 3

Let S be the subset of $\mathbb{R}^3$ consisting of all vectors of the form $(a, b, a + b)$. Let us determine if S is a subspace of $\mathbb{R}^3$.

First, we check condition (i). Is $\mathbf{0} = (0, 0, 0) \in S$? In other words, is $\mathbf{0}$ of the form $(a, b, a + b)$ for some choice of a and b? Yes. If we take $a = 0$ and $b = 0$, then

$$(a, b, a + b) = (0, 0, 0).$$

Next, we check condition (ii). Is $\mathbf{v} + \mathbf{w} \in S$ whenever $\mathbf{v} \in S$ and $\mathbf{w} \in S$? If $\mathbf{v} \in S$, then $\mathbf{v} = (a_1, b_1, a_1 + b_1)$ for some a_1 and b_1. If $\mathbf{w} \in S$, then $\mathbf{w} = (a_2, b_2, a_2 + b_2)$ for some a_2 and b_2. Hence

$$\mathbf{v} + \mathbf{w} = (a_1, b_1, a_1 + b_1) + (a_2, b_2, a_2 + b_2)$$

$$= (a_1 + a_2, b_1 + b_2, a_1 + b_1 + a_2 + b_2).$$

If we set $a = a_1 + a_2$ and $b = b_1 + b_2$, we see that

$$\mathbf{v} + \mathbf{w} = (a, b, a + b),$$

so $\mathbf{v} + \mathbf{w} \in S$. Thus condition (ii) is satisfied. S is closed under addition.

Finally, we check condition (iii). Is $c\mathbf{v} \in S$ whenever $\mathbf{v} \in S$? If $\mathbf{v} \in S$ then $\mathbf{v} = (a_1, b_1, a_1 + b_1)$ for some a_1 and b_1. Hence

$$c\mathbf{v} = c(a_1, b_1, a_1 + b_1) = (ca_1, cb_1, ca_1 + cb_1).$$

If we set $a = ca_1$ and $b = cb_1$ we see that

$$c\mathbf{v} = (a, b, a + b),$$

so $c\mathbf{v} \in S$. Thus condition (iii) is satisfied. S is closed under scalar multiplication.

Since conditions (i), (ii), and (iii) are satisfied, S is a subspace of $\mathbb{R}^3$.

This subspace S is, in fact, a plane in $\mathbb{R}^3$ passing through $\mathbf{0}$. Can you find an equation for this plane? ■

EXAMPLE 4

Let S be the solution set in $\mathbb{R}^2$ of the equation $x_1 x_2 = 0$. Let us determine whether or not S is a subspace of $\mathbb{R}^2$.

First, we check condition (i). Is $\mathbf{0} = (0, 0) \in S$? That is, does the vector $(0, 0)$ satisfy the equation $x_1 x_2 = 0$? Certainly. Condition (i) is satisfied.

Next, we check condition (ii). Is $\mathbf{v} + \mathbf{w} \in S$ whenever $\mathbf{v} \in S$ and $\mathbf{w} \in S$? If $\mathbf{v} = (v_1, v_2) \in S$, then $v_1 v_2 = 0$. If $\mathbf{w} = (w_1, w_2) \in S$, then $w_1 w_2 = 0$. How about $\mathbf{v} + \mathbf{w} = (v_1 + w_1, v_2 + w_2)$? Since

$$(v_1 + w_1)(v_2 + w_2) = v_1 v_2 + v_1 w_2 + w_1 v_2 + w_1 w_2$$

$$= v_1 w_2 + w_1 v_2$$

we see that $(v_1 + w_1)(v_2 + w_2)$ is not necessarily zero. For example, if $v_1 = w_2 = 1$ and $w_1 = v_2 = 0$ then $v_1 v_2 = 0$ and $w_1 w_2 = 0$ but

$$(v_1 + w_1)(v_2 + w_2) = 1 \neq 0.$$

We conclude, then, that condition (ii) is *not* satisfied: $(1, 0) \in S$ and $(0, 1) \in S$, but

$$(1, 0) + (0, 1) = (1, 1) \notin S.$$

Since condition (ii) is not satisfied, S is *not* a subspace of $\mathbb{R}^2$.

This set S is a union of two lines. Can you identify the two lines? ■

EXAMPLE 5

Let us find *all* of the subspaces of $\mathbb{R}^2$.

Suppose S is a subspace of $\mathbb{R}^2$. Then $\mathbf{0} \in S$. Possibly $S = \{\mathbf{0}\}$, since $\{\mathbf{0}\}$ is a subspace of $\mathbb{R}^2$. (Why?)

If S contains a nonzero vector $\mathbf{v}$, then S must contain $c\mathbf{v}$ for each $c \in \mathbb{R}$; that is, S must contain the line $\mathbf{x} = t\mathbf{v}$ through $\mathbf{0}$ in the direction of $\mathbf{v}$. Since this line is a subspace, S is possibly this line.

Suppose S contains a nonzero vector $\mathbf{v}$ and a nonzero vector $\mathbf{w}$ that is not on the line $\mathbf{x} = t\mathbf{v}$. Then, by condition (ii), S must contain $c_1 \mathbf{v}$ for every $c_1 \in \mathbb{R}$ and S must contain $c_2 \mathbf{w}$ for every $c_2 \in \mathbb{R}$. Hence, by condition (iii), S must also contain $c_1 \mathbf{v} + c_2 \mathbf{w}$, for every $c_1, c_2 \in \mathbb{R}$. But *every* vector in $\mathbb{R}^2$ is of the form $c_1 \mathbf{v} + c_2 \mathbf{w}$ for some $c_1, c_2 \in \mathbb{R}$. This can be seen geometrically as follows.

Let $\mathbf{a}$ be any vector in $\mathbb{R}^2$. Consider the lines $\mathbf{x} = \mathbf{a} + t\mathbf{v}$ and $\mathbf{x} = t\mathbf{w}$ (see Figure 4.3). Since these lines are not parallel ($\mathbf{w}$ is not a multiple of $\mathbf{v}$), they must intersect at some point $\mathbf{p}$. Since $\mathbf{p}$ is on the line $\mathbf{x} = \mathbf{a} + t\mathbf{v}$, we see that

$$\mathbf{p} = \mathbf{a} + t_1 \mathbf{v}$$

for some $t_1 \in \mathbb{R}$. Since $\mathbf{p}$ is on the line $\mathbf{x} = t\mathbf{w}$, we see that

$$\mathbf{p} = t_2 \mathbf{w}$$

for some $t_2 \in \mathbb{R}$. Therefore

$$\mathbf{a} + t_1 \mathbf{v} = \mathbf{p} = t_2 \mathbf{w}$$

so

$$\mathbf{a} = (-t_1)\mathbf{v} + t_2 \mathbf{w}.$$

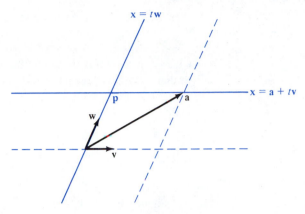

FIGURE 4.3
The vector $\mathbf{a}$ is of the form $\mathbf{a} = c_1\mathbf{v} + c_2\mathbf{w}$ for some c_1 and c_2.

If we set $c_1 = -t_1$ and $c_2 = t_2$, we get that

$$\mathbf{a} = c_1\mathbf{v} + c_2\mathbf{w},$$

as asserted.

Thus, if the subspace S of $\mathbb{R}^2$ contains two vectors neither of which is a scalar multiple of the other, then S must contain every vector in $\mathbb{R}^2$; that is, $S = \mathbb{R}^2$.

We conclude, then, that *the only subspaces of $\mathbb{R}^2$ are $\{\mathbf{0}\}$, lines through $\mathbf{0}$, and $\mathbb{R}^2$ itself.* ■

REMARK In a similar way it can be shown that *the only subspaces of $\mathbb{R}^3$ are $\{\mathbf{0}\}$, lines through $\mathbf{0}$, planes through $\mathbf{0}$, and $\mathbb{R}^3$ itself.*

EXAMPLE 6
Let S be the solution set of the linear system

$$a_{11}x_1 + a_{12}x_2 + \cdots + a_{1n}x_n = b_1$$

$$a_{21}x_1 + a_{22}x_2 + \cdots + a_{2n}x_n = b_2$$

$$\cdots$$

$$a_{m1}x_1 + a_{m2}x_2 + \cdots + a_{mn}x_n = b_m.$$

Then S is a subset of $\mathbb{R}^n$. Let us determine if S is a subspace of $\mathbb{R}^n$.
If we let

$$A = \begin{pmatrix} a_{11} & a_{12} & \cdots & a_{1n} \\ a_{21} & a_{22} & \cdots & a_{2n} \\ & & \cdots & \\ a_{m1} & a_{m2} & \cdots & a_{mn} \end{pmatrix}, \quad \mathbf{x} = \begin{pmatrix} x_1 \\ x_2 \\ \vdots \\ x_n \end{pmatrix}, \quad \text{and} \quad \mathbf{b} = \begin{pmatrix} b_1 \\ b_2 \\ \vdots \\ b_m \end{pmatrix},$$

then we can rewrite this system as the single matrix equation

$$A\mathbf{x} = \mathbf{b}.$$

Clearly, if $\mathbf{x} = \mathbf{0}$ is to satisfy this equation, then the vector $\mathbf{b}$ must be equal to $\mathbf{0}$; that is, the system must be homogeneous. On the other hand, if the system is homogeneous, then S is the set of all $\mathbf{x}$ in $\mathbb{R}^n$ such that $A\mathbf{x} = \mathbf{0}$ and we see that

(i) $\mathbf{0} \in S$ since $A\mathbf{0} = \mathbf{0}$

(ii) If $\mathbf{v} \in S$ and $\mathbf{w} \in S$, then $A\mathbf{v} = \mathbf{0}$ and $A\mathbf{w} = \mathbf{0}$, so

$$A(\mathbf{v} + \mathbf{w}) = A\mathbf{v} + A\mathbf{w} = \mathbf{0} + \mathbf{0} = \mathbf{0};$$

hence $\mathbf{v} + \mathbf{w} \in S$.

(iii) If $\mathbf{v} \in S$ and $c \in \mathbb{R}$, then $A\mathbf{v} = \mathbf{0}$, so

$$A(c\mathbf{v}) = cA\mathbf{v} = c\mathbf{0} = \mathbf{0};$$

hence $c\mathbf{v} \in S$.

We conclude, then, that *the solution set of a linear system of m equations in n unknowns is a subspace of* $\mathbb{R}^n$ *if and only if it is homogeneous.* ■

EXAMPLE 7

Let $f:\mathbb{R}^n \rightarrow \mathbb{R}^m$ be a linear map and let S be the kernel of f. Thus S is the subset of $\mathbb{R}^n$ consisting of all vectors that are mapped to $\mathbf{0}$ by f:

$$S = \{x \in \mathbb{R}^n | f(\mathbf{x}) = \mathbf{0}\}.$$

Then S is a subspace of $\mathbb{R}^n$ because

(i) $f(\mathbf{0}) = \mathbf{0}$ so $\mathbf{0} \in S$.

(ii) If $\mathbf{v} \in S$ and $\mathbf{w} \in S$, then $f(\mathbf{v}) = \mathbf{0}$ and $f(\mathbf{w}) = \mathbf{0}$ so

$$f(\mathbf{v} + \mathbf{w}) = f(\mathbf{v}) + f(\mathbf{w}) = \mathbf{0} + \mathbf{0} = \mathbf{0};$$

hence $\mathbf{v} + \mathbf{w} \in S$.

(iii) If $\mathbf{v} \in S$ and $c \in \mathbb{R}$, then $f(\mathbf{v}) = \mathbf{0}$ so

$$f(c\mathbf{v}) = cf(\mathbf{v}) = c\mathbf{0} = \mathbf{0};$$

hence $c\mathbf{v} \in S$. ■

EXAMPLE 8

Let $f:\mathbb{R}^n \rightarrow \mathbb{R}^m$ be a linear map and let S be the image of $\mathbb{R}^n$ under f. Thus

$$S = \{\mathbf{y} \in \mathbb{R}^m | \mathbf{y} = f(\mathbf{x}) \text{ for some } \mathbf{x} \in \mathbb{R}^n\}.$$

Then S is a subspace of $\mathbb{R}^m$ because:

(i) $\mathbf{0} = f(\mathbf{0})$ so $\mathbf{0} \in S$.

(ii) If $\mathbf{v} \in S$ and $\mathbf{w} \in S$, then $\mathbf{v} = f(\mathbf{a})$ and $\mathbf{w} = f(\mathbf{b})$ for some $\mathbf{a}$ and $\mathbf{b}$ in $\mathbb{R}^n$, so

$$\mathbf{v} + \mathbf{w} = f(\mathbf{a}) + f(\mathbf{b}) = f(\mathbf{a} + \mathbf{b});$$

hence $\mathbf{v} + \mathbf{w} \in S$.

(iii)   If $\mathbf{v} \in S$ and $c \in \mathbb{R}$, then $\mathbf{v} = f(\mathbf{a})$ for some $\mathbf{a} \in \mathbb{R}^n$ so

$$c\mathbf{v} = cf(\mathbf{a}) = f(c\mathbf{a});$$

hence $c\mathbf{v} \in S$.

This subspace,

$$S = \{\mathbf{y} \in \mathbb{R}^m | \mathbf{y} = f(\mathbf{x}) \text{ for some } \mathbf{x} \in \mathbb{R}^n\},$$

is usually called simply *the image of* f, rather than "the image of $\mathbb{R}^n$ under f." ■

EXAMPLE 9

Let $f:\mathbb{R}^2 \to \mathbb{R}^2$ be the linear map defined by

$$f(x_1, x_2) = (2x_1 - 4x_2, -3x_1 + 6x_2).$$

From Example 7 we know that the kernel of f is a subspace of $\mathbb{R}^2$, and from Example 5 we know that this subspace must be either $\{\mathbf{0}\}$, a line passing through $\mathbf{0}$, or all of $\mathbb{R}^2$. Let us determine which of these the kernel of f is. Now $\mathbf{x}$ is in the kernel of f if and only if $f(\mathbf{x}) = \mathbf{0}$; that is, if and only if $\mathbf{x} = (x_1, x_2)$ satisfies the linear system

$$2x_1 - 4x_2 = 0$$

$$-3x_1 + 6x_2 = 0.$$

Since the general solution of this system is $\mathbf{x} = t(2, 1)$, $t \in \mathbb{R}$, we see that the kernel of f is the line $\mathbf{x} = t(2, 1)$ that passes through $\mathbf{0}$ in the direction of the vector $(2, 1)$.

Similarly, we can describe the image of f. It is the set of all vectors of the form

$$\mathbf{y} = f(x_1, x_2) = (2x_1 - 4x_2, -3x_1 + 6x_2)$$

$$= x_1(2, -3) + x_2(-4, 6)$$

$$= (x_1 - 2x_2)(2, -3).$$

Thus every vector $\mathbf{y}$ in the image of f is of the form $\mathbf{y} = t(2, -3)$ for some $t \in \mathbb{R}$. Moreover, every vector of the form $\mathbf{y} = t(2, -3)$, $t \in \mathbb{R}$, is in the image of f (take $x_1 = t$ and $x_2 = 0$). We can conclude, then, that the image of f is the line $\mathbf{x} = t(2, -3)$. ■

EXERCISES

1.   Which of the following are subspaces of $\mathbb{R}^2$?
  (a)   the set of all vectors of the form (a, a)
  (b)   the set of all vectors of the form (a, a^2)
  (c)   the set of all vectors of the form $(a, a + 1)$
  (d)   the solution set of the equation $2x_1 - 3x_2 = 0$
  (e)   the solution set of the equation $2x_1 - 3x_2 = 5$
  (f)   the solution set of the equation $x_1 = x_2^2$

2. Which of the following are subspaces of $\mathbb{R}^3$?
    (a) the set of all vectors of the form $(a, 2a, 3a)$
    (b) the set of all vectors of the form (a, a^2, a^3)
    (c) the set of all vectors of the form $(a, b, a - b)$
    (d) the set of all vectors of the form $(0, a, b)$
    (e) the solution set of the equation $2x_1 - x_2 + x_3 = 0$
    (f) the solution set of the equation $2x_1 - x_2 + x_3 = -1$
    (g) the solution set of the equation $x_1 x_2 x_3 = 0$

3. Which of the following are subspaces of $\mathbb{R}^3$?
    (a) the solution set of the equation $\mathbf{a} \cdot \mathbf{x} = 1$ where $\mathbf{a}$ is some fixed vector in $\mathbb{R}^3$
    (b) the solution set of the equation $\mathbf{a} \cdot \mathbf{x} = 0$ where $\mathbf{a}$ is some fixed vector in $\mathbb{R}^3$
    (c) the solution set of the equation $\mathbf{a} \times \mathbf{x} = \mathbf{0}$ where $\mathbf{a}$ is some fixed vector in $\mathbb{R}^3$
    (d) the set of all vectors of the form $\mathbf{a} \times \mathbf{x}$ where $\mathbf{a}$ is some fixed vector in $\mathbb{R}^3$
    (e) the set of all vectors (1×3 matrices) $\mathbf{x}$ such that $\mathbf{x}A = \mathbf{0}$ where A is some fixed 3×3 matrix
    (f) the set of all vectors of the form $\mathbf{x}A$ where A is some fixed 3×3 matrix

4. Which of the following are subspaces of $\mathbb{R}^4$?
    (a) the solution set of the system

    $$\begin{aligned} x_1 + x_2 + x_3 \quad\;\; &= 0 \\ 2x_1 - x_2 \quad\;\; + x_4 &= 0 \end{aligned}$$

    (b) the solution set of the system

    $$\begin{aligned} x_1 + x_2 + x_2 \quad\;\; &= 3 \\ 2x_1 - x_2 \quad\;\; + x_4 &= -1 \end{aligned}$$

    (c) the set of all (x_1, x_2, x_3, x_4) such that

    $$\det\begin{pmatrix} x_1 & x_2 \\ x_3 & x_4 \end{pmatrix} = 0$$

    (d) the set of all (x_1, x_2, x_3, x_4) such that

    $$\det\begin{pmatrix} x_1 & 1 & 0 & 0 \\ x_2 & 0 & 1 & 0 \\ x_3 & 0 & 0 & 1 \\ x_4 & 1 & 1 & 1 \end{pmatrix} = 0$$

5. The kernel of each of the following linear maps is a line. In each case, find a vector equation describing that line.
    (a) $f:\mathbb{R}^2 \to \mathbb{R}^1$, $f(x_1, x_2) = 2x_1 + 3x_2$
    (b) $f:\mathbb{R}^2 \to \mathbb{R}^2$, $f(x_1, x_2) = (x_1 - x_2, -x_1 + x_2)$
    (c) $f:\mathbb{R}^3 \to \mathbb{R}^2$, $f(x_1, x_2, x_3) = (x_1 - 3x_2 + 2x_3, x_2 + x_3)$
    (d) $f:\mathbb{R}^3 \to \mathbb{R}^3$, $f(x_1, x_2, x_3) = (x_1 + x_2 - x_3, -x_1 + x_2, 2x_1 - x_3)$

6. The image of each of the following linear maps is a line. In each case, find a vector equation describing that line.

(a)　$f:\mathbb{R} \to \mathbb{R}^2$, $f(x) = (2x, 3x)$

(b)　$f:\mathbb{R} \to \mathbb{R}^3$, $f(x) = (x, -x, 2x)$

(c)　$f:\mathbb{R}^2 \to \mathbb{R}^2$, $f(x_1, x_2) = (x_1 - x_2, -x_1 + x_2)$

(d)　$f:\mathbb{R}^2 \to \mathbb{R}^3$, $f(x_1, x_2) = (x_1 + x_2, x_1 + x_2, x_1 + x_2)$

7.　Show that $\{\mathbf{0}\}$ is a subspace of $\mathbb{R}^n$.

8.　Show that $\mathbb{R}^n$ is a subspace of $\mathbb{R}^n$.

9.　Describe all subspaces of $\mathbb{R}^1$.

10.　Let S be a subspace of $\mathbb{R}^n$ and let $f:\mathbb{R}^n \to \mathbb{R}^m$ be a linear map. Show that the image of S under f is a subspace of $\mathbb{R}^m$.

11.　Show that the intersection of any two subspaces of $\mathbb{R}^n$ is a subspace of $\mathbb{R}^n$.

12.　Let S be a subspace of $\mathbb{R}^n$ and let

$$S^\perp = \{\mathbf{w} \in \mathbb{R}^n \mid \mathbf{w} \cdot \mathbf{v} = 0 \text{ for all } \mathbf{v} \in S\}.$$

Show that $S^\perp$ is a subspace of $\mathbb{R}^n$. ($S^\perp$ is called the **orthogonal complement** of S.)

13.　Show that if S is any *nonempty* subset of $\mathbb{R}^n$ that is closed under vector addition and under scalar multiplication, then the zero vector must be in S and hence S is a subspace of $\mathbb{R}^n$.

14.　Show that a subset S of $\mathbb{R}^n$ is a subspace of $\mathbb{R}^n$ if and only if

(i)　S is nonempty, and

(ii)　$c_1\mathbf{v}_1 + c_2\mathbf{v}_2 \in S$ whenever $\mathbf{v}_1, \mathbf{v}_2 \in S$ and $c_1, c_2 \in \mathbb{R}$.

4.2
Spanning Sets

In this section we shall show how each finite set of vectors in $\mathbb{R}^n$ generates, or *spans*, a subspace of $\mathbb{R}^n$. Every subspace of $\mathbb{R}^n$ is spanned by a finite set of vectors, so the ideas of this section will add significantly to our understanding of subspaces of $\mathbb{R}^n$.

Suppose $\{\mathbf{v}_1, \mathbf{v}_2, \ldots, \mathbf{v}_k\}$ is a finite set of vectors in $\mathbb{R}^n$. A vector $\mathbf{v}$ in $\mathbb{R}^n$ is a **linear combination** of the vectors $\mathbf{v}_1, \mathbf{v}_2, \ldots, \mathbf{v}_k$ if $\mathbf{v}$ can be expressed in the form

$$\mathbf{v} = c_1\mathbf{v}_1 + c_2\mathbf{v}_2 + \cdots + c_k\mathbf{v}_k,$$

for some real numbers $c_1, c_2, \ldots, c_k$.

EXAMPLE 1

In $\mathbb{R}^3$, the vector $(-1, -9, 4)$ is a linear combination of the vectors $(1, -1, 2)$ and $(2, 3, 1)$. To show this, we must find real numbers c_1 and c_2 such that

$$c_1(1 -1, 2) + c_2(2, 3, 1) = (-1, -9, 4).$$

Thus, we seek real numbers c_1 and c_2 satisfying

$$c_1 + 2c_2 = -1$$

$$-c_1 + 3c_2 = -9$$

$$2c_1 + c_2 = 4.$$

Since the solution of this linear system is $(c_1, c_2) = (3, -2)$, we see that

$$(-1, -9, 4) = 3(1, -1, 2) - 2(2, 3, 1),$$

so $(-1, -9, 4)$ is a linear combination of $(1, -1, 2)$ and $(2, 3, 1)$, as asserted. ∎

Given any finite set $\{v_1, v_2, \ldots, v_k\}$ of vectors in $\mathbb{R}^n$, let us denote by $\mathcal{L}(v_1, \ldots, v_k)$ the set of *all* linear combinations of the vectors $v_1, v_2, \ldots, v_k$. Thus,

$$\mathcal{L}(v_1, \ldots, v_k) = \{c_1 v_1 + \cdots + c_k v_k \mid c_1, \ldots, c_k \in \mathbb{R}\}.$$

This set is a subspace of $\mathbb{R}^n$ because:

(i) $0 = 0v_1 + 0v_2 + \cdots + 0v_k$ is in $\mathcal{L}(v_1, \ldots, v_k)$.

(ii) If

$$v = c_1 v_1 + c_2 v_2 + \cdots + c_k v_k$$

and

$$w = d_1 v_1 + d_2 v_2 + \cdots + d_k v_k$$

are in $\mathcal{L}(v_1, \ldots, v_k)$, then so is

$$v + w = (c_1 + d_1)v_1 + (c_2 + d_2)v_2 + \cdots + (c_k + d_k)v_k.$$

(iii) If $v = c_1 v_1 + c_2 v_2 + \cdots + c_k v_k$ is in $\mathcal{L}(v_1, \ldots, v_k)$ and $c \in \mathbb{R}$, then

$$cv = (cc_1)v_1 + (cc_2)v_2 + \cdots + (cc_k)v_k$$

is in $\mathcal{L}(v_1, \ldots, v_k)$.

The subspace $\mathcal{L}(v_1, \ldots, v_k)$ is called the **subspace spanned by** the vectors $v_1, v_2, \ldots, v_k$.

EXAMPLE 2

If v is any nonzero vector in $\mathbb{R}^n$, then the subspace $\mathcal{L}(v)$ spanned by the single vector v is the set of all vectors of the form cv, where c is a real number. Thus $\mathcal{L}(v)$ is the line $x = tv$ through 0 in the direction of v (see Figure 4.4). ∎

FIGURE 4.4
The space $\mathcal{L}(v)$ spanned by v is the line $x = tv$ through 0 in the direction of v.

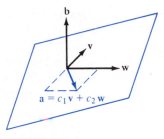

FIGURE 4.5
Each vector **a** in the plane **b** • **x** = 0 is a linear combination of **v** and **w**.

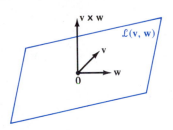

FIGURE 4.6
The space $\mathcal{L}(\mathbf{v}, \mathbf{w})$ spanned by **v** and **w** is the plane through **0** perpendicular to **v** × **w**.

EXAMPLE 3

Suppose **v** and **w** are two vectors in $\mathbb{R}^3$, neither of which is a scalar multiple of the other. Then **v** and **w** are both perpendicular to the vector $\mathbf{b} = \mathbf{v} \times \mathbf{w}$. Each linear combination $c_1\mathbf{v} + c_2\mathbf{w}$ of **v** and **w** is also perpendicular to $\mathbf{b} = \mathbf{v} \times \mathbf{w}$. This can be seen by computing the dot product

$$\mathbf{b} \cdot (c_1\mathbf{v} + c_2\mathbf{w}) = c_1(\mathbf{b} \cdot \mathbf{v}) + c_2(\mathbf{b} \cdot \mathbf{w}) = 0 + 0 = 0.$$

Thus each linear combination of **v** and **w** is in the plane $\mathbf{b} \cdot \mathbf{x} = 0$. Conversely, each vector in the plane $\mathbf{b} \cdot \mathbf{x} = 0$ is a linear combination of **v** and **w** (see Figure 4.5). We see, then, that *the subspace* $\mathcal{L}(\mathbf{v}, \mathbf{w})$ *spanned by two vectors* **v** *and* **w** *in* $\mathbb{R}^3$, *neither of which is a scalar multiple of the other, is a plane, the plane* $\mathbf{b} \cdot \mathbf{x} = 0$ *through* **0** *perpendicular to* $\mathbf{b} = \mathbf{v} \times \mathbf{w}$ (see Figure 4.6).

In particular, if $\mathbf{v} = (1, -1, 2)$ and $\mathbf{w} = (3, 0, 1)$ then $\mathbf{v} \times \mathbf{w} = (-1, 5, 3)$, so the subspace $\mathcal{L}((1, -1, 2), (3, 0, 1))$ of $\mathbb{R}^3$ is the plane whose equation is $-x_1 + 5x_2 + 3x_3 = 0$. ∎

REMARK Notice that the subspace $\mathcal{L}(\mathbf{v}_1, \ldots, \mathbf{v}_k)$ of $\mathbb{R}^n$ spanned by $\mathbf{v}_1, \mathbf{v}_2, \ldots, \mathbf{v}_k$ contains each of the vectors $\mathbf{v}_1, \mathbf{v}_2, \ldots, \mathbf{v}_k$. Indeed,

$$\mathbf{v}_1 = 1\mathbf{v}_1 + 0\mathbf{v}_2 + \cdots + 0\mathbf{v}_k \in \mathcal{L}(\mathbf{v}_1, \ldots, \mathbf{v}_k),$$

$$\mathbf{v}_2 = 0\mathbf{v}_1 + 1\mathbf{v}_2 + \cdots + 0\mathbf{v}_k \in \mathcal{L}(\mathbf{v}_1, \ldots, \mathbf{v}_k),$$

and so on. In fact, it can be shown that $\mathcal{L}(\mathbf{v}_1, \ldots, \mathbf{v}_k)$ is the *smallest* subspace of $\mathbb{R}^n$ that contains $\{\mathbf{v}_1, \mathbf{v}_2, \ldots, \mathbf{v}_k\}$ (see Exercise 20).

If a given subspace S of $\mathbb{R}^n$ can be described as $S = \mathcal{L}(\mathbf{v}_1, \ldots, \mathbf{v}_k)$ for some vectors $\mathbf{v}_1, \mathbf{v}_2, \ldots, \mathbf{v}_k$, then we say that the vectors $\mathbf{v}_1, \mathbf{v}_2, \ldots, \mathbf{v}_k$ *span* S, or that the set $\{\mathbf{v}_1, \mathbf{v}_2, \ldots, \mathbf{v}_k\}$ is a *spanning set* for S. Thus the vectors $\mathbf{v}_1, \mathbf{v}_2, \ldots, \mathbf{v}_k$ span S if and only if (i) each $\mathbf{v}_i \in S$ and (ii) each vector in S is a linear combination of the vectors $\mathbf{v}_1, \mathbf{v}_2, \ldots, \mathbf{v}_k$.

EXAMPLE 4

From Example 2 we know that if S is any line through **0** in $\mathbb{R}^n$, then S is spanned by $\{\mathbf{v}\}$, where **v** is any nonzero vector in S. So $\{\mathbf{v}\}$ is a spanning set for S. In particular, if S is the line $\mathbf{x} = t(5, -1)$ in $\mathbb{R}^2$, then $\{(5, -1)\}$ is a spanning set for S. $\{(10, -2)\}$ is another spanning set for S. The set $\{(-5, 1)\}$ is yet another. But notice that there are also spanning sets for S that contain more than one vector. For example, the set $\{(5, -1), (10, -2)\}$ is a spanning set for S. Why? ∎

EXAMPLE 5

From Example 3 we know that if S is any plane through **0** in $\mathbb{R}^3$, then S is spanned by $\{\mathbf{v}, \mathbf{w}\}$, where **v** and **w** are any two vectors in S neither of which is a scalar multiple of the other. So $\{\mathbf{v}, \mathbf{w}\}$ is a spanning set for S. In particular, if S is the plane $x_1 - x_2 + 2x_3 = 0$, then $\{(1, 1, 0), (-2, 0, 1)\}$ is a spanning set for S because these vectors both lie in the plane and neither is a scalar multiple of the other.

One way to find a spanning set for a plane $a_1x_1 + a_2x_2 + a_3x_3 = 0$ is by inspection. A more systematic way is to solve for the leading variable. For example, the solution set S of the equation $x_1 - x_2 + 2x_3 = 0$ can be obtained by setting $x_2 = c_1$, $x_3 = c_2$, and solving for x_1: $x_1 = c_1 - 2c_2$. Therefore $\mathbf{x}$ is in S if and only if

$$\mathbf{x} = (c_1 - 2c_2, c_1, c_2) = c_1(1, 1, 0) + c_2(-2, 0, 1)$$

and we see that $\{(1, 1, 0), (-2, 0, 1)\}$ is a spanning set for S. ■

EXAMPLE 6

The set

$$\{\mathbf{e}_1, \mathbf{e}_2, \mathbf{e}_3\} = \{(1, 0, 0), (0, 1, 0), (0, 0, 1)\}$$

is a spanning set for $\mathbb{R}^3$ because each vector $\mathbf{a} = (a_1, a_2, a_3)$ in $\mathbb{R}^3$ can be expressed as a linear combination of $\mathbf{e}_1, \mathbf{e}_2, \mathbf{e}_3$:

$$\mathbf{a} = (a_1, a_2, a_3) = a_1(1, 0, 0) + a_2(0, 1, 0) + a_3(0, 0, 1)$$

$$= a_1\mathbf{e}_1 + a_2\mathbf{e}_2 + a_3\mathbf{e}_3. \ ■$$

EXAMPLE 7

The set $\{\mathbf{e}_1, \mathbf{e}_2, \ldots, \mathbf{e}_n\}$ is a spanning set for $\mathbb{R}^n$ because each vector $\mathbf{a} = (a_1, a_2, \ldots, a_n)$ in $\mathbb{R}^n$ can be expressed as a linear combination of $\mathbf{e}_1, \ldots, \mathbf{e}_n$:

$$\mathbf{a} = (a_1, a_2, \ldots, a_n) = a_1\mathbf{e}_1 + a_2\mathbf{e}_2 + \cdots + a_n\mathbf{e}_n.$$

Recall that $\mathbf{e}_1 = (1, 0, \ldots, 0)$, $\mathbf{e}_2 = (0, 1, 0, \ldots, 0), \ldots, \mathbf{e}_n = (0, 0, \ldots, 1)$.] ■

From Example 6 of Section 4.1 we know that the solution space of any homogeneous linear system of m equations in n unknowns is a subspace of $\mathbb{R}^n$. The techniques described in Sections 1.1 and 1.2 for solving these systems lead directly to spanning sets for these solution spaces.

EXAMPLE 8

Consider the system

$$x_1 - x_2 + x_3 + x_4 - 2x_5 = 0$$
$$-2x_1 + 2x_2 - x_3 \qquad + x_5 = 0$$
$$x_1 - x_2 + 2x_3 + 3x_4 - 5x_5 = 0.$$

The matrix of this system is

$$\begin{pmatrix} 1 & -1 & 1 & 1 & -2 & 0 \\ -2 & 2 & -1 & 0 & 1 & 0 \\ 1 & -1 & 2 & 3 & -5 & 0 \end{pmatrix}.$$

Its row echelon matrix is

$$\begin{pmatrix} 1 & -1 & 0 & -1 & 1 & 0 \\ 0 & 0 & 1 & 2 & -3 & 0 \\ 0 & 0 & 0 & 0 & 0 & 0 \end{pmatrix}.$$

The linear system corresponding to this echelon matrix is

$$\begin{aligned} x_1 - x_2 \quad - x_4 + x_5 &= 0 \\ x_3 + 2x_4 - 3x_5 &= 0 \\ 0 &= 0 \end{aligned}$$

with general solution

$$\begin{aligned} \mathbf{x} &= (c_1 + c_2 - c_3, c_1, -2c_2 + 3c_3, c_2, c_3) \\ &= c_1(1, 1, 0, 0, 0) + c_2(1, 0, -2, 1, 0) + c_3(-1, 0, 3, 0, 1) \end{aligned}$$

obtained by setting $x_2 = c_1$, $x_4 = c_2$, $x_5 = c_3$, and solving for x_1 and x_3. Thus, we see that

$$\{(1, 1, 0, 0, 0), (1, 0, -2, 1, 0), (-1, 0, 3, 0, 1)\}$$

is a spanning set for the solution space of the given system, since each solution vector is a linear combination of these three vectors. ■

From Example 7 of Section 4.1, we know that the kernel of any linear map $f:\mathbb{R}^n \to \mathbb{R}^m$ is a subspace of $\mathbb{R}^n$. The kernel of f is the solution set of the vector equation $f(\mathbf{x}) = \mathbf{0}$. This vector equation is equivalent to the homogeneous linear system obtained by setting each entry of $f(\mathbf{x})$ equal to zero. We can, therefore, find a spanning set for the kernel of f by solving this system.

EXAMPLE 9
Consider the linear map $f:\mathbb{R}^4 \to \mathbb{R}^2$ defined by

$$f(x_1, x_2, x_3, x_4) = (x_1 + x_2 - x_3 - x_4, x_1 - x_2 + x_3 + x_4).$$

The kernel of f is the solution set of the vector equation $f(\mathbf{x}) = \mathbf{0}$, that is, of the vector equation

$$(x_1 + x_2 - x_3 - x_4, x_1 - x_2 + x_3 + x_4) = (0, 0).$$

This vector equation is equivalent to the linear system

$$\begin{aligned} x_1 + x_2 - x_3 - x_4 &= 0 \\ x_1 - x_2 + x_3 + x_4 &= 0, \end{aligned}$$

so the kernel of f is the solution space of this system. Since the general solution of this system is

$$\mathbf{x} = c_1(0, 1, 1, 0) + c_2(0, 1, 0, 1),$$

we see that

$$\{(0, 1, 1, 0), (0, 1, 0, 1)\}$$

is a spanning set for the kernel of f. ■

From Example 8 of Section 4.1 we know that the image of any linear map $f : \mathbb{R}^n \rightarrow \mathbb{R}^m$ is a subspace of $\mathbb{R}^m$. It is relatively easy to find a spanning set for this subspace.

Let $f : \mathbb{R}^n \rightarrow \mathbb{R}^m$ be a linear map. The image of f is the set of all vectors of the form $f(\mathbf{x})$ where $\mathbf{x} \in \mathbb{R}^n$. Since $\{\mathbf{e}_1, \mathbf{e}_2, \ldots, \mathbf{e}_n\}$ is a spanning set for $\mathbb{R}^n$, we know that if $\mathbf{x} \in \mathbb{R}^n$ then

$$\mathbf{x} = c_1 \mathbf{e}_1 + c_2 \mathbf{e}_2 + \cdots + c_n \mathbf{e}_n$$

for some real numbers $c_1, c_2, \ldots, c_n$. Hence the image of f consists of all vectors of the form

$$
\begin{aligned}
f(\mathbf{x}) &= f(c_1 \mathbf{e}_1 + c_2 \mathbf{e}_2 + \cdots + c_n \mathbf{e}_n) \\
&= f(c_1 \mathbf{e}_1) + f(c_2 \mathbf{e}_2) + \cdots + f(c_n \mathbf{e}_n) \\
&= c_1 f(\mathbf{e}_1) + c_2 f(\mathbf{e}_2) + \cdots + c_n f(\mathbf{e}_n).
\end{aligned}
$$

In other words, *the set*

$$\{f(\mathbf{e}_1), f(\mathbf{e}_2), \ldots, f(\mathbf{e}_n)\}$$

is a spanning set for the image of f.

EXAMPLE 10

Let $f : \mathbb{R}^3 \rightarrow \mathbb{R}^4$ be the linear map

$$f(x_1, x_2, x_3) = (x_1 + x_2, x_2 + x_3, x_1 + x_2 - x_3, -x_3).$$

Since

$$f(\mathbf{e}_1) = f(1, 0, 0) = (1, 0, 1, 0)$$

$$f(\mathbf{e}_2) = f(0, 1, 0) = (1, 1, 1, 0)$$

$$f(\mathbf{e}_3) = f(0, 0, 1) = (0, 1, -1, -1),$$

we see that

$$\{(1, 0, 1, 0), (1, 1, 1, 0), (0, 1, -1, -1)\}$$

is a spanning set for the image of f. ∎

We close this section with two theorems that describe easy tests to determine whether or not a given set of vectors spans $\mathbb{R}^n$.

Theorem 1 *Let* $\{\mathbf{v}_1, \mathbf{v}_2, \ldots, \mathbf{v}_k\}$ *be a finite set of vectors in* $\mathbb{R}^n$. *Then the vectors* $\mathbf{v}_1, \mathbf{v}_2, \ldots, \mathbf{v}_k$ *span* $\mathbb{R}^n$ *if and only if the matrix whose column vectors are* $\mathbf{v}_1, \mathbf{v}_2, \ldots, \mathbf{v}_k$ *has rank* n.

Proof Consider the map $f : \mathbb{R}^k \rightarrow \mathbb{R}^n$ defined by

$$f(c_1, c_2, \ldots, c_k) = c_1 \mathbf{v}_1 + c_2 \mathbf{v}_2 + \cdots + c_k \mathbf{v}_k.$$

It is easy to verify (see Exercise 21) that this map preserves both vector addition and scalar multiplication and so is a linear map. Its image is $\mathcal{L}(\mathbf{v}_1, \ldots, \mathbf{v}_k)$. So $\mathcal{L}(\mathbf{v}_1, \ldots, \mathbf{v}_k) = \mathbb{R}^n$ if and only if f is onto. But, according to Theorem 4 of Section 2.6, a linear map $f: \mathbb{R}^k \rightarrow \mathbb{R}^n$ is onto if and only if its rank is n.

Now the matrix of f is the matrix whose column vectors are the vectors $f(\mathbf{e}_1), \ldots, f(\mathbf{e}_k)$ (Theorem 2 of Section 2.3). Since

$$f(\mathbf{e}_1) = \mathbf{v}_1, \ldots, f(\mathbf{e}_k) = \mathbf{v}_k,$$

we can conclude that $\mathcal{L}(\mathbf{v}_1, \ldots, \mathbf{v}_k) = \mathbb{R}^n$; that is, the vectors $\mathbf{v}_1, \ldots, \mathbf{v}_k$ span $\mathbb{R}^n$, if and only if the matrix whose column vectors are $\mathbf{v}_1, \ldots, \mathbf{v}_k$ has rank n. ∎

EXAMPLE 11

Let us determine if the vectors $(1, -1, 0)$, $(1, 0, -1)$, $(0, 1, -1)$, $(1, -3, 2)$ span $\mathbb{R}^3$. The matrix with these vectors as column vectors is

$$A = \begin{pmatrix} 1 & 1 & 0 & 1 \\ -1 & 0 & 1 & -3 \\ 0 & -1 & -1 & 2 \end{pmatrix}.$$

Its echelon matrix is

$$B = \begin{pmatrix} 1 & 0 & -1 & 3 \\ 0 & 1 & 1 & -2 \\ 0 & 0 & 0 & 0 \end{pmatrix}.$$

Since the rank of a matrix is the number of corner 1's in its echelon matrix, we see that the rank of A is 2. Since the rank is not 3, the given vectors *do not* span $\mathbb{R}^3$. ∎

One consequence of Theorem 1 is that *it takes at least n vectors to span* $\mathbb{R}^n$. In other words, if $\{\mathbf{v}_1, \ldots, \mathbf{v}_k\}$ is a set of k vectors in $\mathbb{R}^n$, where $k < n$, then $\{\mathbf{v}_1, \ldots, \mathbf{v}_k\}$ cannot be a spanning set for $\mathbb{R}^n$. This is because the rank of a matrix cannot be greater than the number of columns in the matrix. On the other hand, if a subset of $\mathbb{R}^n$ contains exactly n vectors, then there is a simple condition involving the determinant that can be used to determine if the set is a spanning set for $\mathbb{R}^n$.

Theorem 2 *Let $\{\mathbf{v}_1, \mathbf{v}_2, \ldots, \mathbf{v}_n\}$ be a set of n vectors in $\mathbb{R}^n$. Then the vectors $\mathbf{v}_1, \mathbf{v}_2, \ldots, \mathbf{v}_n$ span $\mathbb{R}^n$ if and only if $\det A \neq 0$ where A is the matrix whose column vectors are $\mathbf{v}_1, \mathbf{v}_2, \ldots, \mathbf{v}_n$.*

Equivalently, the vectors $\mathbf{v}_1, \mathbf{v}_2, \ldots, \mathbf{v}_n$ span $\mathbb{R}^n$ if and only if $\det B \neq 0$ where B is the matrix whose row vectors are $\mathbf{v}_1, \mathbf{v}_2, \ldots, \mathbf{v}_n$.

REMARK Notice that the test given by Theorem 2 can be used only when the number of vectors in the set being tested is equal to n.

Proof By Theorem 1, the vectors $\mathbf{v}_1, \ldots, \mathbf{v}_n$ span $\mathbb{R}^n$ if and only if the matrix A whose column vectors are $\mathbf{v}_1, \ldots, \mathbf{v}_n$ has rank n; that is, if and only if the matrix A is invertible (Theorem 2, Section 2.7). But A is invertible if and only if det $A \neq 0$ (Theorem 1, Section 3.4). Hence the vectors $\mathbf{v}_1, \ldots, \mathbf{v}_n$ span $\mathbb{R}^n$ if and only if det $A \neq 0$.

The matrix B is just the transpose of the matrix A. Since det $B =$ det $A' =$ det A, we see that det $B \neq 0$ if and only if det $A \neq 0$, and we have just seen that det $A \neq 0$ if and only if the vectors $\mathbf{v}_1, \mathbf{v}_2, \ldots, \mathbf{v}_n$ span $\mathbb{R}^n$. ■

EXAMPLE 12

Let us determine if

$$\{(1, -1, 1), (1, 1, -1), (1, 1, 1)\}$$

is a spanning set for $\mathbb{R}^3$. Since there are three vectors in this set, we can apply Theorem 2. If we compute the determinant of the matrix whose row vectors are the given vectors, we find

$$\det\begin{pmatrix} 1 & -1 & 1 \\ 1 & 1 & -1 \\ 1 & 1 & 1 \end{pmatrix} = \det\begin{pmatrix} 1 & -1 & 1 \\ 0 & 2 & -2 \\ 0 & 2 & 0 \end{pmatrix} = \det\begin{pmatrix} 2 & -2 \\ 2 & 0 \end{pmatrix} = 4 \neq 0.$$

We can conclude, then, that the given set *is* a spanning set for $\mathbb{R}^3$. ■

EXERCISES

1. Show that the vector $(1, 0)$ in $\mathbb{R}^2$ is a linear combination of the vectors $(1, -1)$ and $(2, 1)$.

2. Which of the given vectors in $\mathbb{R}^3$ are linear combinations of the vectors $(1, -1, 1)$ and $(2, 0, 1)$?
   (a) $(1, 1, 0)$ (b) $(1, 1, 1)$ (c) $(-1, 3, 2)$
   (d) $(1, -3, 2)$ (e) $(1, -1, 1)$ (f) $(0, 0, 0)$

3. Which of the given vectors in $\mathbb{R}^4$ are linear combinations of the vectors $(1, 1, 0, 0)$ and $(0, 0, 1, 1)$?
   (a) $(1, 1, 1, 1)$ (b) $(1, 2, 3, 4)$ (c) $(1, 1, -1, -1)$
   (d) $(-1, 1, 1, -1)$ (e) $(2, 2, 3, 3)$ (f) (a, a, b, b)

4. Find a vector equation that describes the line
   (a) $\mathcal{L}((1, 1))$ in $\mathbb{R}^2$ (b) $\mathcal{L}((-1, 1, 2))$ in $\mathbb{R}^3$
   (c) $\mathcal{L}((3, 4, 1, 2))$ in $\mathbb{R}^4$ (d) $\mathcal{L}(\mathbf{e}_n)$ in $\mathbb{R}^n$

5. Find an equation $\mathbf{b} \cdot \mathbf{x} = 0$ that describes the given plane in $\mathbb{R}^3$.
   (a) $\mathcal{L}((1, 2, 1), (-1, 3, 2))$ (b) $\mathcal{L}((0, 2, 3), (2, 3, 0))$
   (c) $\mathcal{L}((-1, 1, 1), (1, 0, 1))$ (d) $\mathcal{L}((3, -1, 5), (-1, 2, 2))$
   (e) $\mathcal{L}((-1, -1, -3), (2, -3, -1))$ (f) $\mathcal{L}(\mathbf{e}_1, \mathbf{e}_2)$

6. Which of the following are spanning sets for the line $\mathbf{x} = t(-1, 2)$ in $\mathbb{R}^2$?
   (a) $\{(-1, 2)\}$ (b) $\{(2, -1)\}$
   (c) $\{(0, 0)\}$ (d) $\{(2, -4)\}$
   (e) $\{(-1, 2), (2, -4)\}$

7. Which of the following are spanning sets for the plane $x_1 + x_2 - x_3 = 0$ in $\mathbb{R}^3$?
   (a) $\{(1, -1, 0)\}$ (b) $\{(1, -1, 0), (-1, 1, 0)\}$
   (c) $\{(1, -1, 0), (1, 1, 2)\}$ (d) $\{(1, -1, 0), (1, 1, 2), (1, 0, 1)\}$
   (e) $\{(1, 0, 0), (0, 1, 0)\}$

8. Referring to Example 4 of Section 1.2, write down a spanning set for the solution space of the homogeneous linear system

$$x_1 - x_2 + x_3 - x_4 + x_5 = 0$$

$$2x_1 + x_2 - x_3 \qquad - x_5 = 0.$$

9. Find a spanning set for the solution space of each of the given homogeneous linear systems.

(a) $\begin{cases} x_1 + x_2 + x_3 = 0 \\ 2x_1 - x_2 - 7x_3 = 0 \end{cases}$ (b) $\begin{cases} x_1 - 2x_2 + x_3 = 0 \\ -x_1 - 3x_2 + 2x_3 = 0 \end{cases}$

(c) $\begin{cases} x_1 + x_2 - x_3 = 0 \\ x_1 \qquad + 2x_3 = 0 \\ -x_1 + x_2 + 3x_3 = 0 \end{cases}$ (d) $\begin{cases} x_1 + 3x_2 - 4x_3 = 0 \\ -x_1 + 2x_2 - x_3 = 0 \\ x_1 + x_2 - 2x_3 = 0 \end{cases}$

(e) $\begin{cases} x_1 + 2x_2 \qquad - 9x_4 - x_5 = 0 \\ -x_1 - x_2 - x_3 + x_4 - 5x_5 = 0 \\ x_1 + 2x_2 + x_3 - 7x_4 + 4x_5 = 0 \end{cases}$

10. Find a spanning set for each of the following lines through $\mathbf{0}$ in $\mathbb{R}^2$.
    (a) $x_1 + 2x_2 = 0$ (b) $3x_1 - 4x_2 = 0$
    (c) $x_1 - x_2 = 0$ (d) $-5x_1 + 7x_2 = 0$

11. Find a spanning set for each of the following planes through $\mathbf{0}$ in $\mathbb{R}^3$.
    (a) $x_1 + x_2 + x_3 = 0$ (b) $x_2 + x_3 = 0$
    (c) $2x_1 - 3x_2 + 4x_3 = 0$ (d) $x_3 = 0$.

12. Find a spanning set for the hyperplane

$$x_1 + 3x_2 - x_3 - 2x_4 + 4x_5 = 0.$$

13. Find a spanning set for the kernel of each of the following linear maps.
    (a) $f(x_1, x_2) = (x_1 + x_2, x_1 + x_2)$
    (b) $f(x_1, x_2, x_3) = (2x_1 + 3x_2 - x_3, -x_1 + 4x_2 + 6x_3)$
    (c) $f(x_1, x_2, x_3) = (x_1 - x_2 + 2x_3, x_2 + x_3, 2x_1 - x_2 + 5x_3)$
    (d) $f(x_1, x_2, x_3) = (2x_1 - x_2 + 3x_3, -2x_1 + x_2 - 3x_3)$
    (e) $f(x_1, x_2, x_3, x_4) = (x_1 + x_2, x_3 + x_4)$.

14. Find a spanning set for the image of each of the linear maps in Exercise 13.

15. Which of the following sets are spanning sets for $\mathbb{R}^3$?
    (a) $\{(-1, 1, 2), (2, 0, 1)\}$
    (b) $\{(2, 1, 1), (1, 2, 1), (1, 1, 2)\}$
    (c) $\{(0, -1, 1), (-1, 0, 1), (-1, 1, 0)\}$
    (d) $\{(1, 2, 3), (4, 5, 6), (7, 8, 9), (10, 11, 12)\}$
    (e) $\{(1, 1, 1), (1, 2, 4), (1, 3, 9), (1, 4, 16)\}$

16. Which of the following sets are spanning sets for $\mathbb{R}^4$?
    (a) $\{(-1, 1, 0, 2), (1, 3, -1, 1), (2, 5, 4, 0)\}$
    (b) $\{(1, 1, 1, 1), (1, 2, 4, 8), (1, 3, 9, 27), (1, 4, 16, 64)\}$
    (c) $\{(1, 0, 0, 0), (0, 1, 0, 0), (0, 0, 1, 0), (0, 0, 0, 1), (1, 1, 1, 1)\}$
    (d) $\{(1, 1, 0, 0), (-1, 0, 1, 0), (-1, 0, 0, 1), (0, -1, 1, 0), (0, -1, 0, 1), (0, 0, -1, 1)\}$

17. Describe the subspace $\mathcal{L}(\mathbf{0})$, where $\mathbf{0}$ is the zero vector in $\mathbb{R}^n$.

18. Suppose $\mathbf{v}$ and $\mathbf{w}$ are two vectors in $\mathbb{R}^2$, neither of which is a scalar multiple of the other. Describe the subspace $\mathcal{L}(\mathbf{v}, \mathbf{w})$.

19. Show that if A is an $m \times n$ matrix and $f : \mathbb{R}^n \to \mathbb{R}^m$ is the linear map associated with A, then the column vectors of A span the image of f.

20. Let $\{\mathbf{v}_1, \mathbf{v}_2, \ldots, \mathbf{v}_k\}$ be a finite set of vectors in $\mathbb{R}^n$ and let S be a subspace of $\mathbb{R}^n$. Show that if S contains each of the vectors $\mathbf{v}_1, \mathbf{v}_2, \ldots, \mathbf{v}_k$, then S contains $\mathcal{L}(\mathbf{v}_1, \ldots, \mathbf{v}_k)$. Conclude that $\mathcal{L}(\mathbf{v}_1, \ldots, \mathbf{v}_k)$ is the smallest subspace of $\mathbb{R}^n$ that contains $\{\mathbf{v}_1, \mathbf{v}_2, \ldots, \mathbf{v}_k\}$.

21. Let $\{\mathbf{v}_1, \mathbf{v}_2, \ldots, \mathbf{v}_k\}$ be a finite set of vectors in $\mathbb{R}^n$, and let $f : \mathbb{R}^k \to \mathbb{R}^n$ be defined by

$$f(c_1, c_2, \ldots, c_k) = c_1 \mathbf{v}_1 + c_2 \mathbf{v}_2 + \cdots + c_k \mathbf{v}_k.$$

Verify that f is a linear map.

22. Let $f : \mathbb{R}^n \to \mathbb{R}^m$ be a linear map and let S be a subspace of $\mathbb{R}^n$. Show that, if $\{\mathbf{v}_1, \ldots, \mathbf{v}_k\}$ is a spanning set for S, then $\{f(\mathbf{v}_1), \ldots, f(\mathbf{v}_k)\}$ is a spanning set for the image of S under f.

4.3
Bases

If we are given a spanning set $\{\mathbf{v}_1, \ldots, \mathbf{v}_k\}$ for a subspace S of $\mathbb{R}^n$, then we can express each vector $\mathbf{v}$ in S as a linear combination

$$\mathbf{v} = c_1 \mathbf{v}_1 + c_2 \mathbf{v}_2 + \cdots + c_k \mathbf{v}_k$$

of the vectors $\mathbf{v}_1, \ldots, \mathbf{v}_k$. In general there will be many different ways to do this. That is, there will be many solutions $(c_1, c_2, \ldots, c_k)$ of the vector equation

$$c_1 \mathbf{v}_1 + c_2 \mathbf{v}_2 + \cdots + c_k \mathbf{v}_k = \mathbf{v}.$$

For certain special spanning sets, however, this vector equation will have a unique solution, for each $\mathbf{v} \in S$. Such a special spanning set is called a *basis* for S. In this section we shall discuss bases (singular: basis; plural: bases) for subspaces of $\mathbb{R}^n$. In the next section we shall see that, using a basis, we can introduce coordinates for the subspace S. This will enable us to see that each subspace S of $\mathbb{R}^n$ is just a copy of $\mathbb{R}^k$, for some k.

Suppose $\{\mathbf{v}_1, \mathbf{v}_2, \ldots, \mathbf{v}_k\}$ is a finite set of vectors in $\mathbb{R}^n$. Consider the linear map $f : \mathbb{R}^k \to \mathbb{R}^n$ defined by

$$f(c_1, c_2, \ldots, c_k) = c_1 \mathbf{v}_1 + c_2 \mathbf{v}_2 + \cdots + c_k \mathbf{v}_k.$$

The image of this linear map is the subspace $\mathcal{L}(\mathbf{v}_1, \ldots, \mathbf{v}_k)$ of $\mathbb{R}^n$ spanned by the vectors $\mathbf{v}_1, \ldots, \mathbf{v}_k$. Moreover, from Theorem 2 of Section 2.6, we know that f is one-to-one if and only if the kernel of f is $\{\mathbf{0}\}$. In other words, the vector equation

$$f(c_1, \ldots, c_k) = \mathbf{v}$$

has a unique solution for each $\mathbf{v}$ in $\mathcal{L}(\mathbf{v}_1, \ldots, \mathbf{v}_k)$ if and only if the vector equation

$$f(c_1, \ldots, c_k) = \mathbf{0}$$

has the unique solution $(c_1, \ldots, c_k) = (0, \ldots, 0)$. Said yet another way, if $S = \mathcal{L}(\mathbf{v}_1, \ldots, \mathbf{v}_k)$ then the equation

$$c_1\mathbf{v}_1 + \cdots + c_k\mathbf{v}_k = \mathbf{v}$$

has a unique solution for each $\mathbf{v}$ in S, if and only if the only solution of the vector equation

$$c_1\mathbf{v}_1 + \cdots + c_k\mathbf{v}_k = \mathbf{0}$$

is $(c_1, \ldots, c_k) = (0, \ldots, 0)$.

A subset $\{\mathbf{v}_1, \ldots, \mathbf{v}_k\}$ of $\mathbb{R}^n$ is said to be *linearly independent* if the only solution of the vector equation

$$c_1\mathbf{v}_1 + \cdots + c_k\mathbf{v}_k = \mathbf{0}$$

is the zero solution $(c_1, \ldots, c_k) = (0, \ldots, 0)$. If there is a nonzero solution $(c_1, \ldots, c_k) \neq (0, \ldots, 0)$ of this equation then we say that the set $\{\mathbf{v}_1, \ldots, \mathbf{v}_k\}$ is *linearly dependent*. An equation $c_1\mathbf{v}_1 + \cdots + c_k\mathbf{v}_k = \mathbf{0}$, where not all c_i are zero, is called a *dependence relation.*

EXAMPLE 1

Suppose $\{\mathbf{v}_1\}$ is a subset of $\mathbb{R}^n$ consisting of a single vector. Then $\{\mathbf{v}_1\}$ is linearly dependent if and only if there is a real number $c_1 \neq 0$ such that $c_1\mathbf{v}_1 = \mathbf{0}$. This happens precisely when $\mathbf{v}_1 = \mathbf{0}$. Thus $\{\mathbf{v}_1\}$ is linearly dependent if and only if $\mathbf{v}_1 = \mathbf{0}$. In other words, $\{\mathbf{v}_1\}$ is *linearly independent if and only if $\mathbf{v}_1 \neq \mathbf{0}$* (see Figure 4.7). ■

FIGURE 4.7
$\{\mathbf{v}_1\}$ is linearly independent if $\mathbf{v}_1 \neq \mathbf{0}$.

EXAMPLE 2

Suppose $\{\mathbf{v}_1, \mathbf{v}_2\}$ is a subset of $\mathbb{R}^n$ consisting of two vectors. Then $\{\mathbf{v}_1, \mathbf{v}_2\}$ is linearly dependent if and only if the vector equation

$$c_1\mathbf{v}_1 + c_2\mathbf{v}_2 = \mathbf{0}$$

has a nonzero solution $(c_1, c_2) \neq (0, 0)$. If (c_1, c_2) is a solution of this equation with $c_1 \neq 0$, then $\mathbf{v}_1 = -(c_2/c_1)\mathbf{v}_2$. If (c_1, c_2) is a solution with $c_2 \neq 0$, then $\mathbf{v}_2 = -(c_1/c_2)\mathbf{v}_1$. Hence, if $\{\mathbf{v}_1, \mathbf{v}_2\}$ is linearly dependent, then one of the vectors must be a scalar multiple of the other. Conversely, if either of the vectors is a scalar multiple of the other, then the set is linearly dependent (if, for example, $\mathbf{v}_2 = c\mathbf{v}_1$, then $c\mathbf{v}_1 + (-1)\mathbf{v}_2 = \mathbf{0}$ is a dependence relation). We can conclude, therefore, that $\{\mathbf{v}_1, \mathbf{v}_2\}$ *is linearly independent if and only if neither of the vectors is a scalar multiple of the other* (see Figure 4.8). ■

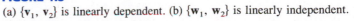

(a)   (b)

FIGURE 4.8
(a) $\{\mathbf{v}_1, \mathbf{v}_2\}$ is linearly dependent. (b) $\{\mathbf{w}_1, \mathbf{w}_2\}$ is linearly independent.

EXAMPLE 3

Suppose $\{\mathbf{v}_1, \mathbf{v}_2, \mathbf{v}_3\}$ is a subset of $\mathbb{R}^n$ consisting of three vectors. If $\{\mathbf{v}_1, \mathbf{v}_2, \mathbf{v}_3\}$ is linearly dependent, then there is a dependence relation

$$c_1\mathbf{v}_1 + c_2\mathbf{v}_2 + c_3\mathbf{v}_3 = \mathbf{0}$$

where $(c_1, c_2, c_3) \neq (0, 0, 0)$. If $c_1 \neq 0$, then

$$\mathbf{v}_1 = \left(-\frac{c_2}{c_1}\right)\mathbf{v}_2 + \left(-\frac{c_3}{c_1}\right)\mathbf{v}_3.$$

If $c_2 \neq 0$, then

$$\mathbf{v}_2 = \left(-\frac{c_1}{c_2}\right)\mathbf{v}_1 + \left(-\frac{c_3}{c_2}\right)\mathbf{v}_3.$$

If $c_3 \neq 0$, then

$$\mathbf{v}_3 = \left(-\frac{c_1}{c_3}\right)\mathbf{v}_1 + \left(-\frac{c_2}{c_3}\right)\mathbf{v}_2.$$

Hence, if $\{\mathbf{v}_1, \mathbf{v}_2, \mathbf{v}_3\}$ is linearly dependent, then at least one of the vectors must be a linear combination of the other two. Conversely, if one of the vectors is a linear combination of the other two then the set $\{\mathbf{v}_1, \mathbf{v}_2, \mathbf{v}_3\}$ is linearly dependent (if, for example, $\mathbf{v}_3 = a\mathbf{v}_1 + b\mathbf{v}_2$, then $a\mathbf{v}_1 + b\mathbf{v}_2 + (-1)\mathbf{v}_3 = \mathbf{0}$ is a dependence relation). Thus, $\{\mathbf{v}_1, \mathbf{v}_2, \mathbf{v}_3\}$ *is linearly independent if and only if none of the vectors is a linear combination of the others* (see Figure 4.9). ■

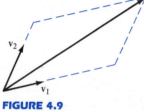

FIGURE 4.9
$\{\mathbf{v}_1, \mathbf{v}_2, \mathbf{v}_3\}$ is linearly dependent if one of the vectors is a linear combination of the others.

REMARK   The discussion in Example 3 generalizes easily to show that a subset $\{\mathbf{v}_1, \ldots, \mathbf{v}_k\}$ of $\mathbb{R}^n$ is linearly independent if and only if none of the vectors $\mathbf{v}_1, \ldots, \mathbf{v}_k$ is a linear combination of the others (see Exercise 13).

EXAMPLE 4

Let us determine if the set

$$\{(1, 2, 3), (4, 5, 6), (7, 8, 9)\}$$

in $\mathbb{R}^3$ is linearly independent. To do this, we must see if there are any nonzero solutions of the vector equation

$$c_1(1, 2, 3) + c_2(4, 5, 6) + c_3(7, 8, 9) = \mathbf{0}.$$

This equation is the same as the equation

$$(c_1 + 4c_2 + 7c_3, 2c_1 + 5c_2 + 8c_3, 3c_1 + 6c_2 + 9c_3) = (0, 0, 0),$$

which is satisfied if and only if

$$c_1 + 4c_2 + 7c_3 = 0$$

$$2c_1 + 5c_2 + 8c_3 = 0$$

$$3c_1 + 6c_2 + 9c_3 = 0.$$

The solution set of this system is the line $(c_1, c_2, c_3) = t(1, -2, 1)$. In particular,

$(1, -2, 1)$ is a nonzero solution, so

$$1(1, 2, 3) + (-2)(4, 5, 6) + 1(7, 8, 9) = \mathbf{0}$$

is a dependence relation. The set $\{(1, 2, 3), (4, 5, 6), (7, 8, 9)\}$ is linearly dependent.

Notice that we can, if we wish, solve this last equation for the vector $(1, 2, 3)$:

$$(1, 2, 3) = 2(4, 5, 6) + (-1)(7, 8, 9),$$

so this vector is a linear combination of the others, in agreement with Example 3. ∎

Testing a finite subset of $\mathbb{R}^n$ for linear independence always leads to a homogeneous system of linear equations. Consider the set

$$\{(a_{11}, \ldots, a_{1n}), (a_{21}, \ldots, a_{2n}), \ldots, (a_{k1}, \ldots, a_{kn})\}$$

of k vectors in $\mathbb{R}^n$. This set is linearly dependent if and only if there exists a nonzero solution $(c_1, c_2, \ldots, c_k)$ of the vector equation

$$c_1(a_{11}, \ldots, a_{1n}) + c_2(a_{21}, \ldots, a_{2n}) + \cdots + c_k(a_{k1}, \ldots, a_{kn}) = \mathbf{0}.$$

This vector equation is equivalent to the linear system

$$a_{11}c_1 + a_{21}c_2 + \cdots + a_{k1}c_k = 0$$

$$\cdots$$

$$a_{1n}c_1 + a_{2n}c_2 + \cdots + a_{kn}c_k = 0$$

with coefficient matrix

$$A = \begin{pmatrix} a_{11} & a_{21} & \cdots & a_{k1} \\ & & \cdots & \\ a_{1n} & a_{2n} & \cdots & a_{kn} \end{pmatrix}.$$

This homogeneous linear system has the zero solution as its only solution if and only if the echelon matrix of A has a corner 1 in every column; that is, if and only if A has rank k. Thus we have proved the following theorem.

Theorem 1   *Suppose $\{\mathbf{v}_1, \ldots, \mathbf{v}_k\}$ is a finite set of vectors in $\mathbb{R}^n$. Then $\{\mathbf{v}_1, \ldots, \mathbf{v}_k\}$ is linearly independent if and only if the matrix whose column vectors are the vectors $\mathbf{v}_1, \ldots, \mathbf{v}_k$ has rank k.*

EXAMPLE 5
The subset $\{(-1, 2, 0, -3), (-2, 1, 1, 2), (2, 4, -1, -3)\}$ of $\mathbb{R}^4$ is linearly independent because the row echelon matrix of

$$A = \begin{pmatrix} -1 & -2 & 2 \\ 2 & 1 & 4 \\ 0 & 1 & -1 \\ -3 & 2 & -3 \end{pmatrix} \quad \text{is} \quad \begin{pmatrix} 1 & 0 & 0 \\ 0 & 1 & 0 \\ 0 & 0 & 1 \\ 0 & 0 & 0 \end{pmatrix}$$

so the rank of A is 3. ∎

REMARK   In using this method to test finite subsets of $\mathbb{R}^n$ for linear independence, it suffices to reduce the matrix A to staircase form. It is not necessary to make the nonzero corner entries into 1's or to introduce stacks of zeros above the corner entries. This is true because the rank of a matrix is equal to the number of nonzero rows in its row echelon matrix, and this is the same as the number of nonzero rows in any staircase matrix that can be obtained from A by a sequence of row operations. Thus in Example 5 we can conclude that the matrix A has rank 3 by observing, for example, that A can be reduced to the staircase matrix

$$\begin{pmatrix} 1 & 2 & -2 \\ 0 & 1 & -1 \\ 0 & 0 & 5 \\ 0 & 0 & 0 \end{pmatrix}.$$

Theorem 1 has two important corollaries.

Corollary 1   *Each set of k vectors in $\mathbb{R}^n$, where $k > n$, is linearly dependent.*

Proof   The rank of a matrix cannot be greater than the number of rows in that matrix. ■

Corollary 2   *A set $\{\mathbf{v}_1, \ldots, \mathbf{v}_n\}$ of n vectors in $\mathbb{R}^n$ is linearly independent if and only if the matrix whose column vectors are the vectors $\mathbf{v}_1, \ldots, \mathbf{v}_n$ has determinant different from zero.*

Equivalently, $\{\mathbf{v}_1, \ldots, \mathbf{v}_n\}$ is linearly independent if and only if the matrix whose row vectors are the vectors $\mathbf{v}_1, \ldots, \mathbf{v}_n$ has determinant different from zero.

Proof   The first statement follows from Theorem 1 and the fact that an $n \times n$ matrix has rank n if and only if its determinant is nonzero. The second statement follows from the first statement and the fact that a square matrix and its transpose have the same determinant.   ■

A spanning set for a subspace S of $\mathbb{R}^n$ that is linearly independent is called a **basis** for S.

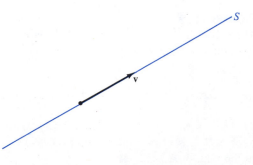

FIGURE 4.10
$\{\mathbf{v}\}$ is a basis for S.

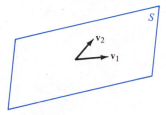

FIGURE 4.11
$\{\mathbf{v}_1, \mathbf{v}_2\}$ is a basis for S.

EXAMPLE 6

If S is a line through $\mathbf{0}$ in $\mathbb{R}^n$ and $\mathbf{v}$ is any nonzero vector in S then the set $\{\mathbf{v}\}$ is a basis for S since it is linearly independent (by Example 1) and it spans S (by Example 4 of Section 4.2). (See Figure 4.10.) ■

EXAMPLE 7

If S is a plane in $\mathbb{R}^3$ and $\mathbf{v}_1, \mathbf{v}_2$ are any two vectors in S neither of which is a scalar multiple of the other, then the set $\{\mathbf{v}_1, \mathbf{v}_2\}$ is a basis for S since it is linearly independent (by Example 2) and it spans S (by Example 5 of Section 4.2). (See Figure 4.11.) ■

EXAMPLE 8

The set

$$\{\mathbf{e}_1, \mathbf{e}_2, \mathbf{e}_3\} = \{(1, 0, 0), (0, 1, 0), (0, 0, 1)\}$$

is a basis for $\mathbb{R}^3$, since it is linearly independent by Theorem 1 (the matrix

$$\begin{pmatrix} 1 & 0 & 0 \\ 0 & 1 & 0 \\ 0 & 0 & 1 \end{pmatrix}$$

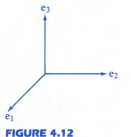

FIGURE 4.12
$\{\mathbf{e}_1, \mathbf{e}_2, \mathbf{e}_3\}$ is a basis for $\mathbb{R}^3$.

has rank 3) and it spans $\mathbb{R}^3$ (by Example 6 of Section 4.2). The set $\{\mathbf{e}_1, \mathbf{e}_2, \mathbf{e}_3\}$ is called the *standard basis* for $\mathbb{R}^3$ (see Figure 4.12). ■

EXAMPLE 9

The set $\{\mathbf{e}_1, \mathbf{e}_2, \ldots, \mathbf{e}_n\}$ is a basis for $\mathbb{R}^n$, since it is linearly independent by Theorem 1 (the matrix

$$\begin{pmatrix} 1 & 0 & & 0 \\ 0 & 1 & & 0 \\ \vdots & \vdots & \vdots & \vdots \\ 0 & 0 & & 1 \end{pmatrix}$$

has rank n) and it spans $\mathbb{R}^n$ (by Example 7 of Section 4.2). The set $\{\mathbf{e}_1, \mathbf{e}_2, \ldots, \mathbf{e}_n\}$ is called the *standard basis* for $\mathbb{R}^n$. ■

EXAMPLE 10

We can find a basis for the solution space of any homogeneous linear system of m equations in n unknowns by solving the system using the techniques described in Sections 1.1 and 1.2. Those techniques lead directly to a spanning set for the solution space which always turns out to be linearly independent. The linear independence is a consequence of the fact that the vectors found always have the property that each has a 1 in some entry where each of the others has a zero. This is sufficient to guarantee linear independence. Thus, in Example 8 of Section 4.2, we found that the vectors

$$(\ \ 1, 1, \ \ \ 0, 0, 0)$$
$$(\ \ 1, 0, -2, 1, 0)$$
$$(-1, 0, \ \ \ 3, 0, 1)$$

form a spanning set for the solution space of the homogeneous linear system considered there. A quick look at the second, fourth, and fifth entries in these vectors should convince you that these vectors form a linearly independent set and hence form a basis for S. ■

EXAMPLE 11

Let us find a basis for the hyperplane

$$x_1 + 2x_2 - x_3 - 3x_4 = 0$$

in $\mathbb{R}^4$. We can solve this equation by setting $x_2 = c_1$, $x_3 = c_2$, $x_4 = c_3$, and solving for x_1 to obtain

$$\mathbf{x} = (x_1, x_2, x_3, x_4) = (-2c_1 + c_2 + 3c_3, c_1, c_2, c_3)$$
$$= c_1(-2, 1, 0, 0) + c_2(1, 0, 1, 0) + c_3(3, 0, 0, 1).$$

The set

$$\{(-2, 1, 0, 0), (1, 0, 1, 0), (3, 0, 0, 1)\}$$

is a basis for the given hyperplane. ■

EXAMPLE 12

Let us find a basis for the kernel of the linear map $f:\mathbb{R}^4 \to \mathbb{R}^2$ defined by

$$f(x_1, x_2, x_3, x_4) = (3x_1 + x_2 - x_4, 2x_1 + x_3 + 3x_4).$$

The kernel of f is the solution set of the vector equation $f(\mathbf{x}) = \mathbf{0}$ which is the same as the solution set of the linear system

$$3x_1 + x_2 \ \ \ \ \ - x_4 = 0$$
$$2x_1 \ \ \ \ \ + x_3 + 3x_4 = 0.$$

The general solution of this system is

$$\mathbf{x} = c_1(-\tfrac{1}{2}, \tfrac{3}{2}, 1, 0) + c_2(-\tfrac{3}{2}, \tfrac{11}{2}, 0, 1).$$

The set $\{(-\tfrac{1}{2}, \tfrac{3}{2}, 1, 0), (-\tfrac{3}{2}, \tfrac{11}{2}, 0, 1)\}$ is a basis for the kernel of f. ■

We shall close this section by stating as a theorem the property that distinguishes bases from arbitrary spanning sets. The proof of this theorem is contained in the discussion that preceded the definition of linear independence at the beginning of this section.

Theorem 2   *Let S be a subspace of $\mathbb{R}^n$ and let $\{\mathbf{v}_1, \ldots, \mathbf{v}_k\}$ be a basis for S. Then each vector $\mathbf{v}$ in S can be expressed in one and only one way as a linear combination*

$$\mathbf{v} = c_1\mathbf{v}_1 + \cdots + c_k\mathbf{v}_k$$

of the vectors $\mathbf{v}_1, \ldots, \mathbf{v}_k$.

EXERCISES

1. Which of the following subsets of $\mathbb{R}^2$ are linearly independent?
   (a) $\{(1, 0), (0, 1)\}$
   (b) $\{(1, 2), (2, 1)\}$
   (c) $\{(-1, 2), (1, -2)\}$
   (d) $\{(-1, 1), (2, 0), (3, 1)\}$
   (e) $\{(1, 1)\}$
   (f) $\{(0, 0)\}$

2. Test the following subsets of $\mathbb{R}^3$ for linear independence.
   (a) $\{(1, 5, -4), (1, 6, -4)\}$
   (b) $\{(1, 5, -4), (1, 6, -4), (1, 7, -4)\}$
   (c) $\{(1, 0, 0), (0, 1, 0), (0, 0, 0)\}$
   (d) $\{(1, 1, 1), (1, 2, 3), (1, 4, 9)\}$
   (e) $\{(1, 2, 3), (1, 4, 9), (1, 8, 27)\}$
   (f) $\{(1, 1, 1), (1, 2, 3), (1, 4, 9), (1, 8, 27)\}$
   (g) $\{(2, -3, 2), (3, -1, 1), (0, -7, 4)\}$
   (h) $\{(-1, 1, 2), (1, 2, 3), (5, 1, 0)\}$
   (i) $\{(1, 1, 0), (0, 1, 1), (3, -5, 12)\}$

3. Test the following subsets of $\mathbb{R}^4$ for linear independence.
   (a) $\{(-1, 0, 0, 1)\}$
   (b) $\{(-1, 0, 0, 1), (0, 1, 1, 0)\}$
   (c) $\{(-1, 0, 0, 1), (0, 1, 1, 0), (1, 0, 0, 1)\}$
   (d) $\{(-1, 0, 0, 1), (0, 1, 1, 0), (1, 0, 0, 1), (0, -1, 1, 0)\}$
   (e) $\{(-1, 0, 0, 1), (0, 1, 1, 0), (1, 0, 0, 1), (0, -1, 1, 0), (1, 0, 1, 0)\}$
   (f) $\{(-1, 1, 0, 1), (1, 2, 3, 3), (2, 1, 3, 2)\}$

4. Which of the following sets are bases for $\mathbb{R}^2$?
   (a) $\{(1, 2)\}$
   (b) $\{(1, 2), (1, 3)\}$
   (c) $\{(1, 2), (1, 3), (1, 4)\}$
   (d) $\{(1, 2), (2, 4)\}$
   (e) $\{(1, 2), (2, 4), (-3, -6)\}$

5. Which of the following are bases for $\mathbb{R}^3$?
   (a) $\{(-1, 0, 0), (0, -1, 0), (0, 0, -1)\}$
   (b) $\{(1, 0, 0), (1, 1, 0), (1, 1, 1)\}$
   (c) $\{(0, 1, -1), (1, 0, -1), (1, -1, 0)\}$
   (d) $\{(1, 2, 3), (1, 4, 9)\}$
   (e) $\{(1, 2, 3), (1, 4, 9), (1, 8, 27)\}$
   (f) $\{(1, 2, 3), (1, 4, 9), (1, 8, 27), (1, 16, 81)\}$
   (g) $\{(1, a, a^2), (1, b, b^2), (1, c, c^2)\}$ where $a, b, c \in \mathbb{R}$ are distinct

6. Which of the sets in Exercise 1 are bases for $\mathbb{R}^2$?

7. Which of the sets in Exercise 2 are bases for $\mathbb{R}^3$?

8. Which of the sets in Exercise 3 are bases for $\mathbb{R}^4$?

9. Find a basis for each of the following subspaces.
   (a) the plane $x_1 + x_2 + x_3 = 0$ in $\mathbb{R}^3$
   (b) the hyperplane $x_1 - 2x_2 + 3x_3 + x_4 = 0$ in $\mathbb{R}^4$
   (c) the hyperplane $3x_1 - 2x_2 - x_4 = 0$ in $\mathbb{R}^4$

(d)   the solution space in $\mathbb{R}^3$ of the linear system

$$x_1 - 5x_2 - 7x_3 = 0$$

$$2x_1 + 2x_2 - 3x_3 = 0$$

(e)   the solution space in $\mathbb{R}^4$ of the linear system

$$2x_1 - x_2 - x_3 + 2x_4 = 0$$

$$3x_1 + 2x_2 - x_3 - x_4 = 0$$

10.   Find a basis for the kernel of each of the following linear maps.

(a) $f(x_1, x_2) = x_1 + x_2$

(b) $f(x_1, x_2, x_3) = x_1 + x_2 + x_3$

(c) $f(x_1, x_2, x_3) = (x_1, x_2 + x_3)$

(d) $f(x_1, x_2, x_3) = (x_1 + 2x_2, 0)$

(e) $f(x_1, x_2, x_3, x_4) = (x_1 - x_2, x_3 - x_4)$

11.   Show that if $\mathbf{v}$ and $\mathbf{w}$ are two vectors in $\mathbb{R}^3$, then $\{\mathbf{v}, \mathbf{w}\}$ is linearly dependent if and only if $\mathbf{v} \times \mathbf{w} = \mathbf{0}$.

12.   Show that if $\mathbf{u}$, $\mathbf{v}$, and $\mathbf{w}$ are three vectors in $\mathbb{R}^3$, then $\{\mathbf{u}, \mathbf{v}, \mathbf{w}\}$ is linearly dependent if and only if $(\mathbf{u} \times \mathbf{v}) \cdot \mathbf{w} = 0$.

13.   Show that the subset $\{\mathbf{v}_1, \mathbf{v}_2, \ldots, \mathbf{v}_k\}$ of $\mathbb{R}^n$ is linearly dependent if and only if one of the vectors is a linear combination of the others.

14.   Explain why, if $\{\mathbf{v}_1, \mathbf{v}_2, \ldots, \mathbf{v}_k\}$ is a basis for $\mathbb{R}^n$, then k must be equal to n. (Why can't k be greater than n? Why can't k be less than n?)

15.   Explain why the condition $\det A \neq 0$ is a necessary and sufficient condition that the column vectors of A form a basis for $\mathbb{R}^n$.

16.   Prove the converse of Theorem 2: if each vector $\mathbf{v}$ in the subspace S of $\mathbb{R}^n$ can be expressed in one and only one way as a linear combination of the vectors $\mathbf{v}_1, \ldots, \mathbf{v}_k \in S$, then $\{\mathbf{v}_1, \ldots, \mathbf{v}_k\}$ is a basis for S.

4.4
Coordinates and Dimension

We saw in Section 4.3 that if $\{\mathbf{v}_1, \ldots, \mathbf{v}_k\}$ is a basis for a subspace S of $\mathbb{R}^n$, then each vector $\mathbf{v}$ in S can be expressed in one and only one way as a linear combination

$$\mathbf{v} = c_1\mathbf{v}_1 + c_2\mathbf{v}_2 + \cdots + c_k\mathbf{v}_k$$

of the vectors $\mathbf{v}_1, \mathbf{v}_2, \ldots, \mathbf{v}_k$. The numbers $c_1, c_2, \ldots, c_k$ are called the *coordinates* of the vector $\mathbf{v}$ relative to the given basis. In this section we shall see that the use of coordinates leads to a complete understanding of subspaces of $\mathbb{R}^n$. They are simply copies of $\mathbb{R}^k$ (for some k) that sit inside $\mathbb{R}^n$.

Our goal, then, is to set up a one-to-one correspondence between the subspace S of $\mathbb{R}^n$ and $\mathbb{R}^k$ for some appropriate k. We shall do this by assigning to each vector $\mathbf{v}$ in S an ordered k-tuple whose entries are the coordinates $c_1, \ldots, c_k$ of $\mathbf{v}$ relative to some basis $\{\mathbf{v}_1, \ldots, \mathbf{v}_k\}$ for S. But we must be careful. The order in which we list the basis vectors is important. For example, if $\{\mathbf{u}, \mathbf{w}\} = \{\mathbf{w}, \mathbf{u}\}$ is a basis for S and if $\mathbf{v} = 2\mathbf{u} + 3\mathbf{w}$, then the coordinates of $\mathbf{v}$ are 2 and 3. Do we assign to $\mathbf{v}$ the ordered pair $(2, 3)$ or the ordered pair

(3, 2)? It depends on whether we think of **u** or **w** as being the first vector in the basis. If **u** is first then we write $\mathbf{v} = 2\mathbf{u} + 3\mathbf{w}$ and assign to **v** the ordered pair (2, 3). If **w** is first then we write $\mathbf{v} = 3\mathbf{w} + 2\mathbf{u}$ and assign to **v** the ordered pair (3, 2). So, to be able to assign to each vector **v** in S exactly one coordinate pair (c_1, c_2) we must specify not only the basis used but also an ordering of the basis vectors. Thus we must specify an *ordered basis* for S.

An **ordered basis** for a subspace S of $\mathbb{R}^n$ is an ordered k-tuple $(\mathbf{v}_1, \ldots, \mathbf{v}_k)$ where $\{\mathbf{v}_1, \ldots, \mathbf{v}_k\}$ is a basis for S. Two ordered bases $(\mathbf{v}_1, \ldots, \mathbf{v}_k)$ and $(\mathbf{w}_1, \ldots, \mathbf{w}_k)$ are equal if and only if $\mathbf{v}_1 = \mathbf{w}_1$, $\mathbf{v}_2 = \mathbf{w}_2, \ldots, \mathbf{v}_k = \mathbf{w}_k$. If $\mathbf{B} = (\mathbf{v}_1, \ldots, \mathbf{v}_k)$ is an ordered basis for a subspace S of $\mathbb{R}^n$ and if **v** is any vector in S, then the numbers $c_1, \ldots, c_k$ such that

$$\mathbf{v} = c_1\mathbf{v}_1 + \cdots + c_k\mathbf{v}_k$$

are called the **B-*coordinates*** of **v**, and the ordered k-tuple $(c_1, \ldots, c_k)$ is called the **B-*coordinate k-tuple*** of **v**.

EXAMPLE 1

Let S be a line in $\mathbb{R}^n$ that passes through **0** and let $\mathbf{v}_1$ be any nonzero vector in S. Then we know, from Example 6 of Section 4.3, that $\{\mathbf{v}_1\}$ is a basis for S. Let $\mathbf{B} = (\mathbf{v}_1)$. Then the **B**-coordinate of any vector $\mathbf{v} \in S$ is the real number c such that $\mathbf{v} = c\mathbf{v}_1$ (see Figure 4.13). The rule that assigns to each vector **v** in S its **B**-coordinate defines a one-to-one correspondence between S and the set of real numbers. In other words, a choice of ordered basis **B** for S makes the line S into a "number line." ■

EXAMPLE 2

Let S be a plane in $\mathbb{R}^3$ that passes through **0** and let $\mathbf{v}_1$ and $\mathbf{v}_2$ be any two vectors in S neither of which is a scalar multiple of the other. Then we know, from Example 7 of Section 4.3, that $\{\mathbf{v}_1, \mathbf{v}_2\}$ is a basis for S. Let $\mathbf{B} = (\mathbf{v}_1, \mathbf{v}_2)$. Then the **B**-coordinate pair of any vector **v** in S is the ordered pair (c_1, c_2) of real numbers such that $\mathbf{v} = c_1\mathbf{v}_1 + c_2\mathbf{v}_2$ (see Figure 4.14). The rule that assigns to each vector **v** in S its **B**-coordinate pair (c_1, c_2) defines a one-to-one correspondence between S and $\mathbb{R}^2$. ■

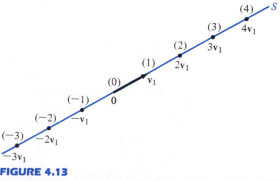

FIGURE 4.13
Points on the line S and their **B**-coordinates, where $\mathbf{B} = (\mathbf{v}_1)$.

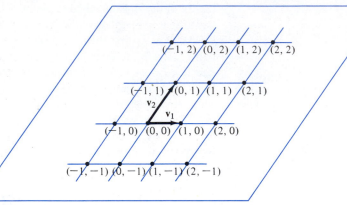

FIGURE 4.14
B-coordinates of points in the plane S, where $\mathbf{B} = (\mathbf{v}_1, \mathbf{v}_2)$.

EXAMPLE 3

Let S be the plane $x_1 - 3x_2 - 2x_3 = 0$ through $\mathbf{0}$ in $\mathbb{R}^3$. Then

$$\{(1, 1, -1), (1, -1, 2)\}$$

is a basis for S, since both of these vectors are in S and neither is a scalar multiple of the other. The vector $\mathbf{v} = (3, 1, 0)$ is also in S. Its coordinates relative to the ordered basis $\mathbf{B} = ((1, 1, -1), (1, -1, 2))$ can be found by solving the vector equation

$$(3, 1, 0) = c_1(1, 1, -1) + c_2(1, -1, 2).$$

This vector equation is equivalent to the linear system

$$c_1 + c_2 = 3$$

$$c_1 - c_2 = 1$$

$$-c_1 + 2c_2 = 0.$$

The unique solution of this system is $(c_1, c_2) = (2, 1)$. Therefore the **B**-coordinate pair of the vector $(3, 1, 0)$ is $(2, 1)$ (see Figure 4.15). ∎

EXAMPLE 4

Let S be any subspace of $\mathbb{R}^n$ that is spanned by two vectors $\mathbf{v}_1$ and $\mathbf{v}_2$, neither of which is a scalar multiple of the other. Then we know, from Example 2 of Section 4.3, that the spanning set $\{\mathbf{v}_1, \mathbf{v}_2\}$ is linearly independent and hence is a basis for S. Let $\mathbf{B} = (\mathbf{v}_1, \mathbf{v}_2)$. Then, as in Example 2 (where $n = 3$), the **B**-coordinate pair of any vector $\mathbf{v}$ in S is the ordered pair (c_1, c_2) of real numbers such that $\mathbf{v} = c_1\mathbf{v}_1 + c_2\mathbf{v}_2$. The rule that assigns to each vector $\mathbf{v}$ in S its **B**-coordinate pair (c_1, c_2) defines a one-to-one correspondence between S and $\mathbb{R}^2$. Thus the subspace S is just a copy of $\mathbb{R}^2$ that sits inside $\mathbb{R}^n$. For this reason, and because $\mathbf{0} \in S$, we call S a *plane through* $\mathbf{0}$ in $\mathbb{R}^n$. ∎

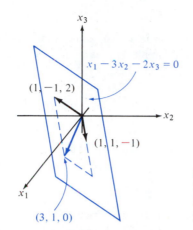

FIGURE 4.15
The **B**-coordinate pair of $(3, 1, 0)$ is $(2, 1)$, where $\mathbf{B} = ((1, 1, -1), (1, -1, 2))$ is an ordered basis for the plane $x_1 - 3x_2 - 2x_3 = 0$.

EXAMPLE 5

Let k be any integer between 1 and n and let S be any subspace of $\mathbb{R}^n$ that has a basis $\{\mathbf{v}_1, \ldots, \mathbf{v}_k\}$ containing k vectors. Let $\mathbf{B} = (\mathbf{v}_1, \ldots, \mathbf{v}_k)$. Then the **B**-coordinate k-tuple of any vector $\mathbf{v}$ in S is the ordered k-tuple $(c_1, \ldots, c_k)$ of real numbers such that $\mathbf{v} = c_1\mathbf{v}_1 + \cdots + c_k\mathbf{v}_k$. The rule that assigns to each vector $\mathbf{v}$ in S its **B**-coordinate k-tuple defines a one-to-one correspondence between S and $\mathbb{R}^k$. Thus the subspace S is just a copy of $\mathbb{R}^k$ that sits inside $\mathbb{R}^n$. For this reason, and because $\mathbf{0} \in S$, we call S a k-***plane through*** $\mathbf{0}$ in $\mathbb{R}^n$. Notice that a 1-plane through $\mathbf{0}$ in $\mathbb{R}^n$ is the same as a line through $\mathbf{0}$, and that a 2-plane through $\mathbf{0}$ in $\mathbb{R}^n$ is the same as a plane through $\mathbf{0}$ (see Example 4). ■

It turns out that *if S is any subspace of $\mathbb{R}^n$ other than the zero subspace* $\{\mathbf{0}\}$, *then S is a k-plane through $\mathbf{0}$ for some $k \leq n$*. Indeed, to find a basis for S all you have to do is pick a nonzero vector $\mathbf{v}_1$ in S, then pick a vector $\mathbf{v}_2$ in S that is not a scalar multiple of $\mathbf{v}_1$, then pick a vector $\mathbf{v}_3$ in S that is not a linear combination of $\mathbf{v}_1$ and $\mathbf{v}_2$, and so on. Eventually this process must stop, because at each step you will have found a subset of $\mathbb{R}^n$ that is linearly independent (see Exercise 14), and we know from Section 4.3 that there are no subsets of $\mathbb{R}^n$ that are linearly independent and contain more than n vectors. But the only way the process *can* stop is if at some point every vector in S is a linear combination of the vectors $\{\mathbf{v}_1, \ldots, \mathbf{v}_k\}$ previously chosen. At this point, then, you will have found a basis for S, since the set $\{\mathbf{v}_1, \ldots, \mathbf{v}_k\}$ that you have found is necessarily linearly independent and it spans S.

It also turns out that, *if S is any subspace of $\mathbb{R}^n$ other than $\{\mathbf{0}\}$, then all bases for S contains the same number of vectors*. We shall prove this fact at the end of this section (see Theorem 2). The number of elements in any basis for S is called the ***dimension*** of S.

EXAMPLE 6

Each line through $\mathbf{0}$ in $\mathbb{R}^n$ has dimension 1 because it has a basis containing one vector (see Example 6 in Section 4.3). ■

EXAMPLE 7

Each plane through $\mathbf{0}$ in $\mathbb{R}^n$ has dimension 2 because it has a basis containing two vectors (see Example 4). ■

EXAMPLE 8

Each k-plane through $\mathbf{0}$ in $\mathbb{R}^n$ has dimension k because it has a basis containing k vectors (see Example 5). ■

REMARK   Although the subspace $\{\mathbf{0}\}$ of $\mathbb{R}^n$ has no basis, it still is assigned a dimension. We say that *the zero subspace* $\{\mathbf{0}\}$ *of* $\mathbb{R}^n$ *has dimension* 0.

EXAMPLE 9

For each integer $n \geq 1$, the space $\mathbb{R}^n$ has dimension n because the standard basis $\{\mathbf{e}_1, \ldots, \mathbf{e}_n\}$ for $\mathbb{R}^n$ contains n vectors. ■

EXAMPLE 10

Let S be the solution space of the homogeneous linear system

$$a_{11}x_1 + \cdots + a_{1n}x_n = 0$$

$$\cdots$$

$$a_{m1}x_1 + \cdots + a_{mn}x_n = 0.$$

We know, from Example 10 of Section 4.3 that we can find a basis for S by solving this system using the techniques described in Sections 1.1 and 1.2. The number of vectors in this basis is the same as the number of arbitrary constants in the general solution. But recall that the arbitrary constants in the general solution correspond to the unknowns that can be specified arbitrarily. These are the unknowns that do not correspond to corner 1's in the echelon matrix of the system. The number of corner 1's is the rank of the matrix of the system. This is the same as the rank of the coefficient matrix, since the system is homogeneous. Since the number of unknowns is n, we can conclude that *the dimension of the solution space of any homogeneous system of m linear equations in n unknowns is equal to $n - r$ where r is the rank of the coefficient matrix.* ■

EXAMPLE 11

Let $f:\mathbb{R}^n \to \mathbb{R}^m$ be a linear map. The kernel of f is the solution space of the vector equation $f(\mathbf{x}) = \mathbf{0}$, which is the same as the solution space of the linear system $A\mathbf{x} = \mathbf{0}$, where A is the matrix of f. Since the coefficient matrix of this system is A, and since the rank of A is equal to the rank of f, we see from Example 10 that *the dimension of the kernel of any linear map $f:\mathbb{R}^n \to \mathbb{R}^m$ is equal to $n - r$, where r is the rank of f.* ■

REMARK   It can be shown (see Exercise 18) that the dimension of the image of a linear map $f:\mathbb{R}^n \to \mathbb{R}^m$ is equal to the rank r of f. Since the dimension of the kernel of f is $n - r$, this says that *the dimension of the kernel of f plus the dimension of the image of f is equal to n, where $f:\mathbb{R}^n \to \mathbb{R}^m$.*

EXAMPLE 12

Let us find the dimension of the kernel of the linear map $f:\mathbb{R}^3 \to \mathbb{R}^3$ defined by

$$f(x_1, x_2, x_3) = (-x_1 + x_2 + 2x_3, 2x_1 + x_2 - x_3, -x_1 + x_3).$$

The matrix of f is

$$\begin{pmatrix} -1 & 1 & 2 \\ 2 & 1 & -1 \\ -1 & 0 & 1 \end{pmatrix}.$$

Its row echelon matrix is

$$\begin{pmatrix} 1 & 0 & -1 \\ 0 & 1 & 1 \\ 0 & 0 & 0 \end{pmatrix}.$$

Therefore the rank of f is 2. By Example 11, the kernel of f has dimension $3 - 2 = 1$.

Can you find a basis for the kernel of f? ■

EXAMPLE 13

The *row space* of an $m \times n$ matrix A is the subspace of $\mathbb{R}^n$ spanned by the row vectors of A. Thus, if

$$A = \begin{pmatrix} a_{11} & a_{12} & \cdots & a_{1n} \\ a_{21} & a_{22} & \cdots & a_{2n} \\ & & \cdots & \\ a_{m1} & a_{m2} & \cdots & a_{mn} \end{pmatrix},$$

then the row space of A is

$$\mathscr{L}((a_{11}, a_{12}, \ldots, a_{1n}), (a_{21}, a_{22}, \ldots, a_{2n}), \ldots, (a_{m1}, a_{m2}, \ldots, a_{mn})).$$

It can be shown (see Exercise 15) that *if B is obtained from A by a sequence of row operations, then A and B have the same row space.* In particular, each matrix has the same row space as its row echelon matrix. But the nonzero row vectors of an echelon matrix span the row space of that matrix and they are linearly independent (each of these vectors has a 1, a corner 1, in a column where each of the other vectors has a zero). Therefore these vectors form a basis for the row space. Since the number of nonzero row vectors in the echelon matrix of A is equal to the rank of A, we can conclude that *the dimension of the row space of A is equal to the rank of A.* ■

Given any finite set $\{\mathbf{v}_1, \ldots, \mathbf{v}_k\}$ of vectors in $\mathbb{R}^n$, we can write down the matrix A whose row vectors are the vectors $\mathbf{v}_1, \ldots, \mathbf{v}_k$. The row space of this matrix A is then the subspace S of $\mathbb{R}^n$ spanned by the given vectors $\mathbf{v}_1, \ldots, \mathbf{v}_k$. The procedure described in Example 13 leads to a basis for S. This procedure may be summarized as follows:

> To find a basis for the subspace S of $\mathbb{R}^n$ spanned by the vectors $\mathbf{v}_1, \ldots, \mathbf{v}_k$.
>
> 1. Write down the matrix A whose row vectors are the vectors $\mathbf{v}_1, \ldots, \mathbf{v}_k$.
> 2. Find the row echelon matrix B of A. Then the nonzero row vectors of B form a basis for S.

EXAMPLE 14

To find a basis for the subspace

$$S = \mathcal{L}((1, -1, 3, 2), (-1, 3, -2, 2), (2, 1, 2, -1), (-1, 0, 2, 7))$$

of $\mathbb{R}^4$, we consider the matrix

$$A = \begin{pmatrix} 1 & -1 & 3 & 2 \\ -1 & 3 & -2 & 2 \\ 2 & 1 & 2 & -1 \\ -1 & 0 & 2 & 7 \end{pmatrix}$$

whose row vectors are the vectors in the given spanning set. The echelon matrix of A is

$$B = \begin{pmatrix} 1 & 0 & 0 & -3 \\ 0 & 1 & 0 & 1 \\ 0 & 0 & 1 & 2 \\ 0 & 0 & 0 & 0 \end{pmatrix}.$$

The set

$$\{(1, 0, 0, -3), (0, 1, 0, 1), (0, 0, 1, 2)\},$$

consisting of the nonzero row vectors of B, is a basis for S. The dimension of S is 3. ∎

We know from Section 4.2 that the image of a linear map $f: \mathbb{R}^n \to \mathbb{R}^m$ is spanned by the vectors $f(\mathbf{e}_1), \ldots, f(\mathbf{e}_n)$. Therefore we can find a basis for the image of f by row reducing the matrix that has $f(\mathbf{e}_1), \ldots, f(\mathbf{e}_n)$ as row vectors.

EXAMPLE 15

Let $f: \mathbb{R}^3 \to \mathbb{R}^3$ be the linear map

$$f(x_1, x_2, x_3) = (-x_1 + x_2 + 2x_3, 2x_1 + x_2 - x_3, -x_1 + x_3),$$

as in Example 12. Then

$$\begin{pmatrix} f(\mathbf{e}_1) \\ f(\mathbf{e}_2) \\ f(\mathbf{e}_3) \end{pmatrix} = \begin{pmatrix} -1 & 2 & -1 \\ 1 & 1 & 0 \\ 2 & -1 & 1 \end{pmatrix} \to \begin{pmatrix} 1 & 0 & \frac{1}{3} \\ 0 & 1 & -\frac{1}{3} \\ 0 & 0 & 0 \end{pmatrix}.$$

The set

$$\{(1, 0, \tfrac{1}{3}), (0, 1, -\tfrac{1}{3})\},$$

consisting of the nonzero row vectors of this echelon matrix, is a basis for the image of f. The dimension of the image of f is 2. ■

We close this section by proving that each basis for a given subspace S of $\mathbb{R}^n$ contains the same number of vectors as every other basis for S. This is a consequence of the following theorem.

Theorem 1 *Suppose $\{v_1, \ldots, v_k\}$ is a spanning set for the subspace S of $\mathbb{R}^n$ and suppose $\{w_1, \ldots, w_m\}$ is a linearly independent set of vectors in S. Then $k \geq m$.*

Proof Since $\{v_1, \ldots, v_k\}$ spans S, each vector w_i can be expressed as a linear combination of $v_1, \ldots, v_k$. Thus there exist real numbers $a_{11}, a_{21}, \ldots, a_{km}$ such that

$$w_1 = a_{11}v_1 + a_{21}v_2 + \cdots + a_{k1}v_k$$

$$w_2 = a_{12}v_1 + a_{22}v_2 + \cdots + a_{k2}v_k$$

(1)

$$\cdots$$

$$w_m = a_{1m}v_1 + a_{2m}v_2 + \cdots + a_{km}v_k.$$

Since $\{w_1, \ldots, w_m\}$ is linearly independent, the only solution $(c_1, \ldots, c_m)$ of the equation

$$(2) \qquad c_1w_1 + \cdots + c_mw_m = 0$$

is the zero solution, $(c_1, \ldots, c_m) = (0, \ldots, 0)$. But by multiplying the ith equation in (1) by $c_i (1 \leq i \leq m)$ and adding the resulting equations, we obtain

$$c_1w_1 + \cdots + c_mw_m = (a_{11}c_1 + \cdots + a_{1m}c_m)v_1$$

$$+ \cdots$$

$$+ (a_{k1}c_1 + \cdots + a_{km}c_m)v_k.$$

Thus $(c_1, \ldots, c_m)$ is a solution of (2) if it is a solution of the system

$$a_{11}c_1 + \cdots + a_{1m}c_m = 0$$

$$\cdots$$

$$a_{k1}c_1 + \cdots + a_{km}c_m = 0.$$

If $k < m$ then this system has more unknowns than equations and hence it must have a nonzero solution. Thus the independence of $\{w_1, \ldots, w_m\}$ implies $k \geq m$. ■

Theorem 2 *Suppose* $\{\mathbf{v}_1, \ldots, \mathbf{v}_k\}$ *and* $\{\mathbf{w}_1, \ldots, \mathbf{w}_m\}$ *are two bases for the same subspace* S *of* $\mathbb{R}^n$. *Then* $k = m$.

Proof   Since $\{\mathbf{v}_1, \ldots, \mathbf{v}_k\}$ spans V and $\{\mathbf{w}_1, \ldots, \mathbf{w}_m\}$ is linearly independent, $k \geq m$ by Theorem 1. On the other hand, since $\{\mathbf{w}_1, \ldots, \mathbf{w}_m\}$ spans V and $\{\mathbf{v}_1, \ldots, \mathbf{v}_k\}$ is linearly independent, $m \geq k$. Thus $k = m$.   ■

EXERCISES

1. Let S be the plane $x_1 - 3x_2 + 5x_3 = 0$ through $\mathbf{0}$ in $\mathbb{R}^3$, and let $\mathbf{B}$ be the ordered basis $((3, 1, 0), (-5, 0, 1))$ for S. Verify that the following vectors are in S and find the $\mathbf{B}$-coordinate pair of each.

    (a) $(1, 2, 1)$ (b) $(-2, 1, 1)$

    (c) $(7, -1, -2)$ (d) $(3, 1, 0)$

    (e) $(-5, 0, 1)$ (f) $(-5, 5, 4)$

2. Let S be the plane $x_1 - 3x_2 + 5x_3 = 0$ through $\mathbf{0}$ in $\mathbb{R}^3$ and let $\mathbf{v} = (-5, 5, 4) \in S$. Find the $\mathbf{B}$-coordinate pair of $\mathbf{v}$, where $\mathbf{B} =$

    (a) $((3, 1, 0), (-5, 0, 1))$ (b) $((-5, 0, 1)), (3, 1, 0))$

    (c) $((3, 1, 0), (1, 2, 1))$ (d) $((7, -1, -2), (1, 2, 1))$

    (e) $((-5, 5, 4), (1, 2, 1))$ (f) $((7, -1, -2), (-5, 5, 4))$

3. Let S be the plane $x_1 - 3x_2 + 5x_3 = 0$ through $\mathbf{0}$ in $\mathbb{R}^3$. Find the vector $\mathbf{v} \in S$ whose $\mathbf{B}$-coordinate pair is $(3, -2)$, where $\mathbf{B} =$

    (a) $((1, 2, 1), (-2, 1, 1))$ (b) $((3, 1, 0), (-5, 0, 1))$

    (c) $((-5, 0, 1), (3, 1, 0))$ (d) $((3, 1, 0), (-2, 1, 1))$

    (e) $((1, \frac{1}{3}, 0), (-2, 1, 1))$ (f) $((-5, 5, 4), (7, -1, -2))$

4. Find the $\mathbf{B}$-coordinate pair of $\mathbf{v} = (3, -4)$ where $\mathbf{B}$ is the given ordered basis for $\mathbb{R}^2$.

    (a) $\mathbf{B} = ((1, 0), (0, 1))$ (b) $\mathbf{B} = ((0, 1), (1, 0))$

    (c) $\mathbf{B} = ((1, 0), (1, 1))$ (d) $\mathbf{B} = ((-1, 1), (1, 1))$

    (e) $\mathbf{B} = ((1, 2), (2, 1))$ (f) $\mathbf{B} = ((2, 1), (1, 2))$

    (g) $\mathbf{B} = ((3, -4), (1, 0))$ (h) $\mathbf{B} = ((\pi, 2\pi), (3, -4))$

5. Find the $\mathbf{B}$-coordinate triple of $\mathbf{v} = (2, 3, 4)$ where $\mathbf{B}$ is the given ordered basis for $\mathbb{R}^3$.

    (a) $((1, 0, 0), (0, 1, 0), (0, 0, 1))$

    (b) $((0, 1, 0), (0, 0, 1), (1, 0, 0))$

    (c) $((1, 1, 0), (1, 0, 1), (0, 1, 1))$

    (d) $((1, 1, 1), (1, 2, 3), (1, 4, 9))$

    (e) $((1, 1, 1), (1, \frac{1}{2}, \frac{1}{3}), (1, \frac{1}{4}, \frac{1}{9}))$

    (f) $((4, 3, 7), (-6, 5, 4), (2, 3, 4))$

6. Let S be the plane through $\mathbf{0}$ in $\mathbb{R}^4$ spanned by the vectors $(-1, 1, 1, 2)$ and $(1, 1, 3, 4)$. Find the $\mathbf{B}$-coordinate pair of each of the following vectors in S, where $\mathbf{B} = ((-1, 1, 1, 2), (1, 1, 3, 4))$.

    (a) $(1, 0, 1, 1)$ (b) $(1, 3, 7, 10)$

    (c) $(1, 5, 10, 16)$ (d) $(3, -1, 1, 0)$

    (e) $(1, 1, 3, 4)$ (f) $(-1, 1, 1, 2)$

7. Let S be the 3-plane through $\mathbf{0}$ in $\mathbb{R}^4$ with ordered basis $\mathbf{B} = ((-1, 1, -1, 1),$ $(1, 1, 1, -3), (1, 0, -2, 1))$. Find the $\mathbf{B}$-coordinate triple of each of the following vectors in S.

   (a) $(1, -1, 0, 0)$ (b) $(1, 0, 0, -1)$

   (c) $(0, 1, -3, 2)$ (d) $(3, -3, 3, -3)$

   (e) $(1, 1, 1, -3)$ (f) $(-3, 1, 1, 1)$

8. Let S and $\mathbf{B}$ be as in Exercise 7. Find the vector $\mathbf{v}$ in S whose $\mathbf{B}$-coordinate triple is

   (a) $(1, 0, 0)$ (b) $(0, 1, 0)$

   (c) $(0, 0, 1)$ (d) $(-1, 2, 0)$

   (e) $(1, 1, -1)$ (f) $(3, -1, 2)$

9. Let $\mathbf{v}$ be a nonzero vector in $\mathbb{R}^n$, let S be the line $\mathbf{x} = t\mathbf{v}$, and let $\mathbf{B} = (2\mathbf{v})$. Find the $\mathbf{B}$-coordinate of

   (a) $\mathbf{v}$ (b) $2\mathbf{v}$

   (c) $-\mathbf{v}$ (d) $6\mathbf{v}$

   (e) $7\mathbf{v}$ (f) $\mathbf{0}$

10. Show that if $\mathbf{B}$ is an ordered basis for a subspace S of $\mathbb{R}^n$, if $\mathbf{v}$ and $\mathbf{w} \in S$ have $\mathbf{B}$-coordinate k-tuples $(a_1, \ldots, a_k)$ and $(b_1, \ldots, b_k)$ respectively, and if $c \in \mathbb{R}$, then

    (a) $\mathbf{v} + \mathbf{w}$ has $\mathbf{B}$-coordinate k-tuple $(a_1 + b_1, \ldots, a_k + b_k)$

    (b) $c\mathbf{v}$ has $\mathbf{B}$-coordinate k-tuple $(ca_1, \ldots, ca_k)$

11. Find a basis for, and the dimension of, each of the following subspaces of $\mathbb{R}^3$.

    (a) the solution space of

    $$x_1 - x_2 + x_3 = 0$$
    $$-x_1 + x_2 - x_3 = 0$$

    (b) the solution space of

    $$x_1 + 2x_2 - x_3 = 0$$
    $$2x_1 + x_2 - x_3 = 0$$

    (c) $\mathscr{L}((-1, 1, 1), (1, -1, 1), (1, 1, -1))$

    (d) $\mathscr{L}((-1, 1, 0), (-1, 0, 1), (0, -1, 1))$

    (e) $\mathscr{L}((2, 1, 0), (-2, -1, 0))$

    (f) the kernel of the linear map $f(x_1, x_2, x_3) = (x_1, x_1, x_1)$

    (g) the image of the linear map $f(x_1, x_2, x_3) = (x_1, x_1, x_1)$

    (h) the kernel of the linear map $f(x_1, x_2, x_3) = (x_1 - x_2, x_1 - x_3, x_2 - x_3)$

    (i) the image of the linear map $f(x_1, x_2, x_3) = (x_1 - x_2, x_1 - x_3, x_2 - x_3)$

    (j) the kernel of the linear map $f(x_1, x_2, x_3) = (x_1, x_1 + x_2, x_1 + x_2 + x_3)$

    (k) the image of the linear map $f(x_1, x_2, x_3) = (x_1, x_1 + x_2, x_1 + x_2 + x_3)$

12. Find a basis for, and the dimension of, each of the given subspaces.

    (a) the solution space in $\mathbb{R}^4$ of the linear system

    $$x_1 + 2x_2 - x_3 - x_4 = 0$$
    $$-x_1 + x_2 - 2x_3 - 3x_4 = 0$$

(b)   the hyperplane

$$x_1 - x_2 - x_3 + x_4 - x_5 = 0$$

in $\mathbb{R}^5$

(c) $\mathcal{L}((-1, 1, 3, 0), (2, -1, 1, 3), (-5, 4, 7, -6))$

(d)   the row space of the matrix

$$\begin{pmatrix} 2 & 1 & -1 & 3 & 6 & 7 \\ -1 & 0 & 2 & 5 & -1 & 3 \\ 4 & 1 & -5 & -7 & 8 & 1 \end{pmatrix}$$

(e)   the kernel of the linear map $f:\mathbb{R}^5 \to \mathbb{R}^3$ given by

$$f(x_1, x_2, x_3, x_4, x_5) = (x_1 + x_2 + x_3, x_2 + x_3 + x_4, x_3 + x_4 + x_5)$$

(f)   the image of the linear map $f:\mathbb{R}^4 \to \mathbb{R}^4$ given by

$$f(x_1, x_2, x_3, x_4) = (x_1 + x_2, x_2 + x_3, x_3 + x_4, x_4 + x_1)$$

13.   Let S be a hyperplane that passes through $\mathbf{0}$ in $\mathbb{R}^n$. What is the dimension of S?

14.   Show that if $\{\mathbf{v}_1, \ldots, \mathbf{v}_k\}$ is a linearly independent set in $\mathbb{R}^n$ and $\mathbf{v}_{k+1}$ is any vector in $\mathbb{R}^n$ that is not in $\mathcal{L}(\mathbf{v}_1, \ldots, \mathbf{v}_k)$ then $\{\mathbf{v}_1, \ldots, \mathbf{v}_k, \mathbf{v}_{k+1}\}$ is linearly independent.

15.   Let A be an $m \times n$ matrix.

(a)   Show that if B is obtained from A by interchanging two rows, then A and B have the same row space.

(b)   Show that if B is obtained from A by multiplying a row of A by a nonzero real number, then A and B have the same row space.

(c)   Show that if B is obtained from A by adding a scalar multiple of one row of A to another, then A and B have the same row space.

(d)   Conclude that if B is obtained from A by any sequence of row operations then A and B have the same row space.

16.   Let k be an integer between 1 and n. A **k-plane** in $\mathbb{R}^n$ is a subset of $\mathbb{R}^n$ of the form

$$P = \{\mathbf{v} + \mathbf{x} \mid \mathbf{x} \in S\}$$

where $\mathbf{v}$ is some fixed vector in $\mathbb{R}^n$ and S is a k-dimensional subspace of $\mathbb{R}^n$ (see Figure 4.16). Explain why the solution set of each linear system of m equations in n unknowns is a k-plane in $\mathbb{R}^n$ for some k.

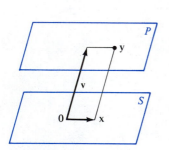

FIGURE 4.16

$\mathbf{y}$ is in the k-plane P if and only if $\mathbf{y} = \mathbf{v} + \mathbf{x}$ where $\mathbf{x}$ is in the subspace S.

17.   Explain why each set of n vectors in $\mathbb{R}^n$ that is linearly independent must span $\mathbb{R}^n$ and hence must be a basis for $\mathbb{R}^n$.

18.   Let $f:\mathbb{R}^n \to \mathbb{R}^m$ be a linear map, let $\{\mathbf{v}_1, \ldots, \mathbf{v}_k\}$ be a basis for the kernel of f, and choose vectors $\mathbf{v}_{k+1}, \ldots, \mathbf{v}_n$ in $\mathbb{R}^n$ so that $\{\mathbf{v}_1, \ldots, \mathbf{v}_k, \mathbf{v}_{k+1}, \ldots, \mathbf{v}_n\}$ is a basis for $\mathbb{R}^n$.

(a)   Show that $\{f(\mathbf{v}_{k+1}), \ldots, f(\mathbf{v}_n)\}$ is a basis for the image of f.

(b)   Conclude that the dimension of the image of f is equal to the rank of f. [*Hint:* Use the fact that the dimension of the kernel of f is equal to $n - r$ where r is the rank of f (see Example 11).]

19.   The **column space** of an $m \times n$ matrix A is the subspace of $\mathbb{R}^m$ spanned by the column vectors of A.

(a)   Show that the equation $A\mathbf{x} = \mathbf{b}$ has a solution if and only if the vector $\mathbf{b}$ is in the column space of A.

(b) Show that the row space of A and the column space of A have the same dimension. [*Hint:* By (a), the column space of A is the image of the linear map $f(\mathbf{x}) = A\mathbf{x}$.]

(c) Let $\mathbf{a}_1, \ldots, \mathbf{a}_n$ be the column vectors of A. Show that

$$\{\mathbf{a}_j | \text{ there is a corner 1 in the } j\text{th column of the echelon matrix of } A\}$$

is a basis for the column space of A. [*Hint:* Use Theorem 1 of Section 4.3 to show that these vectors form a linearly independent set. Why, then, must they form a basis?]

4.5
Orthonormal Bases

We have seen that, by choosing an ordered basis $\mathbf{B} = (\mathbf{v}_1, \ldots, \mathbf{v}_k)$ for a subspace S of $\mathbb{R}^n$, we can define a one-to-one correspondence between S and $\mathbb{R}^k$. Each vector $\mathbf{v}$ in S corresponds to its $\mathbf{B}$-coordinate k-tuple $(c_1, \ldots, c_k)$ in $\mathbb{R}^k$. This process, which assigns to each vector $\mathbf{v}$ in S an ordered k-tuple of real numbers, is a direct generalization of the Cartesian coordinate process that assigns to each point in the Euclidean plane an ordered pair of real numbers. It also generalizes the standard process that assigns to each point in Euclidean 3-space an ordered triple of real numbers.

The lines $\mathbf{x} = t\mathbf{v}_i$ ($i = 1, 2, \ldots, k$) are the ***coordinate axes*** of the $\mathbf{B}$-coordinate system (see Figure 4.17). Each of these axes is a number line because the one-to-one correspondence $t \leftrightarrow t\mathbf{v}_i$ associates with each point on the ith axis a unique real number. However, we do not require, as is usually done with Cartesian coordinates, that the coordinate axes be mutually perpendicular, nor do we require that the unit of distance along the various axes be equal to 1. In other words, we do not require that $\mathbf{v}_i \cdot \mathbf{v}_j = 0$ whenever $i \neq j$, and we do not require that $\|\mathbf{v}_i\| = 1$ for each i.

On the other hand, it is often convenient to impose these requirements on our coordinate system. In this section we shall study the special coordinate systems that arise when we impose these extra conditions.

An ordered basis $(\mathbf{v}_1, \ldots, \mathbf{v}_k)$ for a subspace S of $\mathbb{R}^n$ is said to be ***orthogonal*** if the vectors $\mathbf{v}_1, \ldots, \mathbf{v}_k$ are mutually perpendicular ($\mathbf{v}_i \cdot \mathbf{v}_j = 0$ whenever $i \neq j$). The basis $(\mathbf{v}_1, \ldots, \mathbf{v}_k)$ is ***orthonormal*** if it is orthogonal and

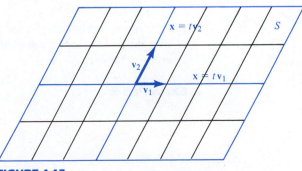

FIGURE 4.17
$\mathbf{B}$-coordinate axes in the subspace S when $\mathbf{B} = (\mathbf{v}_1, \mathbf{v}_2)$.

if, in addition, each $\mathbf{v}_i$ is a unit vector ($\|\mathbf{v}_i\| = 1$ for all i). Similarly, a finite subset $\{\mathbf{v}_1, \ldots, \mathbf{v}_k\}$ of $\mathbb{R}^n$ is said to be **orthogonal** if $\mathbf{v}_i \cdot \mathbf{v}_j = 0$ whenever $i \neq j$, and it is **orthonormal** if, in addition, $\|\mathbf{v}_i\| = 1$ for all i.

EXAMPLE 1

The standard ordered basis $(\mathbf{e}_1, \ldots, \mathbf{e}_n)$ for $\mathbb{R}^n$ is orthonormal because $\mathbf{e}_i \cdot \mathbf{e}_j = 0$ whenever $i \neq j$ and $\|\mathbf{e}_i\| = 1$ for all i. ∎

EXAMPLE 2

The set $\{(1, 0, 2), (0, 1, 0)\}$ is an orthogonal set in $\mathbb{R}^3$, since

$$(1, 0, 1) \cdot (0, 1, 0) = 0,$$

but it is not orthonormal because $\|(1, 0, 1)\| = \sqrt{2} \neq 1$. ∎

The main reason that orthonormal bases for subspaces S of $\mathbb{R}^n$ are useful is that it is easy to compute the **B**-coordinates of vectors in S whenever **B** is orthonormal. The next theorem tells us how to do that.

Theorem 1   *Let S be a subspace of $\mathbb{R}^n$ and let $\mathbf{B} = (\mathbf{v}_1, \ldots, \mathbf{v}_k)$ be an ordered orthonormal basis for S. Then the $\mathbf{B}$-coordinate k-tuple of each vector $\mathbf{v} \in S$ is the k-tuple*

$$(\mathbf{v} \cdot \mathbf{v}_1, \ldots, \mathbf{v} \cdot \mathbf{v}_k).$$

Proof   The **B**-coordinate k-tuple of $\mathbf{v}$ is $(c_1, \ldots, c_k)$ where the real numbers $c_1, \ldots, c_k$ satisfy the equation

$$\mathbf{v} = c_1\mathbf{v}_1 + \cdots + c_k\mathbf{v}_k.$$

Taking the dot product of both sides of this equation with $\mathbf{v}_1$, we obtain

$$\begin{aligned}
\mathbf{v} \cdot \mathbf{v}_1 &= (c_1\mathbf{v}_1 + c_2\mathbf{v}_2 + \cdots + c_k\mathbf{v}_k) \cdot \mathbf{v}_1 \\
&= c_1(\mathbf{v}_1 \cdot \mathbf{v}_1) + c_2(\mathbf{v}_2 \cdot \mathbf{v}_1) + \cdots + c_k(\mathbf{v}_k \cdot \mathbf{v}_1) \\
&= c_1\|\mathbf{v}_1\|^2 + c_2(0) + \cdots + c_k(0) \\
&= c_1,
\end{aligned}$$

so $c_1 = \mathbf{v} \cdot \mathbf{v}_1$ as claimed. To see that $c_2 = \mathbf{v} \cdot \mathbf{v}_2$, simply take the dot product of both sides of the above equation with $\mathbf{v}_2$. Continuing in this way, we see that $c_i = \mathbf{v} \cdot \mathbf{v}_i$ for each i. ∎

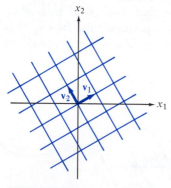

FIGURE 4.18
The ordered basis $(\mathbf{v}_1 \; \mathbf{v}_2)$ for $\mathbb{R}^2$ is orthonormal, where $\mathbf{v}_1 = \left(\dfrac{\sqrt{3}}{2}, \dfrac{1}{2}\right)$ and $\mathbf{v}_2 = \left(-\dfrac{1}{2}, \dfrac{\sqrt{3}}{2}\right)$.

EXAMPLE 3

Let

$$\mathbf{v}_1 = \left(\frac{\sqrt{3}}{2}, \frac{1}{2}\right) \quad \text{and} \quad \mathbf{v}_2 = \left(-\frac{1}{2}, \frac{\sqrt{3}}{2}\right).$$

Then $\mathbf{B} = (\mathbf{v}_1, \mathbf{v}_2)$ is an ordered orthonormal basis for $\mathbb{R}^2$ (see Figure 4.18).

Because **B** is orthonormal, the **B**-coordinate pairs of vectors in $\mathbb{R}^2$ are easy to compute. For example, the **B**-coordinate pair of $\mathbf{v} = (-2, 3)$ is

$$(\mathbf{v} \bullet \mathbf{v}_1, \mathbf{v} \bullet \mathbf{v}_2) = ((-2)\left(\frac{\sqrt{3}}{2}\right) + 3\left(\frac{1}{2}\right), (-2)\left(-\frac{1}{2}\right) + (3)\left(\frac{\sqrt{3}}{2}\right))$$

$$= \left(-\sqrt{3} + \frac{3}{2}, 1 + \frac{3\sqrt{3}}{2}\right).$$

As a check, it is easy to verify that

$$\left(-\sqrt{3} + \frac{3}{2}\right)\mathbf{v}_1 + \left(1 + \frac{3\sqrt{3}}{2}\right)\mathbf{v}_2 = (-2, 3). \quad \blacksquare$$

A nice feature of any orthonormal set is that it forms a basis for the subspace that it spans. This is a consequence of the following theorem.

Theorem 2 *Let* $\{\mathbf{v}_1, \ldots, \mathbf{v}_k\}$ *be an orthonormal set in* $\mathbb{R}^n$. *Then* $\{\mathbf{v}_1, \ldots, \mathbf{v}_k\}$ *is linearly independent.*

Proof Suppose we have a dependence relation

$$c_1\mathbf{v}_1 + \cdots + c_k\mathbf{v}_k = \mathbf{0}.$$

Then,

$$c_1 = \mathbf{v}_1 \bullet (c_1\mathbf{v}_1 + \cdots + c_k\mathbf{v}_k) = \mathbf{v}_1 \bullet \mathbf{0} = 0$$

$$\cdots$$

$$c_k = \mathbf{v}_k \bullet (c_1\mathbf{v}_1 + \cdots + c_k\mathbf{v}_k) = \mathbf{v}_k \bullet \mathbf{0} = 0$$

and hence $c_1 = \cdots = c_k = 0$ as was to be shown. $\blacksquare$

Corollary *Let* $\{\mathbf{v}_1, \ldots, \mathbf{v}_k\}$ *be an orthonormal set in* $\mathbb{R}^n$. *Then* $\{\mathbf{v}_1, \ldots, \mathbf{v}_k\}$ *is a basis for* $\mathcal{L}(\mathbf{v}_1, \ldots, \mathbf{v}_k)$.

Proof The set $\{\mathbf{v}_1, \ldots, \mathbf{v}_k\}$ is linearly independent, by Theorem 2, and it spans $\mathcal{L}(\mathbf{v}_1, \ldots, \mathbf{v}_k)$. $\blacksquare$

EXAMPLE 4

Let $\mathbf{v}_1 = \left(\frac{1}{2}, \frac{1}{2}, \frac{1}{\sqrt{2}}\right)$, $\mathbf{v}_2 = \left(\frac{1}{2}, \frac{1}{2}, -\frac{1}{\sqrt{2}}\right)$, and $S = \mathcal{L}(\mathbf{v}_1, \mathbf{v}_2)$. Then $\{\mathbf{v}_1, \mathbf{v}_2\}$ is an orthonormal set and so by the Corollary it is a basis for S.

The subspace S is a plane through $\mathbf{0}$ in $\mathbb{R}^3$. We can find an equation for this plane by computing the cross product

$$\mathbf{b} = \mathbf{v}_1 \times \mathbf{v}_2 = \left(-\frac{1}{\sqrt{2}}, \frac{1}{\sqrt{2}}, 0\right).$$

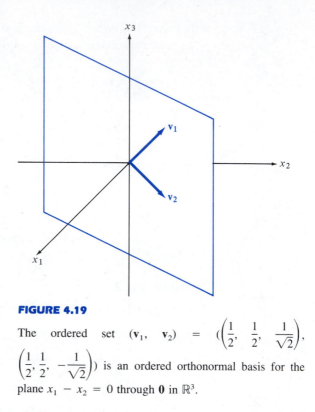

FIGURE 4.19

The ordered set $(\mathbf{v}_1, \ \mathbf{v}_2) = (\left(\dfrac{1}{2}, \ \dfrac{1}{2}, \ \dfrac{1}{\sqrt{2}}\right),$ $\left(\dfrac{1}{2}, \dfrac{1}{2}, \ -\dfrac{1}{\sqrt{2}}\right))$ is an ordered orthonormal basis for the plane $x_1 - x_2 = 0$ through $\mathbf{0}$ in $\mathbb{R}^3$.

This vector $\mathbf{b}$ is perpendicular to S. The vector $\mathbf{0}$ is in S. Therefore S can be described by the equation $\mathbf{b} \cdot (\mathbf{x} - \mathbf{0}) = 0$, or

$$-\frac{1}{\sqrt{2}} x_1 + \frac{1}{\sqrt{2}} x_2 = 0.$$

We can, if we wish, multiply both sides of this equation by $-\sqrt{2}$ to get the equation

$$x_1 - x_2 = 0$$

which also describes S (see Figure 4.19).

The vector $\mathbf{v} = (1, 1, 2\sqrt{2})$ is in S since it satisfies the equation $x_1 - x_2 = 0$. Its $\mathbf{B}$-coordinate pair, where $\mathbf{B}$ is the ordered orthonormal basis $(\mathbf{v}_1, \mathbf{v}_2)$, is

$$(\mathbf{v} \cdot \mathbf{v}_1, \mathbf{v} \cdot \mathbf{v}_2) = ((1, 1, 2\sqrt{2}) \cdot \left(\frac{1}{2}, \frac{1}{2}, \frac{1}{\sqrt{2}}\right), (1, 1, 2\sqrt{2}) \cdot \left(\frac{1}{2}, \frac{1}{2}, -\frac{1}{\sqrt{2}}\right))$$

$$= (3, -1). \quad \blacksquare$$

If $\mathbf{B} = (\mathbf{v}_1, \ldots, \mathbf{v}_k)$ is any ordered basis for a k-dimensional subspace S of $\mathbb{R}^n$, then the one-to-one correspondence between S and $\mathbb{R}^k$ that assigns to each vector $\mathbf{v}$ in S its $\mathbf{B}$-coordinate k-tuple respects both vector addition and scalar multiplication. That is, if $\mathbf{c} = (c_1, \ldots, c_k)$ is the **B**-coordinate k-tuple of $\mathbf{v}$ and $\mathbf{d} = (d_1, \ldots, d_k)$ is the **B**-coordinate k-tuple of $\mathbf{w}$, then $\mathbf{c} + \mathbf{d} = (c_1 + d_1, \ldots, c_k + d_k)$ is the **B**-coordinate k-tuple of $\mathbf{v} + \mathbf{w}$ and $a\mathbf{c} = (ac_1, \ldots, ac_k)$ is the **B**-coordinate k-tuple of $a\mathbf{v}$, for each real number a. When the ordered basis **B** is orthonormal, this one-to-one correspondence also respects dot products and lengths, as the next theorem shows.

Theorem 3 *Let S be a subspace of $\mathbb{R}^n$ and let $\mathbf{B} = (\mathbf{v}_1, \ldots, \mathbf{v}_k)$ be an ordered orthonormal basis for S. Suppose $\mathbf{v} \in S$ and $\mathbf{w} \in S$ have **B**-coordinate k-tuples $\mathbf{c} = (c_1, \ldots, c_k)$ and $\mathbf{d} = (d_1, \ldots, d_k)$ respectively. Then*

$$\mathbf{v} \bullet \mathbf{w} = \mathbf{c} \bullet \mathbf{d} = c_1 d_1 + \cdots + c_k d_k$$

and

$$\|\mathbf{v}\| = \|\mathbf{c}\| = (c_1^2 + \cdots + c_k^2)^{1/2}.$$

Proof Since

$$\mathbf{v} = c_1 \mathbf{v}_1 + \cdots + c_k \mathbf{v}_k$$

and

$$\mathbf{w} = d_1 \mathbf{v}_1 + \cdots + d_k \mathbf{v}_k$$

we find that

$$\mathbf{v} \bullet \mathbf{w} = (c_1 \mathbf{v}_1 + \cdots + c_k \mathbf{v}_k) \bullet (d_1 \mathbf{v}_1 + \cdots + d_k \mathbf{v}_k)$$

$$= c_1 \mathbf{v}_1 \bullet (d_1 \mathbf{v}_1 + \cdots + d_k \mathbf{v}_k)$$

$$+ c_2 \mathbf{v}_2 \bullet (d_1 \mathbf{v}_1 + \cdots + d_k \mathbf{v}_k)$$

$$+ \cdots$$

$$+ c_k \mathbf{v}_k \bullet (d_1 \mathbf{v}_1 + \cdots + d_k \mathbf{v}_k)$$

$$= c_1 d_1 \mathbf{v}_1 \bullet \mathbf{v}_1 + c_2 d_2 \mathbf{v}_2 \bullet \mathbf{v}_2 + \cdots + c_k d_k \mathbf{v}_k \bullet \mathbf{v}_k$$

(since $\mathbf{v}_i \bullet \mathbf{v}_j = 0$ whenever $i \neq j$)

$$= c_1 d_1 + c_2 d_2 + \cdots + c_k d_k$$

(since $\mathbf{v}_i \bullet \mathbf{v}_i = 1$ for all i)

$$= \mathbf{c} \bullet \mathbf{d}$$

and

$$\|\mathbf{v}\| = (\mathbf{v} \bullet \mathbf{v})^{1/2} = (\mathbf{c} \bullet \mathbf{c})^{1/2} = \|\mathbf{c}\|. \quad \blacksquare$$

We have seen some of the advantages of *using* orthonormal bases. Now let us consider the problem of how to *find* an orthonormal basis for a given subspace S of $\mathbb{R}^n$. Since we already know how to find a basis for S (see the discussion following Example 5 in Section 4.4), we need only describe how to replace a given basis by an orthonormal one.

If the dimension k of S is 1, then the given basis will consist of a single vector $\mathbf{v}$. To obtain an orthonormal basis we need only *normalize* $\mathbf{v}$; that is, divide $\mathbf{v}$ by its length. Thus, $\{\mathbf{v}/\|\mathbf{v}\|\}$ is an orthornormal basis for S.

If the dimension of S is 2, then the given basis will consist of two vectors, $\{\mathbf{v}_1, \mathbf{v}_2\}$. First, let $\mathbf{w}_1 = \mathbf{v}_1$. Then replace $\mathbf{v}_2$ by $\mathbf{w}_2 = \mathbf{v}_2 + c\mathbf{w}_1$ where c is chosen so that $\mathbf{w}_2$ is perpendicular to $\mathbf{w}_1$. It is easy to obtain a formula for c. We require that

$$0 = \mathbf{w}_2 \cdot \mathbf{w}_1 = (\mathbf{v}_2 + c\mathbf{w}_1) \cdot \mathbf{w}_1 = \mathbf{v}_2 \cdot \mathbf{w}_1 + c\mathbf{w}_1 \cdot \mathbf{w}_1,$$

so

$$c = -\frac{\mathbf{v}_2 \cdot \mathbf{w}_1}{\mathbf{w}_1 \cdot \mathbf{w}_1}.$$

Thus $\{\mathbf{w}_1, \mathbf{w}_2\}$ is an orthogonal set in S, where

$$\mathbf{w}_1 = \mathbf{v}_1$$

$$\mathbf{w}_2 = \mathbf{v}_2 - \frac{\mathbf{v}_2 \cdot \mathbf{w}_1}{\mathbf{w}_1 \cdot \mathbf{w}_1} \mathbf{w}_1.$$

To obtain an ortho*normal* set, simply normalize $\mathbf{w}_1$ and $\mathbf{w}_2$ to get the orthonormal basis $\left\{\dfrac{\mathbf{w}_1}{\|\mathbf{w}_1\|}, \dfrac{\mathbf{w}_2}{\|\mathbf{w}_2\|}\right\}$ for S. (Note that $\mathbf{w}_2 \neq \mathbf{0}$ since $\{\mathbf{v}_1, \mathbf{v}_2\}$ is linearly independent and that $\left\{\dfrac{\mathbf{w}_1}{\|\mathbf{w}_1\|}, \dfrac{\mathbf{w}_2}{\|\mathbf{w}_2\|}\right\}$ is a basis for S because it is linearly independent [it is orthonormal!] and $S = \mathcal{L}(\mathbf{v}_1, \mathbf{v}_2) = \mathcal{L}(\mathbf{w}_1, \mathbf{w}_2)$.)

If the dimension of S is 3 and $\{\mathbf{v}_1, \mathbf{v}_2, \mathbf{v}_3\}$ is a basis for S, the procedure is similar. First define $\mathbf{w}_1$ and $\mathbf{w}_2$ as before:

$$\mathbf{w}_1 = \mathbf{v}_1$$

$$\mathbf{w}_2 = \mathbf{v}_2 - \frac{\mathbf{v}_2 \cdot \mathbf{w}_1}{\mathbf{w}_1 \cdot \mathbf{w}_1} \mathbf{w}_1.$$

Then take

$$\mathbf{w}_3 = \mathbf{v}_3 + c_1\mathbf{w}_1 + c_2\mathbf{w}_2$$

where c_1 and c_2 are chosen so that $\mathbf{w}_3 \cdot \mathbf{w}_1 = \mathbf{w}_3 \cdot \mathbf{w}_2 = 0$. It is easy to check that

$$c_1 = -\frac{\mathbf{v}_3 \cdot \mathbf{w}_1}{\mathbf{w}_1 \cdot \mathbf{w}_1} \quad \text{and} \quad c_2 = -\frac{\mathbf{v}_3 \cdot \mathbf{w}_2}{\mathbf{w}_2 \cdot \mathbf{w}_2}$$

(see Exercise 8). Then $\{\mathbf{w}_1, \mathbf{w}_2, \mathbf{w}_3\}$ is an orthogonal set in S and $\left\{\dfrac{\mathbf{w}_1}{\|\mathbf{w}_1\|}, \dfrac{\mathbf{w}_2}{\|\mathbf{w}_2\|}, \dfrac{\mathbf{w}_3}{\|\mathbf{w}_3\|}\right\}$ is an orthornormal basis for S.

This process, called the ***Gram-Schmidt orthogonalization process***, generalizes in a straightforward way to subspaces of $\mathbb{R}^n$ of any dimension, as follows.

Theorem 4 *Suppose S is a subspace of $\mathbb{R}^n$ with basis $\{\mathbf{v}_1, \ldots, \mathbf{v}_k\}$. Define $\mathbf{w}_1, \ldots, \mathbf{w}_k \in S$ by*

$$\mathbf{w}_1 = \mathbf{v}_1$$

$$\mathbf{w}_2 = \mathbf{v}_2 - \frac{\mathbf{v}_2 \cdot \mathbf{w}_1}{\mathbf{w}_1 \cdot \mathbf{w}_1} \mathbf{w}_1$$

$$\mathbf{w}_3 = \mathbf{v}_3 - \frac{\mathbf{v}_3 \cdot \mathbf{w}_1}{\mathbf{w}_1 \cdot \mathbf{w}_1} \mathbf{w}_1 - \frac{\mathbf{v}_3 \cdot \mathbf{w}_2}{\mathbf{w}_2 \cdot \mathbf{w}_2} \mathbf{w}_2$$

$$\cdots$$

$$\mathbf{w}_k = \mathbf{v}_k - \frac{\mathbf{v}_k \cdot \mathbf{w}_1}{\mathbf{w}_1 \cdot \mathbf{w}_1} \mathbf{w}_1 - \cdots - \frac{\mathbf{v}_k \cdot \mathbf{w}_{k-1}}{\mathbf{w}_{k-1} \cdot \mathbf{w}_{k-1}} \mathbf{w}_{k-1}.$$

Then $\left\{ \dfrac{\mathbf{w}_1}{\|\mathbf{w}_1\|}, \ldots, \dfrac{\mathbf{w}_k}{\|\mathbf{w}_k\|} \right\}$ is an orthonormal basis for S.

REMARK If, in Theorem 4, we are given only a spanning set $\{\mathbf{v}_1, \ldots, \mathbf{v}_k\}$ for S (not necessarily a basis), then we can still apply the Gram-Schmidt process to obtain an orthonormal basis for S. The only change in the procedure is that we may find that one or more of the vectors $\mathbf{w}_i$ turns out to be zero. If we simply discard any such $\mathbf{w}_i$ and omit the terms $\dfrac{\mathbf{v}_i \cdot \mathbf{w}_i}{\mathbf{w}_i \cdot \mathbf{w}_i} \mathbf{w}_i$ for those i from all subsequent computations, we will still end up with an orthonormal basis for S.

EXAMPLE 5

Let us apply the Gram-Schmidt process to find an orthonormal basis for the subspace $S = \mathscr{L}(\mathbf{v}_1, \mathbf{v}_2)$ of $\mathbb{R}^3$ where $\mathbf{v}_1 = (1, -1, 2)$ and $\mathbf{v}_2 = (2, 1, -1)$. We find

$$\mathbf{w}_1 = \mathbf{v}_1 = (1, -1, 2)$$

$$\mathbf{w}_2 = \mathbf{v}_2 - \frac{\mathbf{v}_2 \cdot \mathbf{w}_1}{\mathbf{w}_1 \cdot \mathbf{w}_1} \mathbf{w}_1 = (2, 1, -1) - \frac{(-1)}{6}(1, -1, 2) = \frac{1}{6}(13, 5, -4)$$

and so

$$\left\{ \frac{\mathbf{w}_1}{\|\mathbf{w}_1\|}, \frac{\mathbf{w}_2}{\|\mathbf{w}_2\|} \right\} = \left\{ \frac{1}{\sqrt{6}}(1, -1, 2), \frac{1}{\sqrt{210}}(13, 5, -4) \right\}$$

is the required orthonormal basis for S. ∎

EXAMPLE 6

Applying the Gram-Schmidt process to find an orthonormal basis for the subspace $S = \mathcal{L}(\mathbf{v}_1, \mathbf{v}_2, \mathbf{v}_3)$ of $\mathbb{R}^4$, where $\mathbf{v}_1 = (2, 0, 1, 2)$, $\mathbf{v}_2 = (-3, 1, 2, 0)$, and $\mathbf{v}_3 = (1, 1, 4, 4)$ we get

$$\mathbf{w}_1 = \mathbf{v}_1 = (2, 0, 1, 2)$$

$$\mathbf{w}_2 = \mathbf{v}_2 - \frac{\mathbf{v}_2 \cdot \mathbf{w}_1}{\mathbf{w}_1 \cdot \mathbf{w}_1} \mathbf{w}_1 = (-3, 1, 2, 0) - \frac{(-4)}{9} (2, 0, 1, 2)$$

$$= \frac{1}{9} (-19, 9, 22, 8)$$

$$\mathbf{w}_3 = \mathbf{v}_3 - \frac{\mathbf{v}_3 \cdot \mathbf{w}_1}{\mathbf{w}_1 \cdot \mathbf{w}_1} \mathbf{w}_1 - \frac{\mathbf{v}_3 \cdot \mathbf{w}_2}{\mathbf{w}_2 \cdot \mathbf{w}_2} \mathbf{w}_2$$

$$= (1, 1, 4, 4) - \frac{14}{9} (2, 0, 1, 2) - \frac{(110/81)}{(990/81)} (-19, 9, 22, 8)$$

$$= (1, 1, 4, 4) - \frac{1}{9} (28, 0, 14, 28) - \frac{1}{9} (-19, 9, 22, 8)$$

$$= (0, 0, 0, 0).$$

Since $\mathbf{w}_3 = \mathbf{0}$ we discard it. Thus

$$\left\{ \frac{\mathbf{w}_1}{\|\mathbf{w}_1\|}, \frac{\mathbf{w}_2}{\|\mathbf{w}_2\|} \right\} = \left\{ \frac{1}{3} (2, 0, 1, 2), \frac{1}{3\sqrt{110}} (-19, 9, 22, 8) \right\}$$

is an orthonormal basis for S. Note that the dimension of S is 2. The given spanning set was linearly dependent! ■

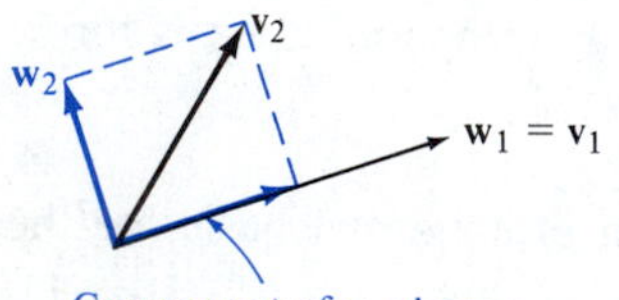

The formulas that occur in Theorem 4 have nice geometric interpretations. The vector $\mathbf{w}_1$ is just $\mathbf{v}_1$. The vector

w_2 v_2

$\mathbf{w}_1 = \mathbf{v}_1$

Component of $\mathbf{v}_2$ along $\mathbf{v}_1$

FIGURE 4.20

The vector $\mathbf{w}_2$ is obtained by subtracting from $\mathbf{v}_2$ its component along $\mathbf{w}_1$.

$$\mathbf{w}_2 = \mathbf{v}_2 - \frac{\mathbf{v}_2 \cdot \mathbf{w}_1}{\mathbf{w}_1 \cdot \mathbf{w}_1} \mathbf{w}_1$$

is the component of $\mathbf{v}_2$ perpendicular to $\mathbf{w}_1$. It is obtained by subtracting from $\mathbf{v}_2$ its component $(\mathbf{v}_2 \cdot \mathbf{w}_1/\mathbf{w}_1 \cdot \mathbf{w}_1)\mathbf{w}_1$ along $\mathbf{w}_1$ (see Figure 4.20). The vector

$$\mathbf{w}_3 = \mathbf{v}_3 - \frac{\mathbf{v}_3 \cdot \mathbf{w}_1}{\mathbf{w}_1 \cdot \mathbf{w}_1} \mathbf{w}_1 - \frac{\mathbf{v}_3 \cdot \mathbf{w}_2}{\mathbf{w}_2 \cdot \mathbf{w}_2} \mathbf{w}_2$$

is obtained by subtracting from $\mathbf{v}_3$ its components along $\mathbf{w}_1$ and $\mathbf{w}_2$ (see Figure 4.21). In general, the vector $\mathbf{w}_i$ is obtained by subtracting from $\mathbf{v}_i$ its component along each of the vectors $\mathbf{w}_1, \mathbf{w}_2, \ldots, \mathbf{w}_{i-1}$.

We can use these same ideas to decompose vectors in $\mathbb{R}^n$ into components relative to any given subspace S of $\mathbb{R}^n$.

Theorem 5 *Let S be a subspace of $\mathbb{R}^n$. Then each vector $\mathbf{v} \in \mathbb{R}^n$ can be decomposed in one and only one way into a sum*

$$\mathbf{v} = \mathbf{v}_1 + \mathbf{v}_2$$

where $\mathbf{v}_1 \in S$ and $\mathbf{v}_2$ is perpendicular to S.

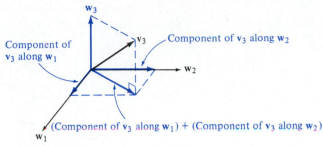

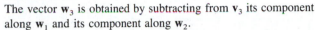

FIGURE 4.21

The vector $\mathbf{w}_3$ is obtained by subtracting from $\mathbf{v}_3$ its component along $\mathbf{w}_1$ and its component along $\mathbf{w}_2$.

Proof Let $\{\mathbf{w}_1, \ldots, \mathbf{w}_k\}$ be an orthonormal basis for S. If there is a decomposition

$$\mathbf{v} = \mathbf{v}_1 + \mathbf{v}_2$$

where $\mathbf{v}_1$ is in S and $\mathbf{v}_2$ is perpendicular to S, then, for each i,

$$\mathbf{v} \cdot \mathbf{w}_i = (\mathbf{v}_1 + \mathbf{v}_2) \cdot \mathbf{w}_i$$

$$= \mathbf{v}_1 \cdot \mathbf{w}_i + \mathbf{v}_2 \cdot \mathbf{w}_i$$

$$= \mathbf{v}_1 \cdot \mathbf{w}_i.$$

This last equality holds because $\mathbf{v}_2$ is perpendicular to each vector in S and $\mathbf{w}_i \in S$ for all i. Since $(\mathbf{v}_1 \cdot \mathbf{w}_1, \ldots, \mathbf{v}_1 \cdot \mathbf{w}_k)$ is the coordinate k-tuple of $\mathbf{v}_1$ relative to the ordered basis $(\mathbf{w}_1, \ldots, \mathbf{w}_k)$ for S, we can conclude that

$$\mathbf{v}_1 = (\mathbf{v}_1 \cdot \mathbf{w}_1)\mathbf{w}_1 + \cdots + (\mathbf{v}_1 \cdot \mathbf{w}_k)\mathbf{w}_k$$

$$= (\mathbf{v} \cdot \mathbf{w}_1)\mathbf{w}_1 + \cdots + (\mathbf{v} \cdot \mathbf{w}_k)\mathbf{w}_k.$$

This formula shows that there is at most one possible choice for $\mathbf{v}_1$, namely

$$\mathbf{v}_1 = (\mathbf{v} \cdot \mathbf{w}_1)\mathbf{w}_1 + \cdots + (\mathbf{v} \cdot \mathbf{w}_k)\mathbf{w}_k$$

and at most one possible choice for $\mathbf{v}_2$, namely,

$$\mathbf{v}_2 = \mathbf{v} - \mathbf{v}_1.$$

So it remains only to show that if we define $\mathbf{v}_1$ and $\mathbf{v}_2$ by the above two formulas, then $\mathbf{v}_1 \in S$, $\mathbf{v}_2$ is perpendicular to S, and $\mathbf{v} = \mathbf{v}_1 + \mathbf{v}_2$. The first and third of these conditions are clear. For the second, simply note that

$$\mathbf{v}_2 \cdot \mathbf{w}_i = (\mathbf{v} - \mathbf{v}_1) \cdot \mathbf{w}_i = \mathbf{v} \cdot \mathbf{w}_i - \mathbf{v}_1 \cdot \mathbf{w}_i = 0$$

for all i, so $\mathbf{v}_2$ is perpendicular to each element of the basis $\{\mathbf{w}_1, \ldots, \mathbf{w}_k\}$ for S and hence is perpendicular to every vector in S. ■

The vector $\mathbf{v}_1$, whose existence is guaranteed by Theorem 5, is called the ***component of* v *parallel to*** S. The vector $\mathbf{v}_2$ is called the ***component of* v *perpendicular to*** S (see Figure 4.22). The vector $\mathbf{v}_1$ is also often called the

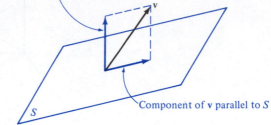

Component of **v** perpendicular to S

S

Component of **v** parallel to S

FIGURE 4.22

Decomposing a vector **v** into its components parallel to S and perpendicular to S.

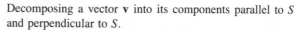

FIGURE 4.23

The orthogonal projection of **v** onto S.

orthogonal projection of **v** onto S and it is denoted $\text{Proj}_S\mathbf{v}$ (see Figure 4.23). From the proof of Theorem 5 we see that this orthogonal projection is given by the formula

$$\text{Proj}_S\mathbf{v} = (\mathbf{v} \bullet \mathbf{w}_1)\mathbf{w}_1 + \cdots + (\mathbf{v} \bullet \mathbf{w}_k)\mathbf{w}_k,$$

where $\{\mathbf{w}_1, \ldots, \mathbf{w}_k\}$ is any orthonormal basis for S. If $\{\mathbf{w}_1, \ldots, \mathbf{w}_k\}$ is an orthogonal, but not necessarily orthonormal, basis for S, this formula becomes

$$\text{Proj}_S\mathbf{v} = \frac{\mathbf{v} \bullet \mathbf{w}_1}{\mathbf{w}_1 \bullet \mathbf{w}_1} \mathbf{w}_1 + \cdots + \frac{\mathbf{v} \bullet \mathbf{w}_k}{\mathbf{w}_k \bullet \mathbf{w}_k} \mathbf{w}_k.$$

EXERCISES

1. Which of the following sets are orthogonal? Which are orthonormal?

  (a) $\{(1, 0, 1, 0), (-1, 0, 1, 0), (-1, 0, -1, 0)\}$

  (b) $(1, 0, 1, 0), (-1, 0, 1, 1), (1, 0, -1, 2)$

  (c) $\left\{\left(\dfrac{1}{\sqrt{3}}, \dfrac{1}{\sqrt{3}}, \dfrac{1}{\sqrt{3}}\right), \left(-\dfrac{1}{\sqrt{6}}, \dfrac{2}{\sqrt{6}}, -\dfrac{1}{\sqrt{6}}\right)\right\}$

  (d) $\left\{\left(\dfrac{1}{2}, -\dfrac{1}{2}, \dfrac{1}{2}, \dfrac{1}{2}\right), \left(\dfrac{1}{2\sqrt{3}}, -\dfrac{1}{2\sqrt{3}}, -\dfrac{3}{2\sqrt{3}}, \dfrac{1}{2\sqrt{3}}\right), \left(-\dfrac{1}{\sqrt{2}}, 0, 0, \dfrac{1}{\sqrt{2}}\right)\right\}$

  (e) $\{(12, 5, 9, -4), (1, 1, -1, 2), (10, -13, -7, -2)\}$

2. Verify that the ordered basis $\mathbf{B} = \left(\left(\dfrac{1}{\sqrt{2}}, -\dfrac{1}{\sqrt{2}}\right), \left(\dfrac{1}{\sqrt{2}}, \dfrac{1}{\sqrt{2}}\right)\right)$ for $\mathbb{R}^2$ is orthonormal, and find the **B**-coordinate pair of each of the following vectors.

  (a) $(1, 0)$ (b) $(0, 1)$

  (c) $(1, 1)$ (d) $(-1, 1)$

  (e) $(-2, 3)$ (f) $(\sqrt{2}, 2\sqrt{2})$

3. Verify that $\mathbf{B} = \left(\left(\dfrac{2}{3}, \dfrac{2}{3}, \dfrac{1}{3}\right), \left(\dfrac{\sqrt{2}}{2}, -\dfrac{\sqrt{2}}{2}, 0\right), \left(\dfrac{\sqrt{2}}{6}, \dfrac{\sqrt{2}}{6}, -\dfrac{2\sqrt{2}}{3}\right)\right)$ is an ordered orthonormal basis for $\mathbb{R}^3$, and find the **B**-coordinate triple of each of the following vectors.

  (a) $\left(\dfrac{\sqrt{2}}{2}, -\dfrac{\sqrt{2}}{2}, 0\right)$ (b) $(1, 1, -1)$

(c) $(1, 0, 0)$ (d) $(1, -1, 2)$

(e) $\left(\frac{2}{3}\sqrt{2}, -\frac{1}{3}\sqrt{2}, -\frac{2}{3}\sqrt{2}\right)$

4.   Let $\mathbf{B} = (\mathbf{v}_1, \mathbf{v}_2, \mathbf{v}_3, \mathbf{v}_4)$, where

$$\mathbf{v}_1 = (\tfrac{1}{2}, \tfrac{1}{2}, -\tfrac{1}{2}, -\tfrac{1}{2})$$

$$\mathbf{v}_2 = (\tfrac{1}{2}, -\tfrac{1}{2}, \tfrac{1}{2}, -\tfrac{1}{2})$$

$$\mathbf{v}_3 = (-\tfrac{1}{2}, \tfrac{1}{2}, \tfrac{1}{2}, -\tfrac{1}{2})$$

$$\mathbf{v}_4 = (\tfrac{1}{2}, \tfrac{1}{2}, \tfrac{1}{2}, \tfrac{1}{2}).$$

Verify that $\mathbf{B}$ is an ordered orthonormal basis for $\mathbb{R}^4$ and find the $\mathbf{B}$-coordinate 4-tuple of each of the following vectors.

(a) $(1, 0, 0, 0)$ (b) $(1, 2, 3, 4)$
(c) $(4, 3, 2, 1)$ (d) $(1, -1, -1, 1)$
(e) $(0, 0, 0, 0)$ (f) $(0, 1, 1, 0)$

5.   Let $\mathbf{v}_1 = \left(\dfrac{1}{\sqrt{6}}, \dfrac{2}{\sqrt{6}}, \dfrac{1}{\sqrt{6}}\right)$, $\mathbf{v}_2 = \left(\dfrac{1}{\sqrt{3}}, -\dfrac{1}{\sqrt{3}}, \dfrac{1}{\sqrt{3}}\right)$, and $S = \mathcal{L}(\mathbf{v}_1, \mathbf{v}_2)$.

(a)   Find an equation $\mathbf{b} \cdot \mathbf{x} = 0$ for S.
(b)   Verify that the vector $\mathbf{v} = (1, 0, 1)$ is in S.
(c)   Verify that $\mathbf{B} = (\mathbf{v}_1, \mathbf{v}_2)$ is an ordered orthonormal basis for S.
(d)   Find the $\mathbf{B}$-coordinate pair of $\mathbf{v} = (1, 0, 1)$, where $\mathbf{B}$ is as in part (c).

6.   Find an orthonormal basis for each of the following subspaces:

(a) $\mathcal{L}((1, 1, 0), (-1, 1, 0))$ in $\mathbb{R}^3$
(b) $\mathcal{L}((1, 1, 1), (-1, 1, 1))$ in $\mathbb{R}^3$
(c) $\mathcal{L}((1, 1, 1, 1), (-2, -1, 1, 2), (1, 0, -1, 2), (0, 0, 1, 5))$ in $\mathbb{R}^4$
(d) $\mathcal{L}((0, 1, 0, \ldots, 0), (0, 1, 2, 0, \ldots, 0), (0, 1, 2, 3, 0, \ldots, 0), \ldots,$
      $(0, 1, 2, \ldots, n))$ in $\mathbb{R}^{n+1}$

7.   Find an orthonormal basis for each of the following subspaces:

(a)   The plane $3x_1 + x_2 - x_3 = 0$ in $\mathbb{R}^3$
(b)   The hyperplane $2x_1 - x_2 + 3x_3 - x_4 = 0$ in $\mathbb{R}^4$
(c)   The solution space in $\mathbb{R}^4$ of the system

$$x_1 + x_2 + x_3 + x_4 = 0$$

$$-x_1 + x_2 \qquad + x_4 = 0$$

8.   Verify that if $\{\mathbf{w}_1, \mathbf{w}_2\}$ is an orthogonal set in $\mathbb{R}^n$ with $\mathbf{w}_1$ and $\mathbf{w}_2$ both nonzero, and if $\mathbf{v}_3 \in \mathbb{R}^n$, then the vector

$$\mathbf{w}_3 = \mathbf{v}_3 + c_1\mathbf{w}_1 + c_2\mathbf{w}_2$$

is perpendicular to both $\mathbf{w}_1$ and $\mathbf{w}_2$ if and only if

$$c_1 = -\frac{\mathbf{v}_3 \cdot \mathbf{w}_1}{\mathbf{w}_1 \cdot \mathbf{w}_1} \quad \text{and} \quad c_2 = -\frac{\mathbf{v}_3 \cdot \mathbf{w}_2}{\mathbf{w}_2 \cdot \mathbf{w}_2}$$

9.   Let $\mathbf{v}_1 = (1, -1, 1)$, $\mathbf{v}_2 = (1, 1, 0)$, and $S = \mathcal{L}(\mathbf{v}_1, \mathbf{v}_2)$.

(a)   Find an orthonormal basis for S.

(b)   Find the orthogonal projection of $\mathbf{v} = (5, 7, -3)$ onto S.

(c)   Find the component of $\mathbf{v} = (5, 7, -3)$ perpendicular to S.

10.   Let $\mathbf{w}_1 = (\frac{3}{5}, \frac{2}{5}, \frac{2}{5}, \frac{2}{5}, \frac{2}{5})$, $\mathbf{w}_2 = (0, \frac{1}{2}, -\frac{1}{2}, \frac{1}{2}, -\frac{1}{2})$, and $\mathbf{w}_3 = (0, \frac{1}{2}, \frac{1}{2}, -\frac{1}{2}, -\frac{1}{2})$.

(a)   Show that $\{\mathbf{w}_1, \mathbf{w}_2, \mathbf{w}_3\}$ is an orthonormal set in $\mathbb{R}^5$.

(b)   Find the orthogonal projection of $\mathbf{v} = (1, 0, -3, 1, 3)$ onto $S = \mathcal{L}(\mathbf{w}_1, \mathbf{w}_2, \mathbf{w}_3)$.

(c)   Find the component of $\mathbf{v}$ perpendicular to S ($\mathbf{v}$ and S as in part (b)).   .

11.   Let $\mathbf{v} \in \mathbb{R}^n$, let $\mathbf{b}$ be a unit vector in $\mathbb{R}^n$, and let H be the hyperplane $\mathbf{b} \cdot \mathbf{x} = 0$. Show that the orthogonal projection of $\mathbf{v}$ onto H is given by the formula

$$\text{Proj}_H \mathbf{v} = \mathbf{v} - (\mathbf{v} \cdot \mathbf{b})\mathbf{b}.$$

[*Hint:* Use Theorem 5.]

12.   Let S be a subspace of $\mathbb{R}^n$ and let $\mathbf{v} \in \mathbb{R}^n$. Show that the orthogonal projection of $\mathbf{v}$ onto S is the vector in S closest to $\mathbf{v}$ by showing that, for each $\mathbf{w} \in S$,

$$\|\mathbf{w} - \mathbf{v}\| \geq \|\text{Proj}_S \mathbf{v} - \mathbf{v}\|$$

and that equality holds if and only if $\mathbf{w} = \text{Proj}_S \mathbf{v}$. [*Hint:* Let $\mathbf{a} = \mathbf{w} - \text{Proj}_S \mathbf{v}$ so that $\mathbf{w} = \mathbf{a} + \text{Proj}_S \mathbf{v}$. Then verify that $\|\mathbf{w} - \mathbf{v}\|^2 = \|\mathbf{a}\|^2 + \|\text{Proj}_S \mathbf{v} - \mathbf{v}\|^2$.]

4.6
Application: Least Squares Approximations

In engineering, science, or social science, theory often predicts that there is a linear relation $y = mx + b$ between two variables x and y, but the constants m and b are not known. In order to find m and b, a sequence of measurements is taken that gives, for several values $(x_1, x_2, \ldots, x_n)$ of x, corresponding experimental values $(y_1, y_2, \ldots, y_n)$ of y. If the points $(x_1, y_1), \ldots, (x_n, y_n)$ all lie on a line in $\mathbb{R}^2$, then it is easy to determine the values of m and b predicted by the experiment: they are the slope and y-intercept of the line through these n points. But usually, due to experimental error, the points $(x_1, y_1), \ldots, (x_n, y_n)$ do not lie on a common line (see Figure 4.24) so the problem is then to find the line that in some sense comes closest to passing through the n points $(x_1, y_1), \ldots, (x_n, y_n)$.

There are various ways to measure how close a point is to a line. Geometrically, the perpendicular distance from the point to the line is probably the best measure. But, from the point of view of analyzing data, vertical distance is usually preferred. If the vertical distance $|y_i - (mx_i + b)|$ is small, then the experimental value y_i of y will be close to the predicted value $mx_i + b$.

A standard measure of the closeness of n points $(x_1, y_1), \ldots, (x_n, y_n)$ in $\mathbb{R}^2$ to the line $y = mx + b$ is the sum S of the squares of the vertical distances from the points to the line:

$$S = (y_1 - (mx_1 + b))^2 + \cdots + (y_n - (mx_n + b))^2$$

(see Figure 4.25). The reason that the sum of the squares of the distances is used rather than simply the sum of the distances is that large errors are less tolerable than small errors. Squaring magnifies the contribution to S of large errors.

The *least squares linear approximation* to the points $(x_1, y_1), \ldots, (x_n, y_n)$ is defined to be the line $y = mx + b$ for which S is as small as possible.

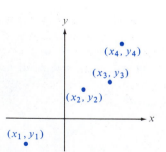

FIGURE 4.24
Experimentally determined points that should, theoretically, lie on a common line.

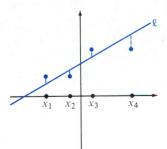

FIGURE 4.25
The closeness of a set of points to a line ℓ is measured by summing the squares of the vertical distances from these points to ℓ.

We shall use linear algebra to help us find a formula for the values of m and b that make S smallest.

Consider first the case where the n points $(x_1, y_1), \ldots, (x_n, y_n)$ are collinear. Then the slope m and y-intercept b of the line through these points must satisfy the linear system

$$mx_1 + b = y_1$$
$$mx_2 + b = y_2$$
$$\cdots$$
$$mx_n + b = y_n.$$

This system is equivalent to the vector equation

$$m\mathbf{x} + b\mathbf{1} = \mathbf{y}$$

where $\mathbf{x} = (x_1, \ldots, x_n)$, $\mathbf{1} = (1, 1, \ldots, 1)$, and $\mathbf{y} = (y_1, \ldots, y_n)$. We see, then, that the n given points lie on a common line ℓ in $\mathbb{R}^2$ if and only if $\mathbf{y}$ is a linear combination of $\mathbf{x}$ and $\mathbf{1}$. The coefficients of this linear combination are the slope and y-intercept of ℓ, respectively.

If the points $(x_1, y_1), \ldots, (x_n, y_n)$ are not collinear, then it is not possible to find m and b such that

$$m\mathbf{x} + b\mathbf{1} = \mathbf{y}.$$

But we can try to find m and b so that $m\mathbf{x} + b\mathbf{1}$ is as close to $\mathbf{y}$ as possible; that is, we can try to find the values of m and b that make $\|\mathbf{y} - (m\mathbf{x} + b\mathbf{1})\|$ as small as possible. But a positive number is smallest when its square is smallest. Since

$$\|\mathbf{y} - (m\mathbf{x} + b\mathbf{1})\|^2 = (y_1 - (mx_1 + b))^2 + \cdots + (y_n - (mx_n + b))^2$$
$$= S$$

we see that finding m and b so that $\|\mathbf{y} - (m\mathbf{x} + b\mathbf{1})\|$ is smallest is exactly the same as finding the least squares linear approximation to the points $(x_1, y_1), \ldots, (x_n, y_n)$.

The geometry of the problem is now clear. The equation $m\mathbf{x} + b\mathbf{1} = \mathbf{y}$ has a solution if and only if the vector $\mathbf{y}$ is in the subspace $\mathcal{L}(\mathbf{x}, \mathbf{1})$ of $\mathbb{R}^n$ spanned by $\{\mathbf{x}, \mathbf{1}\}$. If $\mathbf{y}$ is not in this subspace then the equation $m\mathbf{x} + b\mathbf{1} = \mathbf{y}$ has no solution, so we try to find m and b so that $\|\mathbf{y} - (m\mathbf{x} + b\mathbf{1})\|$ is as small as possible. Since the vector in $\mathcal{L}(\mathbf{x}, \mathbf{1})$ closest to $\mathbf{y}$ is the orthogonal projection of $\mathbf{y}$ onto $\mathcal{L}(\mathbf{x}, \mathbf{1})$ (see Figure 4.26; see also Exercise 12 of Section 4.5 for an outline of an analytic proof of this geometrically evident fact), the solution to the least squares approximation problem will be the solution of the vector equation

$$m\mathbf{x} + b\mathbf{1} = \text{Proj}_{\mathcal{L}(\mathbf{x}, \mathbf{1})}\mathbf{y}.$$

We can solve this equation for m and b as follows:

Since the vector

$$\mathbf{y} - \text{Proj}_{\mathcal{L}(\mathbf{x}, \mathbf{1})}\mathbf{y} = \mathbf{y} - (m\mathbf{x} + b\mathbf{1})$$

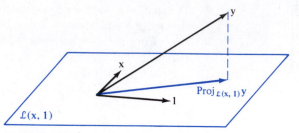

FIGURE 4.26
The vector in $\mathscr{L}(\mathbf{x}, \mathbf{1})$ that is closest to $\mathbf{y}$ is the orthogonal projection $\text{Proj}_{\mathscr{L}(\mathbf{x},\mathbf{1})}\mathbf{y}$ of $\mathbf{y}$ onto $\mathscr{L}(\mathbf{x}, \mathbf{1})$.

is perpendicular to the 2-plane $\mathscr{L}(\mathbf{x}, \mathbf{1})$ (Figure 4.26), and since $\mathbf{x} \in \mathscr{L}(\mathbf{x}, \mathbf{1})$ and $\mathbf{1} \in \mathscr{L}(\mathbf{x}, \mathbf{1})$, we must have

$$\mathbf{x} \cdot [\mathbf{y} - (m\mathbf{x} + b\mathbf{1})] = 0$$

and

$$\mathbf{1} \cdot [\mathbf{y} - (m\mathbf{x} + b\mathbf{1})] = 0.$$

Hence (m, b) must satisfy the linear system

$$(\mathbf{x} \cdot \mathbf{x})m + (\mathbf{x} \cdot \mathbf{1})b = \mathbf{x} \cdot \mathbf{y}$$

$$(\mathbf{1} \cdot \mathbf{x})m + (\mathbf{1} \cdot \mathbf{1})b = \mathbf{1} \cdot \mathbf{y}.$$

We can use Cramer's Rule to solve this linear system. The solution is

$$m = \frac{(\mathbf{1} \cdot \mathbf{1})(\mathbf{x} \cdot \mathbf{y}) - (\mathbf{1} \cdot \mathbf{x})(\mathbf{1} \cdot \mathbf{y})}{(\mathbf{1} \cdot \mathbf{1})(\mathbf{x} \cdot \mathbf{x}) - (\mathbf{1} \cdot \mathbf{x})^2}$$

$$b = \frac{(\mathbf{x} \cdot \mathbf{x})(\mathbf{1} \cdot \mathbf{y}) - (\mathbf{x} \cdot \mathbf{y})(\mathbf{1} \cdot \mathbf{x})}{(\mathbf{1} \cdot \mathbf{1})(\mathbf{x} \cdot \mathbf{x}) - (\mathbf{1} \cdot \mathbf{x})^2}.$$

Notice that the denominator is equal to

$$(\mathbf{1} \cdot \mathbf{1})(\mathbf{x} \cdot \mathbf{x}) - (\mathbf{1} \cdot \mathbf{x})^2 = \|\mathbf{1}\|^2 \|\mathbf{x}\|^2 - \|\mathbf{1}\|^2 \|\mathbf{x}\|^2 \cos^2 \theta$$

$$= \|\mathbf{1}\|^2 \|\mathbf{x}\|^2 \sin^2 \theta$$

where θ is the angle between $\mathbf{1}$ and $\mathbf{x}$. This is zero only when $\sin \theta = 0$, that is, when $\theta = 0$ or $\theta = \pi$. Thus the denominator is zero only when $\mathbf{x}$ is a scalar multiple of $\mathbf{1}$, that is, when $x_1 = x_2 = \cdots = x_n$. In this case the given points are collinear (they all lie on a vertical line).

We have proved the following theorem.

Theorem 1   *The least squares linear approximation to a set $(x_1, y_1), \ldots, (x_n, y_n)$ of noncollinear points in $\mathbb{R}^2$ is the line $y = mx + b$ where*

$$m = \frac{(\mathbf{1} \cdot \mathbf{1})(\mathbf{x} \cdot \mathbf{y}) - (\mathbf{1} \cdot \mathbf{x})(\mathbf{1} \cdot \mathbf{y})}{(\mathbf{1} \cdot \mathbf{1})(\mathbf{x} \cdot \mathbf{x}) - (\mathbf{1} \cdot \mathbf{x})^2} \qquad \mathbf{x} = (x_1, \ldots, x_n)$$

$$\mathbf{y} = (y_1, \ldots, y_n)$$

$$b = \frac{(\mathbf{x} \cdot \mathbf{x})(\mathbf{1} \cdot \mathbf{y}) - (\mathbf{x} \cdot \mathbf{y})(\mathbf{1} \cdot \mathbf{x})}{(\mathbf{1} \cdot \mathbf{1})(\mathbf{x} \cdot \mathbf{x}) - (\mathbf{1} \cdot \mathbf{x})^2} \qquad \mathbf{1} = (1, 1, \ldots, 1)$$

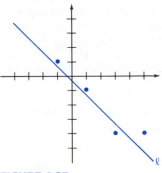

FIGURE 4.27
The line ℓ is the least squares approximation to the set $\{(-1, 1), (1, -1), (3, -4), (5, -4)\}$.

EXAMPLE 1

To find the least squares linear approximation to the set $\{(-1, 1), (1, -1), (3, -4), (5, -4)\}$, we put $\mathbf{x} = (-1, 1, 3, 5)$ and $\mathbf{y} = (1, -1, -4, -4)$, and compute

$$m = \frac{(4)(-34) - (8)(-8)}{(4)(36) - (8)^2} = -\frac{9}{10}$$

$$b = \frac{(36)(-8) - (-34)(8)}{(4)(36) - (8)^2} = -\frac{1}{5}$$

The line is $y = -\frac{9}{10}x - \frac{1}{5}$ (see Figure 4.27). ∎

EXAMPLE 2

The population of a certain small town in southern New Jersey has been growing slowly over a period of 20 years, according to the following table:

YEAR	POPULATION
1960	2000
1970	2050
1980	2080

Let us try to predict the town's population in the year 2000.

Without any additional information available to suggest otherwise, it is reasonable to approximate the population graph by a line (see Figure 4.28). We shall find the least squares linear approximation to the data and use it to predict the population in the year 2000. Setting $\mathbf{x} = (1960, 1970, 1980)$, $\mathbf{y} = (2000, 2050, 2080)$, $\mathbf{1} = (1, 1, 1)$, and using the formulas in Theorem 1, we find that $m = 4$ and, to two decimal places, $b = -5836.67$. So the least squares approximation to the data is

$$y = 4x - 5836.67.$$

Substituting 2000 for x we obtain $y = 2163.33$. We therefore predict that the population in the year 2000 will be approximately 2163. ∎

Least squares linear approximations can also be used to fit certain nonlinear functions to experimental data. The most common examples are power functions and exponentials.

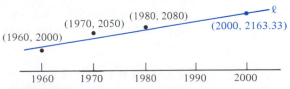

FIGURE 4.28
The line ℓ is the least squares linear approximation to the given population data.

If there is a theoretical relationship between positive variables x and y of the form

$$y = ax^b,$$

then, taking the logarithm (base 10) of both sides, we obtain a linear relation

$$\log y = b \log x + \log a$$

between $\log x$ and $\log y$, so a least squares linear approximation can be used to determine experimental values of b and $\log a$, hence of a and b.

EXAMPLE 3

If we plot the points $(1, 4)$, $(2, 2)$, and $(3, 1)$ we see that they seem to lie on a curve shaped like the graph of $y = c/x^k = cx^{-k}$ for some c and k (see Figure 4.29a). Let us use the least squares formula to find values of c and k that yield a good fit to these data. Taking the logarithm (base 10) of both sides of the equation $y = cx^{-k}$ we get

$$\log y = -k \log x + \log c = m \log x + b$$

where $m = -k$ and $b = \log c$. We want to find m and b that yield the least squares linear approximation to the set $\{(\log 1, \log 4), (\log 2, \log 2), (\log 3, \log 1)\} = \{(0, 2 \log 2), (\log 2, \log 2), (\log 3, 0)\}$ (see Figure 4.29b). Setting $\mathbf{x} = (0, \log 2, \log 3)$, $\mathbf{y} = (2 \log 2, \log 2, 0)$, $\mathbf{1} = (1, 1, 1)$, and applying the formula of Theorem 1 we find that, to two decimal places,

$$m = \frac{3(\log 2)^2 - (\log 2 + \log 3)(3 \log 2)}{3((\log 2)^2 + (\log 3)^2) - (\log 2 + \log 3)^2} = -1.23$$

$$b = \frac{((\log 2)^2 + (\log 3)^2)(3 \log 2) - (\log 2)^2(\log 2 + \log 3)}{3((\log 2)^2 + (\log 3)^2) - (\log 2 + \log 3)^2} = .621$$

Thus, $k = -m = 1.23$, $c = 10^b = 4.18$, and the curve we seek is $y = 4.18x^{-1.23}$. If we compute the values of y as x runs through the values 1, 2, 3 we find that y runs through the values 4.18, 1.78, 1.08, in fair agreement with the given data (see Figure 4.29c). ∎

If there is a theoretical relationship between x and y of the form

$$y = ac^x,$$

where y, a, and c are all positive numbers, then we can take the logarithm of both sides to obtain a linear relation

$$\log y = (\log c)x + \log a$$

between the variables x and $\log y$, so again we can use a least squares linear approximation to determine experimental values of $\log a$ and $\log c$, hence of a and c.

EXAMPLE 4

The uninhibited growth of a bacteria population is described by an equation of the form $N = N_0 c^t$, where N is the number of bacteria at time t, N_0 is the

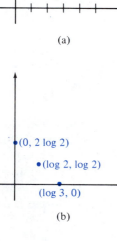

(a)

(b)

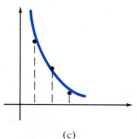

(c)

FIGURE 4.29
The data points, (a), are transformed by the logarithm into points that are nearly collinear, (b). The least squares approximation to these points leads to a curve, (c), that approximates the given data.

number at time $t = 0$, and $c > 0$ is a constant. Population counts of a particular bacteria colony yield the following data:

t	N
3	4.5×10^7
4	7.5×10^7
5	1.2×10^8

We shall use a least squares approximation to determine approximate values of N_0 and c from these data, and we shall use these approximate values to predict the number of bacteria when $t = 10$.

Taking the logarithm of both sides of the equation $N = N_0 c^t$ yields the linear relation

$$\log N = (\log c)t + \log N_0$$

between the variables t and $\log N$. Taking the logarithms of the data in the N column, we get

t	$\log N$
3	7.653
4	7.875
5	8.079

Setting $\mathbf{x} = (3, 4, 5)$, $\mathbf{y} = (7.653, 7.875, 8.079)$, $\mathbf{1} = (1, 1, 1)$, and using the formulas in Theorem 1, we find that

$$\left.\begin{matrix} m(= \log c) = 0.213 \\ b(= \log N_0) = 7.017 \end{matrix}\right\} \quad \text{so} \quad \left\{\begin{matrix} c = 10^m = 1.63 \\ N_0 = 10^b = 1.04 \times 10^7. \end{matrix}\right.$$

Hence the least squares approximation is

$$N = (1.04 \times 10^7)(1.63)^t.$$

When $t = 10$ we obtain, for the predicted value of N,

$$N|_{t=10} = (1.04 \times 10^7)(1.63)^{10} = 1.4 \times 10^9$$

(see Figure (4.30)). ■

The least squares method can be extended to a method for finding approximate solutions for inconsistent systems of linear equations.

Consider the linear system $A\mathbf{x} = \mathbf{b}$, or

$$a_{11}x_1 + \cdots + a_{1n}x_n = b_1$$

$$\cdots$$

$$a_{m1}x_1 + \cdots + a_{mn}x_n = b_m$$

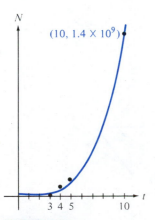

FIGURE 4.30

This curve approximately describes the population growth of a bacteria colony.

where

$$A = \begin{pmatrix} a_{11} & \cdots & a_{1n} \\ & \cdots & \\ a_{m1} & \cdots & a_{mn} \end{pmatrix} \quad \text{and} \quad \mathbf{b} = \begin{pmatrix} b_1 \\ \vdots \\ b_m \end{pmatrix}.$$

This system is equivalent to the vector equation

$$x_1 \mathbf{a}_1 + \cdots + x_n \mathbf{a}_n = \mathbf{b}$$

where $\mathbf{a}_1, \ldots, \mathbf{a}_n$ are the column vectors of the matrix A. It is evident from this vector equation that the system $A\mathbf{x} = \mathbf{b}$ is consistent (has a solution) if and only if $\mathbf{b}$ is in the column space $\mathcal{L}(\mathbf{a}_1, \ldots, \mathbf{a}_n)$ of the matrix A. If $\mathbf{b} \notin \mathcal{L}(\mathbf{a}_1, \ldots, \mathbf{a}_n)$ then the system has no solution. In this case, it is often useful to find an approximate solution. A ***least squares approximate solution*** of the linear system $A\mathbf{x} = \mathbf{b}$ is a vector $(x_1, \ldots, x_n)$ in $\mathbb{R}^n$ with the property that $A\mathbf{x} = x_1\mathbf{a}_1 + \cdots + x_n\mathbf{a}_n$ is as close as possible to $\mathbf{b}$ (that is, such that $\|\mathbf{b} - (x_1\mathbf{a}_1 + \cdots + x_n\mathbf{a}_n)\|$ is as small as possible). Let $S = \mathcal{L}(\mathbf{a}_1, \ldots, \mathbf{a}_n)$. Since the vector in $S = \mathcal{L}(\mathbf{a}_1, \ldots, \mathbf{a}_n)$ closest to $\mathbf{b}$ is the vector $\text{Proj}_S\mathbf{b}$ (see Figure 4.31), *the least squares approximate solutions of the system $A\mathbf{x} = \mathbf{b}$ are the solutions of the system $A\mathbf{x} = \text{Proj}_S\mathbf{b}$.*

Let $\mathbf{v} = \text{Proj}_S\mathbf{b}$. Then $\mathbf{v}$ is the unique vector in S with the property that $\mathbf{b} = \mathbf{v} + \mathbf{w}$ for some vector $\mathbf{w}$ perpendicular to S. In other words, $\mathbf{v}$ is the unique vector in S such that $\mathbf{b} - \mathbf{v}$ is perpendicular to S. Said yet another way, $\mathbf{v}$ is the unique vector in S such that $\mathbf{a}_i \cdot (\mathbf{b} - \mathbf{v}) = 0$ for all i $(1 \leq i \leq n)$ where $\mathbf{a}_1, \ldots, \mathbf{a}_n$ are the column vectors of A.

Now, $\mathbf{x}$ is a least squares approximate solution of $A\mathbf{x} = \mathbf{b}$ if and only if

$$A\mathbf{x} = \mathbf{v},$$

that is, if and only if $A\mathbf{x}$ satisfies the equation

$$\mathbf{a}_i \cdot (\mathbf{b} - A\mathbf{x}) = 0$$

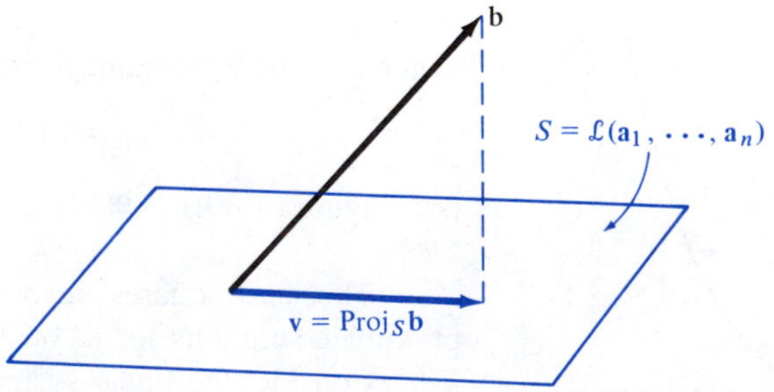

FIGURE 4.31
A least squares approximate solution of the equation $x_1\mathbf{a}_1 + \cdots + x_n\mathbf{a}_n = \mathbf{b}$ is a solution of the equation $x_1\mathbf{a}_1 + \cdots + x_n\mathbf{a}_n = \text{Proj}_S\mathbf{b}$, where $S = \mathcal{L}(\mathbf{a}_1, \ldots, \mathbf{a}_n)$.

for each i ($1 \leq i \leq n$). But the left-hand side of this equation is just the ith row of the matrix $A^t(\mathbf{b} - A\mathbf{x})$. Hence $\mathbf{x}$ is a least squares approximate solution of the equation $A\mathbf{x} = \mathbf{b}$ if and only if

$$A^t(\mathbf{b} - A\mathbf{x}) = 0,$$

that is, if and only if

$$A^tA\mathbf{x} = A^t\mathbf{b}.$$

If the square matrix A^tA is invertible, we can solve this equation explicitly for $\mathbf{x}$.

We have proved the following theorem.

Theorem 2 *The least squares approximate solutions of the linear system* $A\mathbf{x} = \mathbf{b}$ *are the solutions of the equation*

$$A^tA\mathbf{x} = A^t\mathbf{b}.$$

If the matrix A^tA is invertible, then the least squares approximate solution is unique and is given by the formula

$$\mathbf{x} = (A^tA)^{-1}A^t\mathbf{b}.$$

EXAMPLE 5

To find the least squares approximate solution of the system

$$x_1 + x_2 = 9$$

$$3x_1 - x_2 = 1$$

$$x_1 + 2x_2 = 10,$$

we set

$$A = \begin{pmatrix} 1 & 1 \\ 3 & -1 \\ 1 & 2 \end{pmatrix} \quad \text{and} \quad \mathbf{b} = \begin{pmatrix} 9 \\ 1 \\ 10 \end{pmatrix}.$$

Then

$$A^tA = \begin{pmatrix} 1 & 3 & 1 \\ 1 & -1 & 2 \end{pmatrix}\begin{pmatrix} 1 & 1 \\ 3 & -1 \\ 1 & 2 \end{pmatrix} = \begin{pmatrix} 11 & 0 \\ 0 & 6 \end{pmatrix}.$$

Therefore, the least squares approximate solution is

$$\mathbf{x} = (A^tA)^{-1}A^t\mathbf{b} = \begin{pmatrix} \frac{1}{11} & 0 \\ 0 & \frac{1}{6} \end{pmatrix}\begin{pmatrix} 1 & 3 & 1 \\ 1 & -1 & 2 \end{pmatrix}\begin{pmatrix} 9 \\ 1 \\ 10 \end{pmatrix}$$

$$= \begin{pmatrix} \frac{1}{11} & 0 \\ 0 & \frac{1}{6} \end{pmatrix}\begin{pmatrix} 22 \\ 28 \end{pmatrix} = \begin{pmatrix} 2 \\ \frac{14}{3} \end{pmatrix}. \quad \blacksquare$$

EXERCISES

[Note: Use of a hand calculator is recommended for many of these exercises.]

1. Find the least squares linear approximation $y = mx + b$ to the following sets of points.

   (a) $(-1, 0), (1, 3), (3, 4)$

   (b) $(1, 5), (2, 3), (3, 0)$

   (c) $(-1, 0), (0, -1), (1, 1), (2, -1)$

   (d) $(0, -1), (1, 1), (3, 5), (4, 7)$

   (e) $(-1, 3), (0, 1), (1, -1), (2, -2)$

   (f) $(1, 5), (2, 6), (3, 4), (4, 5)$

   (g) $(-1, -4), (1, -3), (2, 0), (3, 0)$

2. Use an appropriate least squares linear approximation to find a curve $y = ax^b$ that provides a good fit with the given set of points.

   (a) $(1, 3), (2, 11), (3, 25)$

   (b) $(1, 2), (2, 1), (3, \frac{1}{2})$

   (c) $(1, 100), (2, 3), (3, \frac{1}{2})$

   (d) $(1, 3), (2, 4), (3, 5)$

   (e) $(1, 10), (2, 75), (3, 260)$

3. Use an appropriate least squares linear approximation to find an exponential curve $y = ac^x$ to fit the following points.

   (a) $(0, 4), (2, 13), (4, 35)$

   (b) $(0, 10), (1, 50), (2, 250)$

   (c) $(-1, \frac{1}{2}), (0, 1), (1, \frac{3}{2})$

   (d) $(-2, 5), (-1, 2), (1, 1)$

   (e) $(1, 2), (2, 4), (3, 8)$

   (f) $(0, 8), (1, 20), (2, 60), (3, 200)$

4. A certain population appears to be growing exponentially with time t. The (t, N) data are $(0, 50), (1, 350), (2, 2400)$. Use the method of least squares to find an exponential function $N = N_0 c^t$ to approximate these data. Then predict the approximate value of N when $t = 3$.

5. Workers on an assembly line are observed to spend a total of 800 worker-hours assembling the first of a new model of automobile, 560 worker-hours assembling the second, 460 assembling the third, and 405 assembling the fourth. Assuming that the total number y of worker-hours required to assemble the xth automobile is given by an equation of the form $y = ax^b$, use the method of least squares to determine values of a and b that provide a good fit to the given data. Then predict the number of worker-hours required to assemble the 100th automobile.

6. Find least squares approximate solutions to the following linear systems.

   (a) $\begin{aligned} 2x_1 + x_2 &= 3 \\ x_1 + 3x_2 &= 4 \\ 5x_1 - x_2 &= 5 \end{aligned}$

   (b) $\begin{aligned} -3x_1 + x_2 &= 0 \\ 2x_1 + x_2 &= 6 \\ x_1 + x_2 &= 5 \end{aligned}$

   (c) $\begin{aligned} x_1 + x_2 - x_3 &= 1 \\ x_1 \phantom{{}+ x_2} - x_3 &= 0 \\ x_1 - x_2 - x_3 &= -1 \\ x_1 \phantom{{}- x_2} + 3x_3 &= -2 \end{aligned}$

7. Find least squares approximate solutions to the following linear systems.

    (a) $2x_1 + x_2 = 1$
         $-x_1 + 3x_2 = 1$
         $x_1 - x_2 = 0$

    (b) $-2x_1 - 3x_2 = -1$
         $3x_1 - 2x_2 = 1$
         $x_1 + x_2 = 1$

    (c) $2x_1 + x_2 = 1$
         $3x_1 + x_2 = 2$
         $5x_1 + x_2 = 3$

    (d) $3x_1 - 4x_2 = 6$
         $-3x_1 + 4x_2 = -5$

    (e) $x_1 + x_2 + x_3 = 1$
         $x_1 \qquad + x_3 = 0$
         $x_2 + x_3 = 0$
         $x_1 + x_2 \qquad = 0$

8. If $\mathbf{x}$ is an approximate solution to the linear system $A\mathbf{x} = \mathbf{b}$, then $E = \|A\mathbf{x} - \mathbf{b}\|$ measures how far $\mathbf{x}$ is from being a true solution. Compute E for the least squares approximate solutions to the systems of Exercise 7.

9. Show that the linear system $A\mathbf{x} = \mathbf{b}$ has a *unique* least squares approximate solution if and only if the columns of A form a linearly independent set.

10. Use the formula of Theorem 2 to find the least squares approximate solution (m, b) of the linear system

    $$mx_1 + b = y_1$$

    $$mx_2 + b = y_2$$

    $$\cdots$$

    $$mx_n + b = y_n$$

    and check that your solution is in agreement with the solution given in Theorem 1.

11. Suppose we know that the data $(x_1, y_1), \ldots, (x_n, y_n)$ should describe a linear relationship of the form $y = mx$. Then we should look, not for the line in $\mathbb{R}^2$ that comes closest to passing through these points, but instead for the line *through* $\mathbf{0}$ that comes closest to passing through these points. Thus we want to choose m to minimize $\|\mathbf{y} - m\mathbf{x}\|$, where $\mathbf{x} = (x_1, \ldots, x_n)$ and $\mathbf{y} = (y_1, \ldots, y_n)$. Show that this m is given by the formula $m = \mathbf{x} \cdot \mathbf{y} / \mathbf{x} \cdot \mathbf{x}$.

12. Suppose we want to find the *horizontal* line $y = b$ that provides the best approximation to the points $(x_1, y_1), \ldots, (x_n, y_n)$. Then we want to choose b to minimize $\|\mathbf{y} - b\mathbf{1}\|$, where $\mathbf{y} = (y_1, \ldots, y_n)$ and $\mathbf{1} = (1, \ldots, 1)$. Show that b is given by the formula $b = (y_1 + \cdots + y_n)/n$.

13. (a) Suppose we want to find a least squares *quadratic* approximation $y = ax^2 + bx + c$ to the set of points $(x_1, y_1), \ldots, (x_n, y_n)$ in $\mathbb{R}^2$. Then we seek a, b, and c that minimize

    $$(y_1 - (ax_1^2 + bx_1 + c))^2 + \cdots + (y_n - (ax_n^2 + bx_n + c))^2$$

    (see Figure 4.32). Show that $(a, b, c) \in \mathbb{R}^3$ is a solution of this least squares quadratic approximation problem if and only if (a, b, c) is a least squares approximate solution of the linear system $A\mathbf{x} = \mathbf{b}$ where

    $$A = \begin{pmatrix} x_1^2 & x_1 & 1 \\ \vdots & \vdots & \vdots \\ x_n^2 & x_n & 1 \end{pmatrix} \quad \text{and} \quad \mathbf{b} = \begin{pmatrix} y_1 \\ \vdots \\ y_n \end{pmatrix}.$$

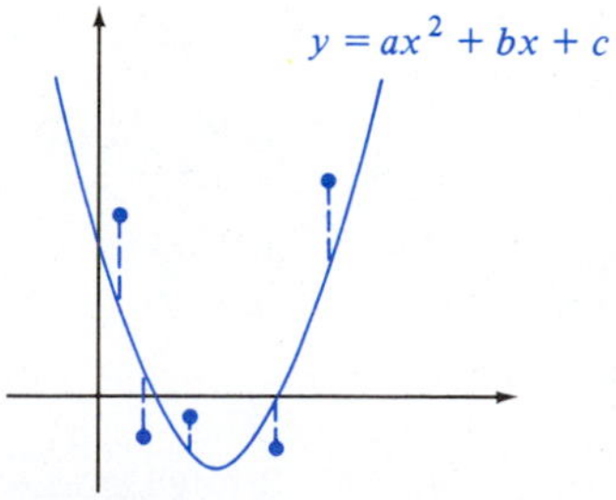

$y = ax^2 + bx + c$

FIGURE 4.32
A least squares quadratic approximation.

(b)   Generalize part (a), replacing "quadratic approximation

$$y = ax^2 + bx + c"$$

by "kth degree polynomial approximation

$$y = a_k x^k + \cdots + a_1 x + a_0."$$

14.   Use the result of Exercise 13a to find the least squares quadratic approximation $y = ax^2 + bx + c$ to the following sets of points.

(a) $\{(-1, 2), (0, 2), (1, 7), (2, 10)\}$

(b) $\{(-2, 7), (-1, 3), (0, 1), (1, 0), (2, 4)\}$

(c) $\{(-1, 6), (0, 2), (1, 0)\}$

REVIEW EXERCISES

1.   Which of the following are subspaces of $\mathbb{R}^n$?

(a)   The set of all $\mathbf{x} \in \mathbb{R}^n$ such that $(1, 2, \ldots, n) \cdot \mathbf{x} = 0$

(b)   The set of all $\mathbf{x} \in \mathbb{R}^n$ such that $(1, 2, \ldots, n) \cdot \mathbf{x} = 1$

(c)   The set of all $\mathbf{x} \in \mathbb{R}^n$ of the form $(a, a^2, \ldots, a^n)$

(d)   The set of all $\mathbf{x} \in \mathbb{R}^n$ of the form $(a, 2a, \ldots, na)$

(e)   The line $\mathbf{x} = \mathbf{a} + t\mathbf{d}$ where $\mathbf{a} \in \mathbb{R}^n$, $\mathbf{d} \in \mathbb{R}^n$, $\mathbf{d} \neq \mathbf{0}$, and $\mathbf{a}$ is not a scalar multiple of $\mathbf{d}$.

2.   Let $f:\mathbb{R}^3 \to \mathbb{R}^2$ be the linear map defined by

$$f(x_1, x_2, x_3) = (x_1 + x_2 - 2x_3, -x_1 - x_2 + 2x_3, x_1 + x_2 + 2x_3).$$

(a)   Find a basis for the kernel of f.

(b)   Find the dimension of the kernel of f.

(c)   Find a basis for the image of f.

(d)   Find the dimension of the image of f.

3.   Repeat Exercise 2 for the linear map $f:\mathbb{R}^4 \to \mathbb{R}^3$ defined by

$$f(x_1, x_2, x_3, x_4) = (x_1 - x_2 + x_3, x_2 - x_3 + x_4, x_1 + x_4).$$

4.   Let $f:\mathbb{R}^2 \to \mathbb{R}^3$ be the linear map defined by

$$f(x_1, x_2) = (2x_1 - 3x_2, x_1 + 4x_2, 2x_1 - x_2).$$

Find an orthonormal basis for the image of f.

5.   Which of the following sets span $\mathbb{R}^3$?

(a) $\{(2, 3, 4), (-1, 0, 1)\}$

(b) $\{(2, 3, 4), (-1, 0, 1), (1, 3, 5)\}$

(c) $\{(2, 3, 4), (-1, 0, 1), (1, 3, 5), (1, 2, 1)\}$

(d) $\{(1, -1, 0), (2, -1, 2), (-1, 2, 2), (1, 0, 2)\}$

6.   Which of the sets in Exercise 5 are linearly independent?

7.   Which of the sets in Exercise 5 are bases for $\mathbb{R}^3$?

8.   Find the **B**-coordinate pair of each of the following vectors, where **B** is the ordered basis $((-1, 3), (3, -4))$ for $\mathbb{R}^2$.

(a) $(5, -5)$ (b) $(-3, 4)$ (c) $(1, 0)$

9. (a) Verify that $\mathbf{B} = (\left(\frac{1}{2}, \frac{1}{2}, \frac{\sqrt{2}}{2}\right), \left(\frac{1}{2}, \frac{1}{2}, -\frac{\sqrt{2}}{2}\right), \left(\frac{\sqrt{2}}{2}, -\frac{\sqrt{2}}{2}, 0\right))$ is an ordered orthonormal basis for $\mathbb{R}^3$.

   (b) Find the $\mathbf{B}$-coordinate triple of the vector $(1, 1, 1)$, where $\mathbf{B}$ is as in part (a).

10. Let $S = \mathcal{L}((1, -1, 2), (-1, 3, 2))$ and let $\mathbf{v} = (1, 2, 3)$.

    (a) Find the component of $\mathbf{v}$ parallel to S.

    (b) Find the component of $\mathbf{v}$ perpendicular to S.

11. Let $S = \{(1, 1), (2, 5), (3, 10)\}$.

    (a) Find the least squares linear approximation $y = mx + b$ to the set S.

    (b) Use a least squares linear approximation to find a curve $y = ax^b$ that provides a good fit with the points in S.

    (c) Use a least squares linear approximation to find a curve $y = ac^x$ that provides a good fit with the points in S.

12. Find all least squares approximate solutions to the inconsistent linear system

$$2x_1 + 3x_2 - 7x_3 = 1$$
$$3x_1 - 5x_2 - x_3 = 3$$
$$2x_1 + 2x_2 - 6x_3 = 1$$

13. True or false?

    (a) Every spanning set for $\mathbb{R}^n$ contains a basis for $\mathbb{R}^n$.

    (b) Every finite subset of $\mathbb{R}^n$ is a basis for some subspace of $\mathbb{R}^n$.

    (c) If S is a subspace of $\mathbb{R}^n$, then the dimension of S must be less than or equal to n.

    (d) If a set of n vectors in $\mathbb{R}^n$ spans $\mathbb{R}^n$, then it must be linearly independent.

    (e) If two matrices have the same echelon matrix, then they have the same row space.

    (f) The nonzero row vectors in a matrix A form a basis for the row space of A.

    (g) If S is an orthogonal set then S is linearly independent.

14. Let $f:\mathbb{R}^n \to \mathbb{R}^m$ be a linear map.

    (a) If f is onto, then what is the dimension of the kernel of f?

    (b) If f is one-to-one, then what is the dimension of the image of f?

5

Linear Maps from $\mathbb{R}^n$ to $\mathbb{R}^n$

In this chapter we shall study linear maps from $\mathbb{R}^n$ to itself. These are the linear maps that have the richest relationship to geometry. Expansions, contractions, reflections, rotations, and shears are examples of linear maps of this type. We shall study subspaces of $\mathbb{R}^n$, called *eigenspaces*, on which a linear map $f:\mathbb{R}^n \to \mathbb{R}^n$ acts as an expansion or as a contraction. Using eigenspaces we will gain insight into the geometry of a large collection of linear maps. We will apply these ideas to the study of conic sections (ellipses, parabolas, and hyperbolas).

5.1
Eigenvectors

The simplest linear maps from $\mathbb{R}^n$ to itself are the maps f_λ ($\lambda \in \mathbb{R}$) defined by

$$f_\lambda(\mathbf{x}) = \lambda\mathbf{x}.$$

Recall, from Section 2.3, that:

(i) If $\lambda > 1$, then f_λ is expansion by the factor λ.
(ii) If $\lambda = 1$, then f_λ is the identity map ($f_1(\mathbf{x}) = \mathbf{x}$ for all $\mathbf{x}$).
(iii) If $0 < \lambda < 1$ then f_λ is contraction by the factor λ.
(iv) If $\lambda = 0$ then f_λ is the zero map ($f_0(\mathbf{x}) = \mathbf{0}$ for all $\mathbf{x}$).
(v) If $\lambda = -1$ then f_λ is reflection in $\mathbf{0}$ ($f_{-1}(\mathbf{x}) = -\mathbf{x}$ for all $\mathbf{x}$).
(vi) If $\lambda < 0$ and $\lambda \neq -1$ then f_λ is a composition

$$f_\lambda = f_{-1} \circ f_{|\lambda|}$$

of an expansion or a contraction followed by reflection in $\mathbf{0}$ (see Figure 5.1).

Most linear maps $f:\mathbb{R}^n \to \mathbb{R}^n$ are not as simple as the maps f_λ. Frequently, however, it is possible to find certain vectors on which a given linear map acts like f_λ for some λ. It is then easy to visualize what the linear map f is doing, geometrically, to these special vectors. These vectors are the *eigenvectors* of the linear map f.

Let $f:\mathbb{R}^n \to \mathbb{R}^n$ be a linear map. An *eigenvector* of f is a nonzero vector $\mathbf{v} \in \mathbb{R}^n$ such that $f(\mathbf{v}) = \lambda\mathbf{v}$ for some $\lambda \in \mathbb{R}$. The real number λ such that $f(\mathbf{v}) = \lambda\mathbf{v}$ is called the *eigenvalue* of f corresponding to the eigenvector $\mathbf{v}$.

EXAMPLE 1
Let $f:\mathbb{R}^2 \to \mathbb{R}^2$ be the linear map $f(x_1, x_2) = (2x_1, -3x_2)$. Then $f(\mathbf{e}_1) = 2\mathbf{e}_1$ and $f(\mathbf{e}_2) = -3\mathbf{e}_2$. Hence $\mathbf{e}_1$ is an eigenvector of f with eigenvalue 2, and $\mathbf{e}_2$

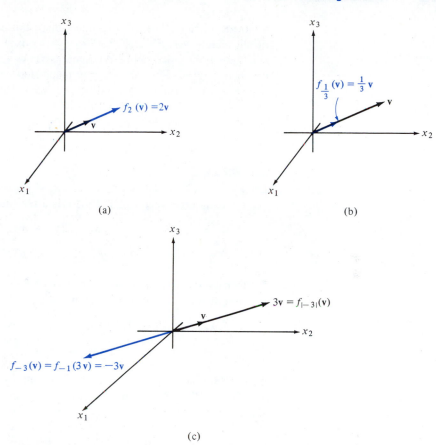

(a)

(b)

(c)

FIGURE 5.1

The effect of the maps $f_\lambda : \mathbb{R}^3 \to \mathbb{R}^3$. (a) The expansion f_2 stretches by a factor of 2. (b) The contraction $f_{1/3}$ shrinks each vector to $\frac{1}{3}$ of its length. (c) The map f_{-3} can be visualized as expansion by a factor of 3 followed by reflection in the origin.

FIGURE 5.2

The linear map $f(x_1, x_2) = (2x_1, -3x_2)$ has eigenvector $\mathbf{e}_1$ with eigenvalue 2 and eigenvector $\mathbf{e}_2$ with eigenvalue -3.

is an eigenvector of f with eigenvalue -3 (see Figure 5.2). So f stretches $\mathbf{e}_1$ by a factor of 2, and it acts on $\mathbf{e}_2$ by stretching by a factor of 3 *and* reversing direction.

Now let us see if there are any other eigenvectors of f. If $\mathbf{v} = (x_1, x_2)$ is any eigenvector of f, with eigenvalue λ, then $f(x_1, x_2) = \lambda(x_1, x_2)$, hence $(2x_1, -3x_2) = (\lambda x_1, \lambda x_2)$. Thus (x_1, x_2) must be a solution of the linear system

$$2x_1 = \lambda x_1$$
$$-3x_2 = \lambda x_2.$$

Since $\mathbf{v} \neq \mathbf{0}$, either $x_1 \neq 0$ or $x_2 \neq 0$. If $x_1 \neq 0$, then $\lambda = 2$. If $x_2 \neq 0$, then $\lambda = -3$. Hence 2 and -3 are the only eigenvalues of f. Moreover, $\lambda = 2$ implies that $x_2 = 0$ so $\mathbf{v} = (x_1, 0) = x_1(1, 0)$ for some x_1 and $\lambda = -3$ implies that $x_1 = 0$ so $\mathbf{v} = (0, x_2) = x_2(0, 1)$ for some x_2. We conclude, then, that the only eigenvectors of f are the nonzero multiples of $\mathbf{e}_1$ (with eigenvalue 2) and the nonzero multiples of $\mathbf{e}_2$ (with eigenvalue -3). ■

REMARK   Let $f:\mathbb{R}^n \to \mathbb{R}^n$ be a linear map. For each real number λ, the set V_λ of all vectors $\mathbf{v} \in \mathbb{R}^n$ such that $f(\mathbf{v}) = \lambda\mathbf{v}$ is a subspace of $\mathbb{R}^n$. Indeed, the condition $f(\mathbf{v}) = \lambda\mathbf{v}$ can be rewritten as $f(\mathbf{v}) - \lambda\mathbf{v} = \mathbf{0}$ or as $(f - \lambda i)(\mathbf{v}) = \mathbf{0}$, where $i:\mathbb{R}^n \to \mathbb{R}^n$ is the identity map. Hence V_λ is the kernel of the linear map $f - \lambda i$. For most values of $\lambda \in \mathbb{R}$ we will find that $V_\lambda = \{\mathbf{0}\}$. But if λ is an eigenvalue of f, then $V_\lambda \neq \{\mathbf{0}\}$. V_λ is called the λ-*eigenspace* of f. Notice that V_λ consists of the eigenvectors of f with eigenvalue λ together with the zero vector.

EXAMPLE 2

Let $f:\mathbb{R}^2 \to \mathbb{R}^2$ be defined by $f(x_1, x_2) = (x_1 + 2x_2, 2x_1 + x_2)$. Then $\mathbf{v} = (x_1, x_2)$ is an eigenvector of f if there is a real number λ for which

$$x_1 + 2x_2 = \lambda x_1$$

$$2x_1 + x_2 = \lambda x_2.$$

In other words, (x_1, x_2) is an eigenvector of f if and only if (x_1, x_2) is a nonzero solution of the homogeneous system

$$(*) \qquad \begin{aligned} (1 - \lambda)x_1 + \qquad\quad 2x_2 &= 0 \\ 2x_1 + (1 - \lambda)x_2 &= 0. \end{aligned}$$

This system has a nonzero solution if and only if its coefficient matrix has rank less than 2; that is, if and only if the determinant of its coefficient matrix is zero. Since

$$\det\begin{pmatrix} 1 - \lambda & 2 \\ 2 & 1 - \lambda \end{pmatrix} = \lambda^2 - 2\lambda - 3,$$

the eigenvalues of f must be the roots of the polynomial equation

$$\lambda^2 - 2\lambda - 3 = 0,$$

namely -1 and 3.

Taking $\lambda = -1$ in $(*)$, we see that the eigenvectors of f with eigenvalue -1 are the nonzero solutions of the homogeneous system

$$2x_1 + 2x_2 = 0$$

$$2x_1 + 2x_2 = 0,$$

namely, the nonzero multiples of the vector $(-1, 1)$. Similarly, taking $\lambda = 3$ in $(*)$, we find that the eigenvectors of f with eigenvalue 3 are the nonzero solutions of the homogeneous system

$$-2x_1 + 2x_2 = 0$$

$$2x_1 - 2x_2 = 0,$$

namely, the nonzero multiples of the vector $(1, 1)$.

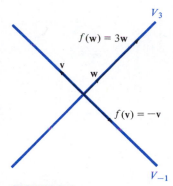

V_3

$f(\mathbf{w}) = 3\mathbf{w}$

$\mathbf{v}$

$\mathbf{w}$

$f(\mathbf{v}) = -\mathbf{v}$

V_{-1}

FIGURE 5.3

The linear map $f(x_1, x_2) = (x_1 + 2x_2, 2x_1 + x_2)$ reflects in $\mathbf{0}$ each vector in V_{-1} and stretches by a factor of 3 each vector in V_3.

Notice that f reflects in $\mathbf{0}$ each vector in the eigenspace $V_{-1} = \mathcal{L}((-1, 1))$, and f stretches by a factor of 3 each vector in the eigenspace $V_3 = \mathcal{L}((1, 1))$ (see Figure 5.3). ∎

The procedure for finding eigenvalues of a linear map $f:\mathbb{R}^n \to \mathbb{R}^n$ always leads to a polynomial whose zeros are the eigenvalues of f. Indeed, $\lambda \in \mathbb{R}$ is an eigenvalue of f if and only if the vector equation $f(\mathbf{x}) = \lambda\mathbf{x}$ has a nonzero solution $\mathbf{x}$. The vector equation

$$f(\mathbf{x}) = \lambda\mathbf{x}$$

is equivalent to the matrix equation

$$A\mathbf{x} = \lambda\mathbf{x}$$

where A is the matrix of f, and this equation is equivalent to the homogeneous linear system

$$(A - \lambda I)\mathbf{x} = \mathbf{0}$$

where I is the $n \times n$ identity matrix. This system has a nonzero solution if and only if the coefficient matrix $A - \lambda I$ has rank less than n; that is, if and only if $\det(A - \lambda I) = 0$.

The function $p(\lambda) = \det(A - \lambda I)$ is a polynomial in λ of degree n. This polynomial is called the **characteristic polynomial** of f.

The preceding discussion proves the following theorem.

Theorem 1 *Let $f:\mathbb{R}^n \to \mathbb{R}^n$ be a linear map and let A be the matrix of f. Then*

(i)   *$\lambda \in \mathbb{R}$ is an eigenvalue of f if and only if*

$$\det(A - \lambda I) = 0$$

where I is the $n \times n$ identity matrix, and

(ii)   *if $\lambda \in \mathbb{R}$ is an eigenvalue of f, then the λ-eigenspace of f is the solution space of the homogeneous linear system*

$$(A - \lambda I)\mathbf{x} = \mathbf{0}.$$

EXAMPLE 3

Let us find the eigenvalues and the eigenspaces of the linear map

$$f(x_1, x_2, x_3) = (2x_1 + x_2 + 2x_3, x_1 + 2x_2 + 2x_3, x_1 + x_2 + 3x_3).$$

The matrix of f is

$$A = \begin{pmatrix} 2 & 1 & 2 \\ 1 & 2 & 2 \\ 1 & 1 & 3 \end{pmatrix}.$$

The characteristic polynomial of f is

$$p(\lambda) = \det(A - \lambda I) = \det\begin{pmatrix} 2 - \lambda & 1 & 2 \\ 1 & 2 - \lambda & 2 \\ 1 & 1 & 3 - \lambda \end{pmatrix}$$

$$= (2 - \lambda)\det\begin{pmatrix} 2 - \lambda & 2 \\ 1 & 3 - \lambda \end{pmatrix} - 1\det\begin{pmatrix} 1 & 2 \\ 1 & 3 - \lambda \end{pmatrix}$$

$$+ 1\det\begin{pmatrix} 1 & 2 \\ 2 - \lambda & 2 \end{pmatrix}$$

$$= -\lambda^3 + 7\lambda^2 - 11\lambda + 5$$

$$= -(\lambda - 1)(\lambda - 1)(\lambda - 5).$$

The eigenvalues of f are the zeros of this polynomial, namely 1 and 5.

The eigenspace V_1 corresponding to the eigenvalue 1 is the solution space of the linear system $(A - 1I)\mathbf{x} = \mathbf{0}$. Since

$$A - 1I = \begin{pmatrix} 1 & 1 & 2 \\ 1 & 1 & 2 \\ 1 & 1 & 2 \end{pmatrix} \longrightarrow \begin{pmatrix} 1 & 1 & 2 \\ 0 & 0 & 0 \\ 0 & 0 & 0 \end{pmatrix}$$

this system has the same solution space as the system

$$x_1 + x_2 + 2x_3 = 0$$

$$0 = 0$$

$$0 = 0.$$

The general solution is

$$\mathbf{x} = c_1(-1, 1, 0) + c_2(-2, 0, 1).$$

Therefore the eigenspace V_1 is the plane $\mathcal{L}((-1, 1, 0), (-2, 0, 1))$ spanned by $(-1, 1, 0)$ and $(-2, 0, 1)$.

The eigenspace V_5 corresponding to the eigenvalue 5 is the solution space of the linear system $(A - 5I)\mathbf{x} = \mathbf{0}$. Since

$$A - 5I = \begin{pmatrix} -3 & 1 & 2 \\ 1 & -3 & 2 \\ 1 & 1 & -2 \end{pmatrix} \longrightarrow \begin{pmatrix} 1 & 0 & -1 \\ 0 & 1 & -1 \\ 0 & 0 & 0 \end{pmatrix}$$

this system has the same solution space as the system

$$x_1 \quad - x_3 = 0$$

$$x_2 - x_3 = 0$$

$$0 = 0.$$

The general solution is

$$\mathbf{x} = c(1, 1, 1).$$

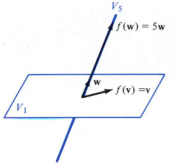

FIGURE 5.4
The linear map f leaves each vector in the eigenspace V_1 fixed and stretches by a factor of 5 each vector in the eigenspace V_5.

Therefore the eigenspace V_5 is the line $\mathcal{L}((1, 1, 1))$ through $\mathbf{0}$ in the direction of the vector $(1, 1, 1)$.

Notice that the results of the above computations give us a good geometric understanding of the linear map f. On the eigenspace $V_1 = \mathcal{L}((-1, 1, 0), (-2, 0, 1))$, f is the identity map ($f(\mathbf{v}) = \mathbf{v}$ for all $\mathbf{v} \in V_1$) and, on the eigenspace $V_5 = \mathcal{L}((1, 1, 1))$, f stretches each vector by a factor of 5 ($f(\mathbf{v}) = 5\mathbf{v}$ for $\mathbf{v} \in V_5$) (see Figure 5.4). ∎

Since any scalar multiple of an eigenvector is also an eigenvector with the same eigenvalue, it is clear that eigenvectors with distinct eigenvalues cannot be scalar multiples of one another. In other words, each pair of eigenvectors with distinct eigenvalues is linearly independent. More generally we can prove the following.

Theorem 2 Let $f:\mathbb{R}^n \to \mathbb{R}^n$ be a linear map and suppose $\mathbf{v}_1, \ldots, \mathbf{v}_k$ are eigenvectors of f with distinct eigenvalues $\lambda_1, \ldots, \lambda_k$. Then $\{\mathbf{v}_1, \ldots, \mathbf{v}_k\}$ is linearly independent.

Proof We shall prove independence by showing that dependence would imply that two of the eigenvalues were equal.

Suppose, then, that $\{\mathbf{v}_1, \ldots, \mathbf{v}_k\}$ is dependent. Let m be the largest integer for which $\{\mathbf{v}_1, \ldots, \mathbf{v}_m\}$ is independent. Thus $m < k$, $\{\mathbf{v}_1, \ldots, \mathbf{v}_m\}$ is independent, and

$$\mathbf{v}_{m+1} = a_1\mathbf{v}_1 + \cdots + a_m\mathbf{v}_m$$

for some $a_1, \ldots, a_m \in \mathbb{R}$. If we apply f to both sides of this equation, we get

$$f(\mathbf{v}_{m+1}) = a_1 f(\mathbf{v}_1) + \cdots + a_m f(\mathbf{v}_m)$$

or

$$\lambda_{m+1}\mathbf{v}_{m+1} = a_1\lambda_1\mathbf{v}_1 + \cdots + a_m\lambda_m\mathbf{v}_m.$$

On the other hand,

$$\lambda_{m+1}\mathbf{v}_{m+1} = \lambda_{m+1}(a_1\mathbf{v}_1 + \cdots + a_m\mathbf{v}_m) = a_1\lambda_{m+1}\mathbf{v}_1 + \cdots + a_m\lambda_{m+1}\mathbf{v}_m.$$

Subtracting, we get

$$\mathbf{0} = a_1(\lambda_{m+1} - \lambda_1)\mathbf{v}_1 + \cdots + a_m(\lambda_{m+1} - \lambda_m)\mathbf{v}_m$$

or, since $\{\mathbf{v}_1, \ldots, \mathbf{v}_m\}$ is independent,

$$a_1(\lambda_{m+1} - \lambda_1) = \cdots = a_m(\lambda_{m+1} - \lambda_m) = 0.$$

But a_i must be different from zero for at least one i (since $\mathbf{v}_{m+1} \neq \mathbf{0}$). For this i we must have $\lambda_i = \lambda_{m+1}$. ∎

Corollary Let $f:\mathbb{R}^n \to \mathbb{R}^n$ be a linear map with n distinct eigenvalues. Then $\mathbb{R}^n$ has a basis consisting of eigenvectors of f.

EXAMPLE 4

Let $f: \mathbb{R}^3 \to \mathbb{R}^3$ be the linear map whose matrix is

$$A = \begin{pmatrix} 0 & -2 & 2 \\ -3 & 1 & 3 \\ -1 & 1 & 3 \end{pmatrix}.$$

Let us find a basis for $\mathbb{R}^3$ consisting of eigenvectors of f.

First we compute the characteristic polynomial:

$$\det(A - \lambda I) = \det \begin{pmatrix} -\lambda & -2 & 2 \\ -3 & 1 - \lambda & 3 \\ -1 & 1 & 3 - \lambda \end{pmatrix}$$

$$= -\lambda[(1 - \lambda)(3 - \lambda) - 3] - (-3)[-2(3 - \lambda) - 2]$$
$$+ (-1)[-6 - 2(1 - \lambda)]$$
$$= -\lambda(-4\lambda + \lambda^2) + 3(2\lambda - 8) - (2\lambda - 8)$$
$$= -\lambda^3 + 4\lambda^2 + 4\lambda - 16 = -(\lambda - 2)(\lambda + 2)(\lambda - 4).$$

Hence f has eigenvalues 2, -2, and 4.

Next, we solve the three homogeneous linear systems whose coefficient matrices are $A - \lambda I$, $\lambda = 2, -2, 4$. We find, by row reduction, that

$$A - 2I = \begin{pmatrix} -2 & -2 & 2 \\ -3 & -1 & 3 \\ -1 & 1 & 1 \end{pmatrix} \longrightarrow \begin{pmatrix} 1 & 0 & -1 \\ 0 & 1 & 0 \\ 0 & 0 & 0 \end{pmatrix}$$

$$A - (-2)I = \begin{pmatrix} 2 & -2 & 2 \\ -3 & 3 & 3 \\ -1 & 1 & 5 \end{pmatrix} \longrightarrow \begin{pmatrix} 1 & -1 & 0 \\ 0 & 0 & 1 \\ 0 & 0 & 0 \end{pmatrix}$$

$$A - 4I = \begin{pmatrix} -4 & -2 & 2 \\ -3 & -3 & 3 \\ -1 & 1 & -1 \end{pmatrix} \longrightarrow \begin{pmatrix} 1 & 0 & 0 \\ 0 & 1 & -1 \\ 0 & 0 & 0 \end{pmatrix}$$

From these echelon matrices we can see that the vectors $(1, 0, 1)$, $(1, 1, 0)$, and $(0, 1, 1)$ are eigenvectors of f with eigenvalues 2, -2, and 4 respectively. Since the eigenvalues are distinct, these vectors form an independent set, hence a basis for $\mathbb{R}^3$. ∎

EXAMPLE 5

Let $f: \mathbb{R}^3 \to \mathbb{R}^3$ be the linear map associated with the matrix

$$A = \begin{pmatrix} 2 & 1 & 2 \\ 1 & 2 & 2 \\ 1 & 1 & 3 \end{pmatrix}.$$

We have already seen in Example 3 that f has eigenvalues 1 and 5 and corresponding eigenspaces

$$V_1 = \mathcal{L}((-1, 1, 0), (-2, 0, 1))$$

and

$$V_5 = \mathcal{L}((1, 1, 1)).$$

This linear map does not have three distinct eigenvalues. Nevertheless, the set $\{(-1, 1, 0), (-2, 0, 1), (1, 1, 1)\}$ of eigenvectors is linearly independent, hence a basis for $\mathbb{R}^3$. ∎

EXAMPLE 6

Let $f : \mathbb{R}^4 \to \mathbb{R}^4$ be the linear map defined by

$$f(x_1, x_2, x_3, x_4) = (x_2, 2x_3, 3x_4, 0).$$

The matrix of f is

$$A = \begin{pmatrix} 0 & 1 & 0 & 0 \\ 0 & 0 & 2 & 0 \\ 0 & 0 & 0 & 3 \\ 0 & 0 & 0 & 0 \end{pmatrix}.$$

The characteristic polynomial of f is

$$\det(A - \lambda I) = \det \begin{pmatrix} -\lambda & 1 & 0 & 0 \\ 0 & -\lambda & 2 & 0 \\ 0 & 0 & -\lambda & 3 \\ 0 & 0 & 0 & -\lambda \end{pmatrix} = \lambda^4.$$

Therefore the only eigenvalue of f is $\lambda = 0$. The 0-eigenspace V_0 of f is the solution space of the homogeneous linear system whose coefficient matrix is $A = A - 0I$, namely $V_0 = \mathcal{L}((1, 0, 0, 0))$.

There is no basis for $\mathbb{R}^4$ consisting of eigenvectors of this linear map f. ∎

One advantage of working with a basis for $\mathbb{R}^n$ consisting of eigenvectors of a linear map f is that, although f may have a very complicated effect on the standard coordinates of a vector, it will have a very simple effect on the **B**-coordinates if each vector in **B** is an eigenvector of f.

EXAMPLE 7

Let $f : \mathbb{R}^n \to \mathbb{R}^n$ be a linear map that has n linearly independent eigenvectors and let $\mathbf{B} = (\mathbf{v}_1, \ldots, \mathbf{v}_n)$ be an ordered basis for $\mathbb{R}^n$ consisting of eigenvectors of f. If

$$\mathbf{v} = c_1\mathbf{v}_1 + c_2\mathbf{v}_2 + \cdots + c_n\mathbf{v}_n$$

is any vector in $\mathbb{R}^n$, then

$$f(\mathbf{v}) = c_1 f(\mathbf{v}_1) + c_2 f(\mathbf{v}_2) + \cdots + c_n f(\mathbf{v}_n)$$

$$= c_1\lambda_1\mathbf{v}_1 + c_2\lambda_2\mathbf{v}_2 + \cdots + c_n\lambda_n\mathbf{v}_n$$

where $\lambda_1, \ldots, \lambda_n$ are the eigenvalues of f corresponding to the eigenvectors $\mathbf{v}_1, \ldots, \mathbf{v}_n$. The **B**-coordinate n-tuple of $\mathbf{v}$ is

$$(c_1, c_2, \ldots, c_n).$$

The **B**-coordinate n-tuple of $f(\mathbf{v})$ is

$$(c_1\lambda_1, c_2\lambda_2, \ldots, c_n\lambda_n).$$

Hence the effect of f on the **B**-coordinates of $\mathbf{v}$ is to multiply the ith coordinate by the ith eigenvalue λ_i, for each $i(1 \leq i \leq n)$. ∎

EXERCISES

1. Find the eigenvalues and the corresponding eigenvectors of the linear map $f:\mathbb{R}^2 \to \mathbb{R}^2$ whose matrix is

(a) $\begin{pmatrix} 2 & 1 \\ -1 & 2 \end{pmatrix}$ (b) $\begin{pmatrix} 5 & -2 \\ -2 & 5 \end{pmatrix}$ (c) $\begin{pmatrix} 1 & -1 \\ 1 & 1 \end{pmatrix}$

(d) $\begin{pmatrix} 33 & -10 \\ 105 & -32 \end{pmatrix}$ (e) $\begin{pmatrix} -11 & -14 \\ 7 & 10 \end{pmatrix}$ (f) $\begin{pmatrix} 0 & -2 \\ -2 & 0 \end{pmatrix}$

(g) $\begin{pmatrix} 13 & -4 \\ 42 & -13 \end{pmatrix}$ (h) $\begin{pmatrix} 0 & -1 \\ 1 & 0 \end{pmatrix}$

2. Find the eigenvalues and the corresponding eigenvectors of the linear map $f:\mathbb{R}^3 \to \mathbb{R}^3$ whose matrix is

(a) $\begin{pmatrix} 1 & -1 & 1 \\ 1 & 1 & 2 \\ 2 & 0 & 3 \end{pmatrix}$ (b) $\begin{pmatrix} 1 & 1 & 1 \\ 1 & 0 & -2 \\ 1 & -1 & 1 \end{pmatrix}$

(c) $\begin{pmatrix} 1 & 0 & 1 \\ 0 & 1 & 0 \\ 1 & 0 & -1 \end{pmatrix}$ (d) $\begin{pmatrix} 7 & -9 & -15 \\ 0 & 4 & 0 \\ 3 & -9 & -11 \end{pmatrix}$

(e) $\begin{pmatrix} 31 & -100 & 70 \\ 18 & -59 & 42 \\ 12 & -40 & 29 \end{pmatrix}$

3. Find the eigenvalues and the corresponding eigenvectors of each of the following linear maps $f:\mathbb{R}^2 \to \mathbb{R}^2$.

   (a) the horizontal shear $f(x_1, x_2) = (x_1 + cx_2, x_2)$, where c is a fixed real number
   (b) the vertical shear $f(x_1, x_2) = (x_1, cx_1 + x_2)$ where c is a fixed real number
   (c) the map $f(x_1, x_2) = (cx_1, cx_2)$, where c is a fixed real number
   (d) the map $f(x_1, x_2) = (x_1 \cos\theta - x_2 \sin\theta, x_1 \sin\theta + x_2 \cos\theta)$ that rotates each vector counterclockwise through the angle θ
   (e) the map $f(x_1, x_2) = (x_1 \cos 2\theta + x_2 \sin 2\theta, x_1 \sin 2\theta - x_2 \cos 2\theta)$ that reflects each vector in the line through $\mathbf{0}$ with inclination angle θ

4. (a) Show that if $f:\mathbb{R}^n \to \mathbb{R}^n$ is a linear map and $\mathbf{v} \in \mathbb{R}^n$ is an eigenvector of f with eigenvalue λ, then $\mathbf{v}$ is also an eigenvector of $f^2 = f \circ f$. What is the corresponding eigenvalue?
   (b) Show that $\mathbf{v} \in \mathbb{R}^n$ may be an eigenvector of f^2 even though it is not an eigenvector of f. [*Hint:* Consider the linear map $f:\mathbb{R}^2 \to \mathbb{R}^2$ that rotates each vector counterclockwise through the angle $\pi/2$.]

5. Show that if $f:\mathbb{R}^n \to \mathbb{R}^n$ is an invertible linear map, then $\mathbf{v} \in \mathbb{R}^n$ is an eigenvector of f if and only if $\mathbf{v}$ is an eigenvector of g, where g is the inverse of f. What is the relationship between the eigenvalues?

6. Show that if n is odd, then each linear map $f:\mathbb{R}^n \to \mathbb{R}^n$ must have at least one eigenvector. [*Hint:* What is the degree of the characteristic polynomial of f?]

7. A linear map $f:\mathbb{R}^n \to \mathbb{R}^n$ is said to be *self-adjoint* if $f(\mathbf{v}) \cdot \mathbf{w} = \mathbf{v} \cdot f(\mathbf{w})$ for all $\mathbf{v}$, $\mathbf{w} \in \mathbb{R}^n$. Show that if $f:\mathbb{R}^n \to \mathbb{R}^n$ is self-adjoint and $\mathbf{v}$, $\mathbf{w} \in \mathbb{R}^n$ are eigenvectors of f with distinct eigenvalues, then $\mathbf{v}$ is perpendicular to $\mathbf{w}$. [*Hint:* Evaluate $f(\mathbf{v}) \cdot \mathbf{w}$ in two different ways.] Conclude that if $f:\mathbb{R}^n \to \mathbb{R}^n$ is self-adjoint and has n distinct eigenvalues, then there is an orthonormal basis for $\mathbb{R}^n$ consisting of eigenvectors of f.

5.2
Change of Basis

We know, from Section 4.4, that each ordered basis for $\mathbb{R}^n$ defines a coordinate system on $\mathbb{R}^n$. Usually the coordinate system on $\mathbb{R}^n$ most convenient to work with is the standard coordinate system, which is defined by the standard ordered basis $(\mathbf{e}_1, \ldots, \mathbf{e}_n)$. Sometimes, however, it is more convenient to work with a different coordinate system. For example, we observed in Section 5.1 (Example 7) that the effect of a linear map $f:\mathbb{R}^n \to \mathbb{R}^n$ on the **B**-coordinate n-tuple of a vector is especially simple if each vector in the basis **B** is an eigenvector of f. In this section we shall discuss the relationship between standard coordinates on $\mathbb{R}^n$ and **B**-coordinates, where **B** is any ordered basis for $\mathbb{R}^n$.

Suppose $\mathbf{B} = (\mathbf{v}_1, \ldots, \mathbf{v}_n)$ is an ordered basis for $\mathbb{R}^n$. Consider the linear map $h:\mathbb{R}^n \to \mathbb{R}^n$ defined by

$$h(c_1, c_2, \ldots, c_n) = c_1\mathbf{v}_1 + c_2\mathbf{v}_2 + \cdots + c_n\mathbf{v}_n.$$

This linear map assigns to each $\mathbf{c} \in \mathbb{R}^n$ the vector $\mathbf{x} \in \mathbb{R}^n$ whose **B**-coordinate n-tuple is $\mathbf{c}$. The map h is invertible. Its inverse is the linear map $h^{-1}:\mathbb{R}^n \to \mathbb{R}^n$ that assigns to each vector $\mathbf{x}$ in $\mathbb{R}^n$ its **B**-coordinate n-tuple $\mathbf{c}$. Thus, by using h and h^{-1} we can switch back and forth between the standard coordinates and the **B**-coordinates of any given vector $\mathbf{x}$ in $\mathbb{R}^n$.

This relationship between a vector $\mathbf{x}$ and its **B**-coordinate n-tuple $\mathbf{c}$ can be restated in terms of matrices as follows. Let H be the matrix of the linear map h. Then H^{-1} is the matrix of h^{-1}. Hence $h(\mathbf{c}) = H\mathbf{c}$ for each $\mathbf{c} \in \mathbb{R}^n$, and $h^{-1}(\mathbf{x}) = H^{-1}\mathbf{x}$ for each $\mathbf{x} \in \mathbb{R}^n$. Thus we can use matrix multiplication to switch back and forth between the standard coordinates and the **B**-coordinates of any vector $\mathbf{x}$ in $\mathbb{R}^n$: the **B**-coordinate n-tuple of $\mathbf{x}$ is $\mathbf{c} = H^{-1}\mathbf{x}$, and if $\mathbf{c}$ is the **B**-coordinate n-tuple of $\mathbf{x}$ then $\mathbf{x} = H\mathbf{c}$. In these equations, of course, the vector $\mathbf{x}$ and its **B**-coordinate n-tuple $\mathbf{c}$ are written vertically rather than horizontally.

The matrix H of h is easy to calculate. It is the matrix whose column vectors are the vectors $h(\mathbf{e}_1), \ldots, h(\mathbf{e}_n)$ (see Theorem 2 of Section 2.3). Since

$$h(c_1, \ldots, c_n) = c_1\mathbf{v}_1 + \cdots + c_n\mathbf{v}_n$$

we see that $h(\mathbf{e}_1) = \mathbf{v}_1, \ldots, h(\mathbf{e}_n) = \mathbf{v}_n$. Hence H is the matrix whose column vectors are the basis vectors $\mathbf{v}_1, \ldots, \mathbf{v}_n$.

We have proved the following theorem.

Theorem 1   *Let* $\mathbf{B} = (\mathbf{v}_1, \ldots, \mathbf{v}_n)$ *be an ordered basis for* $\mathbb{R}^n$ *and let* $\mathbf{x}$ *be any vector in* $\mathbb{R}^n$. *Then the vector* $\mathbf{x}$ *and its* $\mathbf{B}$-*coordinate* n-*tuple* $\mathbf{c}$ *are related by the equation*

$$\mathbf{x} = H\mathbf{c}, \quad \text{or} \quad \mathbf{c} = H^{-1}\mathbf{x},$$

where H *is the matrix whose column vectors are the basis vectors* $\mathbf{v}_1, \ldots, \mathbf{v}_n,$ *and where the vectors* $\mathbf{x}$ *and* $\mathbf{c}$ *are written vertically.*

Often it is convenient to have a special notation for vectors written vertically. If $\mathbf{x}$ is any vector in $\mathbb{R}^n$, then we shall denote by $M(\mathbf{x})$ the vector $\mathbf{x}$ written vertically. Thus, if $\mathbf{x} = (x_1, x_2, \ldots, x_n)$, then

$$M(x) = \begin{pmatrix} x_1 \\ x_2 \\ \vdots \\ x_n \end{pmatrix}.$$

We shall call this matrix $M(\mathbf{x})$ the **standard matrix** of $\mathbf{x}$.

Similarly, if $\mathbf{B}$ is any ordered basis for $\mathbb{R}^n$, then we shall denote by $M_\mathbf{B}(\mathbf{x})$ the $\mathbf{B}$-coordinate n-tuple of $\mathbf{x}$, written vertically. Thus if $\mathbf{B} = (\mathbf{v}_1, \ldots, \mathbf{v}_n)$ and $\mathbf{x} = c_1\mathbf{v}_1 + \cdots + c_n\mathbf{v}_n$, then

$$M_\mathbf{B}(\mathbf{x}) = \begin{pmatrix} c_1 \\ \vdots \\ c_n \end{pmatrix}.$$

The matrix $M_\mathbf{B}(\mathbf{x})$ is called the $\mathbf{B}$-*coordinate matrix* of $\mathbf{x}$. Notice that $M_\mathbf{B}(\mathbf{x}) = M(\mathbf{x})$ when $\mathbf{B} = (\mathbf{e}_1, \ldots, \mathbf{e}_n)$.

Using this matrix notation we can rewrite the formulas of Theorem 1 as follows:

$$M(\mathbf{x}) = HM_\mathbf{B}(\mathbf{x}) \quad \text{and} \quad M_\mathbf{B}(\mathbf{x}) = H^{-1}M(\mathbf{x}).$$

EXAMPLE 1

Let $\mathbf{B} = (\mathbf{v}_1, \mathbf{v}_2)$ where $\mathbf{v}_1 = (1, 2)$ and $\mathbf{v}_2 = (3, 1)$. Then the matrix H, whose column vectors are the basis vectors $\mathbf{v}_1$ and $\mathbf{v}_2$, is

$$H = \begin{pmatrix} 1 & 3 \\ 2 & 1 \end{pmatrix}.$$

The inverse of this matrix is

$$H^{-1} = \begin{pmatrix} -\frac{1}{5} & \frac{3}{5} \\ \frac{2}{5} & -\frac{1}{5} \end{pmatrix}.$$

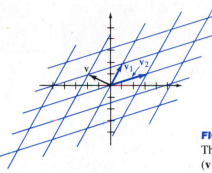

FIGURE 5.5
The **B**-coordinate pair of **v**, where **B** = $(\mathbf{v}_1, \mathbf{v}_2)$, is $(1, -1)$.

We can use the formula $M_{\mathbf{B}}(\mathbf{x}) = H^{-1}M(\mathbf{x})$ to calculate the **B**-coordinates of any vector **x** in $\mathbb{R}^2$. For example, the vector $\mathbf{x} = (-2, 1)$ has **B**-coordinate matrix

$$M_{\mathbf{B}}(\mathbf{x}) = H^{-1}M(\mathbf{x}) = \begin{pmatrix} -\frac{1}{5} & \frac{3}{5} \\ \frac{2}{5} & -\frac{1}{5} \end{pmatrix} \begin{pmatrix} -2 \\ 1 \end{pmatrix} = \begin{pmatrix} 1 \\ -1 \end{pmatrix}.$$

The **B**-coordinate pair of **x** is $(1, -1)$ (see Figure 5.5). ∎

EXAMPLE 2

Let $\mathbf{v}_1$ and $\mathbf{v}_2$ be the vectors in $\mathbb{R}^2$ obtained by rotating $\mathbf{e}_1$ and $\mathbf{e}_2$ counterclockwise through an angle θ. Then

$$\mathbf{v}_1 = (\cos\theta, \sin\theta)$$

$$\mathbf{v}_2 = (-\sin\theta, \cos\theta)$$

(see Figure 5.6). Let $\mathbf{B} = (\mathbf{v}_1, \mathbf{v}_2)$. The matrix whose column vectors are the basis vectors $\mathbf{v}_1$ and $\mathbf{v}_2$ is

$$H = \begin{pmatrix} \cos\theta & -\sin\theta \\ \sin\theta & \cos\theta \end{pmatrix}.$$

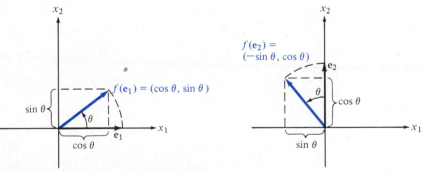

FIGURE 5.6
Rotating $\mathbf{e}_1$ and $\mathbf{e}_2$ counterclockwise through the angle θ.

The inverse of this matrix is

$$H^{-1} = \begin{pmatrix} \cos\theta & \sin\theta \\ -\sin\theta & \cos\theta \end{pmatrix}.$$

We can calculate the **B**-coordinates of any vector $\mathbf{x} \in \mathbb{R}^2$ from the formula

$$M_\mathbf{B}(\mathbf{x}) = H^{-1}M(\mathbf{x}).$$

For any x_1 and x_2, the **B**-coordinate matrix of $\mathbf{x} = (x_1, x_2)$ is

$$M_\mathbf{B}(\mathbf{x}) = \begin{pmatrix} \cos\theta & \sin\theta \\ -\sin\theta & \cos\theta \end{pmatrix}\begin{pmatrix} x_1 \\ x_2 \end{pmatrix} = \begin{pmatrix} x_1\cos\theta + x_2\sin\theta \\ -x_1\sin\theta + x_2\cos\theta \end{pmatrix},$$

and the **B**-coordinate pair of $\mathbf{x}$ is

$$(x_1\cos\theta + x_2\sin\theta, -x_1\sin\theta + x_2\cos\theta)$$

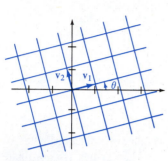

FIGURE 5.7

A rotated coordinate system: The **B**-coordinate pair of (x_1, x_2) is

$(x_1\cos\theta + x_2\sin\theta,$
$-x_1\sin\theta + x_2\cos\theta)$

$[\mathbf{B} = (\mathbf{v}_1, \mathbf{v}_2)].$

(see Figure 5.7). ∎

Just as it is often convenient to represent vectors in terms of their **B**-coordinates, where **B** is some ordered basis for $\mathbb{R}^n$, it is also often helpful to represent linear maps from $\mathbb{R}^n$ to $\mathbb{R}^n$ in terms of a **B**-*coordinate matrix*.

Recall from Section 2.3 that the matrix of a linear map $f:\mathbb{R}^n \to \mathbb{R}^n$ is the matrix A whose column vectors are the vectors $f(\mathbf{e}_1), \ldots, f(\mathbf{e}_n)$. Henceforth we shall call this matrix the ***standard matrix*** of f and we shall denote it by the symbol $M(f)$. Thus $M(f)$ is the matrix whose column vectors are the vectors $f(\mathbf{e}_1), \ldots, f(\mathbf{e}_n)$. Said another way, $M(f)$ is the matrix whose columns are $M(f(\mathbf{e}_1)), \ldots, M(f(\mathbf{e}_n))$.

If $\mathbf{B} = (\mathbf{v}_1, \ldots, \mathbf{v}_n)$ is *any* ordered basis for $\mathbb{R}^n$ and $f:\mathbb{R}^n \to \mathbb{R}^n$ is any linear map, then we define the **B**-*coordinate matrix* of f to be the matrix $M_\mathbf{B}(f)$ whose columns are $M_\mathbf{B}(f(\mathbf{v}_1)), \ldots, M_\mathbf{B}(f(\mathbf{v}_n))$. Thus the **B**-coordinate matrix of f is the matrix

$$M_\mathbf{B}(f) = \begin{pmatrix} c_{11} & c_{12} & \cdots & c_{1n} \\ c_{21} & c_{22} & \cdots & c_{2n} \\ & & \cdots & \\ c_{n1} & c_{n2} & \cdots & c_{nn} \end{pmatrix}$$

where

$$f(\mathbf{v}_1) = c_{11}\mathbf{v}_1 + c_{21}\mathbf{v}_2 + \cdots + c_{n1}\mathbf{v}_n$$

$$f(\mathbf{v}_2) = c_{12}\mathbf{v}_1 + c_{22}\mathbf{v}_2 + \cdots + c_{n2}\mathbf{v}_n$$

$$\cdots$$

$$f(\mathbf{v}_n) = c_{1n}\mathbf{v}_1 + c_{2n}\mathbf{v}_n + \cdots + c_{nn}\mathbf{v}_n.$$

Notice that the matrix $M_\mathbf{B}(f)$ is the *transpose* of the coefficient matrix of this system of vector equations. Notice also that $M_\mathbf{B}(f) = M(f)$ when $\mathbf{B} = (\mathbf{e}_1, \ldots, \mathbf{e}_n)$.

EXAMPLE 3

Let $f:\mathbb{R}^2 \to \mathbb{R}^2$ be the linear map

$$f(x_1, x_2) = (x_1 + x_2, x_1 - x_2).$$

Let us compute the **B**-coordinate matrix $M_\mathbf{B}(f)$ where $\mathbf{B} = ((5, 7), (2, 3))$.
We must find c_{11}, c_{12}, c_{21}, and c_{22} such that

$$f(5, 7) = c_{11}(5, 7) + c_{21}(2, 3)$$

$$f(2, 3) = c_{12}(5, 7) + c_{22}(2, 3).$$

Since $f(5, 7) = (12, -2)$ and $f(2, 3) = (5, -1)$, these equations become

$$c_{11}(5, 7) + c_{21}(2, 3) = (12, -2)$$

$$c_{12}(5, 7) + c_{22}(2, 3) = (5, -1).$$

This pair of vector equations is equivalent to the pair of linear systems

$$5c_{11} + 2c_{21} = 12 \qquad 5c_{12} + 2c_{22} = 5$$
$$\text{and}$$
$$7c_{11} + 3c_{21} = -2 \qquad 7c_{12} + 3c_{22} = -1.$$

The solution of the first of these systems is $(c_{11}, c_{21}) = (40, -94)$. The solution of the second is $(c_{12}, c_{22}) = (17, -40)$. Therefore we can conclude that

$$f(5, 7) = 40(5, 7) + (-94)(2, 3)$$

$$f(2, 3) = 17(5, 7) + (-40)(2, 3)$$

and that

$$M_\mathbf{B}(f) = \begin{pmatrix} 40 & 17 \\ -94 & -40 \end{pmatrix}. \quad \blacksquare$$

EXAMPLE 4

Suppose $f:\mathbb{R}^n \to \mathbb{R}^n$ is a linear map that has n linearly independent eigenvectors and let $\mathbf{B} = (\mathbf{v}_1, \ldots, \mathbf{v}_n)$ be an ordered basis for $\mathbb{R}^n$ consisting of eigenvectors of f. Then

$$f(\mathbf{v}_1) = \lambda_1 \mathbf{v}_1 = \lambda_1 \mathbf{v}_1 + 0\mathbf{v}_2 + \cdots + 0\mathbf{v}_n$$

$$f(\mathbf{v}_2) = \lambda_2 \mathbf{v}_2 = 0\mathbf{v}_1 + \lambda_2 \mathbf{v}_2 + \cdots + 0\mathbf{v}_n$$

$$\cdots$$

$$f(\mathbf{v}_n) = \lambda_n \mathbf{v}_n = 0\mathbf{v}_1 + 0\mathbf{v}_2 + \cdots + \lambda_n \mathbf{v}_n,$$

where $\lambda_1, \ldots, \lambda_n$ are the eigenvalues of f. Therefore the **B**-coordinate matrix of f is the diagonal matrix

$$M_\mathbf{B}(f) = \begin{pmatrix} \lambda_1 & 0 & & 0 \\ 0 & \lambda_2 & & 0 \\ \vdots & \vdots & \vdots & \vdots \\ 0 & 0 & & \lambda_n \end{pmatrix}.$$

Thus, *if* **B** *is an ordered basis for $\mathbb{R}^n$ consisting of eigenvectors of the linear map $f:\mathbb{R}^n \to \mathbb{R}^n$ then the* **B**-*coordinate matrix of f is a diagonal matrix. Moreover, the diagonal entries of this matrix are the eigenvalues of f.* ■

The following theorem tells us why the **B**-coordinate matrix $M_\mathbf{B}(f)$ of a linear map $f:\mathbb{R}^n \to \mathbb{R}^n$ is important: Multiplication by $M_\mathbf{B}(f)$ transforms the **B**-coordinate matrix of **x** into the **B**-coordinate matrix of $f(\mathbf{x})$, for each $\mathbf{x} \in \mathbb{R}^n$.

Theorem 2   *Let $f:\mathbb{R}^n \to \mathbb{R}^n$ be a linear map and let* **B** *be an ordered basis for $\mathbb{R}^n$. Then, for each $\mathbf{x} \in \mathbb{R}^n$,*

$$M_\mathbf{B}(f(\mathbf{x})) = M_\mathbf{B}(f)M_\mathbf{B}(\mathbf{x}).$$

Proof Suppose

$$\mathbf{B} = (\mathbf{v}_1, \ldots, \mathbf{v}_n), \quad M_\mathbf{B}(\mathbf{x}) = \begin{pmatrix} a_1 \\ \vdots \\ a_n \end{pmatrix}, \quad \text{and} \quad M_\mathbf{B}(f) = \begin{pmatrix} c_{11} & \cdots & c_{1n} \\ & \cdots & \\ c_{n1} & \cdots & c_{nn} \end{pmatrix}.$$

Then

$$\mathbf{x} = a_1\mathbf{v}_1 + \cdots + a_n\mathbf{v}_n$$

and

$$f(\mathbf{v}_1) = c_{11}\mathbf{v}_1 + \cdots + c_{n1}\mathbf{v}_n$$

$$\cdots$$

$$f(\mathbf{v}_n) = c_{1n}\mathbf{v}_1 + \cdots + c_{nn}\mathbf{v}_n.$$

Hence

$$f(\mathbf{x}) = f(a_1\mathbf{v}_1 + \cdots + a_n\mathbf{v}_n)$$

$$= a_1 f(\mathbf{v}_1) + \cdots + a_n f(\mathbf{v}_n)$$

$$= a_1(c_{11}\mathbf{v}_1 + \cdots + c_{n1}\mathbf{v}_n) + \cdots + a_n(c_{1n}\mathbf{v}_1 + \cdots + c_{nn}\mathbf{v}_n)$$

$$= (c_{11}a_1 + \cdots + c_{1n}a_n)\mathbf{v}_1 + \cdots + (c_{n1}a_1 + \cdots + c_{nn}a_n)\mathbf{v}_n.$$

Therefore

$$M_\mathbf{B}(f(\mathbf{x})) = \begin{pmatrix} c_{11}a_1 & + \cdots + & c_{1n}a_n \\ & \cdots & \\ c_{n1}a_1 & + \cdots + & c_{nn}a_n \end{pmatrix}$$

$$= \begin{pmatrix} c_{11} & \cdots & c_{1n} \\ & \cdots & \\ c_{n1} & \cdots & c_{nn} \end{pmatrix}\begin{pmatrix} a_1 \\ \vdots \\ a_n \end{pmatrix} = M_\mathbf{B}(f)M_\mathbf{B}(\mathbf{x}). \quad ■$$

EXAMPLE 5

Let $f: \mathbb{R}^n \to \mathbb{R}^n$ be a linear map that has n linearly independent eigenvectors and let $\mathbf{B} = (\mathbf{v}_1, \ldots, \mathbf{v}_n)$ be an ordered basis for $\mathbb{R}^n$ consisting of eigenvectors of f. Then we know from Example 4 that the $\mathbf{B}$-coordinate matrix of f is the diagonal matrix

$$
M_{\mathbf{B}}(f) = \begin{pmatrix} \lambda_1 & 0 & & 0 \\ 0 & \lambda_2 & . & 0 \\ \vdots & \vdots & \vdots & \vdots \\ 0 & 0 & & \lambda_n \end{pmatrix}
$$

where λ_i is the eigenvalue of f corresponding to the eigenvector $\mathbf{v}_i$. If $\mathbf{x} = c_1\mathbf{v}_1 + \cdots + c_n\mathbf{v}_n$ is any vector in $\mathbb{R}^n$ then, according to Theorem 2, the $\mathbf{B}$-coordinate matrix of $f(\mathbf{x})$ is

$$
M_{\mathbf{B}}(f(\mathbf{x})) = M_{\mathbf{B}}(f)M_{\mathbf{B}}(\mathbf{x}) = \begin{pmatrix} \lambda_1 & 0 & & 0 \\ 0 & \lambda_2 & . & 0 \\ \vdots & \vdots & \vdots & \vdots \\ 0 & 0 & & \lambda_n \end{pmatrix} \begin{pmatrix} c_1 \\ c_2 \\ \vdots \\ c_n \end{pmatrix} = \begin{pmatrix} \lambda_1 c_1 \\ \lambda_2 c_2 \\ \vdots \\ \lambda_n c_n \end{pmatrix},
$$

in agreement with Example 7 of Section 5.1. ∎

Theorem 1 in this section describes the relationship between the standard coordinates and the $\mathbf{B}$-coordinates of any vector $\mathbf{x} \in \mathbb{R}^n$. The next theorem describes the relationship between the standard matrix and the $\mathbf{B}$-coordinate matrix of any linear map $f: \mathbb{R}^n \to \mathbb{R}^n$.

Theorem 3 *Let $f: \mathbb{R}^n \to \mathbb{R}^n$ be a linear map and let $\mathbf{B} = (\mathbf{v}_1, \ldots, \mathbf{v}_n)$ be any ordered basis for $\mathbb{R}^n$. Then*

$$
M_{\mathbf{B}}(f) = H^{-1}M(f)H
$$

where H is the matrix whose column vectors are the basis vectors $\mathbf{v}_1, \ldots, \mathbf{v}_n$.

Proof To show that two matrices are equal it suffices to show that all corresponding columns are equal. The jth column of an $n \times n$ matrix A is just

$$
A \begin{pmatrix} 0 \\ \vdots \\ 1 \\ \vdots \\ 0 \end{pmatrix} \quad \leftarrow j\text{th entry} \qquad \text{or} \qquad AM_{\mathbf{B}}(\mathbf{v}_j).
$$

Therefore the jth column of the matrix $M_{\mathbf{B}}(f)$ is

$$M_{\mathbf{B}}(f)M_{\mathbf{B}}(\mathbf{v}_j) = M_{\mathbf{B}}(f(\mathbf{v}_j)) \qquad \text{(by Theorem 2)}$$

$$= H^{-1}M(f(\mathbf{v}_j)) \qquad \text{(by Theorem 1)}$$

$$= H^{-1}M(f)M(\mathbf{v}_j) \qquad \text{(by Theorem 2)}$$

$$= H^{-1}M(f)HM_{\mathbf{B}}(\mathbf{v}_j) \qquad \text{(by Theorem 1).}$$

This last matrix is just the jth column of $H^{-1}M(f)H$. Thus, for each $j(1 \leq j \leq n)$, the jth column of $M_{\mathbf{B}}(f)$ is equal to the jth column of $H^{-1}M(f)H$ and hence

$$M_{\mathbf{B}}(f) = H^{-1}M(f)H. \quad \blacksquare$$

REMARK   The matrix formula

$$M_{\mathbf{B}}(f) = H^{-1}M(f)H$$

simply expresses in symbols the following statement, whose validity is intuitively clear. The linear map (multiplication by $M_{\mathbf{B}}(f)$) that transforms the **B**-coordinate vector of $\mathbf{x}$ into the **B**-coordinate vector of $f(\mathbf{x})$ is the same as the linear map (multiplication by $H^{-1}M(f)H$) that first transforms the **B**-coordinate vector of $\mathbf{x}$ into $\mathbf{x}$ (this is multiplication by H), then transforms $\mathbf{x}$ into $f(\mathbf{x})$ (this is multiplication by $M(f)$), and finally transforms $f(\mathbf{x})$ into the **B**-coordinate vector of $f(\mathbf{x})$ (this is multiplication by H^{-1}). This statement can be represented by the following diagram:

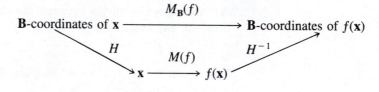

EXAMPLE 6

Let $f:\mathbb{R}^3 \to \mathbb{R}^3$ be the linear map with standard matrix

$$A = \begin{pmatrix} 0 & -2 & 2 \\ -3 & 1 & 3 \\ -1 & 1 & 3 \end{pmatrix}.$$

We observed in Example 4 of Section 5.1 that $\mathbf{B} = ((1, 0, 1), (1, 1, 0), (0, 1, 1))$ is an ordered basis for $\mathbb{R}^3$ consisting of eigenvectors of f. Theorem 3 tells us that the **B**-coordinate matrix of f is

$$M_{\mathbf{B}}(f) = H^{-1}AH,$$

where

$$H = \begin{pmatrix} 1 & 1 & 0 \\ 0 & 1 & 1 \\ 1 & 0 & 1 \end{pmatrix}.$$

Since

$$H^{-1} = \begin{pmatrix} \frac{1}{2} & -\frac{1}{2} & \frac{1}{2} \\ \frac{1}{2} & \frac{1}{2} & -\frac{1}{2} \\ -\frac{1}{2} & \frac{1}{2} & \frac{1}{2} \end{pmatrix}$$

we find that

$$M_{\mathbf{B}}(f) = H^{-1}AH = \begin{pmatrix} 2 & 0 & 0 \\ 0 & -2 & 0 \\ 0 & 0 & 4 \end{pmatrix}.$$

Notice that $M_{\mathbf{B}}(f)$ is the diagonal matrix whose diagonal entries are the eigenvalues of f, in agreement with Example 4. ■

EXERCISES

1. Let $\mathbf{B} = ((2, 3), (1, 2))$. Find the matrix that converts standard coordinates in $\mathbb{R}^2$ to $\mathbf{B}$-coordinates and use this matrix to find the $\mathbf{B}$-coordinate pair of each of the following vectors.

   (a) $(5, -3)$ (b) $(-1, 1)$ (c) $(-3, 4)$

   (d) $(1, 2)$ (e) $(1, 0)$ (f) $(0, 1)$

2. Repeat Exercise 1 for the ordered basis $\mathbf{B} = ((-3, 4), (1, -2))$.

3. Let $\mathbf{v}_1$ and $\mathbf{v}_2$ be the vectors in $\mathbb{R}^2$ obtained by rotating $\mathbf{e}_1$ and $\mathbf{e}_2$ counterclockwise through the angle $\pi/6$ and let $\mathbf{B} = (\mathbf{v}_1, \mathbf{v}_2)$. Find the $\mathbf{B}$-coordinate pair of each of the following vectors.

   (a) $(1, 0)$ (b) $(1, 1)$ (c) $(-1, 1)$

   (d) $(\sqrt{3}, 1)$ (e) $(1, -\sqrt{3})$ (f) $(-1, 2 - \sqrt{3})$

4. Repeat Exercise 3 for the vectors $\mathbf{v}_1$ and $\mathbf{v}_2$ obtained by rotating $\mathbf{e}_1$ and $\mathbf{e}_2$ counterclockwise through the angle $\pi/4$.

5. Let $\mathbf{B} = ((-1, 2, 3), (1, -1, 2), (1, -2, -2))$. Find the matrix that converts standard coordinates in $\mathbb{R}^3$ to $\mathbf{B}$-coordinates and use this matrix to find the $\mathbf{B}$-coordinate triple of each of the following vectors.

   (a) $(1, 1, 1)$ (b) $(-1, 0, 1)$ (c) $(1, -1, 2)$

   (d) $(2, -3, 0)$ (e) $(2, 0, 0)$ (f) $(4, 2, 3)$

6. Repeat Exercise 5 for the ordered basis

   $$\mathbf{B} = ((-1, 2, 3), (1, -1, 2), (1, -2, -1)).$$

7. Let $\mathbf{B} = ((1, 0), (1, 1))$. Find the $\mathbf{B}$-coordinate matrix of each of the following linear maps.

   (a) $f(x_1, x_2) = (x_1 + x_2, x_1 - x_2)$

   (b) $f(x_1, x_2) = (x_2, x_1)$

   (c) $f(x_1, x_2) = (2x_1 - 2x_2, -x_1 + 3x_2)$

   (d) $f(x_1, x_2) = (x_1 + x_2, x_2)$

8. Repeat Exercise 7 for the ordered basis $\mathbf{B} = ((1, -1), (2, 1))$.

9. Let $f:\mathbb{R}^2 \to \mathbb{R}^2$ be the linear map defined by

   $$f(x_1, x_2) = (x_2, x_1).$$

   Find the $\mathbf{B}$-coordinate matrix of f, where $\mathbf{B} =$

   (a) $((1, 0), (0, 1))$ (b) $((0, 1), (1, 0))$ (c) $((1, 1), (-1, 1))$

   (d) $((-1, 1), (1, 1))$ (e) $((3, -1), (-4, 1))$ (f) $((-1, 1), (4, 3))$

10. Repeat Exercise 9 for the linear map $f(x_1, x_2) = (2x_1 + 4x_2, 3x_1 + x_2)$.

11. For each of the given linear maps f, find an ordered basis **B** for $\mathbb{R}^2$ consisting of eigenvectors of f and find the **B**-coordinate matrix of f.

   (a) $f(x_1, x_2) = (2x_1 + x_2, 8x_1)$

   (b) $f(x_1, x_2) = (x_1 + 2x_2, 2x_1 + x_2)$

   (c) $f(x_1, x_2) = (x_1 + x_2, x_1 - x_2)$

   (d) $f(x_1, x_2) = (-3x_1 + 4x_2, -2x_1 + 3x_2)$

   (e) $f(x_1, x_2) = (11x_1 + 21x_2, -10x_1 - 18x_2)$

12. For each of the given linear maps f, find an ordered basis **B** for $\mathbb{R}^3$ consisting of eigenvectors of f and find the **B**-coordinate matrix of f.

   (a) $f(x_1, x_2, x_3) = (5x_1 + x_2 - 2x_3, 6x_1 + 4x_2 - 4x_3, 9x_1 + 3x_2 - 4x_3)$

   (b) $f(x_1, x_2, x_3) = (x_1 - x_2 + x_3, -3x_1 + 5x_2 - x_3, -7x_1 + 11x_2 - 3x_3)$

   (c) $f(x_1, x_2, x_3) = (-4x_1 + 3x_2 - 3x_3, 3x_1 - 4x_2 + 3x_3, 9x_1 - 9x_2 + 8x_3)$

13. Let **B** be the ordered basis $((-1, 2, 3), (1, -1, 2), (1, -2, 2))$ for $\mathbb{R}^3$ and let $f:\mathbb{R}^3 \to \mathbb{R}^3$ be the linear map whose **B**-coordinate matrix is

$$M_\mathbf{B}(f) = \begin{pmatrix} 2 & -1 & 3 \\ 3 & 2 & 4 \\ 1 & -1 & 5 \end{pmatrix}$$

   Find $M_\mathbf{B}(f(\mathbf{v}))$, where $\mathbf{v}$ is the vector in $\mathbb{R}^3$ such that $M_\mathbf{B}(\mathbf{v}) =$

   (a) $\begin{pmatrix} 1 \\ 0 \\ 0 \end{pmatrix}$ (b) $\begin{pmatrix} -1 \\ 1 \\ 1 \end{pmatrix}$ (c) $\begin{pmatrix} 2 \\ 3 \\ -5 \end{pmatrix}$ (d) $\begin{pmatrix} 0 \\ 1 \\ 1 \end{pmatrix}$ (e) $\begin{pmatrix} 3 \\ 3 \\ 2 \end{pmatrix}$

14. Let **B** be an ordered basis for $\mathbb{R}^n$. Show that, if f and g are linear maps from $\mathbb{R}^n$ to $\mathbb{R}^n$ and c is any real number, then

   (a) $M_\mathbf{B}(f + g) = M_\mathbf{B}(f) + M_\mathbf{B}(g)$

   (b) $M_\mathbf{B}(cf) = cM_\mathbf{B}(f)$

   (c) $M_\mathbf{B}(f^{-1}) = [M_\mathbf{B}(f)]^{-1}$, if f is invertible.

15. Let A be any $n \times n$ matrix, let H be any invertible $n \times n$ matrix, and let $C = H^{-1}AH$. Show that there is an ordered basis **B** for $\mathbb{R}^n$ such that $C = M_\mathbf{B}(f)$ where $f:\mathbb{R}^n \to \mathbb{R}^n$ is multiplication by A.

5.3
Diagonalization of Matrices

An $n \times n$ matrix A is said to be ***diagonable*** if there is an invertible matrix H and a diagonal matrix D such that $H^{-1}AH = D$. A matrix H such that $H^{-1}AH$ is a diagonal matrix is said to ***diagonalize*** A. The diagonal matrix D such that $H^{-1}AH = D$ is called a ***diagonalization*** of A.

   In Section 5.2 (see Example 4 and Theorem 3) we saw that if A is the standard matrix of a linear map $f:\mathbb{R}^n \to \mathbb{R}^n$ that has n linearly independent eigenvectors, then A is diagonable. In this section we shall reformulate this result purely in terms of matrices and we shall study a particular class of matrices, symmetric matrices, that are always diagonable.

   In order to diagonalize a diagonable matrix A we must first find the *eigenvectors* and the *eigenvalues* of A.

   A nonzero vector $\mathbf{v} \in \mathbb{R}^n$ is called an ***eigenvector*** of the $n \times n$ matrix A if

$$A\mathbf{v} = \lambda\mathbf{v}$$

for some real number λ. The number λ such that $A\mathbf{v} = \lambda\mathbf{v}$ is the *eigenvalue* of A corresponding to the eigenvector $\mathbf{v}$.

Notice that a nonzero vector $\mathbf{v} \in \mathbb{R}^n$ is an eigenvector of A with eigenvalue λ if and only if

$$(A - \lambda I)\mathbf{v} = \mathbf{0}.$$

This linear system has a nonzero solution if and only if the coefficient matrix $A - \lambda I$ has determinant zero. Hence *the eigenvalues of A are the zeros of the* ***characteristic polynomial*** $p(\lambda) = \det(A - \lambda I)$, *and the eigenvectors of A with eigenvalue λ are the nonzero solutions of the homogeneous linear system*

$$(A - \lambda I)\mathbf{x} = \mathbf{0}.$$

The solution space of the system $(A - \lambda I)\mathbf{x} = \mathbf{0}$ is called the λ-*eigenspace* of A.

EXAMPLE 1

Let us find the eigenvalues and the eigenvectors of the matrix

$$A = \begin{pmatrix} -17 & 30 \\ -10 & 18 \end{pmatrix}.$$

The characteristic polynomial of A is

$$\begin{aligned} p(\lambda) = \det(A - \lambda I) &= \det\begin{pmatrix} -17 - \lambda & 30 \\ -10 & 18 - \lambda \end{pmatrix} \\ &= (-17 - \lambda)(18 - \lambda) + 300 \\ &= \lambda^2 - \lambda - 6 \\ &= (\lambda - 3)(\lambda + 2) \end{aligned}$$

Hence the eigenvalues of A are 3 and -2.

The eigenvectors of A with eigenvalue λ are the nonzero solutions of the linear system

$$(A - \lambda I)\mathbf{x} = \mathbf{0}.$$

Since

$$A - 3I = \begin{pmatrix} -20 & 30 \\ -10 & 15 \end{pmatrix} \longrightarrow \begin{pmatrix} 1 & -\frac{3}{2} \\ 0 & 0 \end{pmatrix}$$

and

$$A - (-2)I = \begin{pmatrix} -15 & 30 \\ -10 & 20 \end{pmatrix} \longrightarrow \begin{pmatrix} 1 & -2 \\ 0 & 0 \end{pmatrix}$$

we can conclude that the eigenvectors of A with eigenvalue 3 are the nonzero scalar multiples of the vector $(\frac{3}{2}, 1)$ and the eigenvectors of A with eigenvalue -2 are the nonzero scalar multiples of the vector $(2, 1)$. ■

Notice that the eigenvectors and eigenvalues of an $n \times n$ matrix A are the same as the eigenvectors and eigenvalues of the linear map $f : \mathbb{R}^n \to \mathbb{R}^n$ defined by $f(\mathbf{x}) = A\mathbf{x}$. Hence everything that we know about eigenvectors and eigenvalues of linear maps applies to eigenvectors and eigenvalues of matrices as well. In particular, we can prove the following theorem.

Theorem 2.   *Let A be an n × n matrix.*

(i)   *If $\mathbf{v}_1, \ldots, \mathbf{v}_k$ are eigenvectors of A with distinct eigenvalues $\lambda_1, \ldots, \lambda_k$, then $\{\mathbf{v}_1, \ldots, \mathbf{v}_k\}$ is linearly independent.*

(ii)   *If A has n distinct eigenvalues, then there is a basis for $\mathbb{R}^n$ consisting of eigenvectors of A.*

(iii)   *A is diagonable if and only if there is a basis for $\mathbb{R}^n$ consisting of eigenvectors of A.*

Proof

(i)   is immediate from Theorem 2 of Section 5.1.

(ii)   follows directly from the Corollary to Theorem 2 of Section 5.1.

(iii)   Suppose there is a basis $\{\mathbf{v}_1, \ldots, \mathbf{v}_k\}$ for $\mathbb{R}^n$ consisting of eigenvectors of A. From Example 4 of Section 5.2 we know that if $\mathbf{B} = (\mathbf{v}_1, \ldots, \mathbf{v}_n)$, then the **B**-coordinate matrix of the linear map $f(\mathbf{x}) = A\mathbf{x}$ is the diagonal matrix

$$D = \begin{pmatrix} \lambda_1 & 0 & & 0 \\ 0 & \lambda_2 & . & 0 \\ \vdots & \vdots & \vdots & \vdots \\ 0 & 0 & & \lambda_n \end{pmatrix}$$

where $\lambda_1, \ldots, \lambda_n$ are the eigenvalues of f corresponding to the eigenvectors $\mathbf{v}_1, \ldots, \mathbf{v}_n$. From Theorem 3 of Section 5.2 it follows that

$$D = H^{-1}AH,$$

where H is the matrix whose column vectors are the vectors $\mathbf{v}_1, \ldots, \mathbf{v}_n$. This shows that A is diagonable.

Conversely, if A is diagonable then $H^{-1}AH$ is a diagonal matrix for some invertible matrix H. It is easy to check (see Exercise 12) that the column vectors of H are eigenvectors of A and that they form a basis for $\mathbb{R}^n$.   ■

From the proof of Theorem 2(iii) we can extract the following algorithm for diagonalizing matrices.

To diagonalize a diagonable matrix A:

1.   Find a basis $\{\mathbf{v}_1, \ldots, \mathbf{v}_n\}$ for $\mathbb{R}^n$ consisting of eigenvectors of A.
2.   Take H to be the matrix whose column vectors are $\mathbf{v}_1, \ldots, \mathbf{v}_n$.

Then H diagonalizes $A : H^{-1}AH = D$ where D is the diagonal matrix whose diagonal entries are the eigenvalues of A corresponding to the eigenvectors $\mathbf{v}_1, \ldots, \mathbf{v}_n$.

EXAMPLE 2

Let us diagonalize the matrix

$$A = \begin{pmatrix} -17 & 30 \\ -10 & 18 \end{pmatrix}.$$

We saw in Example 1 that the vectors $(\frac{3}{2}, 1)$ and $(2, 1)$ are eigenvectors of A, with eigenvalues 3 and -2, respectively. Hence, the matrix

$$H = \begin{pmatrix} \frac{3}{2} & 2 \\ 1 & 1 \end{pmatrix}$$

diagonalizes A, and $H^{-1}AH = D$ where

$$D = \begin{pmatrix} 3 & 0 \\ 0 & -2 \end{pmatrix}.$$

As a check, notice that

$$HD = \begin{pmatrix} \frac{3}{2} & 2 \\ 1 & 1 \end{pmatrix}\begin{pmatrix} 3 & 0 \\ 0 & -2 \end{pmatrix} = \begin{pmatrix} \frac{9}{2} & -4 \\ 3 & -2 \end{pmatrix}$$

and that

$$AH = \begin{pmatrix} -17 & 30 \\ -10 & 18 \end{pmatrix}\begin{pmatrix} \frac{3}{2} & 2 \\ 1 & 1 \end{pmatrix} = \begin{pmatrix} \frac{9}{2} & -4 \\ 3 & -2 \end{pmatrix}.$$

Hence $HD = AH$, or $D = H^{-1}AH$, as required. ■

EXAMPLE 3

Let us diagonalize the matrix

$$A = \begin{pmatrix} 1 & 3 & -3 \\ 15 & -5 & 21 \\ 3 & -3 & 7 \end{pmatrix}.$$

The characteristic polynomial of A is

$$p(\lambda) = \det(A - \lambda I)$$

$$= \det\begin{pmatrix} 1-\lambda & 3 & -3 \\ 15 & -5-\lambda & 21 \\ 3 & -3 & 7-\lambda \end{pmatrix} = (1-\lambda)(\lambda - 4)(\lambda + 2).$$

Hence the eigenvalues of A are 1, 4, and -2. Since

$$A - 1I = \begin{pmatrix} 0 & 3 & -3 \\ 15 & -6 & 21 \\ 3 & -3 & 6 \end{pmatrix} \longrightarrow \begin{pmatrix} 1 & 0 & 1 \\ 0 & 1 & -1 \\ 0 & 0 & 0 \end{pmatrix},$$

$$A - 4I = \begin{pmatrix} -3 & 3 & -3 \\ 15 & -9 & 21 \\ 3 & -3 & 3 \end{pmatrix} \longrightarrow \begin{pmatrix} 1 & 0 & 2 \\ 0 & 1 & 1 \\ 0 & 0 & 0 \end{pmatrix},$$

and

$$A - (-2)I = \begin{pmatrix} 3 & 3 & -3 \\ 15 & -3 & 21 \\ 3 & -3 & 9 \end{pmatrix} \longrightarrow \begin{pmatrix} 1 & 0 & 1 \\ 0 & 1 & -2 \\ 0 & 0 & 0 \end{pmatrix},$$

we see that the vectors

$$(-1, 1, 1), \quad (-2, -1, 1), \quad \text{and} \quad (-1, 2, 1)$$

are eigenvectors of A, with eigenvalues 1, 4, and -2, respectively. Since the eigenvalues are distinct, these eigenvectors form a basis for $\mathbb{R}^3$. Hence $H^{-1}AH = D$ where

$$H = \begin{pmatrix} -1 & -2 & -1 \\ 1 & -1 & 2 \\ 1 & 1 & 1 \end{pmatrix} \quad \text{and} \quad D = \begin{pmatrix} 1 & 0 & 0 \\ 0 & 4 & 0 \\ 0 & 0 & -2 \end{pmatrix}. \quad \blacksquare$$

EXAMPLE 4

Suppose

$$A = \begin{pmatrix} a & b \\ b & c \end{pmatrix},$$

where a, b, and c are any three real numbers. The characteristic polynomial of A is

$$p(\lambda) = \det(A - \lambda I) = \det\begin{pmatrix} a - \lambda & b \\ b & c - \lambda \end{pmatrix}$$

$$= \lambda^2 - (a + c)\lambda + (ac - b^2).$$

Using the quadratic formula we find that the zeros of this polynomial are

$$\frac{(a + c) \pm \sqrt{(a + c)^2 - 4(ac - b^2)}}{2} = \frac{(a + c) \pm \sqrt{(a - c)^2 + 4b^2}}{2}$$

If $b \neq 0$ then

$$(a - c)^2 + 4b^2 > 0$$

so $p(\lambda)$ has two distinct real zeros; that is, A has two distinct eigenvalues. Hence, by Theorem 2, A is diagonable.

If $b = 0$ then A is already a diagonal matrix. We can conclude, then, that every 2×2 matrix of the form

$$A = \begin{pmatrix} a & b \\ b & c \end{pmatrix}$$

is diagonable. $\blacksquare$

A matrix A is said to be **symmetric** if it is a square matrix and it satisfies the equation $A = A^t$. Thus

$$A = \begin{pmatrix} a_{11} & \cdots & a_{1n} \\ & \cdots & \\ a_{n1} & \cdots & a_{nn} \end{pmatrix}$$

is symmetric if $a_{ij} = a_{ji}$ for each i and j ($1 \leq i \leq n$, $1 \leq j \leq n$). Example 4 shows that every symmetric 2×2 matrix is diagonable. More generally, it can be shown that *all* symmetric matrices are diagonable. Actually, symmetric matrices can be diagonalized by matrices of a very special type. This is a consequence of the following theorem.

Theorem 3 *Suppose $f:\mathbb{R}^n \to \mathbb{R}^n$ is a linear map whose standard matrix is symmetric. Then there is an orthonormal basis for $\mathbb{R}^n$ consisting of eigenvectors of f.*

Proof The proof consists of three steps. First we show that f has the property $f(\mathbf{x}) \cdot \mathbf{y} = \mathbf{x} \cdot f(\mathbf{y})$ for all $\mathbf{x}$ and $\mathbf{y}$ in $\mathbb{R}^n$. Then we show that eigenvectors of f with distinct eigenvalues are perpendicular to one another. Finally we show that there is an orthonormal basis for $\mathbb{R}^n$ consisting of eigenvectors of f. Although the theorem is true in general, we will be able to carry out the details of this last step here only in the case where f has n distinct eigenvalues.

Step 1 Suppose $\mathbf{x}$ and $\mathbf{y}$ are any two vectors in $\mathbb{R}^n$. Then the dot product of $\mathbf{x}$ and $\mathbf{y}$ can be expressed in matrix form as

$$\mathbf{x} \cdot \mathbf{y} = [M(\mathbf{x})]^t M(\mathbf{y}).$$

Therefore, for each $\mathbf{x}$ and $\mathbf{y}$ in $\mathbb{R}^n$,

$$
\begin{aligned}
f(\mathbf{x}) \cdot \mathbf{y} &= [M(f(\mathbf{x}))]^t M(\mathbf{y}) \\
&= [M(f)M(\mathbf{x})]^t M(\mathbf{y}) && \text{(by Theorem 2, Section 5.2)} \\
&= [M(\mathbf{x})]^t [M(f)]^t M(\mathbf{y}) && \text{(since } (AB)^t = B^t A^t) \\
&= [M(\mathbf{x})]^t M(f)M(\mathbf{y}) && \text{(since } M(f) \text{ is symmetric)} \\
&= [M(\mathbf{x})]^t M(f(\mathbf{y})) && \text{(by Theorem 2, Section 5.2)} \\
&= \mathbf{x} \cdot f(\mathbf{y}).
\end{aligned}
$$

Thus $f(\mathbf{x}) \cdot \mathbf{y} = \mathbf{x} \cdot f(\mathbf{y})$ *for all $\mathbf{x}$ and $\mathbf{y}$ in $\mathbb{R}^n$.*

Step 2 Suppose $\mathbf{x}$ and $\mathbf{y}$ are eigenvectors of f with eigenvalues λ and μ. Then, using Step 1, we see that

$$\lambda(\mathbf{x} \cdot \mathbf{y}) = (\lambda\mathbf{x}) \cdot \mathbf{y} = f(\mathbf{x}) \cdot \mathbf{y} = \mathbf{x} \cdot f(\mathbf{y}) = \mathbf{x} \cdot (\mu\mathbf{y}) = \mu(\mathbf{x} \cdot \mathbf{y}),$$

so

$$(\lambda - \mu)(\mathbf{x} \cdot \mathbf{y}) = 0.$$

It follows that $\mathbf{x} \cdot \mathbf{y} = 0$ whenever $\lambda \neq \mu$. That is, *eigenvectors of f with distinct eigenvalues are perpendicular to one another.*

Step 3 Suppose f has n distinct eigenvalues $\lambda_1, \ldots, \lambda_n$ with corresponding eigenvectors $\mathbf{v}_1, \ldots, \mathbf{v}_n$. Then by Step 2 $\mathbf{v}_i \cdot \mathbf{v}_j = 0$ whenever $i \neq$

j. Hence $\{\mathbf{v}_1, \ldots, \mathbf{v}_n\}$ is an orthogonal set, and $\{\mathbf{v}_1/\|\mathbf{v}_1\|, \ldots, \mathbf{v}_n/\|\mathbf{v}_1\|\}$ *is an orthonormal basis for* $\mathbb{R}^n$ *consisting of eigenvectors of f.*

If f does not have n distinct eigenvalues, then the proof of this theorem requires techniques that are beyond the scope of this text. What happens is that all the zeros of the characteristic polynomial are real numbers. Hence, if we count multiplicities, there are exactly n eigenvalues. For each eigenvalue of multiplicity k we get an eigenspace of dimension k. Then we can find an orthonormal basis for each of these eigenspaces and put them all together to obtain an orthonormal basis for $\mathbb{R}^n$. ∎

Theorem 4   *Let A be a symmetric matrix. Then A can be diagonalized by a matrix H whose column vectors form an orthonormal set.*

Proof   Let $f:\mathbb{R}^n \to \mathbb{R}^n$ be defined by $f(\mathbf{x}) = A\mathbf{x}$. Then the eigenvectors of A are eigenvectors of f. By Theorem 3, there is an orthonormal basis for $\mathbb{R}^n$ consisting of eigenvectors of f. Take H to be the matrix with these vectors as column vectors. This matrix diagonalizes A. ∎

A square matrix whose column vectors form an orthonormal set is called an ***orthogonal matrix.*** We have just seen that every symmetric matrix can be diagonalized by an orthogonal matrix. In other words, *for every symmetric matrix A there is an orthogonal matrix H such that* $H^{-1}AH$ *is a diagonal matrix.*

One nice property of orthogonal matrices is that their inverses are easy to compute. In fact, there is the following useful characterization of orthogonal matrices.

Theorem 5   *Let A be a square matrix. Then A is orthogonal if and only if* $A^{-1} = A^t$.

Proof   Since the row vectors of A^t are the column vectors of A, and since the entries in the matrix product $A^t A$ are the dot products of the row vectors of A^t and the column vectors of A, the statement that the column vectors of A form an orthonormal set is the same as the statement that $A^t A = I$. Hence A is orthogonal if and only if $A^t = A^{-1}$. ∎

The method for finding an orthogonal matrix that diagonalizes a symmetric matrix may be summarized as follows:

To find an orthogonal matrix H that diagonalizes a symmetric $n \times n$ matrix A:

1.   Find the eigenvalues of A.

If there are n distinct eigenvalues $\lambda_1, \ldots, \lambda_n$, then	If there are only k distinct eigenvalues $\lambda_1, \ldots, \lambda_k$, where $k < n$, then
2. Find corresponding eigenvectors	2. For each eigenvalue λ find the

$$\mathbf{v}_1, \ldots, \mathbf{v}_n.$$

These eigenvectors form an orthogonal set.

3. Normalize these eigenvectors to get an orthonormal eigenvector basis

$$\left\{\frac{\mathbf{v}_1}{\|\mathbf{v}_1\|}, \ldots, \frac{\mathbf{v}_n}{\|\mathbf{v}_n\|}\right\}$$

for $\mathbb{R}^n$.

4. Then the matrix H whose column vectors are the vectors

$$\frac{\mathbf{v}_1}{\|\mathbf{v}_1\|}, \ldots, \frac{\mathbf{v}_n}{\|\mathbf{v}_n\|}$$

is orthogonal and it diagonalizes A.

eigenspace V_λ. The dimension of V_λ equals the multiplicity of λ as a zero of the characteristic polynomial. These eigenspaces are perpendicular to one another.

3. Use the Gram-Schmidt process to find an orthonormal basis for each eigenspace V_λ.

4. Then the matrix H whose column vectors are the vectors found in Step 3 is orthogonal and it diagonalizes A.

EXAMPLE 5

Let us find an orthogonal matrix H that diagonalizes the symmetric matrix

$$A = \begin{pmatrix} -1 & -2 \\ -2 & 2 \end{pmatrix}.$$

The characteristic polynomial of A is

$$p(\lambda) = \det\begin{pmatrix} -1 - \lambda & -2 \\ -2 & 2 - \lambda \end{pmatrix} = \lambda^2 - \lambda - 6 = (\lambda + 2)(\lambda - 3).$$

The eigenvalues of A are -2 and 3. The corresponding eigenvectors are the nonzero multiples of the vectors $(2, 1)$ and $(-1, 2)$. Since

$$\|(2, 1)\| = \sqrt{5} \quad \text{and} \quad \|(-1, 2)\| = \sqrt{5},$$

we obtain the orthonormal eigenvector basis

$$\left\{\left(\frac{2}{\sqrt{5}}, \frac{1}{\sqrt{5}}\right), \left(-\frac{1}{\sqrt{5}}, \frac{2}{\sqrt{5}}\right)\right\}$$

for $\mathbb{R}^2$. The matrix

$$H = \begin{pmatrix} \dfrac{2}{\sqrt{5}} & -\dfrac{1}{\sqrt{5}} \\[2ex] \dfrac{1}{\sqrt{5}} & \dfrac{2}{\sqrt{5}} \end{pmatrix}$$

is orthogonal and it diagonalizes A. Since

$$H^{-1} = H^t = \begin{pmatrix} \dfrac{2}{\sqrt{5}} & \dfrac{1}{\sqrt{5}} \\[2mm] -\dfrac{1}{\sqrt{5}} & \dfrac{2}{\sqrt{5}} \end{pmatrix}$$

it is easy to compute $H^{-1}AH$. We find that

$$H^{-1}AH = \begin{pmatrix} -2 & 0 \\ 0 & 3 \end{pmatrix},$$

as expected. ∎

EXAMPLE 6

Let us find an orthogonal matrix H that diagonalizes the symmetric matrix

$$A = \begin{pmatrix} 3 & 4 & 4 \\ 4 & 3 & 4 \\ 4 & 4 & 3 \end{pmatrix}.$$

The characteristic polynomial of A is

$$p(\lambda) = \det \begin{pmatrix} 3 - \lambda & 4 & 4 \\ 4 & 3 - \lambda & 4 \\ 4 & 4 & 3 - \lambda \end{pmatrix} = -(\lambda - 11)(\lambda + 1)^2$$

so the eigenvalues of A are 11 and -1. The eigenvectors of A with eigenvalue 11 are the nonzero multiples of the vector $(1, 1, 1)$. The eigenvectors of A with eigenvalue -1 are the nonzero vectors of the form $c_1(-1, 0, 1) + c_2(0, -1, 1)$.

We can replace the basis $\{(-1, 0, 1), (0, -1, 1)\}$ for the eigenspace $\mathcal{L}((-1, 0, 1), (0, -1, 1))$ with an orthogonal basis by applying the Gram-Schmidt process. We take

$$\mathbf{w}_1 = (-1, 0, 1)$$

and

$$\mathbf{w}_2 = (0, -1, 1) - \frac{(0, -1, 1) \cdot (-1, 0, 1)}{(-1, 0, 1) \cdot (-1, 0, 1)} (-1, 0, 1)$$

$$= (0, -1, 1) - \tfrac{1}{2}(-1, 0, 1)$$

$$= (\tfrac{1}{2}, -1, \tfrac{1}{2}).$$

The eigenvectors $(1, 1, 1)$, $(0, -1, 1)$, $(\tfrac{1}{2}, -1, \tfrac{1}{2})$ form an orthogonal set. Normalizing each of these vectors we get the orthonormal eigenvector basis

$$\left\{ \left(\frac{1}{\sqrt{3}}, \frac{1}{\sqrt{3}}, \frac{1}{\sqrt{3}} \right), \left(0, -\frac{1}{\sqrt{2}}, \frac{1}{\sqrt{2}} \right), \left(\frac{1}{\sqrt{6}}, -\frac{2}{\sqrt{6}}, \frac{1}{\sqrt{6}} \right) \right\}.$$

Hence the matrix

$$H = \begin{pmatrix} \dfrac{1}{\sqrt{3}} & 0 & \dfrac{1}{\sqrt{6}} \\ \dfrac{1}{\sqrt{3}} & -\dfrac{1}{\sqrt{2}} & -\dfrac{2}{\sqrt{6}} \\ \dfrac{1}{\sqrt{3}} & \dfrac{1}{\sqrt{2}} & \dfrac{1}{\sqrt{6}} \end{pmatrix}$$

is orthogonal and it diagonalizes A. Since

$$H^{-1} = H^t = \begin{pmatrix} \dfrac{1}{\sqrt{3}} & \dfrac{1}{\sqrt{3}} & \dfrac{1}{\sqrt{3}} \\ 0 & -\dfrac{1}{\sqrt{2}} & \dfrac{1}{\sqrt{2}} \\ \dfrac{1}{\sqrt{6}} & -\dfrac{2}{\sqrt{6}} & \dfrac{1}{\sqrt{6}} \end{pmatrix}$$

it is easy to compute $H^{-1}AH$. We find that

$$H^{-1}AH = \begin{pmatrix} 11 & 0 & 0 \\ 0 & -1 & 0 \\ 0 & 0 & -1 \end{pmatrix},$$

as expected. ∎

EXERCISES

1. Find the eigenvalues and the corresponding eigenvectors of each of the following matrices.

   (a) $\begin{pmatrix} 2 & 3 \\ 3 & 2 \end{pmatrix}$ (b) $\begin{pmatrix} 0 & 1 \\ 0 & 0 \end{pmatrix}$

   (c) $\begin{pmatrix} 1 & 1 \\ 0 & 1 \end{pmatrix}$ (d) $\begin{pmatrix} 2 & 1 & 1 \\ 1 & 2 & 1 \\ 2 & 2 & 3 \end{pmatrix}$

2. Decide which of the matrices in Exercise 1 are diagonable. For each matrix A that is diagonable, find an invertible matrix H and a diagonal matrix D such that $H^{-1}AH = D$.

3. Diagonalize each of the following matrices by finding an invertible matrix H and a diagonal matrix D such that $H^{-1}AH = D$.

   (a) $A = \begin{pmatrix} -3 & 4 \\ -2 & 3 \end{pmatrix}$ (b) $A = \begin{pmatrix} 47 & -50 \\ 40 & -43 \end{pmatrix}$ (c) $A = \begin{pmatrix} -8 & -9 \\ 18 & 19 \end{pmatrix}$

   (d) $A = \begin{pmatrix} 1 & 1 \\ 1 & -1 \end{pmatrix}$ (e) $A = \begin{pmatrix} -11 & -21 \\ 10 & 18 \end{pmatrix}$

4. Diagonalize each of the following matrices by finding an invertible matrix H and a diagonal matrix D such that $H^{-1}AH = D$.

(a) $A = \begin{pmatrix} 5 & 1 & -2 \\ 6 & 4 & -4 \\ 9 & 3 & -4 \end{pmatrix}$ (b) $A = \begin{pmatrix} 11 & 6 & 6 \\ 29 & 16 & 16 \\ -48 & -24 & -25 \end{pmatrix}$

(c) $A = \begin{pmatrix} 1 & -1 & 1 \\ -3 & 5 & -1 \\ -7 & 11 & -3 \end{pmatrix}$ (d) $A = \begin{pmatrix} 8 & 11 & -27 \\ 9 & 18 & -39 \\ 5 & 9 & -20 \end{pmatrix}$

5. Which of the following matrices are orthogonal?

(a) $\begin{pmatrix} 1 & -1 \\ 1 & 1 \end{pmatrix}$ (b) $\begin{pmatrix} \frac{1}{2} & \frac{1}{2} \\ -\frac{1}{2} & \frac{1}{2} \end{pmatrix}$ (c) $\begin{pmatrix} 0 & 1 \\ 1 & 0 \end{pmatrix}$

(d) $\begin{pmatrix} \frac{3}{5} & -\frac{4}{5} \\ \frac{4}{5} & \frac{3}{5} \end{pmatrix}$ (e) $\begin{pmatrix} 1 & 0 & 0 \\ 0 & 1 & 0 \\ 0 & 0 & 1 \end{pmatrix}$ (f) $\begin{pmatrix} \frac{1}{9} & 0 & 0 \\ \frac{2}{9} & \frac{1}{\sqrt{2}} & -\frac{1}{\sqrt{2}} \\ \frac{2}{9} & \frac{1}{\sqrt{2}} & \frac{1}{\sqrt{2}} \end{pmatrix}$

6. Diagonalize each of the following symmetric matrices by finding an orthogonal matrix H and a diagonal matrix D such that $H^{-1}AH = D$.

(a) $\begin{pmatrix} 0 & 1 \\ 1 & 0 \end{pmatrix}$ (b) $\begin{pmatrix} 1 & 2 \\ 2 & 1 \end{pmatrix}$ (c) $\begin{pmatrix} 7 & 24 \\ 24 & -7 \end{pmatrix}$

(d) $\begin{pmatrix} 3 & -1 \\ -1 & 1 \end{pmatrix}$ (e) $\begin{pmatrix} 7 & \sqrt{3} \\ \sqrt{3} & 5 \end{pmatrix}$

7. Diagonalize each of the following symmetric matrices by finding an orthogonal matrix H and a diagonal matrix D such that $H^{-1}AH = D$.

(a) $A = \begin{pmatrix} 0 & 1 & 0 \\ 1 & 0 & 0 \\ 0 & 0 & 2 \end{pmatrix}$ (b) $A = \begin{pmatrix} 1 & 0 & 2 \\ 0 & -1 & 2 \\ 2 & 2 & 0 \end{pmatrix}$

(c) $A = \begin{pmatrix} 5 & 2 & 4 \\ 2 & 8 & -2 \\ 4 & -2 & 5 \end{pmatrix}$ (d) $A = \begin{pmatrix} 0 & 1 & 1 \\ 1 & 0 & 1 \\ 1 & 1 & 0 \end{pmatrix}$

8. Show that an $n \times n$ matrix A is orthogonal if and only if its row vectors form an orthonormal set. [*Hint:* Think about the matrix AA'.]

9. Show that if A is an orthogonal matrix then $\det A = \pm 1$. [*Hint:* Think about $\det(A'A)$.]

10. Show that if A and B are orthogonal $n \times n$ matrices then AB is an orthogonal $n \times n$ matrix. [*Hint:* Use Theorem 5.]

11. Show that if A is an orthogonal matrix then A^{-1} is also an orthogonal matrix.

12. Let A be an $n \times n$ matrix. Suppose there is an invertible matrix H such that $H^{-1}AH$ is a diagonal matrix.

(a) Show that the column vectors of H are eigenvectors of A. [*Hint:* Consider the equation $AH = HD$ where $D = H^{-1}AH$.]

(b) Explain why the column vectors of H must form a basis for $\mathbb{R}^n$.

13. Show that if $f:\mathbb{R}^n \to \mathbb{R}^n$ is a linear map such that $f(\mathbf{x}) \cdot \mathbf{y} = \mathbf{x} \cdot f(\mathbf{y})$ for all $\mathbf{x}$ and $\mathbf{y}$ in $\mathbb{R}^n$ then the standard matrix of f is symmetric.

14. Let H be any $n \times n$ orthogonal matrix and let D be any $n \times n$ diagonal matrix. Show that the matrix $H^{-1}DH$ is symmetric.

15. Let A and H be $n \times n$ matrices, with H invertible.

    (a) Show that $(H^{-1}AH)^2 = H^{-1}A^2H$

    (b) Show that $(H^{-1}AH)^k = H^{-1}A^kH$ for each positive integer k.

16. Exercise 15(b) suggests a method for computing powers of diagonable matrices. If $D \doteq H^{-1}AH$ is diagonal, then $A = HDH^{-1}$ and $A^k = HD^kH^{-1}$. Use this formula to find A^5 where $A =$

    (a) $\begin{pmatrix} 1 & 2 \\ 2 & 1 \end{pmatrix}$ (b) $\begin{pmatrix} -17 & 30 \\ -10 & 18 \end{pmatrix}$

    (c) $\begin{pmatrix} 13 & -4 \\ 42 & -13 \end{pmatrix}$ (d) $\begin{pmatrix} 33 & -10 \\ 105 & -32 \end{pmatrix}$

5.4
Isometries

In elementary geometry, the concept of congruence plays a pivotal role. Intuitively, two geometric figures are congruent if one can be superimposed on the other in such a way as to make them coincide. Thus, maps that move figures around but do not change sizes or shapes are important. In this section we shall study such maps. They are called *isometries*.

An *isometry* of $\mathbb{R}^n$ is a map $f:\mathbb{R}^n \to \mathbb{R}^n$ that preserves distances. Thus $f:\mathbb{R}^n \to \mathbb{R}^n$ is an isometry if

$$\|f(\mathbf{b}) - f(\mathbf{a})\| = \|\mathbf{b} - \mathbf{a}\|$$

for all $\mathbf{a}$ and $\mathbf{b}$ in $\mathbb{R}^n$ (see Figure 5.8).

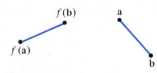

FIGURE 5.8
An isometry $f:\mathbb{R}^n \to \mathbb{R}^n$ is a map that preserves distances:
$\|f(\mathbf{b}) - f(\mathbf{a})\| = \|\mathbf{b} - \mathbf{a}\|$.

EXAMPLE 1
Let $f:\mathbb{R}^2 \to \mathbb{R}^2$ be defined by

$$f(\mathbf{x}) = \mathbf{x} + \mathbf{e}_1.$$

This map moves each point in $\mathbb{R}^2$ one unit to the right (see Figure 5.9). It is called *translation by the vector* $\mathbf{e}_1$. Since

$$\|f(\mathbf{b}) - f(\mathbf{a})\| = \|(\mathbf{b} + \mathbf{e}_1) - (\mathbf{a} + \mathbf{e}_1)\| = \|\mathbf{b} - \mathbf{a}\|$$

for all $\mathbf{a}$ and $\mathbf{b}$ in $\mathbb{R}^2$, we see that f is an isometry. Notice, however, that f is *not* a linear map, since, for example, $f(\mathbf{0}) \neq \mathbf{0}$. ■

EXAMPLE 2
Let $\mathbf{v}$ be any vector in $\mathbb{R}^n$ and let $t_{\mathbf{v}}:\mathbb{R}^n \to \mathbb{R}^n$ be defined by

$$t_{\mathbf{v}}(\mathbf{x}) = \mathbf{x} + \mathbf{v}.$$

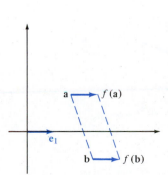

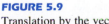

FIGURE 5.9
Translation by the vector $\mathbf{e}_1$ is an isometry.

Then for each pair of vectors $\mathbf{a}$ and $\mathbf{b}$ in $\mathbb{R}^n$ we have

$$\|t_{\mathbf{v}}(\mathbf{b}) - t_{\mathbf{v}}(\mathbf{a})\| = \|(\mathbf{b} + \mathbf{v}) - (\mathbf{a} + \mathbf{v})\| = \|\mathbf{b} - \mathbf{a}\|$$

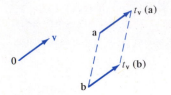

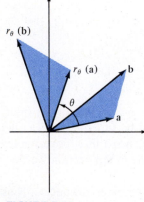

FIGURE 5.10

Translation by the vector **v** is an isometry.

FIGURE 5.11

Rotation r_θ through the angle θ is an isometry. To see that r_θ preserves distances, simply note that the shaded triangles in the figure are congruent.

so f is an isometry. The map $t_\mathbf{v}$ is called **translation by the vector v** (see Figure 5.10. ■

EXAMPLE 3

Let $r_\theta{:}\mathbb{R}^2 \to \mathbb{R}^2$ be the map that rotates each vector in $\mathbb{R}^2$ counterclockwise through the angle θ. Then it is easy to see geometrically that r_θ is an isometry (see Figure 5.11). This can also be checked algebraically (see Exercise 21) using the formula

$$r_\theta(x_1, x_2) = (x_1 \cos \theta - x_2 \sin \theta, x_1 \sin \theta + x_2 \cos \theta),$$

which was derived in Section 2.2, Example 4. ■

EXAMPLE 4

Let ℓ_θ be the line in $\mathbb{R}^2$ that passes through **0** with inclination angle θ and let $s_\theta{:}\mathbb{R}^2 \to \mathbb{R}^2$ be the map that reflects each vector in $\mathbb{R}^2$ in the line ℓ_θ. We can see geometrically that s_θ is an isometry (see Figure 5.12). We can also verify this algebraically (see Exercise 22) using the formula

$$s_\theta(x_1, x_2) = (x_1 \cos 2\theta + x_2 \sin 2\theta, x_1 \sin 2\theta - x_2 \cos 2\theta),$$

which was derived in Section 2.2, Example 5. ■

EXAMPLE 5

Let us determine those real numbers c for which the linear map $f{:}\mathbb{R}^n \to \mathbb{R}^n$ defined by the formula

$$f(\mathbf{x}) = c\mathbf{x}$$

is an isometry. Since

$$\|f(\mathbf{b}) - f(\mathbf{a})\| = \|c\mathbf{b} - c\mathbf{a}\| = |c| \, \|\mathbf{b} - \mathbf{a}\|,$$

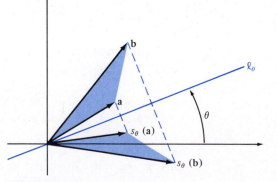

FIGURE 5.12

Reflection in the line ℓ_θ is an isometry. To see that s_θ preserves distances, simply note that the shaded triangles in the figure are congruent. (What does the figure look like if **a** and **b** are on opposite sides of ℓ_θ?)

we see that f is an isometry if and only if $|c| = 1$. The isometry $f(\mathbf{x}) = \mathbf{x}$ ($c = 1$) is the identity map. The isometry $f(\mathbf{x}) = -\mathbf{x}$ ($c = -1$) is rotation through the angle π (see Example 3). ∎

EXAMPLE 6

Suppose $f:\mathbb{R}^n \to \mathbb{R}^n$ and $g:\mathbb{R}^n \to \mathbb{R}^n$ are two isometries of $\mathbb{R}^n$. Then the map $g \circ f:\mathbb{R}^n \to \mathbb{R}^n$ is also an isometry because

$$\|g \circ f(\mathbf{b}) - g \circ f(\mathbf{a})\| = \|g(f(\mathbf{b})) - g(f(\mathbf{a}))\|$$

$$= \|f(\mathbf{b}) - f(\mathbf{a})\| \qquad \text{(since } g \text{ is an isometry)}$$

$$= \|\mathbf{b} - \mathbf{a}\| \qquad \text{(since } f \text{ is an isometry)}$$

for all $\mathbf{a}$ and $\mathbf{b}$ in $\mathbb{R}^n$. ∎

An isometry $f:\mathbb{R}^n \to \mathbb{R}^n$ that is also a linear map is called a ***linear isometry***. Rotations and reflections in lines through $\mathbf{0}$ are examples of linear isometries of $\mathbb{R}^2$. Linear isometries preserve vector addition, scalar multiplication, and distances. They also preserve lengths, dot products, and angles.

Theorem 1   *Let $f:\mathbb{R}^n \to \mathbb{R}^n$ be a linear isometry. Then*

(i) $\|f(\mathbf{a})\| = \|\mathbf{a}\|$ *for all $\mathbf{a}$ in $\mathbb{R}^n$,*

(ii) $f(\mathbf{a}) \cdot f(\mathbf{b}) = \mathbf{a} \cdot \mathbf{b}$ *for all $\mathbf{a}$ and $\mathbf{b}$ in $\mathbb{R}^n$, and*

(iii)   *the angle between $f(\mathbf{a})$ and $f(\mathbf{b})$ is the same as the angle between $\mathbf{a}$ and $\mathbf{b}$, for each pair of nonzero vectors $\mathbf{a}$ and $\mathbf{b}$ in $\mathbb{R}^n$.*

Proof

(i) $\|f(\mathbf{a})\| = \|f(\mathbf{a}) - \mathbf{0}\| = \|f(\mathbf{a}) - f(\mathbf{0})\|$

$$= \|\mathbf{a} - \mathbf{0}\| = \|\mathbf{a}\|.$$

(ii) $f(\mathbf{a}) \cdot f(\mathbf{b}) = \frac{1}{4}[\|f(\mathbf{a}) + f(\mathbf{b})\|^2 - \|f(\mathbf{a}) - f(\mathbf{b})\|^2]$

$$\qquad\qquad \text{(see Section 1.5, Exercise 11)}$$

$$= \tfrac{1}{4}[\|f(\mathbf{a} + \mathbf{b})\|^2 - \|f(\mathbf{a} - \mathbf{b})\|^2] \qquad \text{(since } f \text{ is linear)}$$

$$= \tfrac{1}{4}[\|\mathbf{a} + \mathbf{b}\|^2 - \|\mathbf{a} - \mathbf{b}\|^2] \text{ (by (i))}$$

$$= \mathbf{a} \cdot \mathbf{b}.$$

(iii)   Let θ be the angle between the nonzero vectors $\mathbf{a}$ and $\mathbf{b}$ and let φ be the angle between $f(\mathbf{a})$ and $f(\mathbf{b})$ ($f(\mathbf{a})$ and $f(\mathbf{b})$ are nonzero, by (i)). Then

$$\cos \varphi = f(\mathbf{a}) \cdot f(\mathbf{b})/\|f(\mathbf{a})\| \, \|f(\mathbf{b})\|$$

$$= \mathbf{a} \cdot \mathbf{b}/\|\mathbf{a}\| \, \|\mathbf{b}\| \text{ (by (i) and (ii))}$$

$$= \cos \theta,$$

hence $\varphi = \theta$. ∎

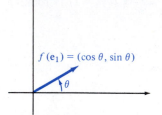

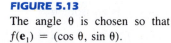

$f(\mathbf{e}_1) = (\cos\theta, \sin\theta)$

FIGURE 5.13
The angle θ is chosen so that $f(\mathbf{e}_1) = (\cos\theta, \sin\theta)$.

Now let us look more closely at linear isometries of the plane.

Suppose $f:\mathbb{R}^2 \to \mathbb{R}^2$ is a linear isometry. Then, by Theorem 1(i), $f(\mathbf{e}_1)$ is a unit vector. Therefore $f(\mathbf{e}_1) = (\cos\theta, \sin\theta)$ for some θ ($0 \le \theta < 2\pi$) (see Figure 5.13). The vector $f(\mathbf{e}_2)$ is also a unit vector, by Theorem 1(i), and it is perpendicular to $f(\mathbf{e}_1) = (\cos\theta, \sin\theta)$, by Theorem 1(iii). Therefore, $f(\mathbf{e}_2)$ is either $(-\sin\theta, \cos\theta)$ or $(\sin\theta, -\cos\theta)$ (see Figure 5.14). The standard matrix $M(f)$ of f is the matrix whose column vectors are $f(\mathbf{e}_1)$ and $f(\mathbf{e}_2)$. Therefore

$$M(f) = \begin{pmatrix} \cos\theta & -\sin\theta \\ \sin\theta & \cos\theta \end{pmatrix} \quad \text{or} \quad M(f) = \begin{pmatrix} \cos\theta & \sin\theta \\ \sin\theta & -\cos\theta \end{pmatrix}$$

Notice that the first matrix has determinant $+1$ whereas the second matrix has determinant -1. If $M(f)$ is the first matrix then f is counterclockwise rotation through the angle θ (see Section 2.2, Example 4). If $M(f)$ is the second matrix then f is reflection in the line $\ell_{\theta/2}$ that passes through $\mathbf{0}$ and has inclination angle $\theta/2$ (see Section 2.2, Example 5). The determinant $\det M(f)$ is also called the **determinant** of f and is written $\det f$.

Thus we have proved the following theorem.

Theorem 2   Let f be a linear isometry of $\mathbb{R}^2$. Then f is either a rotation or a reflection. If $\det f > 0$, then f is counterclockwise rotation through some angle θ. If $\det f < 0$, then f is reflection in a line $\ell_{\theta/2}$ that passes through $\mathbf{0}$ and has inclination angle $\theta/2$. In both cases, θ is the angle such that $f(\mathbf{e}_1) = (\cos\theta, \sin\theta)$ (see Figure 5.15).

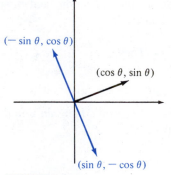

$(-\sin\theta, \cos\theta)$

$(\cos\theta, \sin\theta)$

$(\sin\theta, -\cos\theta)$

FIGURE 5.14
The two unit vectors perpendicular to $(\cos\theta, \sin\theta)$ are $(-\sin\theta, \cos\theta)$ and $(\sin\theta, -\cos\theta)$.

EXAMPLE 7
Consider the map $f = r_{\pi/2} \circ s_{\pi/4}$ from $\mathbb{R}^2$ to $\mathbb{R}^2$ which first reflects in the line $\ell_{\pi/4}$ and then rotates counterclockwise through the angle $\pi/2$. Then f is a composition of linear isometries and hence is a linear isometry. We can identify f as a particular rotation or reflection by computing $\det f$ and $f(\mathbf{e}_1)$. Since

$$\det f = (\det r_{\pi/2})(\det s_{\pi/4}) = (+1)(-1) = -1,$$

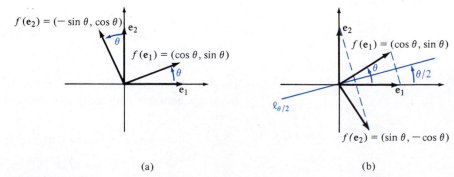

(a)   (b)

FIGURE 5.15
Each linear isometry $f:\mathbb{R}^2 \to \mathbb{R}^2$ is either (a) a counterclockwise rotation, or (b) reflection in a line that passes through $\mathbf{0}$.

we see that f is a reflection. Since (see Figure 5.16)

$$f(\mathbf{e}_1) = r_{\pi/2}(s_{\pi/4}(\mathbf{e}_1)) = r_{\pi/2}(\mathbf{e}_2)$$

$$= -\mathbf{e}_1 = (\cos \pi, \sin \pi),$$

we see that f is reflection in the line $\ell_{\pi/2}$; that is, f is reflection in the x_2-axis. ■

EXAMPLE 8

Consider the map $f = s_{\pi/4} \circ s_{\pi/2}$ that first reflects in the line $\ell_{\pi/2}$ and then reflects in the line $\ell_{\pi/4}$. Then f is a linear isometry. Since

$$\det f = (\det s_{\pi/4})(\det s_{\pi/2}) = (-1)(-1) = +1,$$

we see that f is a rotation. Since (see Figure 5.17)

$$f(\mathbf{e}_1) = s_{\pi/4}(s_{\pi/2}(\mathbf{e}_1)) = s_{\pi/4}(-\mathbf{e}_1)$$

$$= -\mathbf{e}_2 = (\cos 3\pi/2, \sin 3\pi/2),$$

we see that f is counterclockwise rotation through the angle $3\pi/2$. Notice that this is consistent with Theorem 3 of Section 2.5. ■

We have seen that each linear isometry of $\mathbb{R}^2$ is either rotation through some angle or reflection in some line through $\mathbf{0}$. We shall now study linear isometries of $\mathbb{R}^3$, in order to determine how they can be described geometrically.

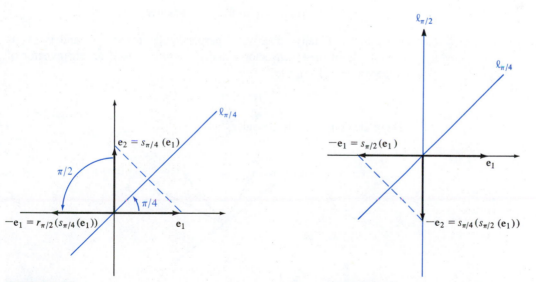

FIGURE 5.16
The linear isometry that first reflects in $\ell_{\pi/4}$ and then rotates counterclockwise through the angle $\pi/2$ sends $\mathbf{e}_1$ to $-\mathbf{e}_1$. This map is the same as reflection in the x_2-axis.

FIGURE 5.17
The linear isometry that first reflects in $\ell_{\pi/2}$ and then reflects in $\ell_{\pi/4}$ sends $\mathbf{e}_1$ to $-\mathbf{e}_2$. This map is the same as counterclockwise rotation through the angle $3\pi/2$.

Suppose $f:\mathbb{R}^3 \to \mathbb{R}^3$ is a linear isometry. Then the characteristic polynomial of f has degree 3 and so it must have at least one real zero. Therefore f must have at least one eigenvalue. If λ is any eigenvalue of f and $\mathbf{v}$ is an eigenvector with eigenvalue λ, then

$$\|\mathbf{v}\| = \|f(\mathbf{v})\| = \|\lambda\mathbf{v}\| = |\lambda|\,\|\mathbf{v}\|,$$

so we see that $|\lambda| = 1$, that is, $\lambda = \pm 1$.

Let $\mathbf{v}$ be any eigenvector of f with eigenvalue $\lambda = \pm 1$, and let W be the subspace of $\mathbb{R}^3$ consisting of all vectors perpendicular to $\mathbf{v}$. Then, for each $\mathbf{w} \in W$,

$$\mathbf{v} \cdot f(\mathbf{w}) = \pm f(\mathbf{v}) \cdot f(\mathbf{w}) = \pm \mathbf{v} \cdot \mathbf{w} = 0$$

so $f(\mathbf{w}) \in W$ whenever $\mathbf{w} \in W$. If we let $\{\mathbf{v}_1, \mathbf{v}_2\}$ be an orthonormal basis for W and let $\mathbf{v}_3 = \mathbf{v}_1 \times \mathbf{v}_2$, then $\{\mathbf{v}_1, \mathbf{v}_2, \mathbf{v}_3\}$ is an orthonormal basis for $\mathbb{R}^3$. We shall compute the matrix of f relative to the ordered orthonormal basis $\mathbf{B} = (\mathbf{v}_1, \mathbf{v}_2, \mathbf{v}_3)$.

Since $f(\mathbf{v}_1)$ is a unit vector in W and $\{\mathbf{v}_1, \mathbf{v}_2\}$ is an orthonormal basis for W, we have

$$f(\mathbf{v}_1) = (\cos \theta)\mathbf{v}_1 + (\sin \theta)\mathbf{v}_2$$

for some θ $(0 \le \theta < 2\pi)$ (see Figure 5.18). Since $f(\mathbf{v}_2)$ is a unit vector in W that is perpendicular to $f(\mathbf{v}_1)$, either

$$f(\mathbf{v}_2) = (-\sin \theta)\mathbf{v}_1 + (\cos \theta)\mathbf{v}_2$$

or

$$f(\mathbf{v}_2) = (\sin \theta)\mathbf{v}_1 - (\cos \theta)\mathbf{v}_2$$

(see Figure 5.19). Finally, since $\mathbf{v}_3$ is perpendicular to both $\mathbf{v}_1$ and $\mathbf{v}_2$, $\mathbf{v}_3$ is a scalar multiple of the eigenvector $\mathbf{v}$ and is therefore itself an eigenvector of f with eigenvalue ± 1. Hence

$$f(\mathbf{v}_3) = \pm \mathbf{v}_3.$$

There are four cases to consider.

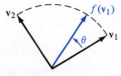

FIGURE 5.18

Since $f(\mathbf{v}_1)$ is a unit vector in W and $\{\mathbf{v}_1, \mathbf{v}_2\}$ is an orthonormal basis for W, we have $f(\mathbf{v}_1) = (\cos \theta)\mathbf{v}_1 + (\sin \theta)\mathbf{v}_2$ for some θ.

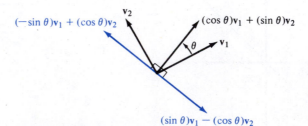

FIGURE 5.19

The two unit vectors in $W = \mathcal{L}(\mathbf{v}_1, \mathbf{v}_2)$ that are perpendicular to $(\cos \theta)\mathbf{v}_1 + (\sin \theta)\mathbf{v}_2$ are $(-\sin \theta)\mathbf{v}_1 + (\cos \theta)\mathbf{v}_2$ and $(\sin \theta)\mathbf{v}_1 - (\cos \theta)\mathbf{v}_2$.

Case 1 Suppose

$$f(\mathbf{v}_1) = (\cos \theta)\mathbf{v}_1 + (\sin \theta)\mathbf{v}_2,$$

$$f(\mathbf{v}_2) = (-\sin \theta)\mathbf{v}_1 + (\cos \theta)\mathbf{v}_2,$$

and

$$f(\mathbf{v}_3) = \mathbf{v}_3.$$

Then

$$M_{\mathbf{B}}(f) = \begin{pmatrix} \cos \theta & -\sin \theta & 0 \\ \sin \theta & \cos \theta & 0 \\ 0 & 0 & 1 \end{pmatrix}$$

and

$$f(a_1\mathbf{v}_1 + a_2\mathbf{v}_2 + a_3\mathbf{v}_3) = (a_1 \cos \theta - a_2 \sin \theta)\mathbf{v}_1$$
$$+ (a_1 \sin \theta + a_2 \cos \theta)\mathbf{v}_2 + a_3\mathbf{v}_3$$

for all real numbers a_1, a_2, and a_3. Thus, given any vector $\mathbf{a} = a_1\mathbf{v}_1 + a_2\mathbf{v}_2 + a_3\mathbf{v}_3$ in $\mathbb{R}^3$, the isometry f leaves fixed the component $a_3\mathbf{v}_3$ of $\mathbf{a}$ along $\mathbf{v}_3$ and rotates the component $a_1\mathbf{v}_1 + a_2\mathbf{v}_2$ of $\mathbf{a}$ perpendicular to $\mathbf{v}_3$ through the angle θ, where the direction of rotation is determined by the right hand rule (see Figure 5.20). This map f is called **rotation about $\mathbf{v}_3$ through the angle θ**. The line $\mathbf{x} = t\mathbf{v}_3$ is called the **axis of rotation**. The subspace W is the **plane of rotation**. Notice that *the angle of rotation is the angle θ $(0 \leq \theta < 2\pi)$ such*

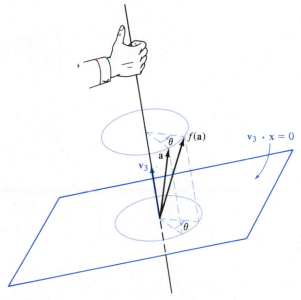

FIGURE 5.20
Rotation about $\mathbf{v}_3$ through the angle θ.

that $f(\mathbf{v}_1) \cdot \mathbf{v}_1 = \cos \theta$ and $f(\mathbf{v}_1) \cdot \mathbf{v}_2 = \sin \theta$. Futhermore, if $\theta \neq 0$, then the axis of rotation is the $+1$ eigenspace of f and the plane of rotation is the row space of the matrix $M(f) - I$. To see this, observe that the equation $(M(f) - I)\mathbf{v}_3 = \mathbf{0}$, which expresses the fact that $\mathbf{v}_3$ is an eigenvector of f with eigenvalue 1, implies that the row vectors of $M(f) - I$ are perpendicular to $\mathbf{v}_3$ and therefore lie in the plane of rotation.

Case 2   Suppose $f(\mathbf{v}_1) = \mathbf{v}_1(\theta = 0)$, $f(\mathbf{v}_2) = \mathbf{v}_2$, and $f(\mathbf{v}_3) = -\mathbf{v}_3$. Then

$$M_{\mathbf{B}}(f) = \begin{pmatrix} 1 & 0 & 0 \\ 0 & 1 & 0 \\ 0 & 0 & -1 \end{pmatrix}$$

and

$$f(a_1\mathbf{v}_1 + a_2\mathbf{v}_2 + a_3\mathbf{v}_3) = a_1\mathbf{v}_1 + a_2\mathbf{v}_2 - a_3\mathbf{v}_3$$

for all real numbers a_1, a_2, and a_3. Thus, given any vector $\mathbf{a} = a_1\mathbf{v}_1 +' a_2\mathbf{v}_2 + a_3\mathbf{v}_3$ in $\mathbb{R}^3$, the isometry f leaves fixed the component of $\mathbf{a}$ perpendicular to $\mathbf{v}_3$ and changes the sign of the component of $\mathbf{a}$ along $\mathbf{v}_3$ (see Figure 5.21). This map f is called *reflection in the plane $\mathbf{v}_3 \cdot \mathbf{x} = 0$*.

Case 3   Suppose

$$f(\mathbf{v}_1) = (\cos \theta)\mathbf{v}_1 + (\sin \theta)\mathbf{v}_2,$$

$$f(\mathbf{v}_2) = (-\sin \theta)\mathbf{v}_1 + (\cos \theta)\mathbf{v}_2,$$

and

$$f(\mathbf{v}_3) = -\mathbf{v}_3.$$

Then

$$M_{\mathbf{B}}(f) = \begin{pmatrix} \cos \theta & -\sin \theta & 0 \\ \sin \theta & \cos \theta & 0 \\ 0 & 0 & -1 \end{pmatrix}$$

$$= \begin{pmatrix} 1 & 0 & 0 \\ 0 & 1 & 0 \\ 0 & 0 & -1 \end{pmatrix} \begin{pmatrix} \cos \theta & -\sin \theta & 0 \\ \sin \theta & \cos \theta & 0 \\ 0 & 0 & 1 \end{pmatrix},$$

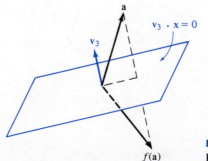

FIGURE 5.21
Reflection in the plane $\mathbf{v}_3 \cdot \mathbf{x} = 0$.

so $f = g \circ h$ where h is rotation about $\mathbf{v}_3$ through the angle θ and g is reflection in the plane $\mathbf{v}_3 \cdot \mathbf{x} = 0$.

Case 4   Suppose

$$f(\mathbf{v}_1) = (\cos \theta)\mathbf{v}_1 + (\sin \theta)\mathbf{v}_2$$

$$f(\mathbf{v}_2) = (\sin \theta)\mathbf{v}_1 - (\cos \theta)\mathbf{v}_2,$$

and

$$f(\mathbf{v}_3) = \pm \mathbf{v}_3.$$

Then

$$M_{\mathbf{B}}(f) = \begin{pmatrix} \cos \theta & \sin \theta & 0 \\ \sin \theta & -\cos \theta & 0 \\ 0 & 0 & \pm 1 \end{pmatrix}.$$

This matrix is symmetric. Therefore the standard matrix of f is also symmetric (see Exercise 23). Hence there is an orthonormal basis for $\mathbb{R}^3$ consisting of eigenvectors of f. Since the only possible eigenvalues of f are $+1$ and -1, f must have a repeated eigenvalue. We can order the basis vectors to get an ordered orthonormal eigenvector basis $\mathbf{B}' = (\mathbf{w}_1, \mathbf{w}_2, \mathbf{w}_3)$ such that the eigenvectors $\mathbf{w}_1$ and $\mathbf{w}_2$ have the same eigenvalue. Then $M_{\mathbf{B}'}(f)$ is one of the following four matrices:

$$\begin{pmatrix} 1 & 0 & 0 \\ 0 & 1 & 0 \\ 0 & 0 & 1 \end{pmatrix}, \qquad \begin{pmatrix} -1 & 0 & 0 \\ 0 & -1 & 0 \\ 0 & 0 & 1 \end{pmatrix},$$

$$\begin{pmatrix} 1 & 0 & 0 \\ 0 & 1 & 0 \\ 0 & 0 & -1 \end{pmatrix}, \quad \text{or} \quad \begin{pmatrix} -1 & 0 & 0 \\ 0 & -1 & 0 \\ 0 & 0 & -1 \end{pmatrix}.$$

In the first two cases, f is a rotation about $\mathbf{w}_3$ (through either the angle 0 or the angle π) and in the last two cases f is a rotation about $\mathbf{w}_3$ (through the angle 0 or π) followed by reflection in the plane $\mathbf{w}_3 \cdot \mathbf{x} = 0$.

This leads us to the following theorem.

Theorem 3   *Let f be a linear isometry of $\mathbb{R}^3$. Then f is either a rotation (if det $f > 0$) or a rotation followed by a reflection (if det $f < 0$).*

Proof   The only thing left to check is that det $f > 0$ if and only if f is a rotation. But

$$\det f = \det M(f) = \det H^{-1}M(f)H$$

for each invertible matrix H, so

$$\det f = \det M_{\mathbf{B}}(f)$$

where $\mathbf{B}$ is any ordered basis for $\mathbb{R}^3$. It is easy to check that the condition det $M_{\mathbf{B}}(f) > 0$ (det $M_{\mathbf{B}'}(f) > 0$ in Case 4) is satisfied precisely in the cases where f is a rotation.   ∎

For any n, we can determine if a linear map from $\mathbb{R}^n$ to $\mathbb{R}^n$ is an isometry by looking at its matrix.

Theorem 4   *Let $f:\mathbb{R}^n \to \mathbb{R}^n$ be a linear map. Then f is a linear isometry if and only if the standard matrix of f is an orthogonal matrix.*

Proof   Suppose $f:\mathbb{R}^n \to \mathbb{R}^n$ is a linear isometry. Since f preserves lengths and angles, the vectors $f(\mathbf{e}_1), \ldots, f(\mathbf{e}_n)$ all have unit length and are perpendicular to one another. Therefore $\{f(\mathbf{e}_1), \ldots, f(\mathbf{e}_n)\}$ is an orthonormal set. The standard matrix of f is precisely the matrix whose column vectors are the vectors $f(\mathbf{e}_1)$, $\ldots, f(\mathbf{e}_n)$. Since these vectors form an orthonormal set, this matrix is an orthogonal matrix.

Conversely, suppose the standard matrix of f is orthogonal. Then its column vectors $f(\mathbf{e}_1), \ldots, f(\mathbf{e}_n)$ form an orthonormal set. If $\mathbf{a} = (a_1, \ldots, a_n)$ is any vector in $\mathbb{R}^n$, then

$$f(\mathbf{a}) = f(a_1\mathbf{e}_1 + \cdots + a_n\mathbf{e}_n)$$
$$= a_1 f(\mathbf{e}_1) + \cdots + a_n f(\mathbf{e}_n)$$

and

$$\|f(\mathbf{a})\| = (a_1^2 + \cdots + a_n^2)^{1/2}$$

$$\text{(since } \{f(\mathbf{e}_1), \ldots, f(\mathbf{e}_n)\} \text{ is orthonormal)}$$

$$= \|\mathbf{a}\|$$

so f preserves lengths of vectors. It follows that

$$\|f(\mathbf{b}) - f(\mathbf{a})\| = \|f(\mathbf{b} - \mathbf{a})\| = \|\mathbf{b} - \mathbf{a}\|$$

for all $\mathbf{a}$ and $\mathbf{b}$ in $\mathbb{R}^n$, so f is an isometry.   ∎

REMARK   Since a linear map is an isometry if and only if its standard matrix is orthogonal, linear isometries are often called *orthogonal linear maps.*

We know from Theorem 3 that every linear isometry of $\mathbb{R}^3$ is either a rotation, or a rotation followed by a reflection. Here is a procedure for describing these maps explicitly. This procedure is based on the facts about linear isometries of $\mathbb{R}^3$ that were derived in the discussion preceding Theorem 3.

To describe a linear isometry f of $\mathbb{R}^3$, first calculate the determinant of f. Then:

(1)   *If det $f > 0$ and f is not the identity map,* find an orthonormal basis $\{\mathbf{v}_1, \mathbf{v}_2\}$ for the row space of $A - I$, where A is the standard matrix of f, and set $\mathbf{v}_3 = \mathbf{v}_1 \times \mathbf{v}_2$. Then f is rotation about $\mathbf{v}_3$ through θ, where θ is the angle such that

$$\cos \theta = f(\mathbf{v}_1) \cdot \mathbf{v}_1$$

$$\sin \theta = f(\mathbf{v}_1) \cdot \mathbf{v}_2.$$

(2) *If f is the identity map*, then *f* can be viewed as a rotation through the angle 0 about **v**, where **v** is any unit vector in $\mathbb{R}^3$.

(3) *If det f < 0*, take $h = g \circ f$ where *g* is reflection in any plane through **0** in $\mathbb{R}^3$. Then det $h > 0$ so *h* is a rotation, as in (1) or (2). The linear isometry *f* is then the composition $f = g \circ h$ of the rotation *h* followed by the reflection *g* (since *g* is its own inverse).

EXAMPLE 9

Let $f:\mathbb{R}^3 \to \mathbb{R}^3$ be the linear map whose standard matrix is

$$A = \begin{pmatrix} \dfrac{3}{4} & \dfrac{1}{4} & -\dfrac{\sqrt{6}}{4} \\[2mm] \dfrac{1}{4} & \dfrac{3}{4} & \dfrac{\sqrt{6}}{4} \\[2mm] \dfrac{\sqrt{6}}{4} & -\dfrac{\sqrt{6}}{4} & \dfrac{1}{2} \end{pmatrix}.$$

This matrix is an orthogonal matrix. Therefore, by Theorem 4, *f* is a linear isometry of $\mathbb{R}^3$. The determinant of *A* is $+1$. Therefore *f* is a rotation of $\mathbb{R}^3$.

The plane of rotation is the row space of the matrix $A - I$. Since

$$A - I = \begin{pmatrix} -\dfrac{1}{4} & \dfrac{1}{4} & -\dfrac{\sqrt{6}}{4} \\[2mm] \dfrac{1}{4} & -\dfrac{1}{4} & \dfrac{\sqrt{6}}{4} \\[2mm] \dfrac{\sqrt{6}}{4} & -\dfrac{\sqrt{6}}{4} & -\dfrac{1}{2} \end{pmatrix} \longrightarrow \begin{pmatrix} 1 & -1 & 0 \\ 0 & 0 & 1 \\ 0 & 0 & 0 \end{pmatrix},$$

we see that the plane of rotation has basis $\{(1, -1, 0), (0, 0, 1)\}$, and orthonormal basis $\{\mathbf{v}_1, \mathbf{v}_2\}$, where $\mathbf{v}_1 = \left(\dfrac{1}{\sqrt{2}}, -\dfrac{1}{\sqrt{2}}, 0\right)$ and $\mathbf{v}_2 = (0, 0, 1)$.

The axis of rotation is the line $\mathbf{x} = t\mathbf{v}_3$, where

$$\mathbf{v}_3 = \mathbf{v}_1 \times \mathbf{v}_2 = \left(-\dfrac{1}{\sqrt{2}}, -\dfrac{1}{\sqrt{2}}, 0\right).$$

Since

$$f(\mathbf{v}_1) = A\mathbf{v}_1 = \begin{pmatrix} \dfrac{3}{4} & \dfrac{1}{4} & -\dfrac{\sqrt{6}}{4} \\[2mm] \dfrac{1}{4} & \dfrac{3}{4} & \dfrac{\sqrt{6}}{4} \\[2mm] \dfrac{\sqrt{6}}{4} & -\dfrac{\sqrt{6}}{4} & \dfrac{1}{2} \end{pmatrix} \begin{pmatrix} \dfrac{1}{\sqrt{2}} \\[2mm] -\dfrac{1}{\sqrt{2}} \\[2mm] 0 \end{pmatrix} = \begin{pmatrix} \dfrac{1}{2\sqrt{2}} \\[2mm] -\dfrac{1}{2\sqrt{2}} \\[2mm] \dfrac{\sqrt{3}}{2} \end{pmatrix},$$

we see that the angle of rotation is the angle θ such that

$$\cos \theta = f(\mathbf{v}_1) \bullet \mathbf{v}_1 = \frac{1}{2}$$

$$\sin \theta = f(\mathbf{v}_1) \bullet \mathbf{v}_2 = \frac{\sqrt{3}}{2};$$

that is, $\theta = \pi/3$.

Thus f is rotation about $\mathbf{v}_3 = \left(-\dfrac{1}{\sqrt{2}}, -\dfrac{1}{\sqrt{2}}, 0 \right)$ through the angle $\pi/3$. ■

The preceding theorems give us a fairly complete understanding of *linear* isometries. The next theorem tells us what nonlinear isometries are like.

Theorem 5   *Let $f:\mathbb{R}^n \to \mathbb{R}^n$ be an isometry.*

(i)   *If $f(\mathbf{0}) = \mathbf{0}$ then f is a linear isometry.*

(ii) *If $f(\mathbf{0}) \neq \mathbf{0}$ then $f = g \circ h$ where g is a translation and h is a linear isometry.*

Proof

(i)   Suppose $f:\mathbb{R}^n \to \mathbb{R}^n$ is an isometry that sends $\mathbf{0}$ to $\mathbf{0}$. Then f *preserves lengths* because

$$\|f(\mathbf{a})\| = \|f(\mathbf{a}) - \mathbf{0}\| = \|f(\mathbf{a}) - f(\mathbf{0})\| = \|\mathbf{a} - \mathbf{0}\| = \|\mathbf{a}\|$$

for all $\mathbf{a}$ in $\mathbb{R}^n$. It follows that f also *preserves dot products:*

$$f(\mathbf{a}) \bullet f(\mathbf{b}) = -\tfrac{1}{2}[\|f(\mathbf{a}) - f(\mathbf{b})\|^2 - \|f(\mathbf{a})\|^2 - \|f(\mathbf{b})\|^2]$$

$$= -\tfrac{1}{2}[\|\mathbf{a} - \mathbf{b}\|^2 - \|\mathbf{a}\|^2 - \|\mathbf{b}\|^2]$$

(since f is an isometry and f preserves lengths)

$$= \mathbf{a} \bullet \mathbf{b}$$

for all $\mathbf{a}$ and $\mathbf{b}$ in $\mathbb{R}^n$. Therefore, the vectors $f(\mathbf{e}_1), \ldots, f(\mathbf{e}_n)$ all have unit length and are perpendicular to one another. Thus $\{f(\mathbf{e}_1), \ldots, f(\mathbf{e}_n)\}$ is an orthonormal set, hence an orthonormal basis for $\mathbb{R}^n$.

To show that f is linear we must verify that

$$f(\mathbf{a} + \mathbf{b}) = f(\mathbf{a}) + f(\mathbf{b})$$

and that

$$f(c\mathbf{a}) = cf(\mathbf{a})$$

for all $\mathbf{a}$ and $\mathbf{b}$ in $\mathbb{R}^n$ and all real numbers c. We will show that $f(\mathbf{a} + \mathbf{b}) = f(\mathbf{a}) + f(\mathbf{b})$ by verifying that the $\mathbf{B}$-coordinate n-tuples of $f(\mathbf{a} + \mathbf{b})$ and of $f(\mathbf{a}) + f(\mathbf{b})$ are equal, where $\mathbf{B} = (f(\mathbf{e}_1), \ldots, f(\mathbf{e}_n))$.

Similarly, we will verify that the **B**-coordinate n-tuples of $f(c\mathbf{a})$ and $cf(\mathbf{a})$ are equal, for all $\mathbf{a} \in \mathbb{R}^n$ and $c \in \mathbb{R}$.

From Section 4.5, Theorem 1, we know that the ith entry in the **B**-coordinate n-tuple of any vector $\mathbf{v}$ in $\mathbb{R}^n$ is $\mathbf{v} \cdot f(\mathbf{e}_i)$. In particular, the ith entry in the **B**-coordinate n-tuple of $f(\mathbf{a} + \mathbf{b})$ is

$$f(\mathbf{a} + \mathbf{b}) \cdot f(\mathbf{e}_i) = (\mathbf{a} + \mathbf{b}) \cdot \mathbf{e}_i = \mathbf{a} \cdot \mathbf{e}_i + \mathbf{b} \cdot \mathbf{e}_i,$$

whereas the ith entry in the **B**-coordinate n-tuple of $f(\mathbf{a}) + f(\mathbf{b})$ is

$$(f(\mathbf{a}) + f(\mathbf{b})) \cdot f(\mathbf{e}_i) = f(\mathbf{a}) \cdot f(\mathbf{e}_i) + f(\mathbf{b}) \cdot f(\mathbf{e}_i) = \mathbf{a} \cdot \mathbf{e}_i + \mathbf{b} \cdot \mathbf{e}_i.$$

Since these are the same for each i ($1 \le i \le n$), we can conclude that $f(\mathbf{a} + \mathbf{b}) = f(\mathbf{a}) + f(\mathbf{b})$, for all $\mathbf{a}$ and $\mathbf{b}$ in $\mathbb{R}^n$.

Similarly, the ith entry in the **B**-coordinate n-tuple of $f(c\mathbf{a})$ is

$$f(c\mathbf{a}) \cdot f(\mathbf{e}_i) = c\mathbf{a} \cdot \mathbf{e}_i$$

whereas the ith entry in the **B**-coordinate n-tuple of $cf(\mathbf{a})$ is

$$(cf(\mathbf{a})) \cdot f(\mathbf{e}_i) = cf(\mathbf{a}) \cdot f(\mathbf{e}_i) = c\mathbf{a} \cdot \mathbf{e}_i.$$

Since these are the same for each i ($1 \le i \le n$) we can conclude that $f(c\mathbf{a}) = cf(\mathbf{a})$ for all $\mathbf{a} \in \mathbb{R}^n$ and each $c \in \mathbb{R}$.

This proves (i).

(ii) Suppose $f : \mathbb{R}^n \to \mathbb{R}^n$ is an isometry such that $f(\mathbf{0}) = \mathbf{v} \ne \mathbf{0}$. Then $t_{-\mathbf{v}} \circ f : \mathbb{R}^n \to \mathbb{R}^n$ is an isometry, where $t_{-\mathbf{v}}$ is translation by $-\mathbf{v}$, and

$$t_{-\mathbf{v}} \circ f(\mathbf{0}) = t_{-\mathbf{v}}(f(\mathbf{0})) = t_{-\mathbf{v}}(\mathbf{v}) = \mathbf{v} - \mathbf{v} = \mathbf{0}.$$

Hence, by (i), the isometry $t_{-\mathbf{v}} \circ f$ is a linear isometry. Set $g = t_{\mathbf{v}}$ and $h = t_{-\mathbf{v}} \circ f$. Then

$$g \circ h = t_{\mathbf{v}} \circ (t_{-\mathbf{v}} \circ f) = (t_{\mathbf{v}} \circ t_{-\mathbf{v}}) \circ f = f.$$

This proves (ii). ∎

EXAMPLE 10

Consider the map $f = r_{\pi/2} \circ t_{\mathbf{e}_1}$ from $\mathbb{R}^2$ to $\mathbb{R}^2$, which first translates by $\mathbf{e}_1$ and then rotates counterclockwise through the angle $\pi/2$. From Theorem 5 we know that f can be described as a linear isometry followed by a translation. We have, in fact, for each $\mathbf{x}$ in $\mathbb{R}^2$,

$$f(\mathbf{x}) = r_{\pi/2}(t_{\mathbf{e}_1}(\mathbf{x})) = r_{\pi/2}(\mathbf{x} + \mathbf{e}_1)$$

$$= r_{\pi/2}(\mathbf{x}) + r_{\pi/2}(\mathbf{e}_1)$$

$$= r_{\pi/2}(\mathbf{x}) + \mathbf{e}_2$$

$$= t_{\mathbf{e}_2}(r_{\pi/2}(\mathbf{x}))$$

so $f = t_{\mathbf{e}_2} \circ r_{\pi/2}$ (see Figure 5.22). ∎

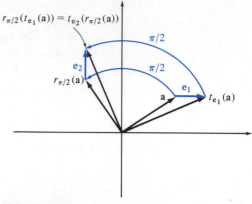

FIGURE 5.22

Translation by e_1 followed by rotation through the angle $\pi/2$ is the same as rotation through $\pi/2$ followed by translation by e_2.

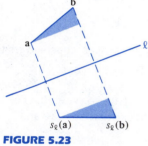

FIGURE 5.23

Reflection in the line ℓ is an isometry. To see that s_ℓ preserves distances, simply note that the shaded triangles are congruent.

EXAMPLE 11

Let ℓ be any line in $\mathbb{R}^2$ (not necessarily passing through **0**) and let $s_\ell : \mathbb{R}^2 \to \mathbb{R}^2$ be reflection in ℓ. Then s_ℓ is an isometry (see Figure 5.23). From Theorem 5 we know that $s_\ell = g \circ h$ where g is a translation and h is a linear isometry. In fact, from the proof of Theorem 5 (ii) we see that $g = t_v$, where $\mathbf{v} = s_\ell(\mathbf{0})$, and $h = t_{-v} \circ s_\ell$. From Figure 5.24a you can see that $\mathbf{v} = 2\mathbf{a}$ where $\mathbf{a}$ is the point of intersection of ℓ and the line through **0** perpendicular to ℓ. From Figure 5.24b you can see that $t_{-2a} \circ s_\ell$ is reflection in the line ℓ_0 that passes through **0** and is parallel to ℓ. Hence

$$s_\ell = t_{2a} \circ s_{\ell_0}$$

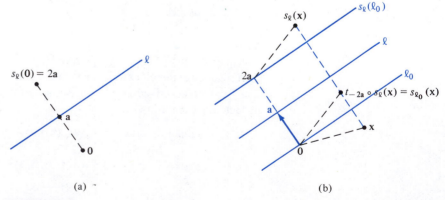

(a)   (b)

FIGURE 5.24

Reflection in an arbitrary line ℓ. (a) $s_\ell(\mathbf{0}) = 2\mathbf{a}$, where $\mathbf{a}$ is the point of intersection of ℓ and the line through **0** perpendicular to ℓ. (b) $(t_{-2a} \circ s_\ell)(\mathbf{x}) = s_\ell(x) - 2\mathbf{a} = s_{\ell_0}(\mathbf{x})$, and hence $s_\ell = t_{2a} \circ s_{\ell_0}$. (What does the figure look like if $\mathbf{x}$ lies between ℓ and ℓ_0?)

where $\mathbf{a}$ is the point of intersection of ℓ and the line perpendicular to ℓ that passes through $\mathbf{0}$, and ℓ_0 is the line parallel to ℓ that passes through $\mathbf{0}$.

In particular, if ℓ is the line $2x_1 + 3x_2 = 13$ then the line perpendicular to ℓ that passes through $\mathbf{0}$ is the line $-3x_1 + 2x_2 = 0$. These two lines intersect at the point $\mathbf{a} = (2, 3)$. Hence

$$s_\ell = t_{(4,6)} \circ s_{\ell_0}$$

where ℓ_0 is the line $2x_1 + 3x_2 = 0$. ∎

We have seen that each isometry of $\mathbb{R}^n$ can be described as a linear isometry followed by a translation. It is also possible to describe isometries as compositions of reflections. We shall now show how to do this for isometries of the plane.

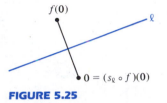

FIGURE 5.25

$s_\ell \circ f(\mathbf{0}) = \mathbf{0}$, where ℓ is the perpendicular bisector of the line segment from $\mathbf{0}$ to $f(\mathbf{0})$.

Theorem 6 *Each isometry of $\mathbb{R}^2$ is a composition of reflections.*

Proof Let $f : \mathbb{R}^2 \to \mathbb{R}^2$ be an isometry.

First we shall find a line ℓ_1 such that $s_{\ell_1} \circ f(\mathbf{0}) = \mathbf{0}$. If $f(\mathbf{0}) \neq \mathbf{0}$, take ℓ_1 to be the perpendicular bisector of the line segment from $\mathbf{0}$ to $f(\mathbf{0})$ (see Figure 5.25). If $f(\mathbf{0}) = \mathbf{0}$, take ℓ_1 to be any line that passes through $\mathbf{0}$. Then, in both cases, $s_{\ell_1} \circ f(\mathbf{0}) = \mathbf{0}$.

Since $s_{\ell_1} \circ f(\mathbf{0}) = \mathbf{0}$, $s_{\ell_1} \circ f$ is a linear isometry, by Theorem 5. If this linear isometry is a reflection, then $s_{\ell_1} \circ f = s_{\ell_2}$ for some line ℓ_2 and hence

$$f = s_{\ell_1}^{-1} \circ s_{\ell_2} = s_{\ell_1} \circ s_{\ell_2}.$$

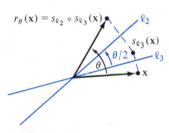

FIGURE 5.26

Rotation through the angle θ is the same as reflection in ℓ_3 followed by reflection in ℓ_2.

If the linear isometry $s_{\ell_1} \circ f$ is a rotation, then Theorem 3 of Section 2.5 tells us that $s_{\ell_1} \circ f = s_{\ell_2} \circ s_{\ell_3}$ where ℓ_2 and ℓ_3 are any two lines through $\mathbf{0}$ such that the angle from ℓ_3 to ℓ_2 is one half the rotation angle (see Figure 5.26). Hence

$$f = s_{\ell_1}^{-1} \circ s_{\ell_2} \circ s_{\ell_2} = s_{\ell_1} \circ s_{\ell_2} \circ s_{\ell_3}.$$

Therefore we can conclude that each isometry f of $\mathbb{R}^2$ can be described as a composition of at most three reflections. ∎

REMARK It can also be shown (see Exercise 26) that each isometry of $\mathbb{R}^3$ can be described as a composition of at most four reflections and, more generally, that each isometry of $\mathbb{R}^n$ can be described as a composition of at most $n + 1$ reflections.

EXERCISES

1. Express each of the following linear isometries of $\mathbb{R}^2$ as either a rotation r_θ or a reflection s_θ.

 (a) $s_{\pi/2} \circ r_{\pi/4}$ (b) $r_{\pi/4} \circ s_{\pi/2}$

 (c) $s_{\pi/2} \circ s_{\pi/4}$ (d) $s_0 \circ s_{3\pi/4}$

 (e) $s_0 \circ r_\pi$ (f) $s_{\pi/4} \circ s_0$

 (g) $s_{\pi/4} \circ s_{\pi/4}$ (h) $s_0 \circ r_{3\pi/2}$

2. Express each of the following linear isometries of $\mathbb{R}^2$ as either a rotation r_θ or a reflection s_θ.

    (a) $s_0 \circ s_{\pi/2}$ (b) $s_{\pi/2} \circ s_0$

    (c) $r_{3\pi/2} \circ s_{3\pi/4}$ (d) $s_{3\pi/4} \circ r_{3\pi/4}$

    (e) $s_{3\pi/2} \circ s_{\pi/4}$ (f) $r_{\pi/4} \circ s_{\pi/4}$

3. Express each of the following linear isometries as either a rotation r_θ or a reflection s_θ.

    (a) $s_{\pi/2} \circ r_{\pi/2} \circ s_{\pi/4}$ (b) $s_{\pi/4} \circ r_\pi \circ s_{\pi/4}$

    (c) $r_{3\pi/4} \circ s_{\pi/2} \circ r_\pi$ (d) $s_{\pi/4} \circ r_{\pi/4} \circ s_{\pi/2} \circ r_{\pi/2}$

    (e) $s_0 \circ r_{\pi/4} \circ s_{\pi/4} \circ r_\pi$ (f) $r_\pi \circ s_{\pi/2} \circ s_{\pi/4} \circ r_{\pi/2}$

4. Express each of the following linear isometries as either a rotation r_θ or a reflection s_θ.

    (a) $r_\alpha \circ r_\beta$ (b) $s_\alpha \circ s_\beta$

    (c) $r_\alpha \circ s_\beta$ (d) $s_\alpha \circ r_\beta$

    (e) $s_\alpha^{-1} \circ r_\beta \circ s_\alpha$ (f) $r_\alpha^{-1} \circ s_\beta \circ r_\alpha$

    (g) $r_\alpha^{-1} \circ s_\beta^{-1} \circ r_\alpha \circ s_\beta$ (h) $s_\alpha^{-1} \circ r_\beta^{-1} \circ s_\alpha \circ r_\beta$

5. Let $f:\mathbb{R}^3 \to \mathbb{R}^3$ be the linear map associated with the matrix

    $$A = \begin{pmatrix} \dfrac{3}{4} & \dfrac{1}{4} & \dfrac{\sqrt{6}}{4} \\[2mm] \dfrac{1}{4} & \dfrac{3}{4} & -\dfrac{\sqrt{6}}{4} \\[2mm] -\dfrac{\sqrt{6}}{4} & \dfrac{\sqrt{6}}{4} & \dfrac{1}{2} \end{pmatrix}.$$

    (a) Verify that f is an isometry.

    (b) Verify that f is a rotation.

    (c) Find the axis of rotation.

    (d) Find a unit vector $\mathbf{u}$ and an angle θ such that f is rotation about $\mathbf{u}$ through the angle θ.

6. Repeat Exercise 5 for the linear map f associated with the matrix

    $$A = \begin{pmatrix} \dfrac{1}{2} & \dfrac{1}{2} & -\dfrac{\sqrt{2}}{2} \\[2mm] \dfrac{1}{2} & \dfrac{1}{2} & \dfrac{\sqrt{2}}{2} \\[2mm] \dfrac{\sqrt{2}}{2} & -\dfrac{\sqrt{2}}{2} & 0 \end{pmatrix}.$$

7. Find a unit vector $\mathbf{u}$ and an angle θ such that f is rotation about $\mathbf{u}$ through the angle θ, where $f:\mathbb{R}^3 \to \mathbb{R}^3$ is the linear isometry whose standard matrix is

    (a) $\begin{pmatrix} \dfrac{1}{\sqrt{2}} & -\dfrac{1}{\sqrt{2}} & 0 \\[2mm] \dfrac{1}{\sqrt{2}} & \dfrac{1}{\sqrt{2}} & 0 \\[2mm] 0 & 0 & 1 \end{pmatrix}$ (b) $\begin{pmatrix} \dfrac{1}{2} & 0 & -\dfrac{\sqrt{3}}{2} \\[2mm] 0 & 1 & 0 \\[2mm] \dfrac{\sqrt{3}}{2} & 0 & \dfrac{1}{2} \end{pmatrix}$

(c) $\begin{pmatrix} 1 & 0 & 0 \\ 0 & -1 & 0 \\ 0 & 0 & -1 \end{pmatrix}$ (d) $\begin{pmatrix} 0 & 0 & 1 \\ 0 & 1 & 0 \\ -1 & 0 & 0 \end{pmatrix}$

8. (a) Show that the standard matrix for rotation about e_1 through the angle θ is

$$\begin{pmatrix} 1 & 0 & 0 \\ 0 & \cos\theta & -\sin\theta \\ 0 & \sin\theta & \cos\theta \end{pmatrix}.$$

[*Hint:* Write down the matrix $M_B(f)$, where $\mathbf{B} = (e_2, e_3, e_1)$, and use the fact that $M_B(f) = H^{-1}M(f)H$ where H is the matrix whose column vectors are e_2, e_3, and e_1.]

(b) Show that the standard matrix for rotation about e_2 through the angle θ is

$$\begin{pmatrix} \cos\theta & 0 & \sin\theta \\ 0 & 1 & 0 \\ -\sin\theta & 0 & \cos\theta \end{pmatrix}.$$

(c) What is the standard matrix for rotation about e_3 through the angle θ?

9. Let $f = g \circ h$ where $h:\mathbb{R}^3 \to \mathbb{R}^3$ is rotation about e_1 through the angle $\pi/2$ and $g:\mathbb{R}^3 \to \mathbb{R}^3$ is rotation about e_2 through the angle $\pi/2$.
   (a) Verify that f is a rotation of $\mathbb{R}^3$.
   (b) Find the axis of rotation.
   (c) Find a unit vector $\mathbf{u}$ and an angle θ such that f is rotation about $\mathbf{u}$ through the angle θ.

10. Repeat Exercise 9 where h is rotation about e_2 through the angle $\pi/2$ and g is rotation about e_1 through the angle $\pi/2$.

11. Find a translation g and a linear isometry h such that $f = g \circ h$, where $f:\mathbb{R}^2 \to \mathbb{R}^2$ is
    (a) $r_\pi \circ t_{(1,2)}$ (b) $r_{\pi/2} \circ t_{(-3,4)}$
    (c) $r_{3\pi/2} \circ t_{(-1,-5)}$ (d) $s_{\pi/4} \circ t_{(1,2)}$
    (e) $s_{\pi/2} \circ t_{(-2,-5)}$ (f) $s_0 \circ t_{(1,-4)}$

12. Let f be any linear isometry of $\mathbb{R}^n$ and let $\mathbf{v}$ be any vector in $\mathbb{R}^n$. Show that

$$f \circ t_\mathbf{v} = t_{f(\mathbf{v})} \circ f.$$

13. Find a translation g and a linear isometry h such that $f = g \circ h$, where $f:\mathbb{R}^2 \to \mathbb{R}^2$ is reflection in the line
    (a) $x_1 = 4$ (b) $x_2 = 7$
    (c) $x_1 + x_2 = 1$ (d) $x_1 - x_2 = 4$
    (e) $x_1 + 2x_2 = 5$ (f) $3x_1 + x_2 = 10$

14. Let g_1 and g_2 be translations of $\mathbb{R}^n$ and let h_1 and h_2 be linear isometries of $\mathbb{R}^n$. Show that $g_1 \circ h_1 = g_2 \circ h_2$ if and only if $g_1 = g_2$ and $h_1 = h_2$. (This exercise shows that, given any isometry of $\mathbb{R}^n$, there is only one translation g and one linear isometry h such that $f = g \circ h$. The translation g is called the ***translational part*** of f and the linear isometry h is the ***linear part*** of f.)

15. Show that $r_\theta \circ t_\mathbf{v} = t_\mathbf{v} \circ r_\theta$ if and only if either $\mathbf{v} = \mathbf{0}$ or θ is an integer multiple of 2π.

16. Show that $s_\theta \circ t_\mathbf{v} = t_\mathbf{v} \circ s_\theta$ if and only if $\mathbf{v} \in \ell_\theta$.

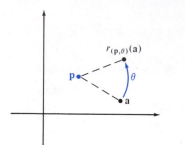

FIGURE 5.27
Rotation about **p** through the angle θ.

17. (a) Show that a linear isometry of $\mathbb{R}^2$ is a rotation if and only if it is a composition of an even number of reflections s_θ.

    (b) Show that a linear isometry of $\mathbb{R}^2$ is a reflection if and only if it is a composition of an odd number of reflections s_θ.

18. Show that if $f:\mathbb{R}^n \to \mathbb{R}^n$ is an isometry, then f is invertible. [*Hint:* Express f as $g \circ h$ where g is a translation and h is a linear isometry.]

19. Let **p** be any point in $\mathbb{R}^2$, let θ be any real number, and let $r_{(\mathbf{p},\theta)}:\mathbb{R}^2 \to \mathbb{R}^2$ be the map that rotates points counterclockwise about **p** through the angle θ (see Figure 5.27).

    (a) Show, geometrically, that $r_{(\mathbf{p},\theta)}$ is an isometry.

    (b) Show that $t_{-\mathbf{p}} \circ r_{(\mathbf{p},\theta)} \circ t_{\mathbf{p}}$ is a linear isometry.

    (c) Show that $t_{-\mathbf{p}} \circ r_{(\mathbf{p},\theta)} \circ t_{\mathbf{p}} = r_\theta$. [*Hint:* Draw a picture.]

    (d) Conclude that $r_{(\mathbf{p},\theta)} = t_{\mathbf{p}} \circ r_\theta \circ t_{-\mathbf{p}} = t_{\mathbf{v}} \circ r_\theta$ where $\mathbf{v} = \mathbf{p} - r_\theta(\mathbf{p})$.

20. Let **v** be a nonzero vector in $\mathbb{R}^2$.

    (a) Show that the perpendicular bisector of the line segment from **0** to $t_{\mathbf{v}}(\mathbf{0})$ is the line ℓ_1 that passes through $\frac{1}{2}\mathbf{v}$ and is perpendicular to **v**.

    (b) Show that if $f = s_{\ell_1} \circ t_{\mathbf{v}}$, then $f(\mathbf{0}) = \mathbf{0}$, $f(-\mathbf{v}) = \mathbf{v}$, and $f(\mathbf{w}) = \mathbf{w}$, whenever **w** is perpendicular to **v**.

    (c) Conclude that $s_{\ell_1} \circ t_{\mathbf{v}} = s_{\ell_2}$ and hence that $t_{\mathbf{v}} = s_{\ell_1} \circ s_{\ell_2}$, where ℓ_2 is the line through **0** perpendicular to **v** and ℓ_1 is the line through $\frac{1}{2}\mathbf{v}$ perpendicular to **v**. Thus, *each translation of $\mathbb{R}^2$ is a composition of two reflections.*

21. Use the formula

$$r_\theta(x_1, x_2) = (x_1 \cos\theta - x_2 \sin\theta, \; x_1 \sin\theta + x_2 \cos\theta)$$

    to give an algebraic proof that r_θ is an isometry.

22. Use the formula

$$s_\theta(x_1, x_2) = (x_1 \cos 2\theta + x_2 \sin 2\theta, \; x_1 \sin 2\theta - x_2 \cos 2\theta)$$

    to give an algebraic proof that s_θ is an isometry.

23. Let $f:\mathbb{R}^n \to \mathbb{R}^n$ be a linear map and let **B** be any ordered orthonormal basis for $\mathbb{R}^n$. Show that the matrix $M_{\mathbf{B}}(f)$ of f relative to **B** is symmetric if and only if the standard matrix $M(f)$ of f is symmetric. [*Hint:* $M_{\mathbf{B}}(f) = H^{-1}M(f)H$, where H is an orthogonal matrix.]

24. Let $f:\mathbb{R}^n \to \mathbb{R}^n$ be a linear map. Show that if f is an isometry then the given statement is true and, conversely, if the given statement is true then f is an isometry.

    (a) $\|f(\mathbf{a})\| = \|\mathbf{a}\|$ for all $\mathbf{a} \in \mathbb{R}^n$

    (b) $f(\mathbf{a}) \cdot f(\mathbf{b}) = \mathbf{a} \cdot \mathbf{b}$ for all **a** and **b** in $\mathbb{R}^n$

    (c) $\{f(\mathbf{v}_1), \ldots, f(\mathbf{v}_n)\}$ is an orthonormal basis for $\mathbb{R}^n$, where $\{\mathbf{v}_1, \ldots, \mathbf{v}_n\}$ is any orthonormal basis for $\mathbb{R}^n$.

    (d) $M_{\mathbf{B}}(f)$ is an orthogonal matrix, where **B** is any ordered orthonormal basis for $\mathbb{R}^n$.

25. Give an alternate proof that each isometry of $\mathbb{R}^2$ is a composition of at most three reflections, as follows.

    (a) If $f(\mathbf{0}) \neq \mathbf{0}$, take ℓ_1 to be the line that passes through $\frac{1}{2}f(\mathbf{0})$ and is perpendicular to $f(\mathbf{0})$. If $f(\mathbf{0}) = \mathbf{0}$, take ℓ_1 to be any line that passes through **0**. Show that $s_1 \circ f(\mathbf{0}) = \mathbf{0}$, where s_1 is reflection in ℓ_1, and conclude that $s_1 \circ f$ is a linear isometry.

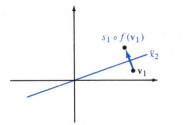

FIGURE 5.28
ℓ_2 is the line that passes through **0** and is perpendicular to
$$s_1 \circ f(\mathbf{v}_1) - \mathbf{v}_1.$$

(b) If $s_1 \circ f$ is the identity, then $f = s_1$. Otherwise choose $\mathbf{v}_1$ so that $s_1 \circ f(\mathbf{v}_1) \neq \mathbf{v}_1$ and take ℓ_2 to be the line that passes through **0** and is perpendicular to $s_1 \circ f(\mathbf{v}_1) - \mathbf{v}_1$ (see Figure 5.28). Show that $s_2 \circ s_1 \circ f$ is a linear isometry and that $s_2 \circ s_1 \circ f(\mathbf{v}_1) = \mathbf{v}_1$, where s_2 is reflection in the line ℓ_2.

(c) If $s_2 \circ s_1 \circ f$ is the identity, then $f = s_1 \circ s_2$. Otherwise choose $\mathbf{v}_2$ perpendicular to $\mathbf{v}_1$ such that $f(\mathbf{v}_2) \neq \mathbf{v}_2$, and take ℓ_3 to be the line that passes through **0** and is perpendicular to $f(\mathbf{v}_2) - \mathbf{v}_2$. Show that $s_3 \circ s_2 \circ s_1 \circ f$ is a linear isometry that leaves both $\mathbf{v}_1$ and $\mathbf{v}_2$ fixed.

(d) Conclude that $s_3 \circ s_2 \circ s_1 \circ f$ is the identity map and hence that $f = s_1 \circ s_2 \circ s_3$.

26. Prove that each isometry of $\mathbb{R}^3$ is a composition of at most four reflections s_i, where each s_i is reflection in a plane. [*Hint:* Imitate the solution of Exercise 25, replacing "line" everywhere by "plane."]

5.5
Application: Conic Sections

In this section we shall apply linear algebra to analytic geometry. We shall see how the technique for diagonalizing symmetric matrices can be used to identify solution sets of quadratic equations in $\mathbb{R}^2$ as circles, ellipses, hyperbolas, and parabolas.

You are probably already familiar with the solution sets of certain quadratic equations in $\mathbb{R}^2$. Recall:

(1) The solution set of the equation $x^2 + y^2 = a^2$ ($a > 0$) is a *circle* (see Figure 5.29a). Notice that this equation can be rewritten as $x^2/a^2 + y^2/a^2 = 1$.

(2) The solution set of the equation $x^2/a^2 + y^2/b^2 = 1$ ($a > 0, b > 0$) is an *ellipse* (see Figure 5.29b). Notice that, if $a = b$, the ellipse is a circle of radius a.

(3) The solution sets of the equations $x^2/a^2 - y^2/b^2 = 1$ and $-x^2/a^2 + y^2/b^2 = 1$ ($a > 0, b > 0$) are *hyperbolas* (see Figures 5.29c and d).

(4) The solution sets of the equations $y = ax^2$ and $x = ay^2$ ($a \neq 0$) are *parabolas* (see Figure 5.29e and f).

These curves are called **conic sections** because each can be obtained as an intersection of a plane with a right circular cone (see Figure 5.30).

The *general quadratic equation* in $\mathbb{R}^2$ is the equation

$$ax^2 + 2bxy + cy^2 + dx + ey + f = 0$$

where a, b, c, d, e, and f are real numbers with at least one of a, b, c nonzero. Notice that the coefficient of the xy term is called $2b$ rather than b. The reason for this is that we shall study this equation in its *matrix form*

$$(x \; y)\begin{pmatrix} a & b \\ b & c \end{pmatrix}\begin{pmatrix} x \\ y \end{pmatrix} + (d \; e)\begin{pmatrix} x \\ y \end{pmatrix} + (f) = (0).$$

You can readily verify that this matrix equation is equivalent to the scalar equation above. By introducing the 2 in the scalar equation we avoid a $\frac{1}{2}$ in the matrix equation.

The left hand side of the equation

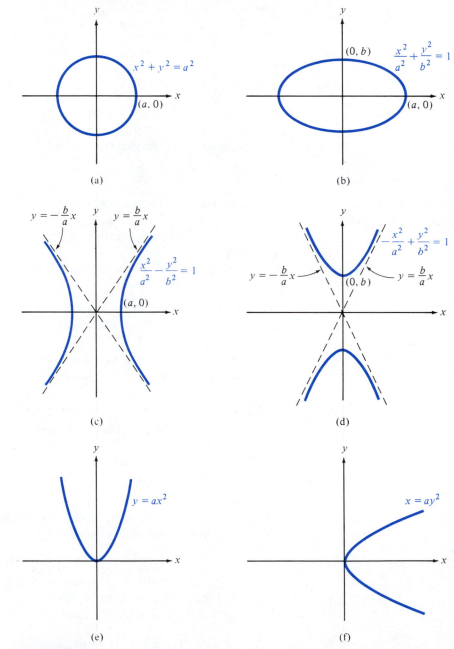

(a)

(b)

(c)

(d)

(e)

(f)

FIGURE 5.29

Quadratic equations in x and y represent conic sections: (a) a circle, (b) an ellipse, (c) and (d) hyperbolas, (e) and (f) parabolas.

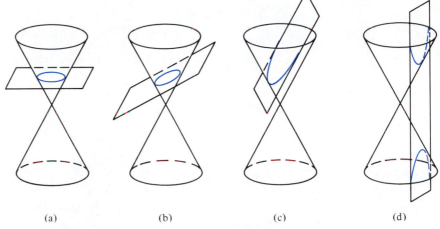

(a)   (b)   (c)   (d)

FIGURE 5.30

Conic sections are curves that can be obtained by cutting an infinite right circular cone with a plane: (a) a circle, (b) an ellipse, (c) a parabola, and (d) a hyperbola.

$$ax^2 + 2bxy + cy^2 + dx + ey + f = 0$$

consists of three parts, a *quadratic* part, $ax^2 + 2bxy + cy^2$, a *linear* part, $dx + ey$, and a *constant* part, f. We shall study first the situation in which the linear part is equal to zero; that is, we shall first study equations of the form

$$ax^2 + 2bxy + cy^2 + f = 0.$$

Notice that if $b = 0$ but a, c, and f are all nonzero then we can readily identify the solution set of this equation as a circle, an ellipse, or a hyperbola. Indeed, if we put the constant term in the equation

$$ax^2 + cy^2 + f = 0$$

on the right-hand side, and then divide through by $-f$, we can convert this equation into one of the standard types (1), (2), or (3) described above.

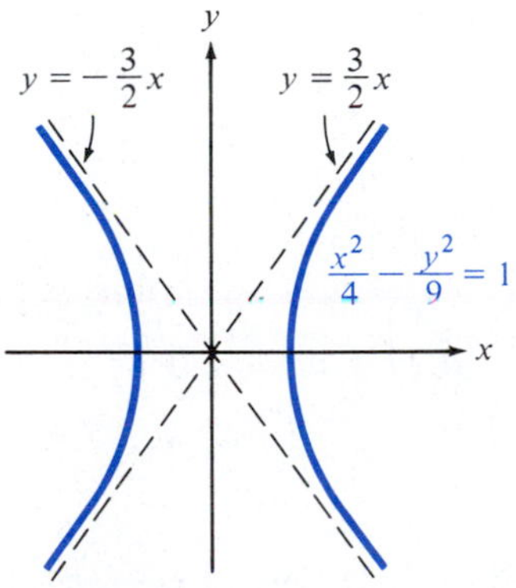

EXAMPLE 1

The equation

$$9x^2 - 4y^2 - 36 = 0$$

describes a hyperbola because it can be rewritten as

$$9x^2 - 4y^2 = 36$$

or as

$$\frac{x^2}{4} - \frac{y^2}{9} = 1$$

(see Figure 5.31). ∎

$y = -\frac{3}{2}x$ $y = \frac{3}{2}x$

$\frac{x^2}{4} - \frac{y^2}{9} = 1$

FIGURE 5.31

The hyperbola $\dfrac{x^2}{4} - \dfrac{y^2}{9} = 1$.

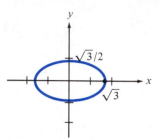

FIGURE 5.32
The ellipse $\dfrac{x^2}{3} + \dfrac{y^2}{\frac{3}{4}} = 1$.

EXAMPLE 2

The equation

$$x^2 + 4y^2 - 3 = 0$$

describes an ellipse because it can be rewritten as

$$x^2 + 4y^2 = 3$$

or as

$$\frac{x^2}{3} + \frac{y^2}{\frac{3}{4}} = 1$$

(see Figure 5.32). ■

When $b \neq 0$, it is not quite so easy to describe the solution set of the equation

$$ax^2 + 2bxy + cy^2 + f = 0.$$

We must, therefore, look at this equation more closely.

The expression

$$Q = ax^2 + 2bxy + cy^2$$

is called a **quadratic form** in $\mathbb{R}^2$. The matrix

$$A = \begin{pmatrix} a & b \\ b & c \end{pmatrix}$$

is called the **matrix of the quadratic form** Q. If the xy-coefficient, b, of Q is equal to zero, then A is a diagonal matrix. If $b \neq 0$ then A is not a diagonal matrix, but A is diagonable. We can use this fact to "diagonalize" the quadratic form Q; that is, we can choose new coordinates on $\mathbb{R}^2$ in such a way that, in this new coordinate system, Q has no xy term. We proceed as follows.

Since A is a symmetric matrix there is an orthogonal matrix H such that

$$H^{-1}AH = D$$

where D is a diagonal matrix. Hence

$$A = HDH^{-1}$$

and

$$Q = (x \; y)A\begin{pmatrix} x \\ y \end{pmatrix} = (x \; y)HDH^{-1}\begin{pmatrix} x \\ y \end{pmatrix}.$$

If we define x' and y' by

$$\begin{pmatrix} x' \\ y' \end{pmatrix} = H^{-1}\begin{pmatrix} x \\ y \end{pmatrix},$$

then, since $H^{-1} = H^t$ (H is orthogonal!), we have

$$(x' \; y') = \begin{pmatrix} x' \\ y' \end{pmatrix}^t = \left[H^{-1}\begin{pmatrix} x \\ y \end{pmatrix}\right]^t = \left[H^t\begin{pmatrix} x \\ y \end{pmatrix}\right]^t = (x \; y)H$$

and so

$$Q = [(x \ y)H]D[H^{-1}\begin{pmatrix} x \\ y \end{pmatrix}] = (x' \ y')D\begin{pmatrix} x' \\ y' \end{pmatrix} = \lambda_1 x'^2 + \lambda_2 y'^2$$

where λ_1 and λ_2 are the diagonal entries in D; that is, λ_1 and λ_2 are eigenvalues of A.

Now, Theorem 1 of Section 5.2 tells us that the matrix

$$\begin{pmatrix} x' \\ y' \end{pmatrix} = H^{-1}\begin{pmatrix} x \\ y \end{pmatrix}$$

is just the **B**-coordinate matrix of the vector (x, y), where **B** is the ordered orthonormal basis for $\mathbb{R}^2$ consisting of the column vectors of H. These column vectors are eigenvectors of the matrix A. Hence we have proved the following theorem.

Theorem 1 *Let*

$$Q = ax^2 + 2bxy + cy^2$$

be the quadratic form in $\mathbb{R}^2$ with matrix

$$A = \begin{pmatrix} a & b \\ b & c \end{pmatrix}$$

and let **B** $= (\mathbf{v}_1, \mathbf{v}_2)$ *be an ordered orthonormal basis for $\mathbb{R}^2$ consisting of eigenvectors of A. Then*

$$Q = \lambda_1 x'^2 + \lambda_2 y'^2$$

where λ_1 and λ_2 are the eigenvalues of A that correspond to the eigenvectors $\mathbf{v}_1$ *and* $\mathbf{v}_2$ *and (x', y') is the* **B***-coordinate pair of the vector (x, y).*

The orthonormal basis **B** in Theorem 1 is said to ***diagonalize*** the quadratic form Q, and the expression $\lambda_1 x'^2 + \lambda_2 y'^2$ is a ***diagonal form*** of Q.

EXAMPLE 3

Let us diagonalize the quadratic form

$$Q = 5x^2 + 6xy + 5y^2.$$

The matrix of Q is

$$A = \begin{pmatrix} 5 & 3 \\ 3 & 5 \end{pmatrix}.$$

(Remember to divide the xy-coefficient by 2.) Since

$$\det (A - \lambda I) = \det\begin{pmatrix} 5 - \lambda & 3 \\ 3 & 5 - \lambda \end{pmatrix}$$
$$= \lambda^2 - 10\lambda + 16 = (\lambda - 8)(\lambda - 2),$$

we see that the eigenvalues of A are 8 and 2. The 8-eigenspace is the line $\mathbf{x} = t(1, 1)$. The 2-eigenspace is the line $\mathbf{x} = t(-1, 1)$. Therefore **B** $=$

$\left(\left(\dfrac{1}{\sqrt{2}}, \dfrac{1}{\sqrt{2}}\right), \left(-\dfrac{1}{\sqrt{2}}, \dfrac{1}{\sqrt{2}}\right)\right)$ is an ordered orthonormal basis for $\mathbb{R}^2$ consisting of eigenvectors of A, and

$$Q = 8x'^2 + 2y'^2$$

where (x', y') is the **B**-coordinate pair of (x, y). ∎

We can use Theorem 1 to identify the solution set of any quadratic equation of the form

$$ax^2 + 2bxy + cy^2 + f = 0.$$

Simply diagonalize the quadratic form

$$Q = ax^2 + 2bxy + cy^2$$

and then rewrite the equation in **B**-coordinates where **B** is the diagonalizing orthonormal basis.

EXAMPLE 4

From Example 3 we see that the quadratic equation

$$5x^2 + 6xy + 5y^2 - 8 = 0$$

can be rewritten as

$$8x'^2 + 2y'^2 - 8 = 0$$

or as

$$x'^2 + \frac{y'^2}{4} = 1$$

where (x', y') is the **B**-coordinate pair of (x, y) and **B** is the ordered orthonormal basis $\left(\left(\dfrac{1}{\sqrt{2}}, \dfrac{1}{\sqrt{2}}\right), \left(-\dfrac{1}{\sqrt{2}}, \dfrac{1}{\sqrt{2}}\right)\right)$. The solution set is therefore an ellipse (see Figure 5.33). ∎

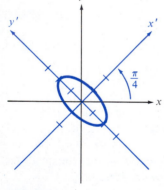

FIGURE 5.33
The ellipse
$5x^2 + 6xy + 5y^2 - 8 = 0$
is described in the rotated (x', y')-coordinate system by the equation $x'^2 + \dfrac{y'^2}{4} = 1$.

REMARK   The ordered orthonormal basis **B** obtained in Example 3 consists of the column vectors of the orthogonal matrix

$$H = \begin{pmatrix} \dfrac{1}{\sqrt{2}} & -\dfrac{1}{\sqrt{2}} \\[2mm] \dfrac{1}{\sqrt{2}} & \dfrac{1}{\sqrt{2}} \end{pmatrix}.$$

Thus **B** $= (\mathbf{v}_1, \mathbf{v}_2)$ where

$$\mathbf{v}_1 = H\mathbf{e}_1 \qquad \text{and} \qquad \mathbf{v}_2 = H\mathbf{e}_2.$$

In other words, $\mathbf{v}_1$ and $\mathbf{v}_2$ are the images of the standard basis vectors $\mathbf{e}_1$ and $\mathbf{e}_2$ under the linear map $f:\mathbb{R}^2 \to \mathbb{R}^2$ defined by $f(\mathbf{x}) = H\mathbf{x}$. Since H is an orthogonal matrix and det $H > 0$, we know from Theorems 2 and 4 of Section 5.4 that f is a rotation. In fact f is counterclockwise rotation through the angle θ, where

$$(\cos\theta, \sin\theta) = f(\mathbf{e}_1) = \mathbf{v}_1 = \left(\frac{1}{\sqrt{2}}, \frac{1}{\sqrt{2}}\right);$$

that is, $\theta = \pi/4$ (see Figure 5.33).

In diagonalizing any quadratic form Q in $\mathbb{R}^2$ we can always arrange that the matrix H, whose columns are the diagonalizing basis vectors $\mathbf{v}_1$ and $\mathbf{v}_2$, has det $H > 0$. To accomplish this, we can interchange the eigenvectors $\mathbf{v}_1$ and $\mathbf{v}_2$ if necessary. Then the basis $\mathbf{B}$ is obtained from the standard basis $(\mathbf{e}_1, \mathbf{e}_2)$ by rotation through the angle θ where θ is the angle such that

$$\mathbf{v}_1 = (\cos\theta, \sin\theta).$$

EXAMPLE 5

Let us analyze the solution set of the equation

$$x^2 + 2\sqrt{3}xy - y^2 + 2 = 0.$$

The quadratic form

$$Q = x^2 + 2\sqrt{3}xy - y^2$$

has matrix

$$A = \begin{pmatrix} 1 & \sqrt{3} \\ \sqrt{3} & -1 \end{pmatrix}.$$

Since

$$\det(A - \lambda I) = \det\begin{pmatrix} 1 - \lambda & \sqrt{3} \\ \sqrt{3} & -1 - \lambda \end{pmatrix} = \lambda^2 - 4,$$

we see that the eigenvalues of A are -2 and 2. The vectors $\left(-\frac{1}{2}, \frac{\sqrt{3}}{2}\right)$ and $\left(\frac{\sqrt{3}}{2}, \frac{1}{2}\right)$ are corresponding unit eigenvectors. Since

$$\det\begin{pmatrix} -\dfrac{1}{2} & \dfrac{\sqrt{3}}{2} \\ \dfrac{\sqrt{3}}{2} & \dfrac{1}{2} \end{pmatrix} < 0$$

we set $\mathbf{B} = (\mathbf{v}_1, \mathbf{v}_2)$ where $\mathbf{v}_1 = \left(\frac{\sqrt{3}}{2}, \frac{1}{2}\right)$ and $\mathbf{v}_2 = \left(-\frac{1}{2}, \frac{\sqrt{3}}{2}\right)$. Then $\lambda_1 = 2$, $\lambda_2 = -2$, and

$$Q = 2x'^2 - 2y'^2$$

where (x', y') is the $\mathbf{B}$-coordinate pair of (x, y). Therefore, the equation

$$x^2 + 2\sqrt{3}xy - y^2 + 2 = 0$$

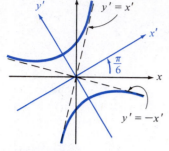

FIGURE 5.34
The hyperbola
$x^2 + 2\sqrt{3}xy - y^2 + 2 = 0$
is described in the rotated (x', y')-
coordinate system by the equation
$-x'^2 + y'^2 = 1$.

can be rewritten as

$$2x'^2 - 2y'^2 + 2 = 0$$

or as

$$-x'^2 + y'^2 = 1.$$

The solution set is a hyperbola. Notice that, since

$$\mathbf{v}_1 = \left(\frac{\sqrt{3}}{2}, \frac{1}{2}\right) = \left(\cos\frac{\pi}{6}, \sin\frac{\pi}{6}\right),$$

the orthonormal basis **B** that determines the (x', y')-coordinate system is obtained by rotating $(\mathbf{e}_1, \mathbf{e}_2)$ counterclockwise through the angle $\pi/6$ (see Figure 5.34). ■

We have seen how to analyze the solution set of the quadratic equation

$$ax^2 + 2bxy + cy^2 + dx + ey + f = 0$$

whenever $d = e = 0$. Now let us consider the case when $b = 0$ but d and e need not be zero. In this case we can rewrite the equation in a more tractible form by completing squares.

Recall that, for any real number p,

$$x^2 + px = \left(x + \frac{p}{2}\right)^2 - \left(\frac{p}{2}\right)^2.$$

The process of writing $x^2 + px$ in this form is called **completing the square** on x. By completing squares, we can rewrite any quadratic equation with no xy term in a form from which we can recognize the solution set as a conic section.

EXAMPLE 6
Consider the quadratic equation

$$2x^2 + y^2 - 12x - 4y + 18 = 0.$$

We can rewrite this equation as

$$2(x^2 - 6x) + (y^2 - 4y) + 18 = 0.$$

Completing the square on both x and y yields

$$2((x - 3)^2 - 9) + ((y - 2)^2 - 4) + 18 = 0$$

or

$$2(x - 3)^2 + (y - 2)^2 = 4$$

or

$$\frac{(x - 3)^2}{2} + \frac{(y - 2)^2}{4} = 1.$$

This equation describes an ellipse, the image under translation by the vector $\mathbf{v} = (3, 2)$ of the ellipse

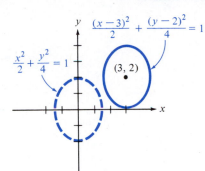

$$\frac{(x-3)^2}{2} + \frac{(y-2)^2}{4} = 1$$

FIGURE 5.35

The ellipse $\dfrac{(x-3)^2}{2} + \dfrac{(y-2)^2}{4} = 1$ is the image under translation by the vector $\mathbf{v} = (3,\ 2)$ of the ellipse $\dfrac{x^2}{2} + \dfrac{y^2}{4} = 1$.

$$\frac{x^2}{2} + \frac{y^2}{4} = 1$$

(see Figure 5.35). To see this, simply note that $(a,\ b)$ satisfies the equation

$$\frac{x^2}{2} + \frac{y^2}{4} = 1$$

if and only if $t_\mathbf{v}(a,\ b) = (a + 3,\ b + 2)$ satisfies the equation

$$\frac{(x-3)^2}{2} + \frac{(y-2)^2}{4} = 1. \quad\blacksquare$$

EXAMPLE 7

Consider the quadratic equation

$$y^2 + x + 8y + 10 = 0.$$

We can rewrite this equation as

$$x + (y^2 + 8y) + 10 = 0.$$

Completing the square on y yields

$$x + ((y + 4)^2 - 16) + 10 = 0$$

or

$$(x - 6) + (y + 4)^2 = 0$$

or

$$x - 6 = -(y + 4)^2.$$

This equation describes a parabola, the image under translation by the vector $\mathbf{v} = (6,\ -4)$ of the parabola

$$x = -y^2$$

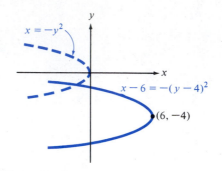

FIGURE 5.36
The parabola $x - 6 = -(y + 4)^2$ is the image under
translation by the vector $\mathbf{v} = (6, -4)$ of the parabola
$x = -y^2$.

(see Figure 5.36). To see this, we note that (a, b) satisfies the equation $x = -y^2$ if and only if $t_v(a, b) = (a + 6, b - 4)$ satisfies the equation $x - 6 = -(y + 4)^2$. ∎

We have seen that the quadratic equation

$$ax^2 + 2bxy + cy^2 + dx + ey + f = 0$$

can be analyzed by rotating coordinates when $b \neq 0$ but $d = e = 0$ or by completing squares when $b = 0$ but d or e is not zero. In the general case, where b, d, and e may all be different from zero, we can first rotate coordinates and then complete squares.

EXAMPLE 8

Consider the equation

$$5x^2 + 6xy + 5y^2 - 4x - 12y = 0.$$

We know, from Example 3, that the orthonormal basis $\mathbf{B} = \left(\left(\dfrac{1}{\sqrt{2}}, \dfrac{1}{\sqrt{2}} \right), \left(-\dfrac{1}{\sqrt{2}}, \dfrac{1}{\sqrt{2}} \right) \right)$ diagonalizes

$$Q = 5x^2 + 6xy + 5y^2$$

and that

$$Q = 8x'^2 + 2y'^2,$$

where (x', y') is the $\mathbf{B}$-coordinate pair of (x, y). Since

$$\binom{x'}{y'} = \begin{pmatrix} \dfrac{1}{\sqrt{2}} & -\dfrac{1}{\sqrt{2}} \\ \dfrac{1}{\sqrt{2}} & \dfrac{1}{\sqrt{2}} \end{pmatrix}^{-1} \binom{x}{y} \quad \text{and} \quad \binom{x}{y} = \begin{pmatrix} \dfrac{1}{\sqrt{2}} & -\dfrac{1}{\sqrt{2}} \\ \dfrac{1}{\sqrt{2}} & \dfrac{1}{\sqrt{2}} \end{pmatrix} \binom{x'}{y'},$$

we can substitute

$$x = \frac{1}{\sqrt{2}} x' - \frac{1}{\sqrt{2}} y'$$

$$y = \frac{1}{\sqrt{2}} x' + \frac{1}{\sqrt{2}} y'$$

into the equation

$$5x^2 + 6xy + 5y^2 - 4x - 12y = 0$$

to obtain

$$8x'^2 + 2y'^2 - 8\sqrt{2}x' - 4\sqrt{2}y' = 0.$$

If we rewrite this equation as

$$8(x'^2 - \sqrt{2}x') + 2(y'^2 - 2\sqrt{2}y') = 0$$

and complete squares, we obtain

$$8\left(\left(x' - \frac{\sqrt{2}}{2}\right)^2 - \frac{1}{2}\right) + 2\left((y' - \sqrt{2})^2 - 2\right) = 0$$

or

$$8\left(x' - \frac{\sqrt{2}}{2}\right)^2 + 2(y' - \sqrt{2})^2 = 8$$

or

$$\left(x' - \frac{\sqrt{2}}{2}\right)^2 + \frac{(y' - \sqrt{2})^2}{4} = 1.$$

This equation describes an ellipse, the image under translation by **v** of the ellipse

$$x'^2 + \frac{y'^2}{4} = 1,$$

where **v** is the vector whose **B**-coordinate pair is $\left(\frac{\sqrt{2}}{2}, \sqrt{2}\right)$ (see Figure 5.37).

■

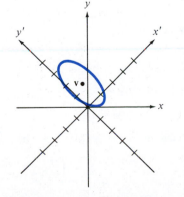

FIGURE 5.37

The ellipse $\left(x' - \frac{\sqrt{2}}{2}\right)^2 + \frac{(y' - \sqrt{2})^2}{4} = 1$ is the image under translation by **v** of the ellipse $x'^2 + \frac{y'^2}{4} = 1$, where **v** is the vector whose (x', y')-coordinate pair is $\left(\frac{\sqrt{2}}{2}, \sqrt{2}\right)$. (Compare this with Figure 5.33.)

We can summarize the method illustrated by Example 8 as follows.

To identify the solution set of the equation

$$ax^2 + 2bxy + cy^2 + dx + ey + f = 0$$

as a particular curve in the plane:

1. Find an orthogonal matrix that diagonalizes the symmetric matrix

$$A = \begin{pmatrix} a & b \\ b & c \end{pmatrix}.$$

2. If necessary, interchange the columns of this orthogonal matrix to obtain an orthogonal matrix H whose determinant is positive.

3. Set

$$\begin{pmatrix} x' \\ y' \end{pmatrix} = H^{-1} \begin{pmatrix} x \\ y \end{pmatrix} \qquad \text{so that} \qquad \begin{pmatrix} x \\ y \end{pmatrix} = H \begin{pmatrix} x' \\ y' \end{pmatrix}.$$

4. Substitute the expressions for x and y obtained in Step 3 into the given equation to obtain an equation of the form

$$\lambda_1 x'^2 + \lambda_2 y'^2 + d'x' + e'y' + f = 0,$$

where λ_1 and λ_2 are the eigenvalues of A corresponding to the eigenvectors that appear as the column vectors of H.

5. Complete the square on x' or y' or both, if necessary, to put the equation into one of the standard forms

$$\lambda_1(x' - h)^2 + \lambda_2(y' - k)^2 = r$$

or

$$y' - k = r(x' - h)^2$$

or

$$x' - h = r(y' - k)^2.$$

6. Sketch the solution set using the fact that the (x', y')-coordinate system is obtained from the standard coordinate system by counterclockwise rotation through the angle θ, where $(\cos \theta, \sin \theta)$ is the first column vector of H.

EXAMPLE 9

Consider the quadratic equation

$$xy + 3x - 2y - 7 = 0.$$

The associated quadratic form is $Q = xy$. The matrix of Q is

$$A = \begin{pmatrix} 0 & \frac{1}{2} \\ \frac{1}{2} & 0 \end{pmatrix}.$$

The eigenvalues of A are the zeros of the polynomial

$$\det(A - \lambda I) = \det\begin{pmatrix} -\lambda & \frac{1}{2} \\ \frac{1}{2} & -\lambda \end{pmatrix} = \lambda^2 - \tfrac{1}{4} = (\lambda + \tfrac{1}{2})(\lambda - \tfrac{1}{2}).$$

Hence the eigenvalues are $-\frac{1}{2}$ and $\frac{1}{2}$. Corresponding unit eigenvectors are $\left(-\dfrac{1}{\sqrt{2}}, \dfrac{1}{\sqrt{2}}\right)$ and $\left(\dfrac{1}{\sqrt{2}}, \dfrac{1}{\sqrt{2}}\right)$. Since

$$\det\begin{pmatrix} -\dfrac{1}{\sqrt{2}} & \dfrac{1}{\sqrt{2}} \\ \dfrac{1}{\sqrt{2}} & \dfrac{1}{\sqrt{2}} \end{pmatrix} < 0,$$

we take

$$H = \begin{pmatrix} \dfrac{1}{\sqrt{2}} & -\dfrac{1}{\sqrt{2}} \\ \dfrac{1}{\sqrt{2}} & \dfrac{1}{\sqrt{2}} \end{pmatrix}.$$

Then $\lambda_1 = \frac{1}{2}$, $\lambda_2 = -\frac{1}{2}$, and

$$\begin{pmatrix} x \\ y \end{pmatrix} = H\begin{pmatrix} x' \\ y' \end{pmatrix} = \begin{pmatrix} \dfrac{1}{\sqrt{2}}x' - \dfrac{1}{\sqrt{2}}y' \\ \dfrac{1}{\sqrt{2}}x' + \dfrac{1}{\sqrt{2}}y' \end{pmatrix}.$$

Substitution into the original quadratic equation then yields

$$\frac{1}{2}x'^2 - \frac{1}{2}y'^2 + \frac{1}{\sqrt{2}}x' - \frac{5}{\sqrt{2}}y' - 7 = 0$$

or

$$x'^2 - y'^2 + \sqrt{2}x' - 5\sqrt{2}y' - 14 = 0.$$

We now rewrite the equation as

$$(x'^2 + \sqrt{2}x') - (y'^2 + 5\sqrt{2}y') - 14 = 0$$

and complete squares to get

$$\left(\left(x' + \frac{\sqrt{2}}{2}\right)^2 - \frac{1}{2}\right) - \left(\left(y' + \frac{5\sqrt{2}}{2}\right)^2 - \frac{25}{2}\right) = 14$$

or

$$\left(x' + \frac{\sqrt{2}}{2}\right)^2 - \left(y' + \frac{5\sqrt{2}}{2}\right)^2 = 2.$$

or

$$\frac{\left(x' + \frac{\sqrt{2}}{2}\right)^2}{2} - \frac{\left(y' + \frac{5\sqrt{2}}{2}\right)^2}{2} = 1.$$

This equation describes a hyperbola, the image under translation by $\mathbf{v}$ of the hyperbola

$$\frac{x'^2}{2} - \frac{y'^2}{2} = 1,$$

where $\mathbf{v}$ is the vector whose **B**-coordinate pair is $\left(-\frac{\sqrt{2}}{2}, -\frac{5\sqrt{2}}{2}\right)$, and where

$\mathbf{B} = \left(\left(\frac{1}{\sqrt{2}}, \frac{1}{\sqrt{2}}\right), \left(-\frac{1}{\sqrt{2}}, \frac{1}{\sqrt{2}}\right)\right)$. Notice that, since $\mathbf{B} = (\mathbf{v}_1, \mathbf{v}_2)$ where

$$\mathbf{v}_1 = \left(\frac{1}{\sqrt{2}}, \frac{1}{\sqrt{2}}\right) = \left(\cos\frac{\pi}{4}, \sin\frac{\pi}{4}\right),$$

the (x', y')-coordinate system is obtained from the standard coordinate system by counterclockwise rotation through the angle $\pi/4$ (see Figure 5.38). ■

REMARK Sometimes the solution set of the equation obtained in Step 5 of this procedure does not describe a circle, an ellipse, a hyperbola, or a parabola. The solution set might be a pair of intersecting lines (for example, $x^2 - y^2 = 0$), a single line (for example, $x^2 = 0$), a point (for example, $x^2 + y^2 = 0$), or even the empty set (for example, $x^2 + y^2 = -1$). These sets are called **degenerate** conic sections (see Figure 5.39).

Sometimes we may be given a quadratic equation

$$ax^2 + 2bxy + cy^2 + dx + ey + f = 0$$

and need only to know which *type* of conic section is described by this equation. From our discussion above we can see that the answer depends solely on the

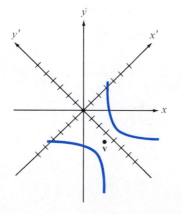

FIGURE 5.38

The hyperbola $\dfrac{\left(x' + \frac{\sqrt{2}}{2}\right)^2}{2} - \dfrac{\left(y' + \frac{5\sqrt{2}}{2}\right)^2}{2} = 1$ is the image under translation by $\mathbf{v}$ of the hyperbola $\dfrac{x'^2}{2} - \dfrac{y'^2}{2} = 1$ where $\mathbf{v}$ is the vector whose (x', y')-coordinate pair is $\left(-\dfrac{\sqrt{2}}{2}, -\dfrac{5\sqrt{2}}{2}\right)$.

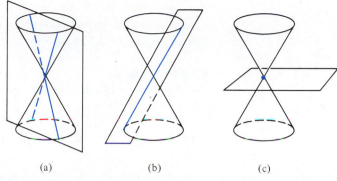

(a)   (b)   (c)

FIGURE 5.39
Degenerate conic sections: (a) a pair of intersecting lines, (b) a single line, (c) a point.

quadratic form $ax^2 + 2bxy + cy^2$ and, in fact, only on the eigenvalues λ_1 and λ_2 of the matrix

$$\begin{pmatrix} a & b \\ b & c \end{pmatrix}.$$

Assuming that the equation describes a nondegenerate conic section, we see that the curve is an ellipse (or a circle, which we view as a special kind of ellipse) if both eigenvalues have the same sign, a hyperbola if the eigenvalues have opposite signs, and a parabola if one of the eigenvalues is zero. Said another way, the curve is an ellipse if $\lambda_1\lambda_2 > 0$, a hyperbola if $\lambda_1\lambda_2 < 0$, and a parabola if $\lambda_1\lambda_2 = 0$. Since

$$\lambda_1\lambda_2 = \det\begin{pmatrix} \lambda_1 & 0 \\ 0 & \lambda_2 \end{pmatrix} = \det H^{-1}\begin{pmatrix} a & b \\ b & c \end{pmatrix} H = \det\begin{pmatrix} a & b \\ b & c \end{pmatrix} = ac - b^2,$$

where H is the orthogonal matrix such that

$$H^{-1}\begin{pmatrix} a & b \\ b & c \end{pmatrix} H = \begin{pmatrix} \lambda_1 & 0 \\ 0 & \lambda_2 \end{pmatrix},$$

we obtain the following theorem.

Theorem 2 *Suppose the quadratic equation*

$$ax^2 + 2bxy + cy^2 + dx + ey + f = 0$$

describes a nondegenerate conic section. Then this conic section is an ellipse if $ac - b^2 > 0$, a hyperbola if $ac - b^2 < 0$, and a parabola if $ac - b^2 = 0$.

The number $D = ac - b^2$ is called the **discriminant** of the quadratic equation

$$ax^2 + 2bxy + cy^2 + dx + ey + f = 0.$$

If this quadratic equation describes a *degenerate* conic section, then we say that the conic section is

(i)   a *degenerate ellipse* if the discriminant D is positive,

(ii) a *degenerate hyperbola* if D is negative, and

(iii) a *degenerate parabola* if $D = 0$.

EXAMPLE 9

The discriminant of the quadratic equation

$$2x^2 + 6xy + 8x^2 - 2x + 5y - 11 = 0$$

is $(2)(8) - (3)^2 = 7$, which is positive, so this equation describes a (possibly degenerate) ellipse.   ■

There is another way to look at the results of this section. We have seen that each equation of the form

(∗) $$ax^2 + 2bxy + cy^2 + f = 0$$

can be rewritten in the form

$$\lambda_1 x'^2 + \lambda_2 y'^2 = r,$$

where $(x'\ y')$ is the **B**-coordinate pair of (x, y) and **B** is obtained by rotating the standard basis counterclockwise through some angle θ. But this is the same as saying that the solution set of (∗) is the image under counterclockwise rotation through the angle θ of the conic section whose equation is

$$\lambda_1 x^2 + \lambda_2 y^2 = r$$

(see Figure 5.40).

Similarly, the solution set of each equation of the form

(1) $$ax^2 + 2bxy + cy^2 + dx + ey + f = 0$$

is the image under counterclockwise rotation through some angle θ of the solution set of an equation of the form

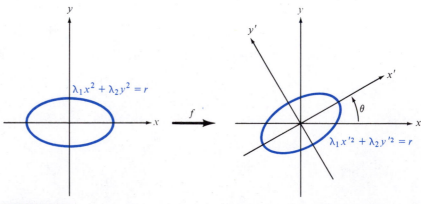

FIGURE 5.40

The conic section $\lambda_1 x'^2 + \lambda_2 y'^2 = r$ is the image under a counterclockwise rotation of the conic section $\lambda_1 x^2 + \lambda_2 y^2 = r$.

(2)
$$\lambda_1 x^2 + \lambda_2 y^2 + d'x + e'y + f = 0$$

and, as we have seen, the solution set of this equation is the image under a translation of the solution set of one of the standard equations

(3)
$$\lambda_1 x^2 + \lambda_2 y^2 = r, \quad y = rx^2, \quad \text{or } x = ry^2.$$

We can conclude, then, that *the solution set of any quadratic equation* (1) *is the image under an isometry (a rotation followed by a translation) of the solution set of one of the standard equations* (3).

EXERCISES

1. Identify the solution set of each of the following quadratic equations as a circle, an ellipse, or a hyperbola. Sketch the solution set.
   (a) $9x^2 + 4y^2 - 36 = 0$ (b) $25x^2 - 4y^2 - 100 = 0$
   (c) $x^2 + y^2 - 9 = 0$ (d) $x^2 + 16y^2 - 4 = 0$
   (e) $x^2 + y^2 - 5 = 0$ (f) $x^2 - 9y^2 + 9 = 0$

2. Sketch the following parabolas.
   (a) $x = 4y^2$ (b) $y = 16x^2$
   (c) $4x - y^2 = 0$ (d) $10y - x^2 = 0$

3. Diagonalize each of the following quadratic forms Q by finding an ordered orthonormal basis $\mathbf{B}$ such that $Q = \lambda_1 x'^2 + \lambda_2 y'^2$ where (x', y') is the $\mathbf{B}$-coordinate pair of the vector (x, y).
   (a) $2x^2 - 4xy + 2y^2$ (b) $2xy$
   (c) $54x^2 - 144xy + 96y^2$ (d) $20x^2 - 120xy + 55y^2$
   (e) $7x^2 - 4\sqrt{3}xy + 3y^2$ (f) $x^2 + 2\sqrt{3}xy - y^2$

4. Repeat Exercise 3 for the following quadratic forms Q.
   (a) $337x^2 + 168xy + 288y^2$ (b) $x^2 - 2\sqrt{3}xy + 3y^2$
   (c) $\frac{3}{2}x^2 + \sqrt{3}xy + \frac{1}{2}y^2$ (d) $21x^2 + 26\sqrt{3}xy + 31y^2$
   (e) $11x^2 + 6\sqrt{3}xy + y^2$ (f) $5x^2 - 6xy + 5y^2$

5. Use the results of Exercise 3 to identify the solution sets of each of the following quadratic equations. Sketch.
   (a) $2x^2 - 4xy + 2y^2 - \dfrac{1}{\sqrt{2}}x - \dfrac{1}{\sqrt{2}}y = 0$
   (b) $2xy + 1 = 0$
   (c) $54x^2 - 144xy + 96y^2 - 20x - 15y = 0$
   (d) $20x^2 - 120xy + 55y^2 - 100 = 0$
   (e) $7x^2 - 4\sqrt{3}xy + 3y^2 - 9 = 0$
   (f) $x^2 + 2\sqrt{3}xy - y^2 + 2 = 0$

6. Use the results of Exercise 4 to identify the solution sets of each of the following quadratic equations. Sketch.
   (a) $337x^2 + 168xy + 288y^2 - 3600 = 0$
   (b) $x^2 - 2\sqrt{3}xy + 3y^2 - 8\sqrt{3}x - 8y = 0$
   (c) $\frac{3}{2}x^2 + \sqrt{3}xy + \frac{1}{2}y^2 - x - \sqrt{3}y = 0$
   (d) $21x^2 + 26\sqrt{3}xy + 31y^2 - 144 = 0$
   (e) $11x^2 + 6\sqrt{3}xy + y^2 - 16 = 0$
   (f) $5x^2 - 6xy + 5y^2 - 2 = 0$

7. Identify the solution sets of the following quadratic equations by completing squares. Sketch.

   (a) $x^2 + 4y^2 - 4x - 32y + 64 = 0$

   (b) $x^2 - 2y^2 - 2x - 8y - 8 = 0$

   (c) $x^2 + 10x - 4y + 25 = 0$

   (d) $x^2 - y^2 + 2x + 6y - 12 = 0$

   (e) $4x^2 + 25y^2 + 16x - 350y + 1141 = 0$

   (f) $4y^2 - 8x - 4y + 1 = 0$

8. Show that the solution set of the equation $\sqrt{x} + \sqrt{y} = 1$, $0 \le x \le 1$, $0 \le y \le 1$ is part (but not all) of a parabola. Sketch.

9. Use the procedure described in this section to identify and sketch the solution sets of the following quadratic equations. [Refer to the corresponding parts of Exercise 3.]

   (a) $2x^2 - 4xy + 2y^2 - \dfrac{8 + \sqrt{2}}{2}x + \dfrac{8 - \sqrt{2}}{2}y + \dfrac{4 + 3\sqrt{2}}{2} = 0$

   (b) $2xy - 2x + 2y - 1 = 0$

   (c) $54x^2 - 144xy + 96y^2 + 196x - 303y + 176 = 0$

   (d) $20x^2 - 120xy + 55y^2 + 480x - 190y - 345 = 0$

   (e) $7x^2 - 4\sqrt{3}xy + 3y^2 + (4\sqrt{3} - 14)x + (4\sqrt{3} - 6)y - 4\sqrt{3} + 1 = 0$

   (f) $x^2 + 2\sqrt{3}xy - y^2 - 6\sqrt{3}x + 6y - 7 = 0$

10. Use the procedure described in this section to identify and sketch the solution sets of the following quadratic equations. [Refer to the corresponding parts of Exercise 4.]

    (a) $337x^2 + 168xy + 288y^2 - 506x + 408y - 2807 = 0$

    (b) $x^2 - 2\sqrt{3}xy + 3y^2 + (2 - 4\sqrt{3})x - (2\sqrt{3} + 4)y + 29 = 0$

    (c) $\frac{3}{2}x^2 + \sqrt{3}xy + \frac{1}{2}y^2 + (2 + 2\sqrt{3})x + 2y + \frac{5}{2} = 0$

    (d) $21x^2 + 26\sqrt{3}xy + 31y^2 - (126 + 104\sqrt{3})x + (26\sqrt{3} - 248)y$
    $+ 541 + 312\sqrt{3} = 0$

    (e) $11x^2 + 6\sqrt{3}xy + y^2 + (66 - 12\sqrt{3})x + (18\sqrt{3} - 4)y$
    $+ 87 - 12\sqrt{3} = 0$

    (f) $5x^2 - 6xy + 5y^2 - 100x + 92y + 578 = 0$

11. Use the discriminant to identify the solution set of each of the following quadratic equations as an ellipse, a hyperbola, or a parabola (possibly degenerate).

    (a) $x^2 - xy + y^2 + 2x - 3y + 7 = 0$

    (b) $x^2 - 2xy + y^2 - x + 2y + 11 = 0$

    (c) $2x^2 + 6xy + 4y^2 - y = 0$

    (d) $3x^2 + 2xy - y^2 + x + y = 0$

    (e) $x^2 - xy - y^2 + 2x + 4y - 9 = 0$

    (f) $3x^2 + 6xy + 3y^2 - x + 2y + 4 = 0$

    (g) $x^2 - 2xy + 5y^2 - 2x - 5y + 1 = 0$

    (h) $4x^2 + 10xy + 9y^2 - 5x + 3y + 10 = 0$

12. Show that:

    (a) Every degenerate ellipse is either a point or the empty set.

    (b) Every degenerate hyperbola is a pair of intersecting lines.

    (c) Every degenerate parabola is a line.

13. Show that the solution set of

$$ax^2 + 2bxy + cy^2 + dx + ey + f = 0$$

is a (possibly degenerate) circle if and only if $a = c \neq 0$ and $b = 0$.

14. Show that if H is an orthogonal matrix that diagonalizes the symmetric matrix

$$\begin{pmatrix} a & b \\ b & c \end{pmatrix},$$

then

$$ax^2 + 2bxy + cy^2 + dx + cy + f$$
$$= \lambda_1 x'^2 + \lambda_2 y'^2 + d'x' + e'y' + f,$$

where $\begin{pmatrix} x' \\ y' \end{pmatrix} = H\begin{pmatrix} x \\ y \end{pmatrix}$, λ_1, λ_2 are eigenvalues of $\begin{pmatrix} a & b \\ b & c \end{pmatrix}$, and $(c'\ d') = (c\ d)H$.
[*Hint:* Write everything in matrix form.]

15. Two plane figures A and B are *similar* if there exists an isometry $f:\mathbb{R}^2 \to \mathbb{R}^2$ and a map $g:\mathbb{R}^2 \to \mathbb{R}^2$, $g(\mathbf{v}) = c\mathbf{v}$ for some $c \neq 0$, such that B is the image of A under $g \circ f$.

   (a) Show that all circles are similar to one another.
   (b) Show that all parabolas are similar to one another.
   (c) Show by example that there exist ellipses that are not similar.
   (d) Show by example that there exist hyperbolas that are not similar.

REVIEW EXERCISES

1. Let

$$A = \begin{pmatrix} 4 & -\sqrt{3} \\ -\sqrt{3} & 2 \end{pmatrix}$$

   (a) Find the characteristic polynomial of A.
   (b) Find the eigenvalues of A.
   (c) Find the eigenspace corresponding to each eigenvalue of A.
   (d) Find an orthonormal basis for $\mathbb{R}^2$ consisting of eigenvectors of A.
   (e) Find an orthogonal matrix H and a diagonal matrix D such that $H^{-1}AH = D$.

2. Let $f:\mathbb{R}^3 \to \mathbb{R}^3$ be the linear map defined by

$$f(x_1, x_2, x_3) = (2x_1 + x_2, 2x_2 + x_3, 2x_3).$$

   (a) Find the characteristic polynomial of f.
   (b) Find the eigenvalues of f.
   (c) Find the eigenspace corresponding to each eigenvalue of f.
   (d) Is there a basis for $\mathbb{R}^3$ consisting of eigenvectors of f? If so, find one.

3. Let **B** be the ordered basis $((-3, 4), (2, -3))$ for $\mathbb{R}^2$.

   (a) Find the matrix that converts standard coordinates to **B**-coordinates and use it to find the **B**-coordinate pair of the vector $(1, 4)$.
   (b) Find the matrix that converts **B**-coordinates to standard coordinates and use it to find the vector whose **B**-coordinate pair is $(1, 4)$.

4. Let **B** be the ordered basis $((-3, 4), (2, -3))$ for $\mathbb{R}^2$ and let $f:\mathbb{R}^2 \to \mathbb{R}^2$ be the linear map defined by

$$f(x_1, x_2) = (x_1 - 5x_2, 3x_1 + 8x_2)$$

(a) Find the standard matrix, $M(f)$, of f.
(b) Find the **B**-coordinate matrix, $M_{\mathbf{B}}(f)$, of f.
(c) Find an invertible matrix H such that $H^{-1}M(f)H = M_{\mathbf{B}}(f)$.

5. Which of the following matrices are diagonable? For each that is, find an invertible matrix H that diagonalizes it and find the associated diagonal matrix D.

(a) $\begin{pmatrix} 0 & -1 \\ 1 & 0 \end{pmatrix}$ (b) $\begin{pmatrix} 0 & 0 \\ 1 & 0 \end{pmatrix}$ (c) $\begin{pmatrix} 0 & 2 \\ 2 & 0 \end{pmatrix}$

6. Which of the following matrices are symmetric? Which are orthogonal?

(a) $\begin{pmatrix} 1 & 1 \\ 1 & -1 \end{pmatrix}$ (b) $\begin{pmatrix} 0 & 1 \\ 1 & 0 \end{pmatrix}$ (c) $\begin{pmatrix} \frac{5}{13} & -\frac{12}{13} \\ \frac{12}{13} & \frac{5}{13} \end{pmatrix}$

(d) $\begin{pmatrix} \frac{1}{\sqrt{2}} & \frac{1}{\sqrt{2}} & 0 \\ \frac{1}{\sqrt{2}} & 0 & \frac{1}{\sqrt{2}} \\ 0 & \frac{1}{\sqrt{2}} & \frac{1}{\sqrt{2}} \end{pmatrix}$ (e) $\begin{pmatrix} \frac{1}{\sqrt{2}} & \frac{1}{\sqrt{2}} & 0 \\ -\frac{1}{\sqrt{2}} & 0 & \frac{1}{\sqrt{2}} \\ 0 & -\frac{1}{\sqrt{2}} & \frac{1}{\sqrt{2}} \end{pmatrix}$

7. Let $f:\mathbb{R}^2 \to \mathbb{R}^2$ be the linear map whose standard matrix is the given matrix A. In each case, decide whether or not f is a linear isometry. If f is a linear isometry, describe f as a particular rotation or reflection of $\mathbb{R}^2$.

(a) $A = \begin{pmatrix} \frac{\sqrt{3}}{2} & -\frac{1}{2} \\ \frac{1}{2} & \frac{\sqrt{3}}{2} \end{pmatrix}$ (b) $A = \begin{pmatrix} \frac{\sqrt{3}}{2} & \frac{1}{2} \\ \frac{1}{2} & \frac{\sqrt{3}}{2} \end{pmatrix}$ (c) $A = \begin{pmatrix} \frac{\sqrt{3}}{2} & \frac{1}{2} \\ \frac{1}{2} & -\frac{\sqrt{3}}{2} \end{pmatrix}$

8. Let $f:\mathbb{R}^3 \to \mathbb{R}^3$ be the linear map whose standard matrix is the given matrix A. In each case, decide whether or not f is a linear isometry. If f is a linear isometry, describe f as a particular rotation, or as a rotation followed by a reflection.

(a) $A = \begin{pmatrix} 0 & 1 & 0 \\ -\frac{4}{5} & 0 & \frac{3}{5} \\ \frac{3}{5} & 0 & \frac{4}{5} \end{pmatrix}$ (b) $A = \begin{pmatrix} \frac{1}{\sqrt{2}} & \frac{1}{\sqrt{3}} & \frac{1}{2} \\ \frac{1}{\sqrt{2}} & -\frac{1}{\sqrt{3}} & \frac{1}{2} \\ 0 & \frac{1}{\sqrt{3}} & -\frac{1}{\sqrt{2}} \end{pmatrix}$.

9. Identify and sketch the solution set of the equation

$$3x^2 + 2xy + 3y^2 - 10x - 14y + 15 = 0.$$

10. Identify the solution set of each of the following equations as an ellipse, a hyperbola, or a parabola (possibly degenerate). Do not sketch.

(a) $3x^2 + 4xy + 5y^2 - 16 = 0$
(b) $x^2 + 5y^2 - 16x + 12y = 0$

(c) $2x^2 + 6xy - y^2 + 11 = 0$

(d) $8x^2 - 8xy + 2y^2 + 6x - 3y + 7 = 0$

(e) $2x^2 + 6xy + y^2 + 2x - y - 13 = 0$

11.   True or false?

(a)   If the 3×3 matrix A has only two distinct eigenvalues, then A cannot be diagonalized.

(b)   The characteristic polynomial of an $n \times n$ matrix has degree n.

(c)   If $f: \mathbb{R}^n \to \mathbb{R}^n$ is an invertible linear map, then the **B**-coordinate matrix $M_{\mathbf{B}}(f)$ is an invertible matrix for every ordered basis **B** for $\mathbb{R}^n$.

(d)   If $f: \mathbb{R}^n \to \mathbb{R}^n$ has n distinct eigenvalues then f is invertible.

(e)   There are exactly two linear isometries of $\mathbb{R}^1$.

(f)   If P and Q are planes through $\mathbf{0}$ in $\mathbb{R}^3$ and if f is reflection in P and g is reflection in Q, then $f \circ g$ is a rotation of $\mathbb{R}^3$.

(g)   In $\mathbb{R}^2$, if f is counterclockwise rotation through the angle θ and g is counterclockwise rotation through the angle ϕ, then $f \circ g = g \circ f$.

(h)   In $\mathbb{R}^3$, if f is rotation about $\mathbf{u}$ through the angle θ and g is rotation about $\mathbf{v}$ through the angle ϕ, then $f \circ g = g \circ f$.

(i)   If $ac - b^2 > 0$ then the equation

$$ax^2 + 2bxy + cy^2 + dx + ey + f = 0$$

describes an ellipse (possibly degenerate).

6 Vector Spaces

In the first five chapters of this book we have studied $\mathbb{R}^n$ and its subspaces. These spaces are examples of a more general kind of space called a *vector space*. In this chapter we shall define the concept of vector space and study the counterparts for vector spaces of the ideas that we have already studied for $\mathbb{R}^n$, such as subspaces, bases, dimension, and linear maps. An an application, we shall learn how to solve an important class of equations called *linear difference equations*.

6.1
Vector Spaces and Subspaces

In Section 1.2 we listed several algebraic properties of vector addition and scalar multiplication in $\mathbb{R}^n$. There are many spaces other than $\mathbb{R}^n$ that share these same properties. Rather than study each of these spaces individually, it makes sense to put them all into a single category, the category of *vector spaces*, so that we can study them simultaneously.

A *vector space* is a set V (whose elements are called *vectors*) together with two operations,

(i) *addition:* To each pair of vectors $\mathbf{u}$ and $\mathbf{v}$ in V there is associated a vector $\mathbf{u} + \mathbf{v}$ in V, called the *sum* of $\mathbf{u}$ and $\mathbf{v}$, and

(ii) *scalar multiplication:* To each vector $\mathbf{v}$ in V and each real number c there is associated a vector $c\mathbf{v}$ in V, called a *scalar multiple* of $\mathbf{v}$.

These operations are required to have the following properties:

(A_1) $\mathbf{u} + \mathbf{v} = \mathbf{v} + \mathbf{u}$ for all $\mathbf{u}$ and $\mathbf{v}$ in V.

(A_2) $(\mathbf{u} + \mathbf{v}) + \mathbf{w} = \mathbf{u} + (\mathbf{v} + \mathbf{w})$ for all $\mathbf{u}$, $\mathbf{v}$, and $\mathbf{w}$ in V.

(A_3) There is an element $\mathbf{0}$ in V, called *zero*, with the property that $\mathbf{0} + \mathbf{v} = \mathbf{v} = \mathbf{v} + \mathbf{0}$ for all $\mathbf{v}$ in V.

(A_4) For each $\mathbf{v}$ in V, there is an element $-\mathbf{v}$ in V such that $\mathbf{v} + (-\mathbf{v}) = \mathbf{0} = (-\mathbf{v}) + \mathbf{v}$.

(S_1) $c(\mathbf{u} + \mathbf{v}) = c\mathbf{u} + c\mathbf{v}$ for all $\mathbf{u}$ and $\mathbf{v}$ in V and c in $\mathbb{R}$.

(S_2) $(c + d)\mathbf{v} = c\mathbf{v} + d\mathbf{v}$ for all c and d in $\mathbb{R}$ and $\mathbf{v}$ in V.

(S_3) $(cd)\mathbf{v} = c(d\mathbf{v})$ for all c and d in $\mathbb{R}$ and $\mathbf{v}$ in V.

(S_4) $1\mathbf{v} = \mathbf{v}$ for all $\mathbf{v}$ in V.

Vector *subtraction* can then be defined by $\mathbf{u} - \mathbf{v} = \mathbf{u} + (-1)\mathbf{v}$.

EXAMPLE 1

The space $\mathbb{R}^n$ is a vector space. Indeed, the properties A_1–A_4 and S_1–S_4 were established in Section 1.2 for vector addition and scalar multiplication in $\mathbb{R}^n$. The zero vector of $\mathbb{R}^n$ is $\mathbf{0} = (0, 0, \ldots, 0)$ and, if $\mathbf{x} = (x_1, x_2, \ldots, x_n)$, then $-\mathbf{x} = (-x_1, -x_2, \ldots, -x_n)$. ∎

EXAMPLE 2

Every subspace V of $\mathbb{R}^n$ is a vector space. The operations of vector addition and scalar multiplication for V are just those of $\mathbb{R}^n$. Since V is a subspace of $\mathbb{R}^n$, V has the following three properties:

  (i) V contains the vector $\mathbf{0}$.

 (ii) V is closed under vector addition.

(iii) V is closed under scalar multiplication.

Properties (ii) and (iii) tell us that the operations of vector addition and scalar multiplication in $\mathbb{R}^n$ actually define operations on V ($\mathbf{u} + \mathbf{v} \in V$ whenever $\mathbf{u} \in V$ and $\mathbf{v} \in V$, and $c\mathbf{v} \in V$ whenever $\mathbf{v} \in V$ and $c \in \mathbb{R}$). Property (i) tells us that $\mathbf{0} \in V$. The condition that $\mathbf{0} + \mathbf{v} = \mathbf{v} = \mathbf{v} + \mathbf{0}$ for all $\mathbf{v}$ in V is true in V because it is true in $\mathbb{R}^n$. This verifies property A_3 for V. Property A_4 holds for V because we can take $-\mathbf{v} = (-1)\mathbf{v}$ and this vector is in V whenever $\mathbf{v}$ is in V, by (iii). The other properties A_1, A_2, and S_1–S_4 are all inherited from $\mathbb{R}^n$. This proves that V is a vector space. ∎

EXAMPLE 3

Let $\mathbb{R}^\infty$ denote the set of all infinite sequences of real numbers. Thus, elements of $\mathbb{R}^\infty$ are of the form $(a_1, a_2, a_3, \ldots)$, where $a_1, a_2, a_3, \ldots$ are real numbers. For example, among the elements of $\mathbb{R}^\infty$ are the sequences $(1, 2, 3, \ldots)$ and $(1, \frac{1}{2}, \frac{1}{3}, \ldots)$.

    We can define addition in $\mathbb{R}^\infty$ by

$$(a_1, a_2, a_3, \ldots) + (b_1, b_2, b_3, \ldots) = (a_1 + b_1, a_2 + b_2, a_3 + b_3, \ldots)$$

and we can define scalar multiplication in $\mathbb{R}^\infty$ by

$$c(a_1, a_2, a_3, \ldots) = (ca_1, ca_2, ca_3, \ldots).$$

Thus, for example,

$$(1, \tfrac{1}{2}, \tfrac{1}{3}, \tfrac{1}{4}, \ldots) + (0, \tfrac{1}{2}, \tfrac{2}{3}, \tfrac{3}{4}, \ldots) = (1, 1, 1, 1, \ldots)$$

and

$$\tfrac{1}{2}(2, -4, 6, -8, \ldots) = (1, -2, 3, -4, \ldots).$$

    The space $\mathbb{R}^\infty$ is a vector space because, with these operations of addition and scalar multiplication, the properties A_1–A_4 and S_1–S_4 all hold. They are easy to verify. For A_1, simply observe that

$$(a_1, a_2, a_3, \ldots) + (b_1, b_2, b_3, \ldots) = (a_1 + b_1, a_2 + b_2, a_3 + b_3, \ldots)$$

$$= (b_1 + a_1, b_2 + a_2, b_3 + a_3, \ldots)$$

$$= (b_1, b_2, b_3, \ldots) + (a_1, a_2, a_3, \ldots).$$

For A_3, notice that

$$(0, 0, 0, \ldots) + (a_1, a_2, a_3, \ldots) = (a_1, a_2, a_3, \ldots)$$

$$= (a_1, a_2, a_3, \ldots) + (0, 0, 0, \ldots).$$

Therefore, A_3 is satisfied if we set $\mathbf{0} = (0, 0, 0, \ldots)$. The other properties are just as easy to check and are left for you to verify as an exercise (Exercise 11). ∎

EXAMPLE 4

Let $\mathcal{M}_{2 \times 2}$ be the set of all 2×2 matrices. Define addition and scalar multiplication of 2×2 matrices as usual:

$$\begin{pmatrix} a_{11} & a_{12} \\ a_{21} & a_{22} \end{pmatrix} + \begin{pmatrix} b_{11} & b_{12} \\ b_{21} & b_{22} \end{pmatrix} = \begin{pmatrix} a_{11} + b_{11} & a_{12} + b_{12} \\ a_{21} + b_{21} & a_{22} + b_{22} \end{pmatrix}$$

and

$$c\begin{pmatrix} a_{11} & a_{12} \\ a_{21} & a_{22} \end{pmatrix} = \begin{pmatrix} ca_{11} & ca_{12} \\ ca_{21} & ca_{22} \end{pmatrix}.$$

Then it is easy to check that with these operations $\mathcal{M}_{2 \times 2}$ is a vector space. The element in $\mathcal{M}_{2 \times 2}$ that is called zero is the 2×2 matrix with all entries equal to zero. ∎

EXAMPLE 5

Let $\mathcal{M}_{m \times n}$ be the set of all $m \times n$ matrices. Define addition and scalar multiplication of $m \times n$ matrices as usual:

$$\begin{pmatrix} a_{11} & a_{12} & \cdots & a_{1n} \\ a_{21} & a_{22} & \cdots & a_{2n} \\ & & \cdots & \\ a_{m1} & a_{m2} & \cdots & a_{mn} \end{pmatrix} + \begin{pmatrix} b_{11} & b_{12} & \cdots & b_{1n} \\ b_{21} & b_{22} & \cdots & b_{2n} \\ & & \cdots & \\ b_{m1} & b_{m2} & \cdots & b_{mn} \end{pmatrix}$$

$$= \begin{pmatrix} a_{11} + b_{11} & a_{12} + b_{12} & \cdots & a_{1n} + b_{1n} \\ a_{21} + b_{21} & a_{22} + b_{22} & \cdots & a_{2n} + b_{2n} \\ & & \cdots & \\ a_{m1} + b_{m1} & a_{m2} + b_{m2} & \cdots & a_{mn} + b_{mn} \end{pmatrix}$$

and

$$c\begin{pmatrix} a_{11} & a_{12} & \cdots & a_{1n} \\ a_{21} & a_{22} & \cdots & a_{2n} \\ & & \cdots & \\ a_{m1} & a_{m2} & \cdots & a_{mn} \end{pmatrix} = \begin{pmatrix} ca_{11} & ca_{12} & \cdots & ca_{1n} \\ ca_{21} & ca_{22} & \cdots & ca_{2n} \\ & & \cdots & \\ ca_{m1} & ca_{m2} & \cdots & ca_{mn} \end{pmatrix}.$$

With these operations, $\mathcal{M}_{m \times n}$ is a vector space. The element of $\mathcal{M}_{m \times n}$ that is called zero is the $m \times n$ matrix with all entries equal to zero. ∎

EXAMPLE 6

Let $\mathcal{P}$ be the set of all polynomials with real coefficients. Thus, elements of $\mathcal{P}$ are functions of the form

$$p(x) = a_n x^n + a_{n-1} x^{n-1} + \cdots + a_1 x + a_0$$

where $a_n, a_{n-1}, \ldots, a_0$ are real numbers, called the *coefficients* of $p(x)$, and n is a nonnegative integer. The integer n is called the *degree* of the polynomial $p(x)$, if $a_n \neq 0$.

We can add polynomials by adding the coefficients of like powers of x. Thus, for example,

$$(x^2 + x - 2) + (x^3 - 2x^2 + 1) = x^3 - x^2 + x - 1.$$

We can multiply a polynomial by a real number by multiplying each coefficient of the polynomial by that real number. Thus, for example,

$$-2(x^4 - x^2 + 3) = -2x^4 + 2x^2 - 6.$$

With these operations of addition and scalar multiplication, $\mathcal{P}$ is a vector space. The element in $\mathcal{P}$ that is called zero is the polynomial all of whose coefficients are zero. ∎

EXAMPLE 7

In Section 2.4 we defined addition and scalar multiplication for linear maps from $\mathbb{R}^n$ to $\mathbb{R}^m$. Recall that, if f and g are two linear maps from $\mathbb{R}^n$ to $\mathbb{R}^m$, then their sum is the linear map $f + g$ defined by

$$(f + g)(\mathbf{x}) = f(\mathbf{x}) + g(\mathbf{x}) \qquad \text{for all } \mathbf{x} \in \mathbb{R}^n.$$

If $f:\mathbb{R}^n \to \mathbb{R}^m$ is a linear map and c is a real number, then $cf:\mathbb{R}^n \to \mathbb{R}^m$ is the linear map defined by

$$(cf)(\mathbf{x}) = cf(\mathbf{x}) \text{ for all } \mathbf{x} \in \mathbb{R}^n.$$

Let $L(\mathbb{R}^n, \mathbb{R}^m)$ denote the set of all linear maps from $\mathbb{R}^n$ to $\mathbb{R}^m$. Then, with the operations of addition and scalar multiplication just described, $L(\mathbb{R}^n, \mathbb{R}^m)$ is a vector space. The element in $L(\mathbb{R}^n, \mathbb{R}^m)$ that is called zero is the map $O:\mathbb{R}^n \to \mathbb{R}^m$ that sends each vector in $\mathbb{R}^n$ to the vector $\mathbf{0}$ in $\mathbb{R}^m$. ∎

We can imitate the definition of subspace of $\mathbb{R}^n$ to define the concept of subspace in an arbitrary vector space.

Suppose V is a vector space. A subset W of V is called a *subspace* of V if

(i) W contains the vector $\mathbf{0}$,

(ii) W is closed under addition, and

(iii) W is closed under scalar multiplication.

An argument almost identical to the one used in Example 2 shows that *every subspace of a vector space is also a vector space.*

EXAMPLE 8

Let $V = \mathbb{R}^\infty$ (see Example 3) and let

$$W = \{(a_1, a_2, a_3, \ldots) \in \mathbb{R}^\infty | a_1 = a_3 = a_5 = \cdots = 0\}.$$

A typical element of W is an infinite sequence of the form

$$(0, a_2, 0, a_4, 0, a_6, \ldots).$$

Since

(i) $\mathbf{0} = (0, 0, 0, \ldots) \in W$,

(ii) W is closed under addition:

$$(0, a_2, 0, a_4, 0, a_6, \ldots) + (0, b_2, 0, b_4, 0, b_6, \ldots)$$

$$= (0, a_2 + b_2, 0, a_4 + b_4, 0, a_6 + b_6, \ldots) \in W,$$

and

(iii) W is closed under scalar multiplication:

$$c(0, a_2, 0, a_4, 0, a_6, \ldots) = (0, ca_2, 0, ca_4, 0, ca_6, \ldots) \in W,$$

we see that W is a subspace of $\mathbb{R}^\infty$.   ∎

EXAMPLE 9

Let V be the vector space of all $n \times n$ matrices, $V = \mathcal{M}_{n \times n}$ (see Example 5), and let W be the subset of $\mathcal{M}_{n \times n}$ consisting of all $n \times n$ diagonal matrices. Thus a typical element of W is an $n \times n$ matrix of the form

$$A = \begin{pmatrix} a_1 & 0 & \cdots & 0 \\ 0 & a_2 & \cdots & 0 \\ & & \cdots & \\ 0 & 0 & \cdots & a_n \end{pmatrix}.$$

Since

(i)   the zero matrix

$$\begin{pmatrix} 0 & 0 & \cdots & 0 \\ 0 & 0 & \cdots & 0 \\ & & \cdots & \\ 0 & 0 & \cdots & 0 \end{pmatrix}$$

is a diagonal matrix,

(ii)   the sum of two diagonal matrices is a diagonal matrix,

$$\begin{pmatrix} a_1 & 0 & \cdots & 0 \\ 0 & a_2 & \cdots & 0 \\ & & \cdots & \\ 0 & 0 & \cdots & a_n \end{pmatrix} + \begin{pmatrix} b_1 & 0 & \cdots & 0 \\ 0 & b_2 & \cdots & 0 \\ & & \cdots & \\ 0 & 0 & \cdots & b_n \end{pmatrix} = \begin{pmatrix} a_1 + b_1 & 0 & \cdots & 0 \\ 0 & a_2 + b_2 & \cdots & 0 \\ & & \cdots & \\ 0 & 0 & \cdots & a_n + b_n \end{pmatrix},$$

and

(iii) each scalar multiple of a diagonal matrix is a diagonal matrix,

$$c\begin{pmatrix} a_1 & 0 & \cdots & 0 \\ 0 & a_2 & \cdots & 0 \\ & & \cdots & \\ 0 & 0 & \cdots & a_n \end{pmatrix} = \begin{pmatrix} ca_1 & 0 & \cdots & 0 \\ 0 & ca_2 & \cdots & 0 \\ & & \cdots & \\ 0 & 0 & \cdots & ca_n \end{pmatrix},$$

we see that W is a subspace of $\mathcal{M}_{n \times n}$. ∎

EXAMPLE 10

Let V be the vector space of all polynomials, $V = \mathcal{P}$ (see Example 6), and let $\mathcal{P}^3$ denote the subset of $\mathcal{P}$ consisting of all polynomials with degree less than or equal to 3. A typical element of $\mathcal{P}^3$ is a polynomial of the form

$$p(x) = a_3x^3 + a_2x^2 + a_1x + a_0.$$

Since

(i) the zero polynomial

$$0x^3 + 0x^2 + 0x + 0$$

is in $\mathcal{P}^3$ (its degree is defined to be zero),

(ii) the sum of two polynomials in $\mathcal{P}^3$ is in $\mathcal{P}^3$,

$$(a_3x^3 + a_2x^2 + a_1x + a_0) + (b_3x^3 + b_2x^2 + b_1x + b_0)$$
$$= (a_3 + b_3)x^3 + (a_2 + b_2)x^2 + (a_1 + b_1)x + (a_0 + b_0),$$

and

(iii) each scalar multiple of a polynomial in $\mathcal{P}^3$ is in $\mathcal{P}^3$,

$$c(a_3x^3 + a_2x^2 + a_1x + a_0) = (ca_3)x^3 + (ca_2)x^2 + (ca_1)x + (ca_0),$$

we see that $\mathcal{P}^3$ is a subspace of $\mathcal{P}$. ∎

REMARK   A similar verification shows that $\mathcal{P}^n$, the subset of $\mathcal{P}$ consisting of all polynomials with degree less than or equal to n, is a subspace of $\mathcal{P}$, for each nonnegative integer n.

EXAMPLE 11

Let V be the vector space of all linear maps from $\mathbb{R}^2$ to $\mathbb{R}^2$, $V = L(\mathbb{R}^2, \mathbb{R}^2)$ (see Example 7), and let

$$W = \{f \in L(\mathbb{R}^2, \mathbb{R}^2) | f(\mathbf{e}_1) = a\mathbf{e}_1 \text{ for some } a \in \mathbb{R}\}.$$

Thus W consists of all linear maps $f:\mathbb{R}^2 \to \mathbb{R}^2$ that have $\mathbf{e}_1$ as an eigenvector. Since

(i)   the zero map is in W—$O(\mathbf{e}_1) = 0\mathbf{e}_1$;

(ii) W is closed under addition—if $f(\mathbf{e}_1) = a\mathbf{e}_1$ and $g(\mathbf{e}_1) = b\mathbf{e}_1$, then

$$(f + g)(\mathbf{e}_1) = f(\mathbf{e}_1) + g(\mathbf{e}_1) = a\mathbf{e}_1 + b\mathbf{e}_1 = (a + b)\mathbf{e}_1;$$

and

(iii) W is closed under scalar multiplication—if $f(\mathbf{e}_1) = a\mathbf{e}_1$, then

$$(cf)(\mathbf{e}_1) = cf(\mathbf{e}_1) = c(a\mathbf{e}_1) = (ca)\mathbf{e}_1,$$

we see that W is a subspace of $L(\mathbb{R}^2, \mathbb{R}^2)$.

Notice that W contains all expansions, contractions, and horizontal shears. W also contains reflection in the line $x_2 = 0$ and reflection in $\mathbf{0}$. ∎

EXAMPLE 12

Let V be the vector space of all 2×2 matrices, $V = \mathcal{M}_{2 \times 2}$ (see Example 4), and let W be the subset of $\mathcal{M}_{2 \times 2}$ consisting of all 2×2 matrices with determinant equal to zero:

$$W = \{A \in \mathcal{M}_{2 \times 2} | \det A = 0\}.$$

Then the zero matrix is in W, since

$$\det\begin{pmatrix} 0 & 0 \\ 0 & 0 \end{pmatrix} = 0.$$

Moreover, W is closed under scalar multiplication, because

$$\det \left(c\begin{pmatrix} a_{11} & a_{12} \\ a_{21} & a_{22} \end{pmatrix}\right) = \det\begin{pmatrix} ca_{11} & ca_{12} \\ ca_{21} & ca_{22} \end{pmatrix}$$

$$= c^2(a_{11}a_{22} - a_{12}a_{21}) = c^2 \det\begin{pmatrix} a_{11} & a_{12} \\ a_{21} & a_{22} \end{pmatrix}$$

so $\det (cA) = 0$ whenever $\det A = 0$. But W is *not closed under addition*. For if

$$A = \begin{pmatrix} 1 & 0 \\ 0 & 0 \end{pmatrix} \quad \text{and} \quad B = \begin{pmatrix} 0 & 0 \\ 0 & 1 \end{pmatrix},$$

then $\det A = 0$ and $\det B = 0$, but

$$\det(A + B) = \det\begin{pmatrix} 1 & 0 \\ 0 & 1 \end{pmatrix} = 1 \neq 0.$$

Hence W is *not* a subspace of $\mathcal{M}_{2 \times 2}$. ∎

EXERCISES

1.   Let $\mathbf{a} = (1, 2, 3, 4, 5, \ldots)$ and $\mathbf{b} = (0, 1, 2, 3, 4, \ldots)$. Using the operations of addition and scalar multiplication in $\mathbb{R}^\infty$, find

(a) $\mathbf{a} + \mathbf{b}$ (b) $\mathbf{a} - \mathbf{b}$

(c) $2\mathbf{a}$ (d) $\frac{1}{2}\mathbf{a} - \frac{3}{2}\mathbf{b}$

(e) $\pi\mathbf{b}$

2. Let $A = \begin{pmatrix} 1 & 0 \\ 1 & 3 \\ 0 & 2 \end{pmatrix}$ and $B = \begin{pmatrix} 1 & 1 \\ -1 & 1 \\ 2 & 4 \end{pmatrix}$. Using the operations of addition and scalar

multiplication in $\mathcal{M}_{3 \times 2}$, find

(a) $A + B$ (b) $A - B$

(c) $\frac{1}{2}B$ (d) $5A + 2B$

(e) $2\pi A$

3. Let $p(x) = x^2 + 3x - 5$ and $q(x) = x^2 - x + 4$. Using the operations of addition and scalar multiplication in $\mathcal{P}$, find

(a) $p(x) + q(x)$ (b) $p(x) - q(x)$

(c) $3p(x)$ (d) $3p(x) - 2q(x)$

(e) $4p(x) + 5q(x)$

4. Let $f(x_1, x_2) = (3x_1 + x_2, -x_1 + 2x_2)$ and $g(x_1, x_2) = (-x_2, x_1)$. Using the operations of addition and scalar multiplication in $L(\mathbb{R}^2, \mathbb{R}^2)$, find

(a) $(f + g)(x_1, x_2)$ (b) $(f - g)(x_1, x_2)$

(c) $(-f)(x_1, x_2)$ (d) $(-2f + 4g)(x_1, x_2)$

5. Which of the following subsets of $\mathbb{R}^\infty$ are subspaces? For those that are not, decide which of the three properties of a subspace are not satisfied.

(a) $\{(a_1, a_2, a_3, \ldots) | a_1 = 0\}$

(b) $\{(a_1, a_2, a_3, \ldots) | a_1 \neq 0\}$

(c) $\{(a_1, a_2, a_3, \ldots) | a_2 = a_4 = a_6 = \cdots = 0\}$

(d) $\{(a_1, a_2, a_3, \ldots) | a_6 = a_7 = a_8 = \cdots = 0\}$

(e) $\{(a_1, a_2, a_3, \ldots) | a_5 \neq 0, a_6 = a_7 = a_8 = \cdots = 0\}$

(f) $\{(a_1, a_2, a_3, \ldots) | a_i \geq 0 \text{ for all } i\}$

(g) $\{(a_1, a_2, a_3, \ldots) | a_2 = -a_1\}$

(h) $\{(a_1, a_2, a_3, \ldots) | a_1, a_2, a_3, \ldots \text{ are integers}\}$

(i) $\{c(1, 2, 3, \ldots) | c \in \mathbb{R}\}$

(j) $\{c_1(1, 0, 0, \ldots) + c_2(0, 1, 0, \ldots) | c_1, c_2 \in \mathbb{R}\}$

6. Which of the following subsets of $\mathcal{M}_{2 \times 2}$ are subspaces? For those that are not, decide which of the three properties of a subspace are not satisfied.

(a) $\{A \in \mathcal{M}_{2 \times 2} | \det A \neq 0\}$

(b) $\{A \in \mathcal{M}_{2 \times 2} | A^t = A\}$

(c) $\{A \in \mathcal{M}_{2 \times 2} | A^t = -A\}$

(d) $\{\begin{pmatrix} a & 0 \\ 0 & a \end{pmatrix} | a \in \mathbb{R}\}$

(e) $\{\begin{pmatrix} a_{11} & a_{12} \\ a_{21} & a_{22} \end{pmatrix} | a_{11} + a_{22} = 0\}$

(f) $\{\begin{pmatrix} a_{11} & a_{12} \\ a_{21} & a_{22} \end{pmatrix} | a_{11} = a_{22} \text{ and } a_{12} = a_{21}\}$

(g) $\{\begin{pmatrix} a_{11} & a_{12} \\ a_{21} & a_{22} \end{pmatrix} | a_{11} + a_{12} + a_{21} + a_{22} = 0\}$.

(h) $\{\begin{pmatrix} a_{11} & a_{12} \\ a_{21} & a_{22} \end{pmatrix} | a_{11} - a_{12} + a_{21} - a_{22} = 0\}$

(i) $\{\begin{pmatrix} a_{11} & a_{12} \\ a_{21} & a_{22} \end{pmatrix} | a_{11}^2 + a_{12}^2 + a_{21}^2 + a_{22}^2 = 0\}$

(j) $\{A | \text{rank } A = 1\}$

7. Which of the following subsets of $\mathcal{P}$ are subspaces? For those that are not subspaces, decide which of the three properties of a subspace are not satisfied.

(a) the set of all polynomials with integer coefficients

(b) the set of all polynomials with rational coefficients

(c) the set of all polynomials with positive coefficients

(d) the set of all polynomials of even degree

(e) the set of all polynomials of odd degree

(f) the set of all polynomials of degree 2

(g) the set of all polynomials with all coefficients of even powers of x equal to zero

(h) the set of all polynomials with all coefficients of odd powers of x equal to zero

(i) the set of all scalar multiples of the polynomial $x^2 + x + 1$

8. Which of the following subsets of $L(\mathbb{R}^2, \mathbb{R}^2)$ are subspaces? For those that are not, decide which of the three properties of a subspace are not satisfied.

(a) the set of all linear maps with kernel equal to $\{0\}$

(b) the set of all linear maps with image equal to $\mathbb{R}^2$

(c) the set of all linear maps with kernel equal to $\mathbb{R}^2$

(d) the set of all expansions

(e) the set of all horizontal shears

(f) the set of all rotations about 0

(g) the set of all f such that $f(\mathbf{e}_2) = a\mathbf{e}_2$ for some real number a

(h) the set of all f such that $f(\mathbf{v}) = a\mathbf{v}$ for some real number a, where $\mathbf{v}$ is some fixed vector in $\mathbb{R}^2$

9. Let V be a vector space and let $\mathbf{0}$ be the zero vector in V. Show that $\{\mathbf{0}\}$ is a subspace of V.

10. Let V be a vector space. Show that V is a subspace of itself.

11. Verify that properties A_2, A_4, and S_1–S_4 hold for $\mathbb{R}^\infty$.

12. Verify that $\mathcal{M}_{2 \times 2}$ is a vector space.

13. Verify that $\mathcal{M}_{m \times n}$ is a vector space.

14. Verify that properties A_1, A_2, A_4, and S_2–S_4 are satisfied by $\mathcal{P}$.

15. Verify that $L(\mathbb{R}^n, \mathbb{R}^m)$ is a vector space.

16. Show that if $m \le n$, then $\mathcal{P}^m$ is a subspace of $\mathcal{P}^n$.

17. Let V be a vector space. Show that if W_1 and W_2 are subspaces of V, then $W_1 \cap W_2$ is a subspace of V. Is $W_1 \cup W_2$ a subspace of V?

18. Let V be a set consisting of a single element a. Define $a + a = a$ and define $ca = a$ for all real numbers c. Show that, with these operations, V is a vector space.

19. Let V and W be vector spaces and let X be the set of all ordered pairs $(\mathbf{v}, \mathbf{w})$ where $\mathbf{v} \in V$ and $\mathbf{w} \in W$. For $(\mathbf{v}, \mathbf{w})$ and $(\mathbf{v}', \mathbf{w}')$ in X and c in $\mathbb{R}$, define

$$(\mathbf{v}, \mathbf{w}) + (\mathbf{v}', \mathbf{w}') = (\mathbf{v} + \mathbf{v}', \mathbf{w} + \mathbf{w}')$$

$$c(\mathbf{v}, \mathbf{w}) = (c\mathbf{v}, c\mathbf{w}).$$

Show that, with these operations, X is a vector space.

20. Let S be a set and let V be the collection of all subsets of S. For A and B in V, define $A + B = A \cup B$. For $A \in V$ and $c \in \mathbb{R}$, define $cA = A$. Which of the properties A_1–A_4 and S_1–S_4 are satisfied by these operations? With these operations, is V a vector space?

21. Define a funny addition $\oplus$ on the set $\mathbb{R}_+$ of positive real numbers by $a \oplus b = ab$ for $a, b \in \mathbb{R}_+$ and define a funny scalar multiplication $\odot$ on this set by $c \odot a = a^c$ for $a \in \mathbb{R}_+$ and $c \in \mathbb{R}$. With these operations, is $\mathbb{R}_+$ a vector space?

22. Let V be a vector space. Show that $\mathbf{z} = \mathbf{0}$ is the only vector in V with the property

$$\mathbf{z} + \mathbf{v} = \mathbf{v} = \mathbf{v} + \mathbf{z}.$$

[*Hint:* Suppose $\mathbf{z}$ has this property. Compute $\mathbf{z} + \mathbf{0}$ in two ways.]

23. Let V be a vector space and let $\mathbf{v} \in V$. Show that $\mathbf{w} = -\mathbf{v}$ is the only vector in V with the property

$$\mathbf{v} + \mathbf{w} = \mathbf{0} = \mathbf{w} + \mathbf{v}.$$

[*Hint:* Suppose $\mathbf{w}$ has this property. Compute $(-\mathbf{v}) + \mathbf{v} + \mathbf{w}$ in two ways.]

24. Let V be a vector space and let $\mathbf{v} \in V$. Show that $0\mathbf{v} = \mathbf{0}$. [*Hint:* Compute $(0 + 0)\mathbf{v}$ in two ways.]

25. Let V be a vector space and let $c \in \mathbb{R}$. Show that $c\mathbf{0} = \mathbf{0}$. [*Hint:* Compute $c(\mathbf{0} + \mathbf{0})$ in two ways.]

26. Let V be a vector space and let $\mathbf{v} \in V$. Show that $-\mathbf{v} = (-1)\mathbf{v}$. [*Hint:* Use the results of Exercises 23 and 24.]

6.2
Basis and Dimension

In earlier chapters of this book, we studied the concepts of spanning set, linear independence, and basis in $\mathbb{R}^n$. We shall now study these concepts in arbitrary vector spaces.

Let V be a vector space and let S be any subset of V. A vector $\mathbf{v}$ in V is a ***linear combination*** of elements of S if $\mathbf{v}$ can be expressed in the form

$$\mathbf{v} = c_1\mathbf{v}_1 + c_2\mathbf{v}_2 + \cdots + c_k\mathbf{v}_k$$

where $\mathbf{v}_1, \ldots, \mathbf{v}_k$ are vectors in S and $c_1, \ldots, c_k$ are real numbers. The set $\mathcal{L}(S)$ of all linear combinations of elements of S is a subspace of V (see Exercise 22). This subspace $\mathcal{L}(S)$ is called the ***subspace of V spanned by S***. If $\mathcal{L}(S) = V$, then we say that S ***spans*** V, or that S is a ***spanning set*** for V.

Notice that spanning sets need not be finite sets, as they were in earlier chapters.

EXAMPLE 1

Let V be the vector space $\mathcal{P}^n$ of all polynomials of degree less than or equal to n and let

$$S = \{1, x, x^2, \ldots, x^n\}.$$

Since each element of $\mathcal{P}^n$ is of the form

$$a_0 \cdot 1 + a_1 x + \cdots + a_n x^n$$

for some real numbers $a_0, a_1, \ldots, a_n$, we see that S is a spanning set for $\mathcal{P}^n$. ■

EXAMPLE 2

Let V be the vector space $\mathcal{P}$ of all polynomials, and let

$$S = \{1, x, x^2, x^3, \ldots\}.$$

Since each element of $\mathcal{P}$ is of the form

$$a_0 \cdot 1 + a_1 x + \cdots + a_n x^n$$

for some positive integer n and some real numbers $a_0, a_1, \ldots, a_n$, we see that S is a spanning set for $\mathcal{P}$. ■

EXAMPLE 3

Let $V = \mathbb{R}^\infty$ and let

$$S = \{\mathbf{e}_1, \mathbf{e}_2, \mathbf{e}_3, \ldots\}$$

where

$$\mathbf{e}_1 = (1, 0, 0, 0, \ldots),$$

$$\mathbf{e}_2 = (0, 1, 0, 0, \ldots),$$

$$\mathbf{e}_3 = (0, 0, 1, 0, \ldots),$$

$$\ldots$$

Thus, for each positive integer i, $\mathbf{e}_i$ is the vector in $\mathbb{R}^\infty$ that has ith entry equal to 1 and all other entries equal to zero.

Let us see if this set S spans $\mathbb{R}^\infty$. Typical elements of $\mathcal{L}(S)$ are

$$2\mathbf{e}_1 + 3\mathbf{e}_2 + 5\mathbf{e}_3 = (2, 3, 5, 0, 0, 0, \ldots),$$

$$(-1)\mathbf{e}_3 + 2\mathbf{e}_6 + \mathbf{e}_7 = (0, 0, -1, 0, 0, 2, 1, 0, 0, 0, \ldots),$$

and

$$4\mathbf{e}_1 + 2\mathbf{e}_2 + (-1)\mathbf{e}_4 + 3\mathbf{e}_5 = (4, 2, 0, -1, 3, 0, 0, 0, \ldots).$$

Notice that each of these vectors has only a finite number of nonzero entries. In fact, if $\mathbf{v}$ is *any* vector in $\mathcal{L}(S)$, then $\mathbf{v}$ has only a finite number of nonzero entries. To see this, simply observe that if $\mathbf{v}$ is a linear combination of k vectors in S, then $\mathbf{v}$ can have at most k entries different from zero. In particular, the vector $(1, 1, 1, 1, \ldots)$, with every entry equal to 1, is not in $\mathcal{L}(S)$. Therefore $\mathcal{L}(S) \neq \mathbb{R}^\infty$; that is, the set

$$S = \{\mathbf{e}_1, \mathbf{e}_2, \mathbf{e}_3, \ldots\}$$

does *not* span $\mathbb{R}^\infty$. ■

Let V be a vector space and let S be a subset of V. A ***dependence relation*** in S is an equation of the form

$$c_1\mathbf{v}_1 + \cdots + c_k\mathbf{v}_k = \mathbf{0}$$

where $\mathbf{v}_1, \ldots, \mathbf{v}_k$ are distinct vectors in S and $c_1, \ldots, c_k$ are real numbers, not all zero. We say that S is ***linearly dependent*** if there is a dependence relation in S, and that S is ***linearly independent*** if there is no dependence relation in S.

EXAMPLE 4

Let $V = \mathcal{P}^n$ and let

$$S = \{1, x, x^2, \ldots, x^n\}.$$

Let us check to see if there is a dependence relation in S. Suppose

$$c_0 + c_1x + c_2x^2 + \cdots + c_nx^n = 0.$$

Since the zero on the right side of this equation is the zero polynomial, this equation can be rewritten as

$$c_0 + c_1x + c_2x^2 + \cdots + c_nx^n = 0 + 0x + 0x^2 + \cdots + 0x^n.$$

But two polynomials are equal if and only if all corresponding coefficients are equal. Thus the only solution $(c_0, c_1, c_2, \ldots, c_n)$ of the polynomial equation

$$c_0 + c_1x + c_2x^2 + \cdots + c_nx^n = 0$$

is the zero solution $(c_0, c_1, \ldots, c_n) = (0, 0, \ldots, 0)$. Hence the set S is linearly independent. ■

EXAMPLE 5

Let $V = \mathcal{P}$ and let

$$S = \{1, x, x^2, x^3, \ldots\}.$$

Then each linear combination of elements of S is a polynomial. The coefficients of this polynomial are the coefficients of the linear combination. A polynomial is equal to the zero polynomial if and only if all its coefficients are equal to zero. Hence there are no dependence relations in S. The set S is linearly independent. ■

Theorem 1   *Let V be a vector space and let S be a subset of V. Then:*

(i)   *If S contains just one vector, then S is linearly dependent if and only if that vector is the zero vector $\mathbf{0}$.*

(ii)   *If S contains exactly two vectors, then S is linearly dependent if and only if one of those vectors is a scalar multiple of the other.*

(iii)   *If S contains more than two vectors, then S is linearly dependent if and only if some vector in S is a linear combination of one or more of the other vectors in S.*

Proof

(i)   Suppose $S = \{\mathbf{v}\}$. Then there is a dependence relation $c\mathbf{v} = \mathbf{0}$ $(c \neq 0)$ in S if and only if $\mathbf{v} = \mathbf{0}$.

(ii) Suppose $S = \{\mathbf{v}, \mathbf{w}\}$. Then there is a dependence relation $c_1\mathbf{v} + c_2\mathbf{w} = \mathbf{0}$ $(c_1$ and c_2 are not both zero) in S if and only if either $\mathbf{w} = -(c_2/c_1)\mathbf{v}$ (if $c_1 \neq 0$) or $\mathbf{v} = -(c_1/c_2)\mathbf{w}$ (if $c_2 \neq 0$).

(iii) Suppose S contains more than two vectors. If S is linearly dependent then there is a dependence relation

$$c_1\mathbf{v}_1 + c_2\mathbf{v}_2 + \cdots + c_k\mathbf{v}_k = \mathbf{0},$$

where $\mathbf{v}_1, \ldots, \mathbf{v}_k$ are vectors in S and $c_i \neq 0$ for some i, $1 \leq i \leq k$. If $k = 1$, then $c_1\mathbf{v}_1 = \mathbf{0}$ where $c_1 \neq 0$. Therefore, $\mathbf{v}_1 = \mathbf{0} = 0\mathbf{v}_2$, where $\mathbf{v}_2$ is any other vector in S. If $k > 1$, we can solve this equation for $\mathbf{v}_i$:

$$-c_i\mathbf{v}_i = c_1\mathbf{v}_1 + \cdots + c_{i-1}\mathbf{v}_{i-1} + c_{i+1}\mathbf{v}_{i+1} + \cdots + c_k\mathbf{v}_k$$

or

$$\mathbf{v}_i = \left(-\frac{c_1}{c_i}\right)\mathbf{v}_1 + \cdots + \left(-\frac{c_{i-1}}{c_i}\right)\mathbf{v}_{i-1}$$
$$+ \left(-\frac{c_{i+1}}{c_i}\right)\mathbf{v}_{i+1} + \cdots + \left(-\frac{c_k}{c_i}\right)\mathbf{v}_k.$$

Hence $\mathbf{v}_i$ is a linear combination of the vectors $\mathbf{v}_1, \ldots, \mathbf{v}_{i-1}, \mathbf{v}_{i+1}, \ldots, \mathbf{v}_k$.

Conversely, suppose some vector $\mathbf{v}$ in S is a linear combination

$$\mathbf{v} = c_1\mathbf{v}_1 + \cdots + c_k\mathbf{v}_k$$

of other vectors in S. Then we have
$$c_1\mathbf{v}_1 + \cdots + c_k\mathbf{v}_k + (-1)\mathbf{v} = \mathbf{0}.$$

This is a dependence relation in S. Hence S is linearly dependent. ■

REMARK   There is a variation on Theorem 1 that is often useful. If S is a linearly independent subset of V and if $\mathbf{v}$ is any vector in V, then the set $S \cup \{\mathbf{v}\}$ is linearly independent if and only if $\mathbf{v} \notin \mathcal{L}(S)$. You will be asked to prove this in Exercise 24.

A subset S of a vector space V is a ***basis*** for V if (i) it is linearly independent and (ii) it spans V.

EXAMPLE 6

The set

$$\{1, x, x^2, \ldots, x^n\}$$

is a basis for $\mathcal{P}^n$ because it is linearly independent (by Example 4) and it spans $\mathcal{P}^n$ (by Example 1). ■

EXAMPLE 7

The set

$$\{1, x, x^2, x^3, \ldots\}$$

is a basis for $\mathcal{P}$ because it is linearly independent (by Example 5) and it spans $\mathcal{P}$ (by Example 2). ∎

EXAMPLE 8

Let V be the vector space of all 2×2 matrices, $V = \mathcal{M}_{2\times2}$, and let

$$S = \left\{ \begin{pmatrix} 1 & 0 \\ 0 & 0 \end{pmatrix}, \begin{pmatrix} 0 & 1 \\ 0 & 0 \end{pmatrix}, \begin{pmatrix} 0 & 0 \\ 1 & 0 \end{pmatrix}, \begin{pmatrix} 0 & 0 \\ 0 & 1 \end{pmatrix} \right\}.$$

Then S is linearly independent because

$$c_1 \begin{pmatrix} 1 & 0 \\ 0 & 0 \end{pmatrix} + c_2 \begin{pmatrix} 0 & 1 \\ 0 & 0 \end{pmatrix} + c_3 \begin{pmatrix} 0 & 0 \\ 1 & 0 \end{pmatrix} + c_4 \begin{pmatrix} 0 & 0 \\ 0 & 1 \end{pmatrix} = \begin{pmatrix} c_1 & c_2 \\ c_3 & c_4 \end{pmatrix}$$

and

$$\begin{pmatrix} c_1 & c_2 \\ c_3 & c_4 \end{pmatrix} = \begin{pmatrix} 0 & 0 \\ 0 & 0 \end{pmatrix}$$

only when $(c_1, c_2, c_3, c_4) = (0, 0, 0, 0)$. Moreover, S spans $\mathcal{M}_{2\times2}$ because each $A \in \mathcal{M}_{2\times2}$ is of the form

$$A = \begin{pmatrix} a_{11} & a_{12} \\ a_{21} & a_{22} \end{pmatrix} = a_{11} \begin{pmatrix} 1 & 0 \\ 0 & 0 \end{pmatrix} + a_{12} \begin{pmatrix} 0 & 1 \\ 0 & 0 \end{pmatrix} + a_{21} \begin{pmatrix} 0 & 0 \\ 1 & 0 \end{pmatrix} + a_{22} \begin{pmatrix} 0 & 0 \\ 0 & 1 \end{pmatrix}.$$

Hence S is a basis for $\mathcal{M}_{2\times2}$. ∎

EXAMPLE 9

Let V be the vector space of all linear maps from $\mathbb{R}^2$ to $\mathbb{R}^2$, $V = L(\mathbb{R}^2, \mathbb{R}^2)$. Since each linear map $f:\mathbb{R}^2 \to \mathbb{R}^2$ corresponds to a 2×2 matrix, you might guess in view of Example 8 that the linear maps that correspond to the matrices

$$\begin{pmatrix} 1 & 0 \\ 0 & 0 \end{pmatrix}, \begin{pmatrix} 0 & 1 \\ 0 & 0 \end{pmatrix}, \begin{pmatrix} 0 & 0 \\ 1 & 0 \end{pmatrix}, \begin{pmatrix} 0 & 0 \\ 0 & 1 \end{pmatrix},$$

form a basis for $L(\mathbb{R}^2, \mathbb{R}^2)$. These linear maps are

$$f_1(x_1, x_2) = (x_1, 0)$$

$$f_2(x_1, x_2) = (x_2, 0)$$

$$f_3(x_1, x_2) = (0, x_1),$$

and

$$f_4(x_1, x_2) = (0, x_2).$$

Let us check to see if the set $S = \{f_1, f_2, f_3, f_4\}$ is linearly independent. Suppose

$$c_1 f_1 + c_2 f_2 + c_3 f_3 + c_4 f_4 = 0.$$

Then

$$(c_1 f_1 + c_2 f_2 + c_3 f_3 + c_4 f_4)(x_1, x_2) = (0, 0)$$

for all $(x_1, x_2) \in \mathbb{R}^2$. Since

$$(c_1 f_1 + c_2 f_2 + c_3 f_3 + c_4 f_4)(x_1, x_2)$$

$$= c_1 f_1(x_1, x_2) + c_2 f_2(x_1, x_2) + c_3 f_3(x_1, x_2) + c_4 f_4(x_1, x_2)$$

$$= c_1(x_1, 0) + c_2(x_2, 0) + c_3(0, x_1) + c_4(0, x_2)$$

$$= (c_1 x_1 + c_2 x_2, c_3 x_1 + c_4 x_2),$$

we see that $c_1 f_1 + c_2 f_2 + c_3 f_3 + c_4 f_4 = 0$ only if

$$(c_1 x_1 + c_2 x_2, c_3 x_1 + c_4 x_2) = (0, 0)$$

for all $(x_1, x_2) \in \mathbb{R}^2$. But if this equation is to hold for all (x_1, x_2), then, in particular, it must hold for $(x_1, x_2) = (1, 0)$ and for $(x_1, x_2) = (0, 1)$. Substituting these values of x_1 and x_2 into the above equation yields

$$(c_1, c_3) = (0, 0) \quad \text{and} \quad (c_2, c_4) = (0, 0).$$

Hence $c_1 f_1 + c_2 f_2 + c_3 f_3 + c_4 f_4 = 0$ only when $(c_1, c_2, c_3, c_4) = (0, 0, 0, 0)$. This shows that $\{f_1, f_2, f_3, f_4\}$ is linearly independent.

It is easy to check that $\{f_1, f_2, f_3, f_4\}$ also spans $L(\mathbb{R}^2, \mathbb{R}^2)$. If $f \in L(\mathbb{R}^2, \mathbb{R}^2)$, then there exist real numbers $a, b, c,$ and d such that

$$f(x_1, x_2) = (ax_1 + bx_2, cx_1 + dx_2)$$

$$= a(x_1, 0) + b(x_2, 0) + c(0, x_1) + d(0, x_2)$$

$$= (af_1 + bf_2 + cf_3 + df_4)(x_1, x_2).$$

Hence

$$f = af_1 + bf_2 + cf_3 + df_4,$$

for some $a, b, c, d \in \mathbb{R}$.

Therefore $\{f_1, f_2, f_3, f_4\}$ is a basis for $L(\mathbb{R}^2, \mathbb{R}^2)$.

Notice that the linear map f_1 is orthogonal projection onto the x_1-axis, f_4 is orthogonal projection onto the x_2-axis, f_2 is reflection in the line $x_1 = x_2$ followed by f_1, and f_3 is reflection in the line $x_1 = x_2$ followed by f_4 (see Figure 6.1). ■

We know from Section 4.4 that each subspace of $\mathbb{R}^n$ has many different bases, but that all of these bases contain the same number of vectors. The number of vectors in any basis for a subspace of $\mathbb{R}^n$ is the dimension of that subspace.

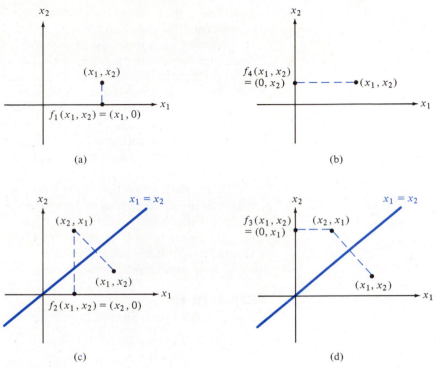

FIGURE 6.1

The set $\{f_1, f_2, f_3, f_4\}$ is a basis for $L(\mathbb{R}^2, \mathbb{R}^2)$, where

(a) f_1 is orthogonal projection onto the x_1-axis
(b) f_4 is orthogonal projection onto the x_2-axis
(c) f_2 is reflection in the line $x_1 = x_2$ followed by f_1
(d) f_3 is reflection in the line $x_1 = x_2$ followed by f_4

We shall now discuss the concept of dimension for arbitrary vector spaces. The key facts that are needed are contained in the following theorem.

Theorem 2 *Let V be a vector space that has a finite basis $\{v_1, \ldots, v_k\}$ containing k vectors. Then*

  (i) *each spanning set for V contains at least k vectors,*

 (ii) *each linearly independent set in V contains at most k vectors,*

and

(iii) *each basis for V contains exactly k vectors.*

Proof First observe that the argument in the proof of Theorem 1 (Section 4.4) can be used, word for word, to show that every *finite* spanning set for V must contain at least as many vectors as any *finite* linearly independent subset of V.

  Suppose S is a spanning set for V. If S is infinite then certainly S contains more than k vectors. On the other hand, if S is finite then the above observation shows that S must contain at least k vectors, since $\{\mathbf{v}_1, \ldots, \mathbf{v}_k\}$ is linearly independent. This proves (i).

Suppose S is a linearly independent set in V. If S is finite, then the above observation shows that S contains at most k vectors, since $\{\mathbf{v}_1, \ldots, \mathbf{v}_k\}$ spans V. On the other hand, S cannot be infinite because if it were it would contain a finite, linearly independent subset with more than k vectors and this is impossible, as we have just observed. This proves (ii).

Finally, suppose S is a basis for V. Then S must contain at least k vectors, by (i), and it can contain at most k vectors, by (ii). Hence S must contain exactly k vectors. This proves (iii). ∎

If a vector space V has a finite basis, we say that V is *finite dimensional* and we define the *dimension* of V to be the number of elements in that basis. If V ($V \neq \{\mathbf{0}\}$) has no finite basis then we say that V is *infinite dimensional*. The zero vector space $V = \{\mathbf{0}\}$ is said to have dimension zero. Notice from Theorem 2 (ii) that *any vector space that contains an infinite linearly independent set is necessarily infinite dimensional.*

EXAMPLE 10

From Examples 6, 7, 8, and 9 we see that

 (i)   the vector space $\mathcal{P}^n$ of all polynomials of degree less than or equal to n has dimension $n + 1$,
 (ii) the vector space $\mathcal{P}$ of all polynomials is infinite dimensional,
 (iii) the vector space $\mathcal{M}_{2\times 2}$ of 2×2 matrices has dimension 4, and
 (iv) the vector space $L(\mathbb{R}^2, \mathbb{R}^2)$ of all linear maps from $\mathbb{R}^2$ to $\mathbb{R}^2$ has dimension 4.

Furthermore,

 (v)   the vector space $\mathbb{R}^\infty$ is infinite dimensional, since it contains the infinite linearly independent set $\{\mathbf{e}_1, \mathbf{e}_2, \ldots\}$ (see Example 3). ∎

Sometimes we may be given a finite spanning set for a vector space and need to find a basis for that space. This can be accomplished by deleting from the given spanning set any vector that is a linear combination of the others to obtain a smaller spanning set, and then repeating this procedure until a linearly independent set is obtained.

EXAMPLE 11

Suppose V is the subspace of $\mathcal{P}^3$ spanned by $S = \{x^3 + 1, (x + 1)^3, x^2 + x\}$. Let us find a basis for V.

First we check to see if S is linearly independent and hence already a basis. Consider the equation

$$c_1(x^3 + 1) + c_2(x + 1)^3 + c_3(x^2 + x) = 0.$$

This equation can be rewritten as

$$(c_1 + c_2)x^3 + (3c_2 + c_3)x^2 + (3c_2 + c_3)x + (c_1 + c_2) = 0.$$

Since a polynomial is equal to the zero polynomial if and only if all its coefficients are zero, this equation is equivalent to the linear system

$$
\begin{aligned}
c_1 + c_2 \quad\quad &= 0 \\
3c_2 + c_3 &= 0 \\
3c_2 + c_3 &= 0 \\
c_1 + c_2 \quad\quad &= 0.
\end{aligned}
$$

The general solution of this system is

$$(c_1, c_2, c_3) = t(\tfrac{1}{3}, -\tfrac{1}{3}, 1), \ t \in \mathbb{R}.$$

In particular, $(c_1, c_2, c_3) = (\tfrac{1}{3}, -\tfrac{1}{3}, 1)$ is a solution so we obtain the dependence relation

$$\tfrac{1}{3}(x^3 + 1) - \tfrac{1}{3}(x + 1)^3 + (x^2 + x) = 0.$$

Thus the given spanning set is linearly dependent.

From the dependence relation we can see that

$$x^2 + x = -\tfrac{1}{3}(x^3 + 1) + \tfrac{1}{3}(x + 1)^3,$$

so we may delete $x^2 + x$ from the given spanning set and still have a spanning set, namely $\{x^3 + 1, (x + 1)^3\}$, for V. Since neither of the polynomials in this set is a scalar multiple of the other, we see that this smaller spanning set, $\{x^3 + 1, (x + 1)^3\}$, is linearly independent and hence is a basis for V. ∎

Sometimes we may be given a linearly independent subset of a finite dimensional vector space V and wish to enlarge this subset to a basis for V. This can be accomplished by adjoining to the given linearly independent set any vector in V that is not a linear combination of the given vectors to obtain a larger linearly independent set, and then repeating this procedure until a spanning set is obtained. In practice, the easiest way to do this is to select the vectors to be adjoined from some convenient basis for V.

EXAMPLE 12
The set

$$S = \left\{ \begin{pmatrix} 1 & 0 \\ 0 & -1 \end{pmatrix}, \begin{pmatrix} 1 & 1 \\ 0 & 1 \end{pmatrix} \right\}$$

is a linearly independent subset of $M_{2 \times 2}$, since neither of these matrices is a scalar multiple of the other. Let us enlarge S to a basis for $M_{2 \times 2}$ by adjoining to it matrices selected from the basis

$$B = \left\{ \begin{pmatrix} 1 & 0 \\ 0 & 0 \end{pmatrix}, \begin{pmatrix} 0 & 1 \\ 0 & 0 \end{pmatrix}, \begin{pmatrix} 0 & 0 \\ 1 & 0 \end{pmatrix}, \begin{pmatrix} 0 & 0 \\ 0 & 1 \end{pmatrix} \right\}.$$

The first matrix in B, $\begin{pmatrix} 1 & 0 \\ 0 & 0 \end{pmatrix}$, is not a linear combination of matrices of S because the equation

$$c_1 \begin{pmatrix} 1 & 0 \\ 0 & -1 \end{pmatrix} + c_2 \begin{pmatrix} 1 & 1 \\ 0 & 1 \end{pmatrix} = \begin{pmatrix} 1 & 0 \\ 0 & 0 \end{pmatrix}$$

is equivalent to the linear system

$$c_1 + c_2 = 1$$
$$c_2 = 0$$
$$0 = 0$$
$$-c_1 + c_2 = 0,$$

and this system has no solution. Hence we may adjoin the matrix $\begin{pmatrix} 1 & 0 \\ 0 & 0 \end{pmatrix}$ to S to obtain the linearly independent set

$$S_1 = \{ \begin{pmatrix} 1 & 0 \\ 0 & -1 \end{pmatrix}, \begin{pmatrix} 1 & 1 \\ 0 & 1 \end{pmatrix}, \begin{pmatrix} 1 & 0 \\ 0 & 0 \end{pmatrix} \}.$$

The set S_1 cannot be a basis for $\mathcal{M}_{2\times 2}$ because it contains only three vectors, and the dimension of $\mathcal{M}_{2\times 2}$ is four. Therefore we must adjoin another matrix to S_1. The second matrix in B, $\begin{pmatrix} 0 & 1 \\ 0 & 0 \end{pmatrix}$, is a linear combination of matrices in S_1,

$$\begin{pmatrix} 0 & 1 \\ 0 & 0 \end{pmatrix} = 1 \begin{pmatrix} 1 & 0 \\ 0 & -1 \end{pmatrix} + 1 \begin{pmatrix} 1 & 1 \\ 0 & 1 \end{pmatrix} + (-2) \begin{pmatrix} 1 & 0 \\ 0 & 0 \end{pmatrix},$$

so we cannot adjoin it to S_1 and still have an independent set. However, the third matrix in B, $\begin{pmatrix} 0 & 0 \\ 1 & 0 \end{pmatrix}$, is not a linear combination of matrices in S_1 because the equation

$$c_1 \begin{pmatrix} 1 & 0 \\ 0 & -1 \end{pmatrix} + c_2 \begin{pmatrix} 1 & 1 \\ 0 & 1 \end{pmatrix} + c_3 \begin{pmatrix} 1 & 0 \\ 0 & 0 \end{pmatrix} = \begin{pmatrix} 0 & 0 \\ 1 & 0 \end{pmatrix}$$

has no solution. Hence we can adjoin it to S_1 to get the linearly independent set

$$S_2 = \{ \begin{pmatrix} 1 & 0 \\ 0 & -1 \end{pmatrix}, \begin{pmatrix} 1 & 1 \\ 0 & 1 \end{pmatrix}, \begin{pmatrix} 1 & 0 \\ 0 & 0 \end{pmatrix}, \begin{pmatrix} 0 & 0 \\ 1 & 0 \end{pmatrix} \}.$$

This set S_2 is a basis for $\mathcal{M}_{2\times 2}$. ∎

REMARK   The reason that the set S_2 in Example 12 is a basis for $\mathcal{M}_{2\times 2}$ is that S_2 is a linearly independent subset of $\mathcal{M}_{2\times 2}$ that contains four vectors. Since $\mathcal{M}_{2\times 2}$ has dimension four, any subset larger than S_2 must be linearly dependent.

It follows that every vector in $\mathcal{M}_{2\times 2}$ must be a linear combination of elements of S_2. Thus S_2 spans $\mathcal{M}_{2\times 2}$ and hence is a basis.

This argument generalizes to prove the following useful theorem.

Theorem 3   *Let V be a finite dimensional vector space.*

(i)   *If S is any linearly independent subset of V that contains n vectors, where n is the dimension of V, then S must span V and hence be a basis for V.*

(ii)   *If S is any spanning set for V that contains n vectors, where n is the dimension of V, then S must be linearly independent and hence a basis for V.*

Proof

(i)   If S did not span V then we could adjoin to S any vector that is not in $\mathcal{L}(S)$ to obtain a linearly independent set containing more than n vectors. But that is impossible, by Theorem 2.

(ii)   If S were not linearly independent then we could delete vectors from S to obtain a basis for V containing fewer than n vectors. But that is impossible, by Theorem 2.   ■

Given an ordered basis $\mathbf{B} = (\mathbf{v}_1, \ldots, \mathbf{v}_n)$ for a finite dimensional vector space V we can write each $\mathbf{v} \in V$ as a linear combination

$$\mathbf{v} = c_1\mathbf{v}_1 + \cdots + c_n\mathbf{v}_n.$$

There is only one ordered n-tuple $(c_1, \ldots, c_n)$ of real numbers that satisfies this equation (see Exercise 27). The n-tuple $(c_1, \ldots, c_n)$ is the **B-*coordinate n-tuple*** of $\mathbf{v}$, and the real numbers $c_1, \ldots, c_n$ are the **B-*coordinates*** of $\mathbf{v}$.

EXAMPLE 13
Let $V = \mathcal{P}^3$ and let $\mathbf{B} = (1, x, x^2, x^3)$. If $p(x) = 2x^3 + x^2 - 3$, then we can write

$$p(x) = (-3)1 + 0x + 1x^2 + 2x^3$$

and see that the **B**-coordinate 4-tuple of $p(x)$ is $(-3, 0, 1, 2)$.   ■

EXERCISES

1.   Which of the following sets are spanning sets for $\mathcal{P}^2$?
   (a) $\{1, x + 1, (x + 1)^2\}$
   (b) $\{x^2 + x + 1, x^2 - x + 1, x^2 + 1\}$
   (c) $\{0, x, x^2\}$
   (d) $\{1, x - 1, (x - 1)^2\}$
   (e) $\{x, 2x, 3x, 4x\}$
   (f) $\mathcal{P}^2$

2. Which of the following sets are spanning sets for $\mathcal{P}^3$?
   (a) $\{1, x + 1, (x + 1)^2, (x + 1)^3\}$
   (b) $\{1, x - 1, (x - 1)^2, (x - 1)^3\}$
   (c) $\{0, x, x^2, x^3\}$
   (d) $\{x^3 + 1, x^3 - 1, x^2 + x + 1, 1\}$
   (e) $\{x, x^2, x^3, x^3 + x^2 + x, x^3 - x^2 + x\}$
   (f) $\mathcal{P}^3$

3. Which of the following sets are spanning sets for $\mathcal{M}_{2 \times 2}$?
   (a) $\left\{ \begin{pmatrix} 1 & 0 \\ 1 & 1 \end{pmatrix}, \begin{pmatrix} 0 & 1 \\ 1 & 1 \end{pmatrix}, \begin{pmatrix} 1 & 1 \\ 0 & 1 \end{pmatrix}, \begin{pmatrix} 1 & 1 \\ 1 & 0 \end{pmatrix} \right\}$
   (b) $\left\{ \begin{pmatrix} 1 & 0 \\ 0 & 1 \end{pmatrix}, \begin{pmatrix} 1 & 0 \\ 0 & 2 \end{pmatrix}, \begin{pmatrix} 1 & 0 \\ 0 & 3 \end{pmatrix}, \begin{pmatrix} 1 & 0 \\ 0 & 4 \end{pmatrix} \right\}$
   (c) $\left\{ \begin{pmatrix} 1 & 0 \\ 0 & 0 \end{pmatrix}, \begin{pmatrix} 1 & 1 \\ 0 & 0 \end{pmatrix}, \begin{pmatrix} 1 & 1 \\ 1 & 0 \end{pmatrix}, \begin{pmatrix} 1 & 1 \\ 1 & 1 \end{pmatrix} \right\}$
   (d) $\left\{ \begin{pmatrix} 1 & 0 \\ 0 & 0 \end{pmatrix}, \begin{pmatrix} 1 & 0 \\ 0 & 1 \end{pmatrix}, \begin{pmatrix} 0 & 2 \\ 1 & 0 \end{pmatrix}, \begin{pmatrix} 0 & 3 \\ 1 & 0 \end{pmatrix} \right\}$
   (e) $\left\{ \begin{pmatrix} 1 & 1 \\ 1 & 1 \end{pmatrix}, \begin{pmatrix} 1 & -1 \\ -1 & 1 \end{pmatrix}, \begin{pmatrix} -1 & 1 \\ 1 & -1 \end{pmatrix}, \begin{pmatrix} -1 & 0 \\ 0 & 1 \end{pmatrix} \right\}$
   (f) $\{A \in \mathcal{M}_{2 \times 2} | \det A = 0\}$

4. Which of the sets $\{f_1, f_2, f_3, f_4\}$ are spanning sets for $L(\mathbb{R}^2, \mathbb{R}^2)$?
   (a) $f_1(x_1, x_2) = (x_1, -x_2), f_2(x_1, x_2) = (x_1, x_2),$
   $f_3(x_1, x_2) = (x_1 + 2x_2, x_2), f_4(x_1, x_2) = (2x_1, x_2)$
   (b) $f_1(x_1, x_2) = (x_1 - x_2, x_1 + x_2), f_2(x_1, x_2) = (x_1 + x_2, x_2),$
   $f_3(x_1, x_2) = (2x_1 - x_2, x_1 + x_2), f_4(x_1, x_2) = (x_1 + x_2, x_1 + x_2)$
   (c) $f_1(x_1, x_2) = (x_1, x_2), f_2(x_1, x_2) = (x_1, 2x_2),$
   $f_3(x_1, x_2) = (x_1, 3x_2), f_4(x_1, x_2) = (x_1, 4x_2)$

5. Which of the sets in Exercise 1 are linearly independent?

6. Which of the sets in Exercise 2 are linearly independent?

7. Which of the sets in Exercise 3 are linearly independent?

8. Which of the sets in Exercise 4 are linearly independent?

9. Show that the set $\{1, 1 + x, 1 + x + x^2, 1 + x + x^2 + x^3, \ldots\}$ is a basis for $\mathcal{P}$.

10. (a) Find a basis for $\mathcal{M}_{2 \times 3}$. What is the dimension of $\mathcal{M}_{2 \times 3}$?
    (b) Find a basis for $\mathcal{M}_{m \times n}$. What is the dimension of $\mathcal{M}_{m \times n}$?

11. Which of the following polynomials are in $\mathcal{L}(x^2 + x + 1, x^2 - 1, x + 2)$?
    (a) x^3 (b) $x^2 + 3x + 5$ (c) $x + 1$

12. Which of the following matrices are in $\mathcal{L}\left(\begin{pmatrix} 1 & 0 \\ 0 & 1 \end{pmatrix}, \begin{pmatrix} 1 & 0 \\ 0 & -1 \end{pmatrix} \right)$?

    (a) $\begin{pmatrix} 1 & 0 \\ 0 & 0 \end{pmatrix}$ (b) $\begin{pmatrix} 0 & 1 \\ 0 & 0 \end{pmatrix}$ (c) $\begin{pmatrix} 0 & 0 \\ 0 & 1 \end{pmatrix}$

13. Find a basis for, and the dimension of, each of the following subspaces of $\mathcal{P}$.
    (a) $\mathcal{L}(x^2 + x + 1, x^2 - x + 1, x^2 + 1)$
    (b) $\mathcal{L}(1, x, 2x, 3x, 4x, 5x)$
    (c) $\mathcal{L}(x, x + x^2, x - x^2, x + x^3, x - x^3)$
    (d) $\mathcal{L}(x + 1, (x + 1)^2, x^2 + 2x + 1, (x + 1)^3, x^3 + 3x^2 + 3x + 1)$
    (e) $\mathcal{L}(1, x, -x, x^2, -x^2, x^3 + x^2 + x + 1)$

14. Find a basis for $\mathcal{P}^2$ that contains the given linearly independent set.
    (a) $\{1, x - 1\}$ (b) $\{x^2 - 1, x^2 + 2x + 1\}$ (c) $\{2, x^2 + 2\}$

15. Find a basis for $\mathcal{M}_{2\times 2}$ that contains the given linearly independent set.

    (a) $\{\begin{pmatrix} 1 & 1 \\ 1 & 1 \end{pmatrix}, \begin{pmatrix} 1 & 0 \\ 0 & 1 \end{pmatrix}\}$

    (b) $\{\begin{pmatrix} 1 & 0 \\ 0 & 1 \end{pmatrix}, \begin{pmatrix} 1 & 1 \\ 0 & 0 \end{pmatrix}\}$

    (c) $\{\begin{pmatrix} 1 & 0 \\ 0 & 2 \end{pmatrix}, \begin{pmatrix} 1 & 2 \\ 0 & 1 \end{pmatrix}, \begin{pmatrix} 1 & 0 \\ 2 & 1 \end{pmatrix}\}$

16. Find the dimension of each of the following vector spaces.
    (a) the subspace of $\mathcal{M}_{2\times 2}$ consisting of all diagonal matrices
    (b) the subspace of $\mathcal{M}_{n\times n}$ consisting of all diagonal matrices

17. Find a basis for $L(\mathbb{R}^2, \mathbb{R}^2)$ that contains the horizontal shear $f(x_1, x_2) = (x_1 + 5x_2, x_2)$.

18. Let $V = \mathcal{P}^2$ and let $\mathbf{B} = (1, (x - 1), (x - 1)^2)$.
    (a) Verify that $\mathbf{B}$ is an ordered basis for $\mathcal{P}^2$.
    (b) Find the $\mathbf{B}$-coordinate triple of each of the following polynomials:
        (i) x^2 (ii) $x^2 + x + 1$ (iii) $x^2 - 2x$

19. Find the $\mathbf{B}$-coordinate 4-tuple of A where $\mathbf{B}$ is the ordered basis

$$\left(\begin{pmatrix} 1 & 0 \\ 0 & 0 \end{pmatrix}, \begin{pmatrix} 0 & 1 \\ 0 & 0 \end{pmatrix}, \begin{pmatrix} 0 & 0 \\ 1 & 0 \end{pmatrix}, \begin{pmatrix} 0 & 0 \\ 0 & 1 \end{pmatrix} \right)$$

    for $\mathcal{M}_{2\times 2}$ and $A = $

    (a) $\begin{pmatrix} 1 & 0 \\ 0 & 1 \end{pmatrix}$ (b) $\begin{pmatrix} 0 & 1 \\ -1 & 0 \end{pmatrix}$ (c) $\begin{pmatrix} 2 & 3 \\ -1 & 4 \end{pmatrix}$

20. Let V be the subset of $\mathbb{R}^\infty$ consisting of all vectors $(a_1, a_2, \ldots)$ such that $a_i = 0$ for all but a finite number of i.
    (a) Show that V is a subspace of $\mathbb{R}^\infty$.
    (b) Show that $V = \mathcal{L}(\mathbf{e}_1, \mathbf{e}_2, \ldots)$, where $\mathbf{e}_i = (0, \ldots, 0, 1, 0, \ldots)$, and where the 1 appears as the ith entry.
    (c) Show that V is infinite dimensional.

21. Show that the set $\{f_{ij} | 1 \leq i \leq m, 1 \leq j \leq n\}$ is a basis for $L(\mathbb{R}^n, \mathbb{R}^m)$, where
    $$f_{ij}(x_1, \ldots, x_n) = (0, \ldots, 0, \overset{\text{ith entry}}{x_j}, 0, \ldots, 0).$$ What is the dimension of $L(\mathbb{R}^n, \mathbb{R}^m)$?

22. Verify that if S is any subset of a vector space V, then $\mathcal{L}(S)$ is a subspace of V.

23. Show that if V is a finite dimensional vector space and W is a subspace of V, then W is finite dimensional and has dimension less than or equal to the dimension of V.

24. Let S be a linearly independent subset of a vector space V and let $\mathbf{v} \in V$. Prove that $S \cup \{\mathbf{v}\}$ is linearly dependent if and only if $\mathbf{v} \in \mathcal{L}(S)$.

25. Explain why every minimal spanning set for a vector space V is a basis for V. That is, explain why, if S spans V and no proper subset of S spans V, then S must be a basis for V.

26. Explain why every maximal linearly independent set in a vector space V is a basis for V. That is, explain why, if S is a linearly independent subset of V and every larger subset of V is linearly dependent, then S must be a basis for V.

27. Let $\{\mathbf{v}_1, \ldots, \mathbf{v}_n\}$ be a basis for the finite dimensional vector space V. Show that, if

$$c_1\mathbf{v}_1 + \cdots + c_n\mathbf{v}_n = d_1\mathbf{v}_1 + \cdots + d_n\mathbf{v}_n,$$

then $c_1 = d_1, \ldots, c_n = d_n$.

28. Let V be a vector space and let S be a subset of V. Show that if W is any subspace of V that contains S, then W contains $\mathcal{L}(S)$. [Thus $\mathcal{L}(S)$ is the *smallest* subspace of V that contains S.]

29. (a) Let $p(x)$ and $q(x)$ be polynomials such that

$$\det\begin{pmatrix} p(0) & q(0) \\ p(1) & q(1) \end{pmatrix} \neq 0.$$

Show that $\{p(x), q(x)\}$ is linearly independent.

(b) Let $p_1(x), \ldots, p_k(x)$ be polynomials. Suppose

$$\det\begin{pmatrix} p_1(a_1) & p_2(a_1) & \cdots & p_k(a_1) \\ p_1(a_2) & p_2(a_2) & \cdots & p_k(a_2) \\ & & \cdots & \\ p_1(a_k) & p_2(a_k) & \cdots & p_k(a_k) \end{pmatrix} \neq 0$$

for some real numbers $a_1, a_2, \ldots, a_k$. Show that $\{p_1(x), \ldots, p_k(x)\}$ is linearly independent.

30. Suppose $\mathbf{B} = (\mathbf{v}_1, \ldots, \mathbf{v}_n)$ and $\mathbf{B}' = (\mathbf{w}_1, \ldots, \mathbf{w}_n)$ are two ordered bases for the same vector space V. Show that, for each $\mathbf{v} \in V$, the $\mathbf{B}$ and $\mathbf{B}'$ coordinate n-tuples of $\mathbf{v}$ are related by the equations

$$M_{\mathbf{B}}(\mathbf{v}) = HM_{\mathbf{B}'}(\mathbf{v}), \qquad M_{\mathbf{B}'}(\mathbf{v}) = H^{-1}M_{\mathbf{B}}(\mathbf{v})$$

where H is the matrix whose column vectors are the $\mathbf{B}$-coordinate n-tuples of the $\mathbf{B}'$-basis vectors $\mathbf{w}_1, \ldots, \mathbf{w}_n$, and the matrices $M_{\mathbf{B}}(\mathbf{v})$ and $M_{\mathbf{B}'}(\mathbf{v})$ are the $\mathbf{B}$- and $\mathbf{B}'$-coordinate n-tuples of $\mathbf{v}$, written vertically. [*Hint:* Imitate the proof of Theorem 1 in Section 5.2.]

6.3

Linear Maps

In Chapter 2 we studied linear maps from $\mathbb{R}^n$ to $\mathbb{R}^m$. Now we shall study linear maps between arbitrary vector spaces. Recall that linear maps from $\mathbb{R}^n$ to $\mathbb{R}^m$ can be described as those functions from $\mathbb{R}^n$ to $\mathbb{R}^m$ that preserve both vector addition and scalar multiplication. We shall use this characterization to define linearity for maps between vector spaces in general.

Let V and W be any two vector spaces. A **linear map** from V to W is a function $f: V \to W$ that satisfies

(i) $f(\mathbf{u} + \mathbf{v}) = f(\mathbf{u}) + f(\mathbf{v})$, for all $\mathbf{u}$ and $\mathbf{v}$ in V, and

(ii) $f(c\mathbf{v}) = cf(\mathbf{v})$, for all $c \in \mathbb{R}$ and all $\mathbf{v} \in V$.

Property (i) states that *f preserves addition* and property (ii) states that *f preserves scalar multiplication*.

We have already seen some examples of linear maps in Chapter 2. Let us look now at some others.

EXAMPLE 1

Let $f:\mathcal{P} \to \mathbb{R}$ be defined by $f(p(x)) = p(0)$. Then f preserves addition because

$$f(p(x) + q(x)) = p(0) + q(0) = f(p(x)) + f(q(x))$$

and f preserves scalar multiplication because

$$f(cp(x)) = cp(0) = cf(p(x)).$$

Therefore f is a linear map from $\mathcal{P}$ to $\mathbb{R}$. ∎

EXAMPLE 2

More generally, if a is any fixed real number, then the map $f:\mathcal{P} \to \mathbb{R}$ defined by

$$f(p(x)) = p(a)$$

is a linear map (see Exercise 9). This linear map is called an ***evaluation map*** since it evaluates the polynomial $p(x)$ at the real number a. ∎

EXAMPLE 3

Let $f:\mathbb{R}^{n+1} \to \mathcal{P}^n$ be defined by

$$f(a_1, a_2, \ldots, a_{n+1}) = a_1 + a_2x + \cdots + a_{n+1}x^n.$$

Then f preserves addition because

$$f((a_1, \ldots, a_{n+1}) + (b_1, \ldots, b_{n+1})) = f((a_1 + b_1, \ldots, a_{n+1} + b_{n+1}))$$

$$= (a_1 + b_1) + (a_2 + b_2)x + \cdots + (a_{n+1} + b_{n+1})x^n$$

$$= (a_1 + a_2x + \cdots + a_{n+1}x^n) + (b_1 + b_2x + \cdots + b_{n+1}x^{n+1})$$

$$= f(a_1, \ldots, a_{n+1}) + f(b_1, \ldots, b_{n+1}).$$

Also, f preserves scalar multiplication because

$$f(c(a_1, \ldots, a_{n+1})) = f(ca_1, \ldots, ca_{n+1})$$

$$= ca_1 + ca_2x + \cdots + ca_{n+1}x^{n+1}$$

$$= c(a_1 + a_2x + \cdots + a_{n+1}x^{n+1})$$

$$= cf(a_1, \ldots, a_{n+1}).$$

Thus f is a linear map from $\mathbb{R}^{n+1}$ to $\mathcal{P}^n$. Notice that this linear map is both one-to-one and onto. ∎

EXAMPLE 4

Let $f:\mathcal{P} \to \mathbb{R}^{\infty}$ be defined by

$$f(a_0 + a_1x + \cdots + a_nx^n) = (a_0, a_1, \ldots, a_n, 0, 0, 0, \ldots).$$

Then f is a linear map (see Exercise 20). Notice that this linear map is one-to-one but it is not onto. ∎

Linear maps from $\mathbb{R}^n$ to $\mathbb{R}^m$ always send $\mathbf{0}$ to $\mathbf{0}$ and they preserve linear combinations. Linear maps between arbitrary vector spaces also have these properties.

Theorem 1 *Let V and W be any two vector spaces and let $f:V \to W$ be a linear map. Then*

(i) $f(\mathbf{0}) = \mathbf{0}$, *and*

(ii) $f(c_1\mathbf{v}_1 + \cdots + c_k\mathbf{v}_k) = c_1f(\mathbf{v}_1) + \cdots + c_kf(\mathbf{v}_k)$

for all $\mathbf{v}_1, \ldots, \mathbf{v}_k$ in V and all real numbers $c_1, \ldots, c_k$.

Proof   To prove (i) notice that

$$f(\mathbf{0}) = f(\mathbf{0} + \mathbf{0}) = f(\mathbf{0}) + f(\mathbf{0}).$$

If we add $-f(\mathbf{0})$ to both sides of this equation we get $\mathbf{0} = f(\mathbf{0})$.

Property (ii) follows from a direct calculation:

$$f(c_1\mathbf{v}_1 + \cdots + c_k\mathbf{v}_k) = f(c_1\mathbf{v}_1) + f(c_2\mathbf{v}_2 + \cdots + c_k\mathbf{v}_k)$$

$$= \cdots$$

$$= f(c_1\mathbf{v}_1) + f(c_2\mathbf{v}_2) + \cdots + f(c_k\mathbf{v}_k)$$

$$= c_1f(\mathbf{v}_1) + c_2f(\mathbf{v}_2) + \cdots + c_kf(\mathbf{v}_k). \quad \blacksquare$$

Theorem 1 is sometimes useful in showing that a map is *not* linear.

EXAMPLE 5

The map $f:\mathbb{R}^\infty \to \mathbb{R}^\infty$ defined by

$$f(x_1, x_2, \ldots) = (x_1 + 1, x_2 + 1, \ldots)$$

is not linear, because $f(0, 0, \ldots) = (1, 1, \ldots) \neq (0, 0, \ldots)$.

On the other hand, the map $\det: \mathcal{M}_{2 \times 2} \to \mathbb{R}$ does send the zero matrix to the number $0 \in \mathbb{R}$, since

$$\det\begin{pmatrix} 0 & 0 \\ 0 & 0 \end{pmatrix} = 0 - 0 = 0,$$

but it nevertheless is not linear since it does not preserve scalar multiplication. For example,

$$\det\left((-1)\begin{pmatrix} 1 & 0 \\ 0 & 1 \end{pmatrix}\right) = \det\begin{pmatrix} -1 & 0 \\ 0 & -1 \end{pmatrix} = +1$$

whereas

$$(-1)\det\begin{pmatrix} 1 & 0 \\ 0 & 1 \end{pmatrix} = -1. \quad \blacksquare$$

The next theorem tells us that each linear map from V to W is uniquely determined by its values on any basis for V. For simplicity we consider first the case when V is finite dimensional.

Theorem 2 *Let V and W be vector spaces, with V finite dimensional. Let $\{\mathbf{v}_1, \ldots, \mathbf{v}_n\}$ be any basis for V and let $\mathbf{w}_1, \ldots, \mathbf{w}_n$ be any n vectors in W. Then there is one and only one linear map $f:V \to W$ for which*

$$f(\mathbf{v}_1) = \mathbf{w}_1, f(\mathbf{v}_2) = \mathbf{w}_2, \ldots, f(\mathbf{v}_n) = \mathbf{w}_n.$$

Proof Define $f:V \to W$ as follows. Each vector $\mathbf{v} \in V$ is uniquely expressible as a linear combination

$$\mathbf{v} = c_1\mathbf{v}_1 + \cdots + c_n\mathbf{v}_n$$

where $c_1, \ldots, c_n$ are real numbers. Set

$$f(\mathbf{v}) = c_1\mathbf{w}_1 + \cdots + c_n\mathbf{w}_n.$$

If

$$\mathbf{u} = b_1\mathbf{v}_1 + \cdots + b_n\mathbf{v}_n$$

and

$$\mathbf{v} = c_1\mathbf{v}_1 + \cdots + c_n\mathbf{v}_n,$$

then

$$
\begin{aligned}
f(\mathbf{u} + \mathbf{v}) &= f((b_1\mathbf{v}_1 + \cdots + b_n\mathbf{v}_n) + (c_1\mathbf{v}_1 + \cdots + c_n\mathbf{v}_n)) \\
&= f((b_1 + c_1)\mathbf{v}_1 + \cdots + (b_n + c_n)\mathbf{v}_n) \\
&= (b_1 + c_1)\mathbf{w}_1 + \cdots + (b_n + c_n)\mathbf{w}_n \\
&= (b_1\mathbf{w}_1 + \cdots + b_n\mathbf{w}_n) + (c_1\mathbf{w}_1 + \cdots + c_n\mathbf{w}_n) \\
&= f(\mathbf{u}) + f(\mathbf{v}).
\end{aligned}
$$

Thus f preserves addition. A similar calculation shows that f preserves scalar multiplication. Therefore f is a linear map and for each i

$$
\begin{aligned}
f(\mathbf{v}_i) &= f(0\mathbf{v}_1 + \cdots + 0\mathbf{v}_{i-1} + 1\mathbf{v}_i + 0\mathbf{v}_{i+1} + \cdots + 0\mathbf{v}_n) \\
&= 0\mathbf{w}_1 + \cdots + 0\mathbf{w}_{i-1} + 1\mathbf{w}_i + 0\mathbf{w}_{i+1} + \cdots + 0\mathbf{w}_n = \mathbf{w}_i.
\end{aligned}
$$

We still must check that f is the *only* linear map from V to W that sends $\mathbf{v}_i$ to $\mathbf{w}_i$ for each i. Suppose $g:V \to W$ is another linear map such that

$$g(\mathbf{v}_1) = \mathbf{w}_1, \ldots, g(\mathbf{v}_n) = \mathbf{w}_n.$$

Then for each vector $\mathbf{v} \in V$, $\mathbf{v} = c_1\mathbf{v}_1 + \cdots + c_n\mathbf{v}_n$, we have

$$
\begin{aligned}
g(\mathbf{v}) &= g(c_1\mathbf{v}_1 + \cdots + c_n\mathbf{v}_n) \\
&= c_1 g(\mathbf{v}_1) + \cdots + c_n g(\mathbf{v}_n) \\
&= c_1\mathbf{w}_1 + \cdots + c_n\mathbf{w}_n \\
&= f(\mathbf{v}),
\end{aligned}
$$

hence $g = f$ and the theorem is proved. ∎

The following theorem extends Theorem 2 to the case where V is infinite dimensional. The proof is quite similar to the proof of Theorem 2. We shall not carry out the details here.

Theorem 3   *Let V and W be any two vector spaces. Suppose B is a basis for V, and suppose $h:B \rightarrow W$ is any function that assigns to each vector $\mathbf{v}$ in B a vector $h(\mathbf{v})$ in W. Then there is one and only one linear map $f:V \rightarrow W$ such that $f(\mathbf{v}) = h(\mathbf{v})$ for each $\mathbf{v} \in B$.*

EXAMPLE 6

The set $\{1, x, x^2, x^3, \ldots\}$ is a basis for $\mathcal{P}$. Theorem 3 tells us that there is a unique linear map $D:\mathcal{P} \rightarrow \mathcal{P}$ such that

$$D(1) = 0$$

and

$$D(x^n) = nx^{n-1}$$

for each integer $n \geq 1$. Using the knowledge of what D does to the polynomials $1, x, x^2, \ldots$, it is easy to compute what D does to any polynomial. We see, for example, that

$$D(3x^5 - 5x^2 + 1) = 3D(x^5) - 5D(x^2) + D(1)$$
$$= 3(5x^4) - 5(2x) + 0$$
$$= 15x^4 - 10x$$

and that

$$D((2x + 1)^3) = D(8x^3 + 12x^2 + 6x + 1)$$
$$= 8D(x^3) + 12D(x^2) + 6D(x) + D(1)$$
$$= 24x^2 + 24x + 6.$$

The polynomial $D(p(x))$, $p(x) \in \mathcal{P}$, is called the **derivative** of the polynomial $p(x)$. The linear map D, which assigns to each polynomial its derivative, has many important applications (see for example Exercises 15 and 16). This linear map is studied in great detail and in a more general setting in calculus courses. ∎

REMARK   Notice that the derivative of each polynomial $p(x)$ of degree $n \geq 1$ is a polynomial of degree $n - 1$. Hence, if we are interested only in polynomials of degree less than or equal to n, we may view D as a linear map from $\mathcal{P}^n$ to $\mathcal{P}^{n-1}$ or, if we wish, from $\mathcal{P}^n$ to $\mathcal{P}^n$.

The **kernel** of a linear map $f:V \rightarrow W$ is the solution set of the equation $f(\mathbf{v}) = \mathbf{0}$. In other words, the kernel of f is the set of all vectors $\mathbf{v}$ in V that are mapped by f to the zero vector in W. The kernel of each linear map $f:V \rightarrow W$ is a subspace of V (see Exercise 21).

EXAMPLE 7

Let $f:\mathcal{P} \to \mathbb{R}$ be the evaluation map $f(p(x)) = p(1)$. Then $x - 1$, $x^2 - 1$, and $x^2 - 3x + 2$ are all in the kernel of f because the value of each of these polynomials at $x = 1$ is 0:

$$1 - 1 = 0$$
$$1^2 - 1 = 0$$

and

$$1^2 - 3 \cdot 1 + 2 = 0.$$

Recall from algebra that if $p(x)$ is a polynomial and r is any real number, then $p(r) = 0$ if and only if $x - r$ is a factor of $p(x)$. That is, $p(r) = 0$ if and only if $p(x) = (x - r)q(x)$ for some polynomial $q(x)$. Using this fact, we see that $p(x)$ is in the kernel of f if and only if $x - 1$ is a factor of $p(x)$. ∎

EXAMPLE 8

Let $D:\mathcal{P} \to \mathcal{P}$ be the linear map that assigns to each polynomial its derivative (see Example 6). Then

$$D(a_n x^n + \cdots + a_2 x^2 + a_1 x + a_0)$$

$$= a_n D(x^n) + \cdots + a_2 D(x^2) + a_1 D(x) + a_0 D(1)$$

$$= n a_n x^{n-1} + \cdots + 2a_2 x + a_1.$$

Hence

$$D(a_n x^n + \cdots + a_1 x + a_0) = 0$$

if and only if

$$n a_n x^{n-1} + \cdots + 2a_2 x + a_1 = 0x^{n-1} + \cdots + 0x + 0,$$

that is, if and only if $a_n = \cdots = a_2 = a_1 = 0$. Therefore, the only polynomials that are in the kernel of D are the "constant" polynomials, $p(x) = a_0$ for some $a_0 \in \mathbb{R}$. ∎

The *image* of a linear map $f:V \to W$ is the set of all $\mathbf{w} \in W$ such that $\mathbf{w} = f(\mathbf{v})$ for some $\mathbf{v} \in V$. The image of each linear map $f:V \to W$ is a subspace of W (see Exercise 22).

EXAMPLE 9

Let $f:\mathcal{P} \to \mathbb{R}^\infty$ be the linear map

$$f(a_0 + a_1 x + \cdots + a_n x^n) = (a_0, a_1, \ldots, a_n, 0, 0, 0, \ldots),$$

as in Example 4. The image of this linear map is the set of all vectors in $\mathbb{R}^\infty$ of the form

$$(a_0, a_1, \ldots, a_n, 0, 0, 0, \ldots).$$

In other words, the image of f is the set of all sequences $(a_0, a_1, a_2, \ldots)$ such that, for some integer n, $a_k = 0$ for all $k > n$. ∎

EXAMPLE 10

Let $D:\mathcal{P} \to \mathcal{P}$ be the linear map that assigns to each polynomial its derivative (see Example 6). Every polynomial in $\mathcal{P}$ is of the form

$$p(x) = a_n x^n + \cdots + a_1 x + a_0.$$

Since

$$a_n x^n + \cdots + a_1 x + a_0 = D\left(\frac{a_{n+1}}{n+1} x^{n+1} + \cdots + \frac{a_1}{2} x^2 + a_0 x\right),$$

we see that every polynomial in $\mathcal{P}$ is in the image of D. In other words, the image of D is the whole vector space $\mathcal{P}$. ∎

REMARK   The reader who has studied calculus will notice some familiar concepts appearing here. The derivative D is the same derivative that you studied in calculus, and the polynomial that appeared in Example 10 is an integral:

$$\frac{a_{n+1}}{n+1} x^{n+1} + \cdots + \frac{a_1}{2} x^2 + a_0 x = \int_0^x (a_n t^n + \cdots + a_1 t + a_0)\, dt.$$

The concepts of linear algebra appear repeatedly in calculus. For example, the set $\mathcal{F}$, consisting of all real valued functions with domain $\mathbb{R}$, forms a vector space, with addition and scalar multiplication of functions defined in the standard way. The subset $\mathcal{C}$, consisting of those functions in $\mathcal{F}$ that are continuous, is a subspace of $\mathcal{F}$, as is the subset $\mathcal{D}$ consisting of all $f \in \mathcal{F}$ that are differentiable. The map $\int:\mathcal{C} \to \mathcal{D}$, which assigns to each $f \in \mathcal{C}$ the function $F(x) = \int_0^x f(t)dt$, is a linear map, as is the function $D:\mathcal{D} \to \mathcal{F}$ defined by $D(f) = f'$. Notice that $(D \circ \int)(f) = f$ for each $f \in \mathcal{C}$ (this is the fundamental theorem of calculus!) but that D and $\int$ are not inverse to each other. (Why?)

The following theorem states the relationship of the kernel and the image of a linear map to the properties one-to-one and onto. The theorem is proved in this general setting in exactly the same way that it was proved in Section 2.6 (see Theorem 2) for linear maps from $\mathbb{R}^n$ to $\mathbb{R}^m$ so we shall not reproduce the proof here.

Theorem 4   *Let V and W be vector spaces and let $f:V \to W$ be a linear map. Then*

   (i)   *f is one-to-one if and only if its kernel is $\{0\}$, and*
   (ii)   *f is onto if and only if its image is W.*

EXAMPLE 11

From Examples 8 and 10 we see that the linear map $D:\mathcal{P} \to \mathcal{P}$ that assigns to each polynomial its derivative is not one-to-one but that it is onto. ∎

When the vector space V is finite dimensional, there is a nice relationship between the dimension of the kernel and the dimension of the image of any linear map $f:V \rightarrow W$.

Theorem 5 *Let V and W be vector spaces, with V finite dimensional, and let $f:V \rightarrow W$ be a linear map. Then*

$$\dim(\ker f) + \dim(\operatorname{im} f) = \dim V$$

where $\ker f$ *is the kernel of* f, $\operatorname{im} f$ *is the image of* f, *and* $\dim$ *stands for dimension.*

Proof Let $\{v_1, \ldots, v_k\}$ be a basis for $\ker f$. By adjoining vectors we can extend this basis to a basis $\{v_1, \ldots, v_k, v_{k+1}, \ldots, v_n\}$ for V. If we can show that $\{f(v_{k+1}), \ldots, f(v_n)\}$ is a basis for $\operatorname{im} f$, then we will be able to conclude that

$$\dim(\ker f) + \dim(\operatorname{im} f) = k + (n - k) = n = \dim V,$$

as asserted.

So we must show that $\{f(v_{k+1}), \ldots, f(v_n)\}$ is a basis for the image of f. This set certainly spans the image of f since each vector v in V is a linear combination

$$v = c_1 v_1 + \cdots + c_n v_n$$

and hence each vector $f(v)$ in the image of f is a linear combination

$$f(v) = c_1 f(v_1) + \cdots + c_k f(v_k) + c_{k+1} f(v_{k+1}) + \cdots + c_n f(v_n)$$

$$= 0 + \cdots + 0 + c_{k+1} f(v_{k+1}) + \cdots + c_n f(v_n)$$

$$= c_{k+1} f(v_{k+1}) + \cdots + c_n f(v_n).$$

So we need only show that the set $\{f(v_{k+1}), \ldots, f(v_n)\}$ is linearly independent. Suppose we have a dependence relation

$$c_{k+1} f(v_{k+1}) + \cdots + c_n f(v_n) = 0.$$

Then

$$f(c_{k+1} v_{k+1} + \cdots + c_n v_n) = c_{k+1} f(v_{k+1}) + \cdots + c_n f(v_n) = 0,$$

so $c_{k+1} v_{k+1} + \cdots + c_n v_n$ is in the kernel of f. Since $\{v_1, \ldots, v_k\}$ is a basis for the kernel of f, it follows that

$$c_{k+1} v_{k+1} + \cdots + c_n v_n = c_1 v_1 + \cdots + c_k v_k$$

for some $c_1, \ldots, c_k$. But then we would have

$$(-c_1) v_1 + \cdots + (-c_k) v_k + c_{k+1} v_{k+1} + \cdots + c_n v_n = 0.$$

Since $\{v_1, \ldots, v_n\}$ is linearly independent, this equation implies that

$$-c_1 = \cdots = -c_k = c_{k+1} = \cdots = c_n = 0.$$

In particular, $c_{k+1} = \cdots = c_n = 0$. $\{f(\mathbf{v}_{k+1}), \ldots, f(\mathbf{v}_n)\}$ is linearly independent, as asserted. ■

A linear map $f:V \rightarrow W$ is ***invertible*** if there is a linear map $g:W \rightarrow V$ such that

$$g \circ f(\mathbf{v}) = \mathbf{v} \qquad \text{for all } \mathbf{v} \, \epsilon \, V$$

and

$$f \circ g(\mathbf{w}) = \mathbf{w} \qquad \text{for all } \mathbf{w} \, \epsilon \, W.$$

The linear map g, if it exists, is unique (see Exercise 26) and is called the ***inverse*** of f.

EXAMPLE 12

Suppose $F:\mathcal{M}_{2\times2} \rightarrow L(\mathbb{R}^2, \mathbb{R}^2)$ assigns to each 2×2 matrix the linear map defined by that matrix. For example,

$$F\begin{pmatrix} 1 & 2 \\ -1 & 1 \end{pmatrix} = f$$

where $f:\mathbb{R}^2 \rightarrow \mathbb{R}^2$ is defined by

$$f(x_1, x_2) = (x_1 + 2x_2, -x_1 + x_2).$$

The map F preserves addition because if

$$A = \begin{pmatrix} a_{11} & a_{12} \\ a_{21} & a_{22} \end{pmatrix} \qquad \text{and} \qquad B = \begin{pmatrix} b_{11} & b_{12} \\ b_{21} & b_{22} \end{pmatrix}$$

then

$$F(A) = f \qquad \text{and} \qquad F(B) = g$$

where

$$f(x_1, x_2) = (a_{11}x_1 + a_{12}x_2, a_{21}x_1 + a_{22}x_2)$$

$$g(x_1, x_2) = (b_{11}x_1 + b_{12}x_2, b_{21}x_1 + b_{22}x_2).$$

Furthermore,

$$F(A + B) = h$$

where

$$h(x_1, x_2) = ((a_{11} + b_{11})x_1 + (a_{12} + b_{12})x_2, (a_{21} + b_{21})x_1 + (a_{22} + b_{22})x_2)$$

$$= (a_{11}x_1 + a_{12}x_2, a_{21}x_1 + a_{22}x_2) + (b_{11}x_1 + b_{12}x_2, b_{21}x_1 + b_{22}x_2)$$

$$= f(x_1, x_2) + g(x_1, x_2)$$

$$= (f + g)(x_1, x_2).$$

Therefore,

$$F(A + B) = F(A) + F(B).$$

In a similar way, you can check that F preserves scalar multiplication. Hence F is a linear map.

This linear map $F:\mathcal{M}_{2\times 2} \to L(\mathbb{R}^2, \mathbb{R}^2)$ is invertible because, if $G:L(\mathbb{R}^2, \mathbb{R}^2) \to \mathcal{M}_{2\times 2}$ is defined by

$$G(f) = A,$$

where A is the matrix of f, then G is a linear map (see Section 2.5, Theorem 1) and

$$G \circ F(A) = A \qquad \text{for all } A \in \mathcal{M}_{2\times 2}$$

$$F \circ G(f) = f \qquad \text{for all } f \in L(\mathbb{R}^2, \mathbb{R}^2). \qquad \blacksquare$$

REMARK An invertible linear map $f:V \to W$ from one vector space to another is a one-to-one map from V onto W that preserves both addition and scalar multiplication. Its inverse $g:W \to V$ is also one-to-one and onto and it too preserves addition and scalar multiplication. So these maps define a one-to-one correspondence between V and W that respects the vector space operations. Such a correspondence is called an ***isomorphism.*** Two vector spaces are said to be ***isomorphic*** if there is an isomorphism between them. According to Example 12, the vector spaces $\mathcal{M}_{2\times 2}$ and $L(\mathbb{R}^2, \mathbb{R}^2)$ are isomorphic. Similarly it can be shown that $\mathcal{M}_{m\times n}$ and $L(\mathbb{R}^n, \mathbb{R}^m)$ are isomorphic vector spaces, for each m and n. Another isomorphism with which you are familiar is the one that assigns to each vector $\mathbf{v}$ in an n-dimensional vector space its coordinate n-tuple. This isomorphism shows that each n-dimensional vector space is isomorphic to $\mathbb{R}^n$.

EXERCISES

1. Determine if $f:\mathcal{P}^2 \to \mathcal{P}^2$ is a linear map, where $f(ax^2 + bx + c) =$
   (a) $cx^2 + bx + a$ (b) $(a + b + c)x^2$
   (c) $(a^2 + b^2 + c^2)x$ (d) $abcx$
   (e) $ax^2 - bx + c$ (f) $(a - b)x^2 + (b - c)x + (c - a)$

2. Determine if $f:\mathbb{R}^\infty \to \mathbb{R}^\infty$ is a linear map, where $f(x_1, x_2, x_3, \ldots) =$
   (a) $(0, x_1, 0, x_2, 0, x_3, \ldots)$
   (b) $(x_1, x_2 + 1, x_3 + 2, x_4 + 3, \ldots)$
   (c) $(x_1 - x_2, x_2 - x_3, x_3 - x_4, \ldots)$
   (d) $(x_1^2, x_2^2, x_3^2, \ldots)$
   (e) $(0, x_1, x_2, x_3, \ldots)$
   (f) $(x_1, -x_2, x_3, -x_4, \ldots)$

3. Determine if $f:\mathcal{P} \to \mathcal{P}$ is a linear map, where
   (a) $f(p(x)) = p(-x)$ (b) $f(p(x)) = -p(x)$
   (c) $f(p(x)) = p(x + 1)$ (d) $f(p(x)) = p(x) + 1$
   (e) $f(p(x)) = p(2x)$ (f) $f(p(x)) = 2p(x)$
   (g) $f(p(x)) = p(x^2)$ (h) $f(p(x)) = (p(x))^2$

4. Determine if $f:\mathcal{M}_{2\times 2} \to \mathbb{R}$ is a linear map, where $f\begin{pmatrix} a & b \\ c & d \end{pmatrix} =$
   (a) $a + d$ (b) $a^2 + b^2 + c^2 + d^2$
   (c) $abcd$ (d) $a + b - c + d$

5. Determine if $f:\mathcal{M}_{n\times n} \to \mathcal{M}_{n\times n}$ is a linear map where
   (a) $f(A) = 2A$ (b) $f(A) = A^2$ (c) $f(A) = A^t$
   (d) $f(A) = BA$, where B is some fixed $n \times n$ matrix

6. Determine if $T:L(\mathbb{R}^2, \mathbb{R}^2) \to \mathbb{R}^2$ is a linear map, where
   (a) $T(f) = f(\mathbf{e}_1)$, for all $f \in L(\mathbb{R}^2, \mathbb{R}^2)$
   (b) $T(f) = \mathbf{e}_1$, for all $f \in L(\mathbb{R}^2, \mathbb{R}^2)$
   (c) $T(f) = f(f(\mathbf{e}_1))$, for all $f \in L(\mathbb{R}^2, \mathbb{R}^2)$
   (d) $T(f) = f(\mathbf{e}_1) + \mathbf{e}_1$, for all $f \in L(\mathbb{R}^2, \mathbb{R}^2)$

7. Let V and W be any two vector spaces and let $f:V \to W$ be defined by $f(\mathbf{v}) = \mathbf{0}$ for all $\mathbf{v} \in V$. Show that f is a linear map. This linear map is called the **zero map.**

8. Let V be any vector space and let $f:V \to V$ be defined by $f(\mathbf{v}) = \mathbf{v}$ for all $\mathbf{v} \in V$. Show that f is a linear map. This linear map is called the **identity map**.

9. Let a be a fixed real number. Define $f_a:\mathcal{P} \to \mathbb{R}$ by $f_a(p(x)) = p(a)$ for all $p(x) \in \mathcal{P}$. Show that f_a is a linear map.

10. Let $f:\mathcal{P}^3 \to \mathcal{P}^3$ be the linear map such that $f(1) = x^2 + 1$, $f(x) = -x$, $f(x^2) = x^3$, $f(x^3) = x^2 + x - 1$. Find
   (a) $f(x^2 + 2x + 1)$ (b) $f((x - 1)^3)$
   (c) $f(4x^3 + x^2 + 2)$ (d) $f(x^3 - x^2 + x - 1)$
   (e) $f(1 + 2x + 3x^2 + 4x^3)$ (f) $f((x - 2)^2 + x^3)$

11. Describe the kernel and the image of the linear map $f:\mathcal{P}^2 \to \mathcal{P}^2$ defined by
   (a) $f(1) = x + 1$, $f(x) = x - 1$, $f(x^2) = x^2 + 1$
   (b) $f(1) = x + 1$, $f(x) = x^2$, $f(x^2) = x^2 + x + 1$
   (c) $f(1) = f(x) = f(x^2) = x^2$

12. Describe the image of the evaluation map $f:\mathcal{P} \to \mathbb{R}$ defined by $f(p(x)) = p(a)$, where a is some fixed real number.

13. Let $D:\mathcal{P}^n \to \mathcal{P}^n$ be the linear map that assigns to each polynomial its derivative. Find $\dim(\ker D)$ and $\dim(\operatorname{im} D)$. Check that your answers are consistent with Theorem 5.

14. Find the derivative of each of the given polynomials.
   (a) x^2 (b) x^3
   (c) x^5 (d) $x^4 + x + 1$
   (e) $x^{10} + x^5 - 5x$ (f) $3x^2 + 7x - 3$

15. Let $p(x) = ax^2 + bx + c$, where $a > 0$.
   (a) Find the derivative of $p(x)$.
   (b) Show that $x = -b/2a$ is the only zero of the polynomial $D(p(x))$.
   (c) Show that

$$p(x) = a\left(x + \frac{b}{2a}\right)^2 - \frac{b^2 - 4ac}{4a}$$

and conclude that $p(x)$ is smallest when $x = -b/2a$.

REMARK This exercise shows that the real number x that makes $p(x)$ smallest can be found by setting $D(p(x))$ equal to zero and solving for x.

16. Let a be any real number, let $\mathbf{B} = (1, x - a, (x - a)^2)$, and let $p(x) \in \mathcal{P}^2$.
    (a) Verify that $\mathbf{B}$ is an ordered basis for $\mathcal{P}^2$.
    (b) Show that the $\mathbf{B}$-coordinate triple of $p(x)$ is $(p(a), p_1(a), p_2(a))$, where $p_1(x) = D(p(x))$ and $p_2(x) = \frac{1}{2}D(p_1(x))$. [*Hint:* Write $p(x)$ as $p(x) = c_1 + c_2(x - a) + c_3(x - a)^2$, expand, and compute derivatives.]

17. Let $I{:}\mathcal{P} \to \mathcal{P}$ be the linear map such that $I(1) = x$ and $I(x^n) = \dfrac{x^{n+1}}{n+1}$ for all $n \geq 1$. Find
    (a) $I(x^3 + x^2 + x)$ (b) $I(2x^4 - x^3 + x)$
    (c) $I((3 - x)^3)$ (d) $I((x^2 - 1)^2)$
    (e) $I(1 + x + x^2 + \cdots + x^n)$ (f) $I(1 + 2x + 3x^2 + \cdots + (n + 1)x^n$

18. Let $I{:}\mathcal{P} \to \mathcal{P}$ be the linear map defined in Exercise 17.
    (a) Describe the kernel of I.
    (b) Describe the image of I.
    (c) Show that $D \circ I(p(x)) = p(x)$ for all $p(x) \in \mathcal{P}$.
    (d) For what polynomials $p(x) \in \mathcal{P}$ is $I \circ D(p(x)) = p(x)$?

19. The *trace* of an $n \times n$ matrix A is the real number $T(A)$ defined by

$$T(A) = T\begin{pmatrix} a_{11} & a_{12} & \cdots & a_{1n} \\ a_{21} & a_{22} & \cdots & a_{2n} \\ & & \cdots & \\ a_{n1} & a_{n2} & \cdots & a_{nn} \end{pmatrix} = a_{11} + a_{22} + \cdots + a_{nn}.$$

    (a) Show that $T{:}\mathcal{M}_{n \times n} \to \mathbb{R}$ is a linear map.
    (b) Describe the kernel of T.
    (c) Describe the image of T.
    (d) Find $\dim(\operatorname{im} T)$ and $\dim(\ker T)$.
    (e) Find a basis for $\ker T$.

20. Show that the map $f{:}\mathcal{P} \to \mathbb{R}^\infty$ defined by

$$f(a_0 + a_1 x + \cdots + a_n x^n) = (a_0, a_1, \ldots, a_n, 0, 0, 0, \ldots)$$

    is a linear map.

21. Let $f{:}V \to W$ be a linear map. Show that the kernel of f is a subspace of V.

22. Let $f{:}V \to W$ be a linear map. Show that the image of f is a subspace of W.

23. Let V and W be any two vector spaces, let f and g be linear maps from V into W, and let c be any real number. Show that
    (a) $f + g$ is a linear map
    (b) cf is a linear map

24. Let V and W be vector spaces and let $L(V, W)$ be the set of all linear maps from V into W. Show that, with the operations of addition and scalar multiplication given in this section, $L(V, W)$ is a vector space.

25. Let V, W, and U be vector spaces and let $f{:}V \to W$ and $g{:}W \to U$ be linear maps. Show that $g \circ f$ is a linear map from V into U.

26. Let V and W be vector spaces and let $f{:}V \to W$ be a linear map. Show that there is at most one linear map $g{:}W \to V$ such that $g \circ f(\mathbf{v}) = \mathbf{v}$ for all $\mathbf{v} \in V$ and $f \circ g(\mathbf{w}) = \mathbf{w}$ for all $\mathbf{w} \in W$.

27. (a) Show that the linear map

$$f(x_1, x_2, x_3, x_4) = \begin{pmatrix} x_1 & x_2 \\ x_3 & x_4 \end{pmatrix}$$

defines an isomorphism between $\mathbb{R}^4$ and $\mathcal{M}_{2 \times 2}$.

(b) Describe an isomorphism between $\mathbb{R}^{mn}$ and $\mathcal{M}_{m \times n}$.

28. Show that if V and W are any two n-dimensional vector spaces, then V and W are isomorphic.

6.4
The Matrix of a Linear Map

We know from Chapter 2 that each linear map $f:\mathbb{R}^n \to \mathbb{R}^m$ corresponds to an $m \times n$ matrix, the *standard matrix* of f. We can calculate $f(\mathbf{x})$ for each vector $\mathbf{x} \in \mathbb{R}^n$ by multiplying $\mathbf{x}$ by this matrix.

In Chapter 5 we saw that there are many different ways to associate an $n \times n$ matrix with a linear map $f:\mathbb{R}^n \to \mathbb{R}^n$. Given any ordered basis $\mathbf{B}$ for $\mathbb{R}^n$, we can associate with each linear map $f:\mathbb{R}^n \to \mathbb{R}^n$ its $\mathbf{B}$-*coordinate matrix* $M_\mathbf{B}(f)$. Then we can calculate the $\mathbf{B}$-coordinate n-tuple of $f(\mathbf{x})$ for each vector $\mathbf{x} \in \mathbb{R}^n$ by multiplying the $\mathbf{B}$-coordinate n-tuple of $\mathbf{x}$ by this matrix.

In this section we will see how to associate a matrix with each linear map $f:V \to W$, where V and W are any two finite dimensional vector spaces. In order to assign a matrix to the linear map $f:V \to W$ we will first need to choose an ordered basis $\mathbf{B}$ for V and an ordered basis $\mathbf{B}'$ for W. The matrix associated with f will have the property that multiplication by this matrix transforms the $\mathbf{B}$-coordinate n-tuple of each vector $\mathbf{v} \in V$, where n is the dimension of V, into the $\mathbf{B}'$-coordinate m-tuple of the vector $f(\mathbf{v})$, where m is the dimension of W.

Recall that the $\mathbf{B}$-coordinate matrix of $f:\mathbb{R}^n \to \mathbb{R}^n$, where $\mathbf{B} = (\mathbf{v}_1, \ldots, \mathbf{v}_n)$, is the matrix whose columns are the $\mathbf{B}$-coordinate n-tuples of the vectors $f(\mathbf{v}_1), \ldots, f(\mathbf{v}_n)$. Coordinate matrices of linear maps $f:V \to W$ are defined in a similar way.

Suppose $\mathbf{B} = (\mathbf{v}_1, \ldots, \mathbf{v}_n)$ is an ordered basis for the finite dimensional vector space V, and $\mathbf{B}' = (\mathbf{w}_1, \ldots, \mathbf{w}_m)$ is an ordered basis for the finite dimensional vector space W. The $(\mathbf{B}, \mathbf{B}')$-*coordinate matrix* $M(f)$ of a linear map $f:V \to W$ is the matrix whose column vectors are the $\mathbf{B}'$-coordinate n-tuples of the vectors $f(\mathbf{v}_1), \ldots, f(\mathbf{v}_n)$. Thus the $(\mathbf{B}, \mathbf{B}')$-coordinate matrix of f is the matrix

$$M(f) = \begin{pmatrix} c_{11} & c_{12} & \cdots & c_{1n} \\ c_{21} & c_{22} & \cdots & c_{2n} \\ & & \cdots & \\ c_{m1} & c_{m2} & \cdots & c_{mn} \end{pmatrix},$$

where

$$f(\mathbf{v}_1) = c_{11}\mathbf{w}_1 + c_{21}\mathbf{w}_2 + \cdots + c_{m1}\mathbf{w}_m$$

$$f(\mathbf{v}_2) = c_{12}\mathbf{w}_1 + c_{22}\mathbf{w}_2 + \cdots + c_{m2}\mathbf{w}_m$$

$$\cdots$$

$$f(\mathbf{v}_n) = c_{1n}\mathbf{w}_1 + c_{2n}\mathbf{w}_2 + \cdots + c_{mn}\mathbf{w}_m.$$

Notice that the $(\mathbf{B}, \mathbf{B}')$-coordinate matrix $M(f)$ is the *transpose* of the coefficient matrix of this system of vector equations.

Also notice that *the $(\mathbf{B}, \mathbf{B}')$-coordinate matrix $M(f)$ depends on the choice of ordered bases $\mathbf{B}$ and $\mathbf{B}'$ for V and W.* If you change $\mathbf{B}$ or $\mathbf{B}'$ (or both) then the matrix $M(f)$ will change accordingly. Sometimes the notation $M_{(\mathbf{B}, \mathbf{B}')}(f)$ is used instead of $M(f)$ to show explicitly that this matrix does depend on the choices of $\mathbf{B}$ and of $\mathbf{B}'$, but this notation is somewhat cumbersome and we will use it only when necessary to avoid confusion.

Finally, notice that if V has dimension n and W has dimension m, then the $(\mathbf{B}, \mathbf{B}')$-coordinate matrix of each linear map $f: V \to W$ is an $m \times n$ matrix.

EXAMPLE 1

Let $D: \mathcal{P}^3 \to \mathcal{P}^2$ be the linear map that assigns to each polynomial $p(x)$ in $\mathcal{P}^3$ its derivative. Let us calculate the $(\mathbf{B}, \mathbf{B}')$-coordinate matrix of D, where $\mathbf{B}$ is the ordered basis $(x^3, x^2, x, 1)$ for $\mathcal{P}^3$ and $\mathbf{B}'$ is the ordered basis $(x^2, x, 1)$ for $\mathcal{P}^2$. Since

$$D(x^3) = 3x^2 = 3x^2 + 0x + 0 \cdot 1$$
$$D(x^2) = 2x = 0x^2 + 2x + 0 \cdot 1$$
$$D(x) = 1 = 0x^2 + 0x + 1 \cdot 1$$
$$D(1) = 0 = 0x^2 + 0x + 0 \cdot 1,$$

we see that the $(\mathbf{B}, \mathbf{B}')$-coordinate matrix of D is the matrix

$$M(D) = \begin{pmatrix} 3 & 0 & 0 & 0 \\ 0 & 2 & 0 & 0 \\ 0 & 0 & 1 & 0 \end{pmatrix}. \quad \blacksquare$$

EXAMPLE 2

Let $f: \mathcal{P}^2 \to \mathbb{R}^4$ be the linear map defined by

$$f(p(x)) = (p(0), p(1), p(2), p(3)).$$

Let us find the $(\mathbf{B}, \mathbf{B}')$-coordinate matrix of f, where $\mathbf{B}$ is the ordered basis $(1, x, x^2)$ for $\mathcal{P}^2$ and $\mathbf{B}'$ is the ordered basis $(\mathbf{e}_1, \mathbf{e}_2, \mathbf{e}_3, \mathbf{e}_4)$ for $\mathbb{R}^4$. Since

$$f(1) = (1, 1, 1, 1) = 1\mathbf{e}_1 + 1\mathbf{e}_2 + 1\mathbf{e}_3 + 1\mathbf{e}_4$$
$$f(x) = (0, 1, 2, 3) = 0\mathbf{e}_1 + 1\mathbf{e}_2 + 2\mathbf{e}_3 + 3\mathbf{e}_4$$
$$f(x^2) = (0, 1, 4, 9) = 0\mathbf{e}_1 + 1\mathbf{e}_2 + 4\mathbf{e}_3 + 9\mathbf{e}_4$$

we see that the $(\mathbf{B}, \mathbf{B}')$-coordinate matrix of f is the matrix

$$M(f) = \begin{pmatrix} 1 & 0 & 0 \\ 1 & 1 & 1 \\ 1 & 2 & 4 \\ 1 & 3 & 9 \end{pmatrix}. \quad \blacksquare$$

The next theorem tells us why the $(\mathbf{B}, \mathbf{B}')$-coordinate matrix of a linear map $f:V \rightarrow W$ is important: multiplication by this matrix transforms the $\mathbf{B}$-coordinates of each vector $\mathbf{v} \in V$ into the $\mathbf{B}'$-coordinates of $f(\mathbf{v})$.

Given an ordered basis $\mathbf{B}$ for an n-dimensional vector space V and any vector $\mathbf{v}$ in V, we call the matrix $M(\mathbf{v})$, obtained by writing the $\mathbf{B}$-coordinate n-tuple of $\mathbf{v}$ vertically, the $\mathbf{B}$-*coordinate matrix* of $\mathbf{v}$.

Theorem 1 *Let V and W be finite dimensional vector spaces and let $f:V \rightarrow W$ be a linear map. If $\mathbf{B}$ is any ordered basis for V and $\mathbf{B}'$ is any ordered basis for W, then for each $\mathbf{v} \in V$*

$$M(f(\mathbf{v})) = M(f)M(\mathbf{v}),$$

where $M(\mathbf{v})$ is the $\mathbf{B}$-coordinate matrix of $\mathbf{v}$, $M(f(\mathbf{v}))$ is the $\mathbf{B}'$-coordinate matrix of $f(\mathbf{v})$, and $M(f)$ is the $(\mathbf{B}, \mathbf{B}')$-coordinate matrix of f.

The proof of this theorem is nearly identical to the proof of Theorem 2 in Section 5.2 so we will not reproduce it here.

EXAMPLE 3

Let $D:\mathcal{P}^3 \rightarrow \mathcal{P}^2$ be the linear map that assigns to each polynomial $p(x)$ in $\mathcal{P}^3$ its derivative. From Example 1 we know that the $(\mathbf{B}, \mathbf{B}')$-coordinate matrix of D, where $\mathbf{B} = (x^3, x^2, x, 1)$ and $\mathbf{B}' = (x^2, x, 1)$, is

$$M(D) = \begin{pmatrix} 3 & 0 & 0 & 0 \\ 0 & 2 & 0 & 0 \\ 0 & 0 & 1 & 0 \end{pmatrix}.$$

We can use this matrix to compute the derivative of any polynomial of degree less than or equal to 3. For example, if

$$p(x) = x^3 + 3x^2 - 5x + 7,$$

then by Theorem 1,

$$M(D(p(x)) = M(D)M(p(x))$$

$$= \begin{pmatrix} 3 & 0 & 0 & 0 \\ 0 & 2 & 0 & 0 \\ 0 & 0 & 1 & 0 \end{pmatrix} \begin{pmatrix} 1 \\ 3 \\ -5 \\ 7 \end{pmatrix} = \begin{pmatrix} 3 \\ 6 \\ -5 \end{pmatrix}$$

so

$$D(p(x)) = 3x^2 + 6x - 5. \quad \blacksquare$$

When $f:V \rightarrow V$ is a linear map from a finite dimensional vector space to itself, then it is usually most convenient to take $\mathbf{B} = \mathbf{B}'$ when we need a coordinate matrix for f. We call the $(\mathbf{B}, \mathbf{B}')$-coordinate matrix of f simply the $\mathbf{B}$-*coordinate matrix* of f. In the next theorem we shall deal only with this situation in order to keep the discussion simple. The more general situation is dealt with in the exercises.

The next theorem relates the algebraic properties of linear maps to the algebraic properties of matrices. Given any two linear maps $f:V \to V$ and $g:V \to V$ and any real number c, we can define $(f + g):V \to V$, $cf:V \to V$ and $f \circ g:V \to V$ by

$$(f + g)(\mathbf{x}) = f(\mathbf{x}) + g(\mathbf{x})$$

$$(cf)(\mathbf{x}) = cf(\mathbf{x})$$

and

$$(f \circ g)(\mathbf{x}) = f(g(\mathbf{x})),$$

for all $\mathbf{x} \in V$. You can verify that each of these maps is linear (see Exercises 16 and 17).

Theorem 2   *Let V be a finite dimensional vector space and let $\mathbf{B}$ be an ordered basis for V. Suppose $f:V \to V$ and $g:V \to V$ are linear maps and c is any real number. Then the $\mathbf{B}$-coordinate matrices of $f + g$, cf, and $f \circ g$ are related to the $\mathbf{B}$-coordinate matrices of f and g by*

$$M(f + g) = M(f) + M(g)$$

$$M(cf) = cM(f)$$

$$M(f \circ g) = M(f)M(g).$$

Furthermore, $M(f)$ is invertible if and only if f is invertible, in which case

$$M(f^{-1}) = [M(f)]^{-1},$$

where f^{-1} is the inverse of f.

The proofs of these formulas are nearly identical to the proofs given in Chapter 2 for the case when $V = \mathbb{R}^n$.

EXAMPLE 4

Let $D:\mathcal{P}^3 \to \mathcal{P}^3$ be the linear map that assigns to each polynomial in $\mathcal{P}^3$ its derivative. Then

$$D(x^3) = 3x^2 = 0x^3 + 3x^2 + 0x + 0 \cdot 1$$

$$D(x^2) = 2x = 0x^3 + 0x^2 + 2x + 0 \cdot 1$$

$$D(x) = 1 = 0x^3 + 0x^2 + 0x + 1 \cdot 1$$

$$D(1) = 0 = 0x^3 + 0x^2 + 0x + 0 \cdot 1$$

so the $\mathbf{B}$-coordinate matrix of D, where $\mathbf{B} = (x^3, x^2, x, 1)$, is

$$M(D) = \begin{pmatrix} 0 & 0 & 0 & 0 \\ 3 & 0 & 0 & 0 \\ 0 & 2 & 0 & 0 \\ 0 & 0 & 1 & 0 \end{pmatrix}.$$

By Theorem 2, the **B**-coordinate matrix of $D^2 = D \circ D$ is

$$[M(D)]^2 = \begin{pmatrix} 0 & 0 & 0 & 0 \\ 3 & 0 & 0 & 0 \\ 0 & 2 & 0 & 0 \\ 0 & 0 & 1 & 0 \end{pmatrix}^2 = \begin{pmatrix} 0 & 0 & 0 & 0 \\ 0 & 0 & 0 & 0 \\ 6 & 0 & 0 & 0 \\ 0 & 2 & 0 & 0 \end{pmatrix}.$$

It follows from Theorem 1 that the **B**-coordinate matrix of the polynomial $D^2(p(x))$, where $p(x) = ax^3 + bx^2 + cx + d$, is

$$M(D^2(p(x)) = M(D^2)M(p(x))$$

$$= \begin{pmatrix} 0 & 0 & 0 & 0 \\ 0 & 0 & 0 & 0 \\ 6 & 0 & 0 & 0 \\ 0 & 2 & 0 & 0 \end{pmatrix} \begin{pmatrix} a \\ b \\ c \\ d \end{pmatrix} = \begin{pmatrix} 0 \\ 0 \\ 6a \\ 2b \end{pmatrix}.$$

Hence

$$D^2(p(x)) = 0x^3 + 0x^2 + (6a)x + (2b) \cdot 1;$$

that is,

$$D^2(ax^3 + bx^2 + cx + d) = 6ax + 2b. \quad \blacksquare$$

EXAMPLE 5

Let $f : \mathcal{P}^2 \to \mathcal{P}^2$ be the linear map defined by

$$f(p(x)) = p(x + 1).$$

Then

$$f(x^2) = (x + 1)^2 = 1x^2 + 2x + 1$$
$$f(x) = x + 1 \quad = 0x^2 + 1x + 1$$
$$f(1) = 1 \quad\quad = 0x^2 + 0x + 1$$

so the **B**-coordinate matrix of f, where $\mathbf{B} = (x^2, x, 1)$, is

$$M(f) = \begin{pmatrix} 1 & 0 & 0 \\ 2 & 1 & 0 \\ 1 & 1 & 1 \end{pmatrix}.$$

This matrix $M(f)$ is invertible. Its inverse is

$$[M(f)]^{-1} = \begin{pmatrix} 1 & 0 & 0 \\ -2 & 1 & 0 \\ 1 & -1 & 1 \end{pmatrix}.$$

By Theorem 2, the linear map f is also invertible and the **B**-coordinate matrix of the inverse f^{-1} of f is

$$M(f^{-1}) = [M(f)]^{-1} = \begin{pmatrix} 1 & 0 & 0 \\ -2 & 1 & 0 \\ 1 & -1 & 1 \end{pmatrix}.$$

It follows from Theorem 1 that the **B**-coordinate matrix of $f^{-1}(p(x))$, where $p(x) = ax^2 + bx + c$, is

$$M(f^{-1}(p(x)) = M(f^{-1})M(p(x))$$

$$= \begin{pmatrix} 1 & 0 & 0 \\ -2 & 1 & 0 \\ 1 & -1 & 1 \end{pmatrix}\begin{pmatrix} a \\ b \\ c \end{pmatrix} = \begin{pmatrix} a \\ -2a + b \\ a - b + c \end{pmatrix}.$$

Hence

$$f^{-1}(p(x)) = ax^2 + (-2a + b)x + (a - b + c)1,$$

that is,

$$f^{-1}(ax^2 + bx + c) = ax^2 + (b - 2a)x + (a - b + c). \quad \blacksquare$$

The next theorem is a straightforward generalization of Theorem 3 (Section 5.2).

Theorem 3   *Let V be a finite dimensional vector space and let $f:V \to V$ be a linear map. Suppose $\mathbf{B} = (\mathbf{v}_1, \ldots, \mathbf{v}_n)$ and $\mathbf{B}' = (\mathbf{w}_1, \ldots, \mathbf{w}_n)$ are two ordered bases for V. Then the $\mathbf{B}'$-coordinate matrix of f, $M_{\mathbf{B}'}(f)$, is related to the $\mathbf{B}$-coordinate matrix of f, $M_{\mathbf{B}}(f)$, by the formula*

$$M_{\mathbf{B}'}(f) = H^{-1}M_{\mathbf{B}}(f)H,$$

where H is the matrix whose columns are the $\mathbf{B}$-coordinate matrices $M_{\mathbf{B}}(\mathbf{w}_1)$, $\ldots$, $M_{\mathbf{B}}(\mathbf{w}_n)$ of the $\mathbf{B}'$-basis vectors $\mathbf{w}_1, \ldots, \mathbf{w}_n$.

EXAMPLE 6
Let V be a 2-dimensional vector space with ordered basis $\mathbf{B} = (\mathbf{v}_1, \mathbf{v}_2)$ and let $\mathbf{B}' = (\mathbf{v}_2, \mathbf{v}_1)$. Suppose $f:V \to V$ is a linear map whose $\mathbf{B}$-coordinate matrix is

$$M_{\mathbf{B}}(f) = \begin{pmatrix} a & b \\ c & d \end{pmatrix}.$$

Then we can use Theorem 3 to calculate the $\mathbf{B}'$-coordinate matrix of f. First notice that $\mathbf{B}' = (\mathbf{w}_1, \mathbf{w}_2)$, where

$$\mathbf{w}_1 = \mathbf{v}_2 = 0\mathbf{v}_1 + 1\mathbf{v}_2$$

$$\mathbf{w}_2 = \mathbf{v}_1 = 1\mathbf{v}_1 + 0\mathbf{v}_2.$$

Therefore

$$M_{\mathbf{B}}(\mathbf{w}_1) = \begin{pmatrix} 0 \\ 1 \end{pmatrix} \quad \text{and} \quad M_{\mathbf{B}}(\mathbf{w}_2) = \begin{pmatrix} 1 \\ 0 \end{pmatrix}.$$

Let H be the matrix whose columns are $M_\mathbf{B}(\mathbf{w}_1)$ and $M_\mathbf{B}(\mathbf{w}_2)$,

$$H = \begin{pmatrix} 0 & 1 \\ 1 & 0 \end{pmatrix}.$$

Then, by Theorem 3,

$$M_{\mathbf{B}'}(f) = H^{-1}M_\mathbf{B}(f)H$$

$$= \begin{pmatrix} 0 & 1 \\ 1 & 0 \end{pmatrix}\begin{pmatrix} a & b \\ c & d \end{pmatrix}\begin{pmatrix} 0 & 1 \\ 1 & 0 \end{pmatrix}$$

$$= \begin{pmatrix} 0 & 1 \\ 1 & 0 \end{pmatrix}\begin{pmatrix} b & a \\ d & c \end{pmatrix} = \begin{pmatrix} d & c \\ b & a \end{pmatrix}. \quad \blacksquare$$

EXERCISES

1. Find the $(\mathbf{B}, \mathbf{B}')$-coordinate matrix, where $\mathbf{B} = \mathbf{B}' = (x^2, x, 1)$, for each of the given linear maps $f:\mathcal{P}^2 \to \mathcal{P}^2$.

    (a) $f(p(x)) = p(-x)$ (b) $f(p(x)) = p(x - 1)$
    (c) $f(p(x)) = p(x + 2)$ (d) $f(p(x)) = xD(p(x))$.

2. Find the $(\mathbf{B}, \mathbf{B}')$-coordinate matrix, where $\mathbf{B} = (x^3, x^2, x, 1)$ and $\mathbf{B}' = (\mathbf{e}_1, \mathbf{e}_2, \mathbf{e}_3)$, for each of the following linear maps $f:\mathcal{P}^3 \to \mathbb{R}^3$.

    (a) $f(p(x)) = (p(0), p(1), p(2))$
    (b) $f(p(x)) = (p(-1), p(0), p(1))$
    (c) $f(ax^3 + bx^2 + cx + d) = (a, b, c)$
    (d) $f(ax^3 + bx + cx + d) = (a + b, b + c, c + d)$

3. Let $f:\mathcal{P}^2 \to \mathcal{P}^2$ be the linear map

    $$f(p(x)) = p(x + 1).$$

    Find the $(\mathbf{B}, \mathbf{B}')$-coordinate matrix of f, where

    (a) $\mathbf{B} = \mathbf{B}' = (1, x, x^2)$
    (b) $\mathbf{B} = (1, x, x^2)$, $\mathbf{B}' = (x^2, x, 1)$
    (c) $\mathbf{B} = (x^2, x, 1)$, $\mathbf{B}' = (1, x, x^2)$
    (d) $\mathbf{B} = \mathbf{B}' = (1, x - 1, (x - 1)^2)$

4. Let $f:\mathcal{P}^2 \to \mathcal{P}^2$ be the linear map whose $(\mathbf{B}, \mathbf{B}')$-coordinate matrix is

    $$M(f) = \begin{pmatrix} 1 & 2 & 0 \\ 0 & -1 & 1 \\ 1 & 3 & 2 \end{pmatrix},$$

    where $\mathbf{B} = \mathbf{B}' = (x^2, x, 1)$. Find

    (a) $f(x^2 + 1)$ (b) $f(x^2 + x + 1)$
    (c) $f((x + 1)^2)$ (d) $f(1 + x)$
    (e) $f(x^2)$ (f) $f(2x - 3)$

5. Let $f:\mathcal{P}^2 \to \mathcal{P}^3$ be the linear map whose $(\mathbf{B}, \mathbf{B}')$-coordinate matrix is

    $$M(f) = \begin{pmatrix} -1 & 0 & 1 \\ 2 & 3 & 1 \\ -1 & 2 & 2 \\ 0 & -1 & 1 \end{pmatrix},$$
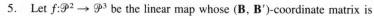

where $\mathbf{B} = (x^2, x, 1)$ and $\mathbf{B}' = (x^3, x^2, x, 1)$. Find

(a) $f(x^2 + x + 1)$ (b) $f((x + 1)^2)$

(c) $f((1 - x)^2)$ (d) $f(1 - x^2)$

(e) $f(x^2 - 1)$ (f) $f(x^2 - x + 1)$

6.   (a) Find the $(\mathbf{B}, \mathbf{B}')$-coordinate matrix of the linear map $f:\mathscr{P}^3 \to \mathscr{P}^3$, where $f(p(x)) = p(x + 1)$ and $\mathbf{B} = \mathbf{B}' = (x^3, x^2, x, 1)$.

(b) Find the $(\mathbf{B}, \mathbf{B}')$-coordinate matrix of the linear map $f:\mathscr{P}^n \to \mathscr{P}^n$ defined by $f(p(x)) = p(x + 1)$, where $\mathbf{B} = \mathbf{B}' = (x^n, x^{n-1}, \dots, x^2, x, 1)$.

7.   (a) Find the $(\mathbf{B}, \mathbf{B}')$-coordinate matrix of $D:\mathscr{P}^4 \to \mathscr{P}^4$ where D is the linear map that assigns to each polynomial in $\mathscr{P}^4$ its derivative, and where $\mathbf{B} = \mathbf{B}' = (x^4, x^3, x^2, x, 1)$.

(b) Use the result of part (a) to compute the derivative of

$$p(x) = x^4 + 2x^3 - 2x^2 + x + 1.$$

8.   Let $D:\mathscr{P}^3 \to \mathscr{P}^3$ be the linear map that assigns to each polynomial in $\mathscr{P}^3$ its derivative and let $M(D)$ be the $\mathbf{B}$-coordinate matrix of D, where $\mathbf{B} = (x^3, x^2, x, 1)$.

(a) Calculate $[M(D)]^3$.

(b) Calculate $[M(D)]^4$. What does your answer imply about

$$D \circ D \circ D \circ D(p(x)) = D^4(p(x))$$

when $p(x) \in \mathscr{P}^3$?

9.   Let $f:\mathscr{P}^3 \to \mathscr{P}^3$ be the linear map $f(p(x)) = p(x + 1)$.

(a) Find a formula for $f \circ f(p(x))$.

(b) Find the $\mathbf{B}$-coordinate matrix of $f \circ f$, where $B = (x^3, x^2, x, 1)$, in two different ways: (i) by direct computation, and (ii) by using Theorem 2.

10.   Let $f:\mathscr{P}^2 \to \mathscr{P}^2$ be defined by $f(p(x)) = p(x + 1)$ and let $g:\mathscr{P}^2 \to \mathscr{P}^2$ be defined by $g(p(x)) = p(x - 1)$.

(a) Verify that f and g are inverse to each other.

(b) Calculate the $\mathbf{B}$-coordinate matrix of g, where $\mathbf{B} = (x^2, x, 1)$, and compare your answer with the matrix $M(f^{-1})$ found in Example 5.

11.   Let V be a 3-dimensional vector space with ordered basis $\mathbf{B} = (\mathbf{v}_1, \mathbf{v}_2, \mathbf{v}_3)$ and suppose $f:V \to V$ is the linear map whose $\mathbf{B}$-coordinate matrix is

$$M_{\mathbf{B}}(f) = \begin{pmatrix} a_{11} & a_{12} & a_{13} \\ a_{21} & a_{22} & a_{23} \\ a_{31} & a_{32} & a_{33} \end{pmatrix}.$$

Using Theorem 3, find $M_{\mathbf{B}'}(f)$, where

(a) $\mathbf{B}' = (\mathbf{v}_2, \mathbf{v}_1, \mathbf{v}_3)$

(b) $\mathbf{B}' = (\mathbf{v}_2, \mathbf{v}_3, \mathbf{v}_1)$

(c) $\mathbf{B}' = (\mathbf{v}_3, \mathbf{v}_2, \mathbf{v}_1)$

12.   Let $f:\mathscr{P}^1 \to \mathscr{P}^1$ be the linear map defined by

$$f(ax + b) = bx + a.$$

(a) Find $M_{\mathbf{B}}(f)$ where $\mathbf{B} = (x, 1)$.

(b) Find $M_{\mathbf{B}'}(f)$ where $\mathbf{B}' = (x + 1, x - 1)$.

(c) Find H such that $M_{\mathbf{B}'}(f) = H^{-1}M_{\mathbf{B}}(f)H$.

13. Let $f:\mathcal{P}^2 \to \mathcal{P}^2$ be the linear map such that $f(1) = 1$, $f(x) = 2x + 1$, and $f(x^2) = -x^2 - 6x - 4$.

(a) Find $M_{\mathbf{B}}(f)$ where $\mathbf{B} = (x^2, x, 1)$.

(b) Find $M_{\mathbf{B}'}(f)$ where $\mathbf{B}' = (1, x + 1, (x + 1)^2)$.

(c) Find a matrix H such that $M_{\mathbf{B}'}(f) = H^{-1}M_{\mathbf{B}}(f)H$.

14. Let V be any vector space and let $f:V \to V$ be a linear map. A real number λ is an *eigenvalue* of f if there is a nonzero vector $\mathbf{v} \in V$ that such that $f(\mathbf{v}) = \lambda\mathbf{v}$. The vector $\mathbf{v}$ is an *eigenvector* of f. If λ is an eigenvalue of f, then the set of all $\mathbf{v} \in V$ such that $f(\mathbf{v}) = \lambda\mathbf{v}$ is the λ-*eigenspace* of f.

(a) Show that if V is finite dimensional and if $\mathbf{B}$ is an ordered basis for V, then the $\mathbf{B}$-coordinate matrix $M_{\mathbf{B}}(f)$ is a diagonal matrix if and only if $\mathbf{B}$ consists of eigenvectors of f.

(b) Find the eigenvalues and the corresponding eigenspaces of the linear map $f:\mathcal{P}^n \to \mathcal{P}^n$ defined by $f(p(x)) = p(-x)$.

15. Let V and W be finite dimensional vector spaces of the same dimension and let $f:V \to W$ be any linear map.

(a) Show that there exist ordered bases $\mathbf{B}$ for V and $\mathbf{B}'$ for W such that the $(\mathbf{B}, \mathbf{B}')$-coordinate matrix of f is a diagonal matrix. [*Hint:* You can take $\mathbf{B}$ to be any ordered basis for V. How must $\mathbf{B}'$ be chosen?]

(b) Show that $f:V \to W$ is invertible if and only if there exist ordered bases $\mathbf{B}$ for V and $\mathbf{B}'$ for W such that the $(\mathbf{B}, \mathbf{B}')$-coordinate matrix of f is the identity matrix.

16. Let V and W be any two finite dimensional vector spaces. Let $f:V \to W$ and $g:V \to W$ be linear maps.

(a) Verify that the map $f + g:V \to W$ defined by

$$(f + g)(\mathbf{x}) = f(\mathbf{x}) + g(\mathbf{x})$$

is a linear map.

(b) Verify that, for each real number c, the map $cf:V \to W$ defined by

$$(cf)(\mathbf{x}) = cf(\mathbf{x})$$

is a linear map.

(c) Show that if $\mathbf{B}$ and $\mathbf{B}'$ are ordered bases for V and for W respectively, then the $(\mathbf{B}, \mathbf{B}')$-coordinate matrices of f, g, $f + g$, and cf are related by

$$M(f + g) = M(f) + M(g)$$

and

$$M(cf) = cM(f).$$

17. Let V, W, and U be finite dimensional vector spaces with ordered bases $\mathbf{B}$, $\mathbf{B}'$, and $\mathbf{B}''$ respectively. Suppose $f:V \to W$ and $g:W \to U$ are linear maps.

(a) Verify that $g \circ f:V \to U$ is a linear map.

(b) Show that

$$M_{(\mathbf{B},\mathbf{B}'')}(g \circ f) = M_{(\mathbf{B}',\mathbf{B}'')}(g)M_{(\mathbf{B},\mathbf{B}')}(f).$$

(c) Show that $f:V \to W$ is invertible if and only if $M_{(\mathbf{B},\mathbf{B}')}(f)$ is invertible.

(d) Show that if $f:V \to W$ is invertible with inverse f^{-1}, then

$$M_{(\mathbf{B}',\mathbf{B})}(f^{-1}) = [M_{(\mathbf{B},\mathbf{B}')}(f)]^{-1}.$$

6.5

Application: Linear Difference Equations

Sequences $\mathbf{x} = (x_1, x_2, x_3, \ldots)$ of numbers arise in all branches of mathematics. Sometimes they are specified by a formula for the nth term. For example, the geometric sequence $(1, \frac{1}{2}, \frac{1}{4}, \frac{1}{8}, \ldots)$ can be described by the formula $x_n = (\frac{1}{2})^{n-1}$, $n = 1, 2, 3, \ldots$. In other cases, a sequence may be specified by a relationship between the nth term and one or more of the previous terms. For example, the sequence $\mathbf{x}$ defined by the equation

$$x_n - 2x_{n-1} - x_{n-2} = 0, \quad n > 2$$

and satisfying the conditions $x_1 = 0$ and $x_2 = 1$ can be calculated to as many terms as desired by solving for x_n. For each $n > 2$ we have

$$x_n = 2x_{n-1} + x_{n-2},$$

so

$$x_3 = 2x_2 + x_1 = 2$$
$$x_4 = 2x_3 + x_2 = 5$$
$$x_5 = 2x_4 + x_3 = 12,$$

and so on.

The equation $x_n - 2x_{n-1} - x_{n-2} = 0$, $n > 2$, is an example of a linear difference equation. A **linear difference equation** is an equation of the form

$$a_0 x_n + a_1 x_{n-1} + \cdots + a_k x_{n-k} = 0, \quad n > k$$

relating, for all $n > k$, the nth term x_n of a sequence $\mathbf{x} = (x_1, x_2, x_3, \ldots)$ to the k previous terms. The integer k is called the **order** of the difference equation. The coefficients $a_0, \ldots a_k$ are real numbers, with $a_0 \neq 0$ and $a_k \neq 0$. A **solution** of this equation is any vector $\mathbf{x} \in \mathbb{R}^\infty$ that satisfies it.

It may appear, at first glance, that the equation

$$a_0 x_n + a_1 x_{n-1} + \cdots + a_k x_{n-k} = 0, \quad n > k,$$

is not *one equation* but rather a *sequence of equations*. That is technically correct. However, we can rewrite this sequence of *scalar* equations as the single *vector* equation

$$(a_0 S^k + a_1 S^{k-1} + \cdots + a_k)(\mathbf{x}) = \mathbf{0},$$

where $S:\mathbb{R}^\infty \to \mathbb{R}^\infty$ is the **shift operator** defined by

$$S(x_1, x_2, x_3, \ldots) = (x_2, x_3, x_4, \ldots).$$

In reality it is the vector equation that is being discussed when we talk about *the* difference equation

$$a_0 x_n + a_1 x_{n-1} + \cdots + a_k x_{n-k} = 0, \quad n > k.$$

The shift operator $S:\mathbb{R}^\infty \to \mathbb{R}^\infty$ is a linear map. Hence the map $f:\mathbb{R}^\infty \to \mathbb{R}^\infty$ defined by

$$f(\mathbf{x}) = (a_0 S^k + a_1 S^{k-1} + \cdots + a_k)(\mathbf{x})$$

is also a linear map. The solution set of the linear difference equation

$$a_0 x_n + a_1 x_{n-1} + \cdots + a_k x_{n-k} = 0, \qquad n > k,$$

is just the kernel of this linear map f. We can conclude, therefore, that the solution set of this difference equation is a subspace of $\mathbb{R}^\infty$.

For an explanation of why these equations are called difference equations, see Exercise 5.

We shall study here only linear difference equations of order 2. These are equations of the form

$$ax_n + bx_{n-1} + cx_{n-2} = 0, \qquad n > 2,$$

where a, b, and c are real numbers with $a \neq 0$ and $c \neq 0$. To solve an equation of this type, it is convenient to rewrite the equation in the form

$$x_n = -\frac{b}{a} x_{n-1} - \frac{c}{a} x_{n-2}, \qquad n > 2.$$

If we let $y_n = x_{n-1}$, $(n > 1)$, this equation can be described by the pair of equations

$$x_n = -\frac{b}{a} x_{n-1} - \frac{c}{a} y_{n-1}$$

$$y_n = x_{n-1} \quad n > 2$$

or as the single matrix equation

$$\begin{pmatrix} x_n \\ y_n \end{pmatrix} = \begin{pmatrix} -b/a & -c/a \\ 1 & 0 \end{pmatrix} \begin{pmatrix} x_{n-1} \\ y_{n-1} \end{pmatrix}, \qquad n > 2.$$

By setting

$$A = \begin{pmatrix} -b/a & -c/a \\ 1 & 0 \end{pmatrix}$$

and applying this matrix equation repeatedly for various values of n, we see that

$$\begin{pmatrix} x_n \\ y_n \end{pmatrix} = A \begin{pmatrix} x_{n-1} \\ y_{n-1} \end{pmatrix} = A^2 \begin{pmatrix} x_{n-2} \\ y_{n-2} \end{pmatrix} = \cdots = A^{n-2} \begin{pmatrix} x_2 \\ y_2 \end{pmatrix}$$

for all $n > 2$. Since $y_n = x_{n-1}$, it follows that

$$\begin{pmatrix} x_n \\ x_{n-1} \end{pmatrix} = A^{n-2} \begin{pmatrix} x_2 \\ x_1 \end{pmatrix}$$

for all $n > 2$. This formula is also valid when $n = 2$ since, by definition, $A^0 = I$. Working backwards, we can also see that each $\mathbf{x} = (x_1, x_2, x_3, \ldots)$ $\in \mathbb{R}^\infty$ given by this formula, where x_1 and x_2 are arbitrary real numbers, is a solution of the given difference equation.

We have proved the following theorem.

Theorem 1 *The solution set of the linear difference equation*

$$ax_n + bx_{n-1} + cx_{n-2} = 0, \qquad n > 2,$$

is the set of all vectors $\mathbf{x} = (x_1, x_2, x_3, \ldots) \in \mathbb{R}^\infty$ such that, for all $n \geq 2$,

$$\begin{pmatrix} x_n \\ x_{n-1} \end{pmatrix} = A^{n-2} \begin{pmatrix} x_2 \\ x_1 \end{pmatrix}, \text{ where } A = \begin{pmatrix} -b/a & -c/a \\ 1 & 0 \end{pmatrix}.$$

In order to use Theorem 1 to calculate an explicit formula for the nth term x_n of a sequence $\mathbf{x}$ satisfying the difference equation

$$ax_n + bx_{n-1} + cx_{n-2} = 0, \qquad n > 2,$$

we must be able to compute the matrix power A^{n-2}, where

$$A = \begin{pmatrix} -b/a & -c/a \\ 1 & 0 \end{pmatrix}.$$

When A is diagonable, we can do this by diagonalizing A, as follows.

Since

$$\det(A - \lambda I) = \lambda^2 + \frac{b}{a}\lambda + \frac{c}{a} = \frac{1}{a}(a\lambda^2 + b\lambda + c),$$

the eigenvalues of A are the zeros of the polynomial $a\lambda^2 + b\lambda + c$. This polynomial is called the **auxiliary polynomial** of the difference equation $ax_n + bx_{n-1} + cx_{n-2} = 0, n > 2$. When $b^2 - 4ac > 0$ the zeros of this polynomial, $\lambda_1 = (-b + \sqrt{b^2 - 4ac})/2a$ and $\lambda_2 = (-b - \sqrt{b^2 - 4ac})/2a$, are real and distinct. Hence, in this case, A is diagonable. It is easy to see that $(\lambda_1, 1)$ and $(\lambda_2, 1)$ are eigenvectors corresponding to the eigenvalues λ_1 and λ_2. Hence, if we let

$$H = \begin{pmatrix} \lambda_1 & \lambda_2 \\ 1 & 1 \end{pmatrix}, \qquad \text{then} \qquad H^{-1}AH = D = \begin{pmatrix} \lambda_1 & 0 \\ 0 & \lambda_2 \end{pmatrix}.$$

It follows that $A = HDH^{-1}$ and

$$A^{n-2} = (HDH^{-1})(HDH^{-1}) \cdots (HDH^{-1}) = HD^{n-2}H^{-1}$$

so, by Theorem 1,

$$\begin{pmatrix} x_n \\ x_{n-1} \end{pmatrix} = A^{n-2} \begin{pmatrix} x_2 \\ x_1 \end{pmatrix} = HD^{n-2}H^{-1} \begin{pmatrix} x_2 \\ x_1 \end{pmatrix}$$

$$= \begin{pmatrix} \lambda_1 & \lambda_2 \\ 1 & 1 \end{pmatrix} \begin{pmatrix} \lambda_1^{n-2} & 0 \\ 0 & \lambda_2^{n-2} \end{pmatrix} \begin{pmatrix} c_1 \\ c_2 \end{pmatrix}, \qquad \text{for } n \geq 2,$$

where

$$\begin{pmatrix} c_1 \\ c_2 \end{pmatrix} = H^{-1} \begin{pmatrix} x_2 \\ x_1 \end{pmatrix}.$$

Note that (c_1, c_2) is an arbitrary vector in $\mathbb{R}^2$, since (x_2, x_1) is arbitrary and H^{-1} is invertible.

If we carry out the matrix multiplication we find that

$$x_n = c_1\lambda_1^{n-1} + c_2\lambda_2^{n-1}$$

$$x_{n-1} = c_1\lambda_1^{n-2} + c_2\lambda_2^{n-2}, \qquad n \geq 2,$$

where c_1 and c_2 are arbitrary constants. In other words,

$$\mathbf{x} = c_1(1, \lambda_1, \lambda_1^2, \ldots) + c_2(1, \lambda_2, \lambda_2^2, \ldots).$$

In summary, we have shown the following:

> If $b^2 - 4ac > 0$, then the solution set of the difference equation $ax_n + bx_{n-1} + cx_{n-2} = 0$, $n > 2$, is the subspace of $\mathbb{R}^\infty$ spanned by $\{(1, \lambda_1, \lambda_1^2, \ldots), (1, \lambda_2, \lambda_2^2, \ldots)\}$, where λ_1 and λ_2 are the zeros of the auxiliary polynomial $a\lambda^2 + b\lambda + c$.

EXAMPLE 1

Consider the difference equation

$$6x_n - 5x_{n-1} + x_{n-2} = 0, \qquad n > 2.$$

The auxiliary polynomial is

$$6\lambda^2 - 5\lambda + 1 = (2\lambda - 1)(3\lambda - 1).$$

Its zeros are $\lambda_1 = \frac{1}{2}$ and $\lambda_2 = \frac{1}{3}$. Hence the solution set of this difference equation is the set of all $\mathbf{x} \in \mathbb{R}^\infty$ of the form

$$\mathbf{x} = c_1(1, \tfrac{1}{2}, \tfrac{1}{4}, \tfrac{1}{8}, \ldots) + c_2(1, \tfrac{1}{3}, \tfrac{1}{9}, \tfrac{1}{27}, \ldots).$$

If we want the solution of this difference equation with $x_1 = 1$ and $x_2 = 2$, we must choose c_1 and c_2 so that

$$1 = x_1 = c_1 + c_2$$

$$2 = x_2 = \tfrac{1}{2}c_1 + \tfrac{1}{3}c_2.$$

The solution of this pair of equations is $(c_1, c_2) = (10, -9)$, so the particular solution we seek is

$$\mathbf{x} = 10(1, \tfrac{1}{2}, \tfrac{1}{4}, \tfrac{1}{8}, \ldots) - 9(1, \tfrac{1}{3}, \tfrac{1}{9}, \tfrac{1}{27}, \ldots)$$

$$= (1, 2, \tfrac{3}{2}, \tfrac{11}{12}, \ldots). \quad \blacksquare$$

EXAMPLE 2

Let us find all sequences with the property that each term beyond the second in the sequence is the average of the preceding two terms. This condition may be expressed as $x_n = \frac{1}{2}(x_{n-1} + x_{n-2})$, $n > 2$, or as

$$2x_n - x_{n-1} - x_{n-2} = 0, \qquad n > 2.$$

Since the auxiliary polynomial $2\lambda^2 - \lambda - 1$ has zeros $\lambda_1 = 1$ and $\lambda_2 = -\frac{1}{2}$, the solution set is the set of all $\mathbf{x} \in \mathbb{R}$ of the form

$$\mathbf{x} = c_1(1, 1, 1, \ldots) + c_2(1, -\tfrac{1}{2}, \tfrac{1}{4}, -\tfrac{1}{8}, \ldots). \quad \blacksquare$$

EXAMPLE 3

The *Fibonacci sequence* is the sequence $\mathbf{x}$ with $x_1 = x_2 = 1$ and with the property that each term beyond the second is the sum of the preceding two. Thus

$$x_1 = 1, x_2 = 1, x_3 = 2, x_4 = 3, x_5 = 5, x_6 = 8,$$

and so on. Let us find a formula for the nth term x_n of the Fibonacci sequence.

The property defining the Fibonacci sequence may be expressed as $x_n = x_{n-1} + x_{n-2}$, $n > 2$, or as

$$x_n - x_{n-1} - x_{n-2} = 0, \qquad n > 2.$$

The auxiliary polynomial $\lambda^2 - \lambda - 1$ has zeros $(1 \pm \sqrt{5})/2$, and hence this difference equation has solution set consisting of those $\mathbf{x} \in \mathbb{R}^\infty$ of the form

$$\mathbf{x} = c_1\left(1, \frac{1 + \sqrt{5}}{2}, \left(\frac{1 + \sqrt{5}}{2}\right)^2, \ldots\right)$$

$$+ c_2\left(1, \frac{1 - \sqrt{5}}{2}, \left(\frac{1 - \sqrt{5}}{2}\right)^2, \ldots\right).$$

The nth term of this sequence is

$$x_n = c_1\left(\frac{1 + \sqrt{5}}{2}\right)^{n-1} + c_2\left(\frac{1 - \sqrt{5}}{2}\right)^{n-1}, \, n \geq 1.$$

The particular solution with $x_1 = x_2 = 1$ must satisfy

$$1 = x_1 = c_1 + c_2$$

$$1 = x_2 = c_1\left(\frac{1 + \sqrt{5}}{2}\right) + c_2\left(\frac{1 - \sqrt{5}}{2}\right).$$

Hence, for the Fibonacci sequence, we must have

$$c_1 = \frac{1 + \sqrt{5}}{2\sqrt{5}} \quad \text{and} \quad c_2 = -\frac{1 - \sqrt{5}}{2\sqrt{5}}.$$

Therefore, the nth term of the Fibonacci sequence is

$$x_n = \frac{1}{\sqrt{5}}\left[\left(\frac{1 + \sqrt{5}}{2}\right)^n - \left(\frac{1 - \sqrt{5}}{2}\right)^n\right], \, n \geq 1.$$

Notice that $|(1 - \sqrt{5})/2| \approx 0.62 < 1$ so, for large n, x_n is approximately equal to $\left(\frac{1 + \sqrt{5}}{2}\right)^n / \sqrt{5}. \quad \blacksquare$

So far we have seen how to obtain explicit solutions of the difference equation $ax_n + bx_{n-1} + cx_{n-2} = 0$, $n > 2$, only when $b^2 - 4ac > 0$. In this case we were able to express A in the form $A = HDH^{-1}$, where D is a diagonal matrix, and hence we were able to compute the powers of A and apply the formula of Theorem 1. In the remaining two cases, when $b^2 - 4ac = 0$ and when $b^2 - 4ac < 0$, we cannot diagonalize A. However, we can express A in the form $A = HBH^{-1}$, where B is not diagonal but nevertheless is so simple that its powers are easy to compute.

Consider first the case where $b^2 - 4ac = 0$. Then the auxiliary polynomial $a\lambda^2 + b\lambda + c$ has a repeated zero $\lambda_1 = -b/2a$. We can express the entries of A in terms of λ_1. Indeed, since λ_1 is a repeated zero, the auxiliary polynomial must be of the form

$$a\lambda^2 + b\lambda + c = a(\lambda - \lambda_1)^2 = a\lambda^2 - 2a\lambda_1\lambda + a\lambda_1^2.$$

Hence $b = -2a\lambda_1$, $c = a\lambda_1^2$ and

$$A = \begin{pmatrix} -b/a & -c/a \\ 1 & 0 \end{pmatrix} = \begin{pmatrix} 2\lambda_1 & -\lambda_1^2 \\ 1 & 0 \end{pmatrix}.$$

If we let

$$H = \begin{pmatrix} \lambda_1 & 1 \\ 1 & 0 \end{pmatrix} \quad \text{and} \quad B = \begin{pmatrix} \lambda_1 & 1 \\ 0 & \lambda_1 \end{pmatrix},$$

then it is easy to check that $A = HBH^{-1}$ (see Exercise 9a). The matrix B is called the "Jordan canonical form" of A. Its powers are easy to compute. We find that

$$B^2 = \begin{pmatrix} \lambda_1^2 & 2\lambda_1 \\ 0 & \lambda_1^2 \end{pmatrix}, B^3 = \begin{pmatrix} \lambda_1^3 & 3\lambda_1^2 \\ 0 & \lambda_1^3 \end{pmatrix}, B^4 = \begin{pmatrix} \lambda_1^4 & 4\lambda_1^3 \\ 0 & \lambda_1^4 \end{pmatrix}.$$

and so on. In general,

$$B^k = \begin{pmatrix} \lambda_1^k & k\lambda_1^{k-1} \\ 0 & \lambda_1^k \end{pmatrix} \text{ for } k \geq 1.$$

Hence, using Theorem 1, we see that the solutions of the difference equation $ax_n + bx_{n-1} + cx_{n-2} = 0$, $n > 2$, when $b^2 - 4ac = 0$, are given by the formula

$$\begin{pmatrix} x_n \\ x_{n-1} \end{pmatrix} = A^{n-2}\begin{pmatrix} x_2 \\ x_1 \end{pmatrix} = HB^{n-2}H^{-1}\begin{pmatrix} x_2 \\ x_1 \end{pmatrix}$$

$$= \begin{pmatrix} \lambda_1 & 1 \\ 1 & 0 \end{pmatrix}\begin{pmatrix} \lambda_1^{n-2} & (n-2)\lambda_1^{n-3} \\ 0 & \lambda_1^{n-2} \end{pmatrix}\begin{pmatrix} c_1 \\ c_2 \end{pmatrix}, n \geq 2,$$

where

$$\begin{pmatrix} c_1 \\ c_2 \end{pmatrix} = H^{-1}\begin{pmatrix} x_2 \\ x_1 \end{pmatrix}.$$

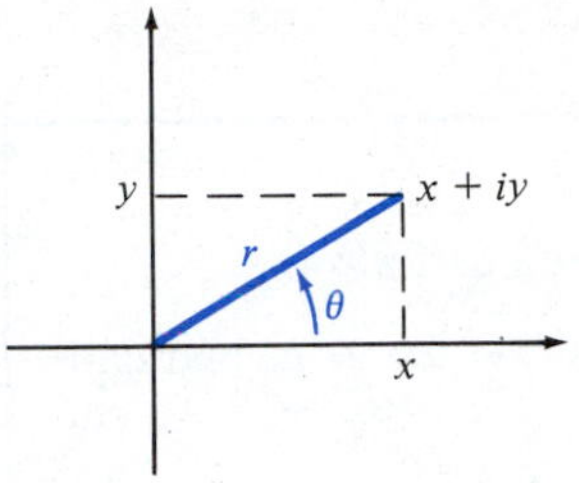

FIGURE 6.2
The polar form of the complex number $x + iy$ is
$$r(\cos \theta + i \sin \theta).$$

Carrying out the matrix multiplication, we find that

$$x_n = c_1\lambda_1^{n-1} + c_2(n-1)\lambda_1^{n-2}, \, n \geq 1.$$

The case when $b^2 - 4ac < 0$ can be handled in the same way. In this case the zeros of the auxiliary polynomial are complex numbers, which can be expressed in polar form as $r(\cos \theta \pm i \sin \theta)$ for some $r > 0$ and some θ not a multiple of π (see Figure 6.2). Since

$$a\lambda^2 + b\lambda + c = a(\lambda - r(\cos \theta + i \sin \theta))(\lambda - r(\cos \theta - i \sin \theta))$$

$$= a\lambda^2 - (2ar \cos \theta)\lambda + ar^2,$$

we see that $b = -2ar \cos \theta$ and $c = ar^2$. Hence

$$A = \begin{pmatrix} -b/a & -c/a \\ 1 & 0 \end{pmatrix} = \begin{pmatrix} 2r \cos \theta & -r^2 \\ 1 & 0 \end{pmatrix}.$$

If we set

$$H = \begin{pmatrix} r \cos \theta & r \sin \theta \\ 1 & 0 \end{pmatrix} \quad \text{and} \quad B = \begin{pmatrix} r \cos \theta & r \sin \theta \\ -r \sin \theta & r \cos \theta \end{pmatrix},$$

then it is easy to check that $A = HBH^{-1}$ (see Exercise 9b). The powers of B are once again easy to compute. We find that

$$B^2 = \begin{pmatrix} r^2 \cos 2\theta & r^2 \sin 2\theta \\ -r^2 \sin 2\theta & r^2 \cos 2\theta \end{pmatrix}, B^3 = \begin{pmatrix} r^3 \cos 3\theta & r^3 \sin 3\theta \\ -r^3 \sin 3\theta & r^3 \sin 3\theta \end{pmatrix},$$

and so on. In general,

$$B^k = \begin{pmatrix} r^k \cos k\theta & r^k \sin k\theta \\ -r^k \sin k\theta & r^k \cos k\theta \end{pmatrix} \text{ for } k \geq 1.$$

Hence, using Theorem 1, we see that the general solution of the difference equation $ax_n + bx_{n-1} + cx_{n-2} = 0$, $n > 2$, when $b^2 - 4ac < 0$, is given by

$$\begin{pmatrix} x_n \\ x_{n-1} \end{pmatrix} = A^{n-2}\begin{pmatrix} x_2 \\ x_1 \end{pmatrix} = HB^{n-2}H^{-1}\begin{pmatrix} x_2 \\ x_1 \end{pmatrix}$$

$$= \begin{pmatrix} r \cos \theta & r \sin \theta \\ 1 & 0 \end{pmatrix}\begin{pmatrix} r^{n-2} \cos (n-2)\theta & r^{n-2} \sin (n-2)\theta \\ -r^{n-2} \sin (n-2)\theta & r^{n-2} \cos (n-2)\theta \end{pmatrix}\begin{pmatrix} c_1 \\ c_2 \end{pmatrix},$$

$$n \geq 2,$$

where

$$\begin{pmatrix} c_1 \\ c_2 \end{pmatrix} = H^{-1}\begin{pmatrix} x_2 \\ x_1 \end{pmatrix}.$$

Carrying out the matrix multiplication, we find that

$$x_n = c_1 r^{n-1} \cos (n-1)\theta + c_2 r^{n-1} \sin (n-1)\theta, \, n \geq 1.$$

Here is a summary of the results of this section:

To solve a second order linear difference equation
$$ax_n + bx_{n-1} + cx_{n-2} = 0, \quad n > 2, \text{ proceed as follows:}$$

(1) Find the zeros of the auxiliary polynomial $a\lambda^2 + b\lambda + c$, and call these zeros λ_1 and λ_2.

(2) If λ_1 and λ_2 are real and unequal, then

$$x_n = c_1\lambda_1^{n-1} + c_2\lambda_2^{n-1}$$

for all $n \geq 1$, where c_1 and c_2 are arbitrary constants.

(3) If $\lambda_1 = \lambda_2$, then

$$x_n = c_1\lambda_1^{n-1} + c_2(n-1)\lambda_1^{n-2}$$

for all $n \geq 1$, where c_1 and c_2 are arbitrary constants.

(4) If λ_1 and λ_2 are the complex numbers $r(\cos\theta \pm i\sin\theta)$ ($r \neq 0$, $\theta \neq k\pi$ for any integer k), then

$$x_n = c_1 r^{n-1}\cos(n-1)\theta + c_2 r^{n-1}\sin(n-1)\theta$$

for all $n \geq 1$, where c_1 and c_2 are arbitrary constants.

EXAMPLE 4

Let us find all sequences with the property that each term beyond the first in the sequence is the average of the terms immediately preceding and following it. This condition can be expressed as

$$x_{n-1} = \tfrac{1}{2}(x_n + x_{n-2}), \qquad n > 2,$$

or as

$$x_n - 2x_{n-1} + x_{n-2} = 0, \qquad n > 2.$$

Since the auxiliary polynomial $\lambda^2 - 2\lambda + 1 = (\lambda - 1)^2$ has the repeated zero $\lambda_1 = \lambda_2 = 1$, the solution is given by

$$x_n = c_1 1^{n-1} + c_2(n-1)1^{n-2} = c_1 + c_2(n-1), \, n \geq 1.$$

In other words, the solution set is the set of all $\mathbf{x} \in \mathbb{R}^\infty$ of the form

$$\mathbf{x} = c_1(1, 1, 1, 1, \ldots) + c_2(0, 1, 2, 3, \ldots). \quad \blacksquare$$

EXAMPLE 5

The sequence that begins with $x_1 = 0$ and $x_2 = 1$ and that satisfies the difference equation

$$x_n - 2x_{n-1} + 2x_{n-2} = 0, \qquad n > 2$$

starts out $\mathbf{x} = (0, 1, 2, 2, 0, -4, -8, \ldots)$. Let us find a formula for the nth

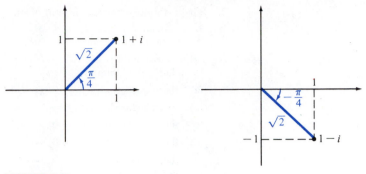

FIGURE 6.3

The polar forms of the complex numbers $1 \pm i$ are $\sqrt{2}\left(\cos \dfrac{\pi}{4} \pm i \sin \dfrac{\pi}{4}\right)$.

term of this sequence. Since the auxiliary polynomial $\lambda^2 - 2\lambda + 2$ has the complex zeros

$$\frac{2 \pm \sqrt{4 - 8}}{2} = 1 \pm i = \sqrt{2}\left(\cos \frac{\pi}{4} \pm i \sin \frac{\pi}{4}\right)$$

(see Figure 6.3), the solution of this difference equation is given by

$$x_n = c_1(\sqrt{2})^{n-1} \cos(n - 1)\frac{\pi}{4} + c_2(\sqrt{2})^{n-1} \sin(n - 1)\frac{\pi}{4}, \ n \geq 1.$$

The particular solution with $x_1 = 0$ and $x_2 = 1$ must satisfy

$$0 = x_1 = c_1$$

$$1 = x_2 = c_1\sqrt{2}\cos\frac{\pi}{4} + c_2\sqrt{2}\sin\frac{\pi}{4} = c_1 + c_2$$

Hence $c_1 = 0$ and $c_2 = 1$. Therefore

$$x_n = (\sqrt{2})^{n-1} \sin(n - 1)\frac{\pi}{4}$$

for all $n \geq 1$. ∎

EXERCISES

1. Find the solution set of each of the following difference equations:
   (a) $5x_n - 6x_{n-1} + x_{n-2} = 0, \ n > 2$
   (b) $x_n - 6x_{n-1} + 5x_{n-2} = 0, \ n > 2$
   (c) $x_n + 6x_{n-1} + 8x_{n-2} = 0, \ n > 2$
   (d) $x_n - 2x_{n-1} - 2x_{n-2} = 0, \ n > 2$
   (e) $x_n + 4x_{n-1} + 4x_{n-2} = 0, \ n > 2$
   (f) $4x_n + 4x_{n-1} + x_{n-2} = 0, \ n > 2$
   (g) $x_n + 6x_{n-1} + 9x_{n-2} = 0, \ n > 2$
   (h) $x_n + x_{n-1} + x_{n-2} = 0, \ n > 2$
   (i) $x_n + 2x_{n-1} + 2x_{n-2} = 0, \ n > 2$
   (j) $x_n + x_{n-2} = 0, \ n > 2$

2. Find a formula for the nth term of the sequence satisfying $x_1 = 0$, $x_2 = 1$, and $x_n - 2x_{n-1} - x_{n-2} = 0$, $n > 2$. (This is the sequence described in the first paragraph of this section.)

3. Write down the first five terms of each of the following sequences, and then find a formula for the nth term x_n.

   (a) the sequence with $x_1 = 0$, $x_2 = 1$, and $x_n = x_{n-1} - x_{n-2}$ for $n > 2$

   (b) the sequence with $x_1 = -1$, $x_2 = 1$, and $x_n = -2x_{n-2}$ for $n > 2$

   (c) the sequence with $x_1 = 0$, $x_2 = 1$, and $x_n = 2(x_{n-1} - x_{n-2})$ for $n > 2$

   (d) the sequence with $x_1 = 0$, $x_2 = 1$, and $x_n = 4(x_{n-1} - x_{n-2})$ for $n > 2$

   (e) the sequence with $x_1 = 0$, $x_2 = 1$, and $x_n = 5(x_{n-1} - x_{n-2})$ for $n > 2$

4. (a) Show that the solution set of the first order linear difference equation

   $$ax_n + bx_{n-1} = 0, n > 1,$$

   is the set of all $\mathbf{x} \in \mathbb{R}^\infty$ of the form $\mathbf{x} = c(1, \lambda, \lambda^2, \lambda^3, \ldots)$, where $\lambda = -b/a$ and $c \in \mathbb{R}$ is arbitrary.

   (b) If an amount P_0 is deposited in a savings account at an interest rate of 6% compounded annually, then the amount P_n at the end of the nth year satisfies $P_n = 1.06P_{n-1}$, $n \geq 1$. Use the result of (a) to find a formula for P_n in terms of P_0 and n.

5. Let $\Delta : \mathbb{R}^\infty \to \mathbb{R}^\infty$ be defined by

   $$\Delta(x_1, x_2, x_3, \ldots) = (x_2 - x_1, x_3 - x_2, x_4 - x_3, \ldots).$$

   (a) Show that Δ is a linear map.

   (b) Compute $\Delta^2\mathbf{x}$ where $\mathbf{x} = (x_1, x_2, x_3, \ldots)$.

   (c) Show that each second order linear difference equation

   $$ax_n + bx_{n-1} + cx_{n-2} = 0, \qquad n > 2,$$

   can be rewritten in the form

   $$(p\Delta^2 + q\Delta + r)\mathbf{x} = \mathbf{0}$$

   where p, q, and $r \in \mathbb{R}$, and $\mathbf{x} = (x_1, x_2, x_3, \ldots)$.

   REMARK The sequence $\Delta\mathbf{x}$ is called the *first difference* of $\mathbf{x}$, and $\Delta^2\mathbf{x}$ is the *second difference* of $\mathbf{x}$. The linear map $p\Delta^2 + q\Delta + r$ is a second order *difference operator*. The equivalence established in (c) is the reason that the equation $ax_n + bx_{n-1} + cx_{n-2} = 0$, $n > 2$, is called a *difference equation.*

6. Consider the linear difference equation

   $$a_0x_n + a_1x_{n-1} + \cdots + a_kx_{n-k} = 0, \qquad n > k.$$

   (a) Show that for each i, $1 \leq i \leq k$, there is a unique solution $\mathbf{y}_i = (y_{i1}, y_{i2}, y_{i3}, \ldots)$ of this equation that satisfies the conditions $y_{ii} = 1$ and $y_{ij} = 0$ for $j \neq i$, $1 \leq j \leq k$.

   (b) Verify that the set $\{\mathbf{y}_1, \ldots, \mathbf{y}_k\}$ is a basis for the solution space of this equation.

   (c) Conclude that the dimension of the solution space is k.

7. (a) Show that all solutions of the third order linear difference equation

   $$ax_n + bx_{n-1} + cx_{n-2} + dx_{n-3} = 0, \qquad n > 3,$$

   can be calculated from the formula

$$\begin{pmatrix} x_n \\ x_{n-1} \\ x_{n-2} \end{pmatrix} = A^{n-3} \begin{pmatrix} x_3 \\ x_2 \\ x_1 \end{pmatrix}, \qquad n \geq 3,$$

where x_1, x_2, and x_3 are arbitrary, and where

$$A = \begin{pmatrix} -b/a & -c/a & -d/a \\ 1 & 0 & 0 \\ 0 & 1 & 0 \end{pmatrix}.$$

[*Hint:* Let $y_n = x_{n-1}$ and $z_n = x_{n-2}$.]

   (b) Generalize part (a) to describe the solution set of the kth order linear difference equation

$$a_0 x_n + a_1 x_{n-1} + \cdots + a_k x_{n-k} = 0, \qquad n > k.$$

8.   (a) Use the result of Exercise 7(a) to show that if the zeros λ_1, λ_2, λ_3 of the auxiliary polynomial $a\lambda^3 + b\lambda^2 + c\lambda + d$ are real and distinct, then the solution space of the equation

$$a x_n + b x_{n-1} + c x_{n-2} + d x_{n-3} = 0, \qquad n > 3,$$

is spanned by

$$\{(1, \lambda_1, \lambda_1^2, \ldots), (1, \lambda_2, \lambda_2^2, \ldots), (1, \lambda_3, \lambda_3^2, \ldots)\}.$$

   (b) Generalizing part (a), describe a basis for the solution space of the difference equation

$$a_0 x_n + a_1 x_{n-1} + \cdots + a_k x_{n-k} = 0, \qquad n > k,$$

in the case where the zeros of the auxiliary polynomial $a_0\lambda^k + a_1\lambda^{k-1} + \cdots + a_k$ are real and distinct.

9.   (a) Let $A = \begin{pmatrix} 2\lambda_1 & -\lambda_1^2 \\ 1 & 0 \end{pmatrix}$, $H = \begin{pmatrix} \lambda_1 & 1 \\ 1 & 0 \end{pmatrix}$, and $B = \begin{pmatrix} \lambda_1 & 1 \\ 0 & \lambda_1 \end{pmatrix}$, where $\lambda_1 \in \mathbb{R}$.

     Show that $AH = HB$, and hence that $A = HBH^{-1}$.

   (b) Let $A = \begin{pmatrix} 2r\cos\theta & -r^2 \\ 1 & 0 \end{pmatrix}$, $H = \begin{pmatrix} r\cos\theta & r\sin\theta \\ 1 & 0 \end{pmatrix}$, and $B =$

$$\begin{pmatrix} r\cos\theta & r\sin\theta \\ -r\sin\theta & r\cos\theta \end{pmatrix}.$$ Show that $AH = HB$, and hence that $A = HBH^{-1}$.

REVIEW EXERCISES

1.   Which of the following are subspaces?

   (a) the subset of $\mathbb{R}^\infty$ consisting of all $(a_1, a_2, a_3, \ldots)$ with $a_n = a_1^n$ for all n.

   (b) the subset of $M_{2\times 2}$ consisting of all 2×2 matrices A with $A^2 = 0$.

   (c) the subset of $\mathcal{P}$ consisting of all polynomials $p(x)$ with $p(1) = 0$.

   (d) the subset of $\mathcal{P}$ consisting of all polynomials $p(x)$ with $p(0) = 1$.

   (e) the subset of $M_{2\times 2}$ consisting of all 2×2 matrices with all diagonal entries equal to zero.

   (f) the subset of $\mathbb{R}^\infty$ consisting of all $(a_1, a_2, a_3, \ldots)$ with $a_n = 0$ for some n.

2. Which of the following sets are bases for the indicated vector spaces?
   (a) $\{x - 2, (x - 2)^2, (x - 2)^3\}$, for $\mathcal{P}^3$
   (b) $\{x, (x - 1)^2, (x + 1)^2\}$, for $\mathcal{P}^2$
   (c) $\{x - 1, x + 1, (x - 1)(x + 1)\}$, for $\mathcal{P}^2$
   (d) $\{x, x + 1, x + 2\}$, for $\mathcal{P}^1$

3. Find a basis for and the dimension of the subspace $\mathcal{L}(S)$ of $\mathcal{P}$ spanned by S, where $S =$
   (a) $\{x^3 + x^2 + x + 1, x^3 - x^2 + x - 1, x^3 + x, x^2 + 1\}$
   (b) $\{x^2 + 3x + 2, x^2 + 2x - 1, x^2 + 5x + 8\}$
   (c) $\{x^3, (x + 1)^3, (x + 2)^3\}$.

4. Find a basis for $\mathcal{P}^3$ that contains the given linearly independent set.
   (a) $\{x^3 + 1, x^3 - 1\}$ (b) $\{(x + 1)^3, (x - 1)^3\}$
   (c) $\{x^3 + 2x^2 + 3x - 1, x^3 - 2x^2 + 3x + 1, x^3 + x^2 + 2\}$

5. Which of the following maps are linear maps?
   (a) $f:\mathbb{R}^\infty \to \mathbb{R}^\infty$ defined by $f(x_1, x_2, x_3, \ldots) = (x_1, x_2^2, x_3^3, \ldots)$
   (b) $f: \mathcal{M}_{2 \times 3} \to \mathcal{M}_{2 \times 2}$ defined by

   $$f\begin{pmatrix} a_{11} & a_{12} & a_{13} \\ a_{21} & a_{22} & a_{23} \end{pmatrix} = \begin{pmatrix} a_{11} + a_{12} & a_{12} + a_{13} \\ a_{21} + a_{22} & a_{22} + a_{23} \end{pmatrix}$$

   (c) $f:\mathcal{M}_{n \times n} \to \mathbb{R}$ defined by $f(A) = \det A$
   (d) $f:\mathcal{P} \to \mathbb{R}$ defined by $f(a_n x^n + \cdots + a_1 x + a_0) = a_n + \cdots + a_1 + a_0$.

6. Let $f:\mathcal{P}^3 \to \mathcal{P}^3$ be the linear map such that $f(1) = 1$, $f(x) = x + 1$, $f(x^2) = x^2 + x + 1$, and $f(x^3) = x^3 + x^2 + x + 1$. Let $\mathbf{B} = (1, x, x^2, x^3)$.
   (a) Find the $\mathbf{B}$-coordinate matrix of f.
   (b) Find the polynomial $f(x^3 + 3x^2 + 4x - 2)$.
   (c) Show that f is invertible.
   (d) Find the $\mathbf{B}$-coordinate matrix for the inverse map f^{-1}.

7. Describe the kernel and the image of each of the following linear maps.
   (a) $f:\mathbb{R}^\infty \to \mathbb{R}^2$, $f(x_1, x_2, x_3, \ldots) = (x_1 - x_2, -x_1 + x_2)$
   (b) $f:\mathcal{P}^3 \to \mathbb{R}^3$, $f(p(x)) = (p(0), p(1), p(-1))$
   (c) $f:\mathcal{M}_{2 \times 2} \to \mathcal{M}_{2 \times 2}$, $f(A) = AB$ where

   $$B = \begin{pmatrix} 1 & -1 \\ -1 & 1 \end{pmatrix}$$

   (d) $f:\mathcal{P} \to \mathcal{P}$, $f(a_n x^n + \cdots + a_1 x + a_0) = a_n x^{n-1} + \cdots + a_2 x + a_1$.

8. Find the derivative of each of the given polynomials.
   (a) $x^2 + 5x - 6$ (b) $x^3 - x$
   (c) $x^{10} + x^5 + 1$ (d) $x^n + x^{n-1} + \cdots + x + 1$.

9. Suppose V is a vector space with ordered basis $\mathbf{B} = (\mathbf{v}, \mathbf{w})$ and suppose $\mathbf{B}' = (\mathbf{v} + \mathbf{w}, \mathbf{v} - \mathbf{w})$. Find the $\mathbf{B}'$-coordinate matrix $M_{\mathbf{B}'}(f)$, if the $\mathbf{B}$-coordinate matrix of f is

   $$M_{\mathbf{B}}(f) = \begin{pmatrix} 2 & 0 \\ -1 & 3 \end{pmatrix}.$$

10. If f is a linear map from $\mathcal{P}^3$ to $\mathbb{R}^2$ which is onto, then what is the dimension of the kernel of f?

11. True or false?

(a) If $\{\mathbf{v}_1, \mathbf{v}_2, \mathbf{v}_3\}$ is a linearly independent subset of the vector space V then the dimension of V must be equal to 3.

(b) If $\{p_1(x), p_2(x), p_3(x)\}$ spans $\mathcal{P}^n$ then n must be equal to 2.

(c) If V and W are finite dimensional vector spaces and the linear map $f:V \rightarrow W$ is onto, then the dimension of V must be greater than or equal to the dimension of W.

(d) The linear map $f:\mathcal{P} \rightarrow \mathbb{R}^\infty$ defined by

$$f(a_0 + a_1x + \cdots + a_nx^n) = (a_0, a_1, \ldots, a_n, 0, 0, \ldots)$$

is onto.

(e) If $f:\mathcal{M}_{3 \times 4} \rightarrow \mathbb{R}^7$ is a linear map whose kernel has dimension 9, then the image of f must have dimension 3.

(f) If A is the **B**-coordinate matrix of $f:V \rightarrow V$, then A^2 is the **B**-coordinate matrix of $f \circ f$.

(g) If the $(\mathbf{v}_1, \mathbf{v}_2)$-coordinate matrix of $f:V \rightarrow V$ is

$$\begin{pmatrix} a & b \\ c & d \end{pmatrix},$$

then the $(\mathbf{v}_2, \mathbf{v}_1)$-coordinate matrix of f is

$$\begin{pmatrix} d & c \\ b & a \end{pmatrix}.$$

12. Find the solution set of each of the following linear difference equations.

(a) $x_n - 6x_{n-1} - 25x_{n-2} = 0, n > 2$

(b) $3x_n + 7x_{n-1} + 2x_{n-2} = 0, n > 2.$

(c) $9x_n + 12x_{n-1} + 4x_{n-2} = 0, n > 2.$

13. Find all sequences with the property that each term beyond the second is one-sixth the sum of the preceding two terms.

APPENDIX
Linear Algebra and
the Computer

The computational techniques that have been discussed in this book are efficient for computations by hand when the matrices involved are not too large and when the entries in the matrices are integers. However, when solving linear algebra problems that involve large matrices, or matrices whose entries have many significant figures, then the solutions are difficult, if not impossible, to obtain by hand. In such situations, it is very helpful to have a computer available to carry out the computations.

It is easy to write a computer program that will row reduce matrices and that will therefore solve most of the problems in this book. However, it is important to realize that the computer works not with real numbers but with rational approximations. For example, the irrational number π is stored in the computer as a rational number, the rational number obtained by rounding off (or perhaps truncating) the decimal expansion of π to a specified number of decimal places, possibly 8 or 16. Because of this, the computer sometimes cannot distinguish between zero and a number that is not zero but is very small. For example, row reduction of the matrix

$$\begin{pmatrix} 1 & \frac{1}{3} \\ 2 & \frac{2}{3} \end{pmatrix} \qquad \text{yields} \qquad \begin{pmatrix} 1 & \frac{1}{3} \\ 0 & 0 \end{pmatrix},$$

whereas if we first express all fractions in their decimal expansions, rounding off to two places, and then row reduce, we get

$$\begin{pmatrix} 1 & \frac{1}{3} \\ 2 & \frac{2}{3} \end{pmatrix} \approx \begin{pmatrix} 1 & .33 \\ 2 & .67 \end{pmatrix} \longrightarrow \begin{pmatrix} 1 & .33 \\ 0 & .01 \end{pmatrix} \longrightarrow \begin{pmatrix} 1 & .33 \\ 0 & 1 \end{pmatrix} \longrightarrow \begin{pmatrix} 1 & 0 \\ 0 & 1 \end{pmatrix}.$$

Thus the process of rounding off has caused a matrix of rank one to be perceived as a matrix of rank two!

In general, unless extreme care is taken in the programming, the results of computer calculations on matrices will be unreliable whenever the matrix has less than maximal rank. (A matrix has **maximal rank** if its rank is equal to the number of rows in the matrix, or the number of columns in the matrix, whichever is smaller.) Even when the rank is maximal, roundoff errors can accumulate and magnify to the point where the computer solution to a problem has little meaning.

In this appendix, we shall first describe a modification of Gaussian elimination that will reduce errors due to roundoff and truncation. Then we shall describe some other algorithms that are useful when using computers to solve linear algebra problems.

A.1
Pivoting and Rescaling

If a is any real number, and if $\tilde{a}$ is obtained from a by rounding off (or by truncating), then

$$a = \tilde{a} + \varepsilon$$

where ε is the **roundoff error.** When we multiply both sides of this equation by c we find that

$$ca = c\tilde{a} + c\varepsilon$$

Thus multiplication of a number by c multiplies the absolute value of the round-off error by $|c|$. Whenever $|c| > 1$, roundoff errors are magnified. But smaller values of $|c|$ yield smaller magnification of roundoff error. If $|c| < 1$, then roundoff error is actually reduced! Therefore, it makes sense to try to arrange the computer calculations involved in Gaussian elimination in such a way that whenever a row vector is multiplied by a nonzero real number c then the absolute value $|c|$ of that number is as small as possible.

In Gaussian elimination, row vectors are multiplied by real numbers only when introducing a new corner 1 into the matrix. At this stage in the process, the partially reduced matrix looks like this:

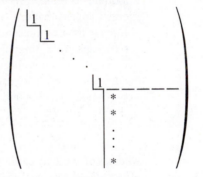

The position in the matrix occupied by the top asterisk is called the **pivot position.**

Ordinarily, the next step in the algorithm is to use a row interchange, if necessary, to *bring any nonzero entry c in the column of asterisks to the pivot position.* Then we multiply the row in which c appears by $1/c$. A preferred strategy, from the point of view of reducing roundoff error, is to use a row interchange to *bring the entry with the largest absolute value to the pivot position* before multiplying to introduce the corner 1.

The modified Gaussian elimination process that at each stage brings the entry with largest absolute value into the pivot position is called **pivoting on the largest available entry.**

EXAMPLE 1
Let us solve the linear system

$$\tfrac{1}{7}x_1 + \tfrac{4}{3}x_2 + \tfrac{2}{13}x_3 = 0$$

$$\tfrac{19}{6}x_1 + \tfrac{1}{3}x_2 + \tfrac{8}{7}x_3 = 1$$

$$\tfrac{11}{3}x_1 + \tfrac{2}{7}x_2 + \tfrac{1}{2}x_3 = 2$$

by Gaussian elimination, rounding off each calculation to two decimal places. If we pivot at each stage on the largest available entry, the calculation proceeds as follows, where an asterisk above an arrow indicates a pivoting step:

$$\begin{pmatrix} 0.14 & 1.33 & 0.15 & 0.00 \\ 3.17 & 0.33 & 1.14 & 1.00 \\ 3.67 & 0.29 & 0.50 & 2.00 \end{pmatrix} \overset{*}{\longrightarrow} \begin{pmatrix} 3.67 & 0.29 & 0.50 & 2.00 \\ 3.17 & 0.33 & 1.14 & 1.00 \\ 0.14 & 1.33 & 0.15 & 0.00 \end{pmatrix}$$

$$\longrightarrow \begin{pmatrix} 1 & 0.08 & 0.14 & 0.54 \\ 3.17 & 0.33 & 1.14 & 1.00 \\ 0.14 & 1.33 & 0.15 & 0.00 \end{pmatrix} \longrightarrow \begin{pmatrix} 1 & 0.08 & 0.14 & 0.54 \\ 0 & 0.08 & 0.70 & -0.71 \\ 0 & 1.32 & 0.13 & -0.08 \end{pmatrix}$$

$$\overset{*}{\longrightarrow} \begin{pmatrix} 1 & 0.08 & 0.14 & 0.54 \\ 0 & 1.32 & 0.13 & -0.08 \\ 0 & 0.08 & 0.70 & -0.71 \end{pmatrix} \longrightarrow \begin{pmatrix} 1 & 0.08 & 0.14 & 0.54 \\ 0 & 1 & 0.10 & -0.06 \\ 0 & 0.08 & 0.70 & -0.71 \end{pmatrix}$$

$$\longrightarrow \begin{pmatrix} 1 & 0 & 0.13 & 0.54 \\ 0 & 1 & 0.10 & -0.06 \\ 0 & 0 & 0.69 & -0.71 \end{pmatrix} \longrightarrow \begin{pmatrix} 1 & 0 & 0.13 & 0.54 \\ 0 & 1 & 0.10 & -0.06 \\ 0 & 0 & 1 & -1.03 \end{pmatrix}$$

$$\longrightarrow \begin{pmatrix} 1 & 0 & 0 & 0.67 \\ 0 & 1 & 0 & 0.04 \\ 0 & 0 & 1 & -1.03 \end{pmatrix}$$

Hence we find the (approximate) solution

$$(x_1, x_2, x_3) = (0.67, 0.04, -1.03).$$

If we had used standard Gaussian elimination, without pivoting on largest available entries, we would have found the approximate solution

$$(x_1, x_2, x_3) = (0.74, 0.09, -1.45).$$

The correct solution, rounded to two decimal places, is

$$(x_1, x_2, x_3) = (0.68, 0.05, -1.03).$$

Clearly we have obtained a better approximate solution by pivoting on the largest available entry. ∎

REMARK 1 Computers do not actually round off to a fixed number of decimal places. Rather, they round off to a fixed number of *significant figures;* that is, they store only the first n digits (possibly $n = 8$, or $n = 16$), beginning the count with the first nonzero digit, and they record the location of the decimal point. Thus, for example, a computer storing two significant figures would store $\tfrac{4}{3}$ as 1.3, $\tfrac{1}{3}$ as .33, and $\tfrac{1}{13}$ as .077. But either type of roundoff introduces errors. In both cases, pivoting on the largest available entry will help to minimize the roundoff error.

REMARK 2   For some linear systems, the largest available entry may not be the best choice of pivot entry because all of the coefficients in that equation may be abnormally large when compared to the coefficients that occur in the other equations. For example, in the matrix

$$\begin{pmatrix} 2 & 2 & -1 \\ 10 & 20 & 10 \end{pmatrix},$$

the largest entry in the first column is the one in the second row, but *all* the entries in the second row are large compared to the entries in the first row. If we multiply the first row by $\frac{1}{2}$ and the second row by $\frac{1}{20}$, we obtain the matrix

$$\begin{pmatrix} 1 & 1 & -\frac{1}{2} \\ \frac{1}{2} & 1 & \frac{1}{2} \end{pmatrix},$$

in which the entries in the two rows have comparable size. Now the largest entry in the first column is in the *first* row, and indeed the 1 that appears there is the better choice of pivot entry. In programming Gaussian elimination on a computer, it is a good idea to **rescale** the rows of the given matrix at the outset to make them comparable. One way to do this is to multiply each row in the matrix by the reciprocal of the entry in that row with the largest absolute value.

REMARK 3   Some linear systems are so sensitive to roundoff error that even rescaling and pivoting on the largest available entry will not help. Such systems are called *ill-conditioned*. You can expect that computers will not be of much help in trying to solve such systems.

A.2
Jacobi and Gauss-Seidel Iteration

The method of Gaussian elimination that has been described in this book gives a *direct method* for solving linear systems. It is a method that prescribes a finite (although possibly very large) number of steps, the execution of which will lead to the exact solution of the given system (assuming no round off). When using a computer to solve linear systems, it is often preferable to use an *iterative method*. An iterative method provides a way of passing from one approximate solution of the system to another approximate solution. In favorable situations, the second approximate solution will be closer to the true solution than the first one was, and by applying this method repeatedly one eventually obtains an approximate solution that is as close as desired to the true solution.

The principle that underlies the iterative methods that we will describe here is the following. Suppose we have a linear system that can be written in the form

$$\mathbf{x} = B\mathbf{x} + \mathbf{c}$$

where B is an $n \times n$ matrix and $\mathbf{c}$ is a vector in $\mathbb{R}^n$. Let $f:\mathbb{R}^n \to \mathbb{R}^n$ be the map defined by

$$f(\mathbf{x}) = B\mathbf{x} + \mathbf{c}.$$

Then $\mathbf{x}$ satisfies the given linear system if and only if $\mathbf{x}$ is a **fixed point** of f; that is, if and only if $f(\mathbf{x}) = \mathbf{x}$. One method of searching for a fixed point of the map $f:\mathbb{R}^n \to \mathbb{R}^n$ is to pick any vector $\mathbf{x}_0 \in \mathbb{R}^n$ and then define vectors $\mathbf{x}_1$, $\mathbf{x}_2$, $\mathbf{x}_3$, . . . by the formulas

$$\mathbf{x}_1 = f(\mathbf{x}_0), \ \mathbf{x}_2 = f(\mathbf{x}_1), \ \mathbf{x}_3 = f(\mathbf{x}_2), \text{ and so on.}$$

Notice that *if the vectors $\mathbf{x}_k$ approach, or **converge to,** a vector $\mathbf{x}$ (written* $\mathbf{x} = \lim_{k \to \infty} \mathbf{x}_k$) *then that vector $\mathbf{x}$ must be a fixed point of f.* This is because

$$f(\mathbf{x}) = f(\lim_{k \to \infty} \mathbf{x}_k) = \lim_{k \to \infty} f(\mathbf{x}_k) = \lim_{k \to \infty} \mathbf{x}_{k+1} = \mathbf{x}.$$

The vector $\mathbf{x} = \lim_{k \to \infty} \mathbf{x}_k$ is called the **limit** of the sequence $(\mathbf{x}_1, \mathbf{x}_2, \ldots)$.

The most straightforward way to convert a linear system $A\mathbf{x} = \mathbf{b}$, where A is an $n \times n$ matrix with all diagonal entries nonzero, into an equation of the form $\mathbf{x} = B\mathbf{x} + \mathbf{c}$ is to solve the ith equation for x_i, for each $i(1 \le i \le n)$. If we do this, the system

$$a_{11}x_1 + a_{12}x_2 + \cdots + a_{1n}x_n = b_1$$

$$a_{21}x_1 + a_{22}x_2 + \cdots + a_{2n}x_n = b_2$$

$$\cdots$$

$$a_{n1}x_1 + a_{n2}x_2 + \cdots + a_{nn}x_n = b_n$$

becomes

$$x_1 = \qquad -\frac{a_{12}}{a_{11}}x_2 \ - \cdots - \ \frac{a_{1n}}{a_{11}}x_n + \frac{b_1}{a_{11}}$$

$$x_2 = -\frac{a_{21}}{a_{22}}x_1 \qquad - \cdots - \ \frac{a_{2n}}{a_{22}}x_n + \frac{b_2}{a_{22}}$$

$$\cdots$$

$$x_n = -\frac{a_{n1}}{a_{nn}}x_1 - \frac{a_{n2}}{a_{nn}}x_2 - \cdots \qquad + \frac{b_n}{a_{nn}}$$

or

$$\mathbf{x} = B\mathbf{x} + \mathbf{c}$$

where

$$B = \begin{pmatrix} 0 & -\dfrac{a_{12}}{a_{11}} & \cdots & -\dfrac{a_{1n}}{a_{11}} \\ -\dfrac{a_{21}}{a_{22}} & 0 & \cdots & -\dfrac{a_{2n}}{a_{22}} \\ & & \cdots & \\ -\dfrac{a_{n1}}{a_{nn}} & -\dfrac{a_{n2}}{a_{nn}} & \cdots & 0 \end{pmatrix} \quad \text{and} \quad \mathbf{c} = \begin{pmatrix} \dfrac{b_1}{a_{11}} \\ \dfrac{b_2}{a_{22}} \\ \cdots \\ \dfrac{b_n}{a_{nn}} \end{pmatrix}$$

The method that attempts to find a solution of the system $A\mathbf{x} = \mathbf{b}$ by picking a vector $\mathbf{x}_0$, computing the vectors

$$\mathbf{x}_1 = B\mathbf{x}_0 + \mathbf{c}$$

$$\mathbf{x}_2 = B\mathbf{x}_1 + \mathbf{c}$$

$$\mathbf{x}_3 = B\mathbf{x}_2 + \mathbf{c}$$

$$\cdots$$

and identifying the vector $\mathbf{x} = \lim_{k \to \infty} \mathbf{x}_k$ (if it exists) is called **Jacobi iteration.** The vectors $\mathbf{x}_k$ are the **Jacobi iterates** of $\mathbf{x}_0$.

EXAMPLE 2

Let us use Jacobi iteration to find an approximate solution of the linear system

$$2x_1 + x_2 = 1$$

$$x_1 + 4x_2 = 3$$

Solving the first equation for x_1 and the second for x_2 yields

$$x_1 = \qquad -\tfrac{1}{2}x_2 + \tfrac{1}{2}$$

$$x_2 = -\tfrac{1}{4}x_1 \qquad + \tfrac{3}{4}$$

or

$$\begin{pmatrix} x_1 \\ x_2 \end{pmatrix} = \begin{pmatrix} 0 & -\tfrac{1}{2} \\ -\tfrac{1}{4} & 0 \end{pmatrix} \begin{pmatrix} x_1 \\ x_2 \end{pmatrix} + \begin{pmatrix} \tfrac{1}{2} \\ \tfrac{3}{4} \end{pmatrix}.$$

If we take $\mathbf{x}_0 = (0,0)$ and calculate the Jacobi iterates of $\mathbf{x}_0$, rounding off at each stage to three decimal places, we obtain

$$\mathbf{x}_1 = \begin{pmatrix} .500 \\ .750 \end{pmatrix}, \qquad \mathbf{x}_2 = \begin{pmatrix} .125 \\ .625 \end{pmatrix}, \qquad \mathbf{x}_3 = \begin{pmatrix} .188 \\ .719 \end{pmatrix},$$

$$\mathbf{x}_4 = \begin{pmatrix} .141 \\ .703 \end{pmatrix}, \qquad \mathbf{x}_5 = \begin{pmatrix} .149 \\ .715 \end{pmatrix}, \qquad \mathbf{x}_6 = \begin{pmatrix} .143 \\ .713 \end{pmatrix},$$

and so on. These vectors appear to be approaching a vector $\mathbf{x}$ whose entries, rounded to two decimal places, are .14 and .71.

The exact solution of the given system is $\mathbf{x} = (\tfrac{1}{7}, \tfrac{5}{7})$, which, expressed in decimal form and rounded to two places, is $\mathbf{x} = (.14, .71)$. ■

The Jacobi iteration technique does not always yield a sequence of vectors that approaches some limiting vector $\mathbf{x}$ (see Exercise 7). However, if the given system does have a unique solution, and if the Jacobi iterates do approach some vector $\mathbf{x}$, then we can be sure that the limiting vector $\mathbf{x}$ is the unique solution of the given system. Furthermore, if the diagonal entries of the coefficient matrix are large compared to the other entries in the matrix then the sequence of Jacobi iterates will converge (see Exercise 10).

It is easy to write a computer program that will carry out Jacobi iteration. Be sure to instruct the computer to stop the iteration process whenever two successive answers are within a prescribed tolerance of one another, or as soon as a specified number of iterations have been carried out. Otherwise, the computer will run indefinitely!

There is a variation on the Jacobi iteration technique, called **Gauss-Seidel iteration,** that uses less computer memory for storage and that usually converges more rapidly to the solution of the given system. In the Jacobi technique, all of the entries in $\mathbf{x}_k$ are used to compute each entry in $\mathbf{x}_{k+1}$. In the Gauss-Seidel technique, we use the entries of $\mathbf{x}_k$ to compute the first entry in $\mathbf{x}_{k+1}$. Then, having the first entry in $\mathbf{x}_{k+1}$ available, we use it together with the other entries in $\mathbf{x}_k$ to compute the second entry in $\mathbf{x}_{k+1}$. We then use the first and second entries of $\mathbf{x}_{k+1}$ together with the remaining entries in $\mathbf{x}_k$ to compute the third entry in $\mathbf{x}_{k+1}$, and so on.

EXAMPLE 3

Let us use Gauss-Seidel iteration to find an approximate solution of the linear system

$$
\begin{aligned}
3x_1 + x_2 - x_3 &= 1 \\
x_1 - 4x_2 + 2x_3 &= 0 \\
x_1 + x_2 + 5x_3 &= 0
\end{aligned}
$$

First we rewrite the system in the form

$$
\begin{aligned}
x_1 &= \qquad -\tfrac{1}{3}x_2 + \tfrac{1}{3}x_3 + \tfrac{1}{3} \\
x_2 &= \tfrac{1}{4}x_1 \qquad\qquad + \tfrac{1}{2}x_3 \\
x_3 &= -\tfrac{1}{5}x_1 - \tfrac{1}{5}x_2
\end{aligned}
$$

If we denote the kth Gauss-Seidel iterate by $\mathbf{x}_k = (x_1^{(k)}, x_2^{(k)}, x_3^{(k)})$ then we have, for each $k \geq 0$,

$$
\begin{aligned}
x_1^{(k+1)} &= \qquad -\tfrac{1}{3}x_2^{(k)} + \tfrac{1}{3}x_3^{(k)} + \tfrac{1}{3} \\
x_2^{(k+1)} &= \tfrac{1}{4}x_1^{(k+1)} \qquad\qquad + \tfrac{1}{2}x_3^{(k)} \\
x_3^{(k+1)} &= -\tfrac{1}{5}x_1^{(k+1)} - \tfrac{1}{5}x_2^{(k+1)}
\end{aligned}
$$

If we take $\mathbf{x}_0 = (0,0,0)$ and use these equations, rounding off at each stage to two decimal places, we obtain

$$
\mathbf{x}_1 = \begin{pmatrix} .33 \\ .08 \\ -.08 \end{pmatrix}, \quad
\mathbf{x}_2 = \begin{pmatrix} .28 \\ .03 \\ -.06 \end{pmatrix}, \quad
\mathbf{x}_3 = \begin{pmatrix} .30 \\ .05 \\ -.07 \end{pmatrix}, \quad
\mathbf{x}_4 = \begin{pmatrix} .29 \\ .04 \\ -.07 \end{pmatrix},
$$

and so on. The exact solution is $\mathbf{x} = (\tfrac{22}{74}, \tfrac{3}{74}, -\tfrac{5}{74})$, which, expressed in decimal form and rounded to two places, is $\mathbf{x} = (.30, .04, -.07)$. ∎

A.3

The Power Method for Finding an Eigenvector

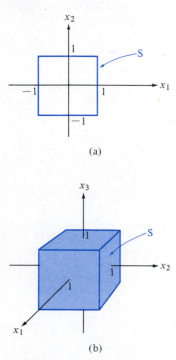

(a)

(b)

Figure A.1
The set S of all vectors $\mathbf{v}$ in $\mathbb{R}^n$ with $\ell(\mathbf{v}) = 1$. (a) $n = 2$. (b) $n = 3$.

The procedure described in Chapter 5 for finding eigenvectors and eigenvalues of linear map $f:\mathbb{R}^n \to \mathbb{R}^n$ is not well suited for use in computer calculations. A fixed point principle analogous to that described in Section A.2 can be used to derive an alternate method that will find an eigenvector for many linear maps. This method is an iterative method and hence is well suited for computer use.

The method is based on the following observation. For each vector $\mathbf{v}$ in $\mathbb{R}^n$ define $\ell(\mathbf{v})$ to be the entry in $\mathbf{v}$ with the largest absolute value. If there is more than one entry in $\mathbf{v}$ with this absolute value, take $\ell(\mathbf{v})$ to be the first such entry. Thus

$$\ell(-1,1,2) = 2, \quad \ell(-3,2,1) = -3, \quad \text{and} \quad \ell(0,-1,1) = -1.$$

Let S be the set of all vectors in $\mathbb{R}^n$ with $\ell(\mathbf{v}) = 1$. If $n = 2$ then S is the boundary of a square (see Figure A1a). If $n = 3$ then S is the boundary of a cube (see Figure A1b). Notice that, whenever $\mathbf{v}$ is a nonzero vector in $\mathbb{R}^n$ then $\mathbf{v}/\ell(\mathbf{v}) \in S$.

Now observe that *a vector $\mathbf{v} \in S$ is an eigenvector of f with nonzero eigenvalue if and only if $g(\mathbf{v}) = \mathbf{v}$, where $g(\mathbf{v}) = f(\mathbf{v})/\ell(f(\mathbf{v}))$.*

To see that this statement is correct, notice first that if $\mathbf{v} \in S$ and $f(\mathbf{v}) = \lambda\mathbf{v}$ where $\lambda \neq 0$ then

$$g(\mathbf{v}) = f(\mathbf{v})/\ell(f(\mathbf{v})) = \lambda\mathbf{v}/\ell(\lambda\mathbf{v}) = \lambda\mathbf{v}/\lambda\ell(\mathbf{v}) = \mathbf{v}.$$

Then notice that, conversely, if $\mathbf{v} \in S$ and $g(\mathbf{v}) = \mathbf{v}$ then $f(\mathbf{v})/\ell(f(\mathbf{v})) = \mathbf{v}$ so

$$f(\mathbf{v}) = \ell(f(\mathbf{v}))\mathbf{v}$$

and hence $\mathbf{v}$ is an eigenvector of f with eigenvalue $\lambda = \ell(f(\mathbf{v}))$.

Thus, if $\mathbf{v}$ is a vector in S that is a fixed point of g then $\mathbf{v}$ is an eigenvector of f. If we construct a sequence $(\mathbf{x}_0, \mathbf{x}_1, \mathbf{x}_2, \ldots)$ of vectors in S by choosing any $\mathbf{x}_0 \in S$ and defining $\mathbf{x}_{k+1} = g(\mathbf{x}_k)$ for each $k \geq 0$, then the limit of this sequence, if it exists, will be a fixed point of g:

$$g(\mathbf{v}) = g(\lim_{k\to\infty} \mathbf{x}_k) = \lim_{k\to\infty} g(\mathbf{x}_k) = \lim_{k\to\infty} \mathbf{x}_{k+1} = \mathbf{v}.$$

Hence this vector $\mathbf{v}$ will be an eigenvector of f.

In other words, *if $\mathbf{x}_0$ is any vector in S, if vectors $\mathbf{x}_1, \mathbf{x}_2, \mathbf{x}_3, \ldots$ in S are defined by*

$$\mathbf{x}_{k+1} = f(\mathbf{x}_k)/\ell(f(\mathbf{x}_k)),$$

and if these vectors converge to some vector $\mathbf{v}$, then the vector $\mathbf{v}$ is an eigenvector of f. The corresponding eigenvalue is $\ell(f(\mathbf{v}))$.

REMARK   If $\ell(f(\mathbf{x}_k)) = 0$ for some k then $\mathbf{x}_{k+1}$ is not defined. However, if this happens then $f(\mathbf{x}_k) = \mathbf{0}$ so $\mathbf{x}_k$ itself is an eigenvector of f, with eigenvalue zero. In every other case, an infinite sequence $(\mathbf{x}_1, \mathbf{x}_2, \mathbf{x}_3, \ldots)$ is defined by the formula $\mathbf{x}_{k+1} = f(\mathbf{x}_k)/\ell(f(\mathbf{x}_k))$ $(k \geq 0)$. In particular, we will obtain an infinite sequence, for every choice of $\mathbf{x}_0$, whenever the linear map f is invertible, for then the only vector $\mathbf{w}$ with $f(\mathbf{w}) = \mathbf{0}$ is the zero vector $\mathbf{w} = \mathbf{0}$.

EXAMPLE 4

Let $f: \mathbb{R}^2 \to \mathbb{R}^2$ be the linear map whose standard matrix is

$$A = \begin{pmatrix} -1 & 2 \\ -2 & 7 \end{pmatrix}.$$

If we take $\mathbf{x}_0 = (0,1)$ and define $\mathbf{x}_1, \mathbf{x}_2, \mathbf{x}_3, \ldots$ by the formula

$$\mathbf{x}_{k+1} = f(\mathbf{x}_k)/\ell(f(\mathbf{x}_k)) \ (k \geq 0),$$

we find, rounding off at each stage to three decimal places, that

$$\mathbf{x}_0 = (0,1)$$

$$\mathbf{x}_1 = (2,7)/(7) = (.286,1)$$

$$\mathbf{x}_2 = (1.714,6.428)/(6.428) = (.267,1)$$

$$\mathbf{x}_3 = (1.733,6.466)/(6.466) = (.268,1)$$

$$\mathbf{x}_4 = (1.732,6.464)/(6.464) = (.268,1).$$

If we compute additional iterations we will get $\mathbf{x}_4 = \mathbf{x}_5 = \mathbf{x}_6 = \ldots$, so we expect that, correct to three decimal places, the vector $\mathbf{v} = (.268,1)$ is an eigenvector of f.

If $\mathbf{v} = (.268,1)$ is a good approximation to an eigenvector of f, then the number

$$\ell(f(\mathbf{v})) = \ell(1.732,6.464) = 6.464$$

must be a good approximation to the corresponding eigenvalue λ.

The exact eigenvectors of f are the nonzero scalar multiples of $(2 - \sqrt{3},1) \approx (.268,1)$ and the nonzero scalar multiples of $(1,2 - \sqrt{3}) \approx (1,.268)$. The corresponding eigenvalues are $3 + 2\sqrt{3} \approx 6.464$ and $3 - 2\sqrt{3} \approx -.464$. ■

This iteration technique for finding an eigenvector of a linear map f is called the ***power method*** because, for each $k > 0$,

$$\mathbf{x}_k = f^k(\mathbf{x}_0)/\ell(f^k(\mathbf{x}_0))$$

(see Exercise 8). The iterates $\mathbf{x}_k$ do not always converge to an eigenvector of f, but it can be shown that if f is diagonable and has a ***dominant eigenvalue*** (an eigenvalue that is strictly greater in absolute value than all the other eigenvalues of f), then the sequence $(\mathbf{x}_0, \mathbf{x}_1, \mathbf{x}_2, \ldots)$ will converge to an eigenvector of f, associated with this dominant eigenvalue, for almost every choice of $\mathbf{x}_0$ (see Exercise 9).

There are many iterative techniques that are available for solving linear algebra problems on the computer. We have been able to present only a few of the simplest ones here. Discussions of more sophisticated techniques can be found in books on numerical analysis.

EXERCISES

1. Reduce each of the following matrices to a row echelon matrix, rounding off each calculation to two decimal places and pivoting at each stage on the largest available entry.

    (a) $\begin{pmatrix} \frac{1}{5} & \frac{1}{2} & 1 \\ \frac{1}{4} & \frac{2}{3} & -1 \\ \frac{1}{3} & -1 & \frac{1}{6} \end{pmatrix}$
    (b) $\begin{pmatrix} 1 & \frac{1}{3} & \frac{2}{3} \\ 2 & \frac{2}{3} & \frac{4}{3} \\ \frac{3}{7} & \frac{1}{7} & \frac{2}{7} \end{pmatrix}$
    (c) $\begin{pmatrix} \frac{1}{5} & \frac{1}{7} & \frac{2}{3} \\ \frac{2}{7} & \frac{1}{4} & \frac{3}{14} \\ \frac{1}{8} & \frac{2}{11} & \frac{1}{3} \end{pmatrix}$

2. Rescale the rows of each of the following matrices. Then reduce each to a row echelon matrix, rounding off each calculation to two decimal places and pivoting at each stage on the largest available entry.

    (a) $\begin{pmatrix} \frac{1}{4} & \frac{2}{3} & \frac{1}{5} \\ \frac{8}{3} & \frac{11}{4} & -\frac{16}{3} \\ 1 & \frac{1}{2} & \frac{1}{4} \end{pmatrix}$
    (b) $\begin{pmatrix} -1 & 15 & 2 \\ 3 & -45 & -6 \\ \frac{1}{2} & -\frac{1}{10} & \frac{1}{15} \end{pmatrix}$
    (c) $\begin{pmatrix} \frac{1}{2} & \frac{1}{3} & \frac{1}{4} \\ 2 & 3 & 4 \\ -\frac{1}{3} & \frac{2}{5} & \frac{1}{7} \end{pmatrix}$

3. Use Jacobi iteration to find an approximate solution of the given linear system. In each case, take $\mathbf{x}_0 = \mathbf{0}$ and calculate the first four Jacobi iterates, rounding off all calculations to three decimal places. Then find the exact solution of the system and compare it with these iterates.

    (a) $3x_1 - 2x_2 = 1$
        $x_1 + 3x_2 = -1$

    (b) $3x_1 + x_2 \quad\quad = 1$
        $\quad\quad 6x_2 + x_3 = 0$
        $x_1 \quad\quad - 5x_3 = 1$

    (c) $5x_1 + 2x_2 + 2x_3 = 0$
        $-x_1 + 7x_2 + 3x_3 = 0$
        $\quad\quad 3x_2 + 4x_3 = 1$

    (d) $9x_1 + 4x_2 - 2x_3 = 1$
        $x_1 - 7x_2 \quad\quad = 2$
        $x_1 - x_2 - 8x_3 = 3$

4. Repeat Exercise 3, replacing Jacobi iteration by Gauss-Seidel iteration.

5. Use the power method with $\mathbf{x}_0 = (0,1)$ to find an approximate eigenvector and the corresponding approximate eigenvalue for each of the following matrices. Calculate the first five iterates, rounding off each calculation to three decimal places. Then find the exact dominant eigenvalue and the corresponding eigenvector $\mathbf{v}$ with $\ell(\mathbf{v}) = 1$ and compare them with your approximate eigenvalue and eigenvector.

    (a) $\begin{pmatrix} -8 & -9 \\ 18 & 19 \end{pmatrix}$
    (b) $\begin{pmatrix} 2 & 3 \\ 3 & 2 \end{pmatrix}$
    (c) $\begin{pmatrix} 1 & \frac{5}{2} \\ \frac{1}{2} & 1 \end{pmatrix}$
    (d) $\begin{pmatrix} 2 & -\frac{1}{2} \\ 1 & 0 \end{pmatrix}$

6. Repeat Exercise 5 with $\mathbf{x}_0 = (1,0)$.

7. Consider the linear system

    $$x_1 + x_2 = 0$$
    $$x_1 - x_2 = 1$$

    (a) Show that this system has a unique solution.
    (b) Show that the Jacobi iterates, with $\mathbf{x}_0 = (0,0)$, do not converge.
    (c) Show that the Gauss-Seidel iterates, with $\mathbf{x}_0 = (0,0)$, do not converge.

8. Verify that if $f:\mathbb{R}^n \to \mathbb{R}^n$ is a linear map, if $\mathbf{x}_0$ is a vector with $\ell(\mathbf{x}_0) = 1$, and if $\ell(f^i(\mathbf{x}_0)) \neq 0$ for $1 \leq i \leq k - 1$, then the kth iterate of $\mathbf{x}_0$ obtained by the power method is given by

    $$\mathbf{x}_k = f^k(\mathbf{x}_0)/\ell(f^k(\mathbf{x}_0)).$$

9. Suppose $f:\mathbb{R}^n \to \mathbb{R}^n$ is a diagonable linear map. Let $\{\mathbf{v}_1, \ldots, \mathbf{v}_n\}$ be a basis for $\mathbb{R}^n$ consisting of eigenvectors of f and let $\lambda_1, \ldots, \lambda_n$ be the corresponding eigenvalues.

    (a) Show that if $\mathbf{x}_0 = a_1\mathbf{v}_1 + \cdots + a_n\mathbf{v}_n$ and $\lambda_1 \neq 0$ then

    $$\frac{1}{\lambda_1^k} f^k(\mathbf{x}_0) = a_1\mathbf{v}_1 + a_2\left(\frac{\lambda_2}{\lambda_1}\right)^k \mathbf{v}_2 + \cdots + a_n\left(\frac{\lambda_n}{\lambda_1}\right)^k \mathbf{v}_n.$$

(b)   Conclude that, if $|\lambda_1| > |\lambda_i|$ for all i > 1 then

$$\lim_{k \to \infty} \frac{1}{\lambda_1^k} f^k(\mathbf{x}_0) = a_1\mathbf{v}_1$$

and that

$$\lim_{k \to \infty} f^k(\mathbf{x}_0)/\ell(f^k(\mathbf{x}_0)) = \mathbf{v}_1/\ell(\mathbf{v}_1)$$

provided $a_1 \neq 0$.

REMARK   This exercise shows that, unless $\mathbf{x}_0$ is in the subspace of $\mathbb{R}^n$ spanned by $\{\mathbf{v}_2, \ldots, \mathbf{v}_n\}$, the sequence $(\mathbf{x}_1, \mathbf{x}_2, \ldots)$ constructed by the power method will converge to an eigenvector of f whose eigenvalue is the dominant eigenvalue $\lambda_d = \lambda_1$.

10.   For $\mathbf{v} \in \mathbb{R}^n$, define $|\!|\!| \mathbf{v} |\!|\!|$ to be the largest of the absolute values of the entries in $\mathbf{v}$. Thus $|\!|\!| \mathbf{v} |\!|\!| = |\ell(\mathbf{v})|$.

(a)   Suppose B is an $n \times n$ matrix with the property that the sum of the absolute values of the entries in each row is less than 1 and suppose $\mathbf{c} \in \mathbb{R}^n$. Show that if $f:\mathbb{R}^n \to \mathbb{R}^n$ is the map defined by

$$f(\mathbf{x}) = B\mathbf{x} + \mathbf{c},$$

then

$$|\!|\!| f(\mathbf{x}) - f(\mathbf{y}) |\!|\!| < |\!|\!| \mathbf{x} - \mathbf{y} |\!|\!|$$

whenever $\mathbf{x} \neq \mathbf{y}$, $\mathbf{x}, \mathbf{y} \in \mathbb{R}^n$. [*Hint:* Do it first for $n = 2$.]

(b)   Show that if $f:\mathbb{R}^n \to \mathbb{R}^n$ is as in part (a), then f can have at most one fixed point. [*Hint:* If $\mathbf{x}$ and $\mathbf{y}$ are two fixed points then what can you say about $|\!|\!| f(\mathbf{x}) - f(\mathbf{y}) |\!|\!|$?]

(c)   Show that if $f:\mathbb{R}^n \to \mathbb{R}^n$ is as in part (a) and $\mathbf{v} \in \mathbb{R}^n$ is a fixed point of f then, for each $\mathbf{x} \in \mathbb{R}^n$, $\mathbf{x} \neq \mathbf{v}$,

$$|\!|\!| f(\mathbf{x}) - \mathbf{v} |\!|\!| < |\!|\!| \mathbf{x} - \mathbf{v} |\!|\!|.$$

(d)   An $n \times n$ matrix A is **diagonally dominant** if the absolute value of each diagonal entry is greater than the sum of the absolute values of all the other entries in the same row. Show that if A is diagonally dominant then for each $\mathbf{b} \in \mathbb{R}^n$ the linear system $A\mathbf{x} = \mathbf{b}$ has a unique solution. [*Hint:* Prove it first for $\mathbf{b} = \mathbf{0}$ by applying part (b) to the function $f(\mathbf{x}) = B\mathbf{x} + \mathbf{c}$ where B and $\mathbf{c}$ are defined as in the Jacobi iteration process.]

(e)   Show that if A is diagonally dominant, if $\mathbf{b} \in \mathbb{R}^n$, and if $\mathbf{x}$ is the unique solution of the equation $A\mathbf{x} = \mathbf{b}$ then, for each choice of $\mathbf{x}_0$ in $\mathbb{R}^n$, the Jacobi iterates $\mathbf{x}_k$ of the system $A\mathbf{x} = \mathbf{b}$ get closer to $\mathbf{v}$ as k gets larger; that is, show that

$$|\!|\!| \mathbf{x}_{k+1} - \mathbf{v} |\!|\!| < |\!|\!| x_k - \mathbf{v} |\!|\!|$$

for all $k \geq 0$. [*Hint:* Use part (c).]

REMARK   It can be shown, by a slightly more delicate argument, that the Jacobi iterates actually converge to the solution of $A\mathbf{x} = \mathbf{b}$ whenever the matrix A is diagonally dominant.

ANSWERS AND HINTS

Section 1.1 (page 9)

1. (a) $(x_1, x_2) = (0, 1)$ (b) $(x_1, x_2) = (1, 0)$
   (c) no solutions

3. $(\frac{13}{5} + \frac{4}{5}c, \frac{4}{5} - \frac{3}{5}c, c)$, where c is any real number

5. $(11 + c, 5 + c, 3 + c, c)$, where c is any real number

7. $(\frac{15}{8} - \frac{3}{8}c_1 + \frac{3}{8}c_2, \frac{11}{8} + \frac{17}{8}c_1 - \frac{25}{8}c_2, c_1, c_2)$, where c_1 and c_2 are any real numbers.

9. $(1, \frac{1}{2}, -1, \frac{3}{2}, -\frac{1}{2})$

11. *Hint:* Look at the equation corresponding to the row in which the last corner 1 occurs.

13. *Hint:* Look at the system of equations corresponding to the echelon matrix.

Section 1.2 (page 16)

1. (a) $(0, 0, 0, 0, 0)$ (b) $(1, 6, 0, 14)$
   (c) $(2, 2, -2)$ (d) $(-1, 0, 0, 1)$
   (e) $(-1, 2, 9, 24)$ (f) $(0, 4, 1, 1)$

3. #3 $(\frac{13}{5}, \frac{4}{5}, 0) + c(\frac{4}{5}, -\frac{3}{5}, 1)$
   #5 $(11, 5, 3, 0) + c(1, 1, 1, 1)$
   #7 $(\frac{15}{8}, \frac{11}{8}, 0, 0) + c_1(-\frac{3}{8}, \frac{17}{8}, 1, 0)$
   $+ c_2(\frac{3}{8}, -\frac{25}{8}, 0, 1)$
   #9 $(1, \frac{1}{2}, -1, \frac{3}{2}, -\frac{1}{2})$

5. $(5, -8)$

9. *Hint:* The first part is straightforward. Once completed, assume **u** is a particular solution of the system, and that **w** is any other solution of the system. Set $\mathbf{v} = \mathbf{w} - \mathbf{u}$.

Section 1.3 (page 23)

1.

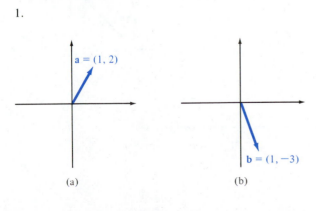

(a)   (b)

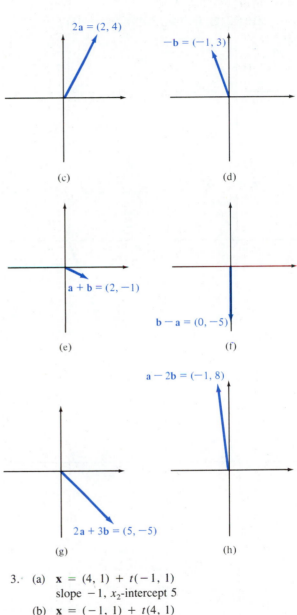

3. (a) $\mathbf{x} = (4, 1) + t(-1, 1)$
      slope -1, x_2-intercept 5
   (b) $\mathbf{x} = (-1, 1) + t(4, 1)$
      slope $\frac{1}{4}$, x_2-intercept $\frac{5}{4}$
   (c) $\mathbf{x} = (4, 1) + t(5, 0)$
      slope 0, x_2-intercept 1
   (d) $\mathbf{x} = (-4, -1) + t(2, 3)$
      slope $\frac{3}{2}$, x_2-intercept 5

5. (a) $(\frac{7}{3}, \frac{2}{3})$ (b) $(-\frac{3}{7}, -\frac{12}{7})$ (c) $(2, 1)$

7. (a) $(\frac{3}{2}, 15)$

    (b) $(\frac{4}{3}, \frac{40}{3})$ and $(\frac{5}{3}, \frac{50}{3})$

    (c) $(\frac{6}{5}, 12)$, $(\frac{7}{5}, 14)$, $(\frac{8}{5}, 16)$, and $(\frac{9}{5}, 18)$

9.

Section 1.4 (page 32)

1.

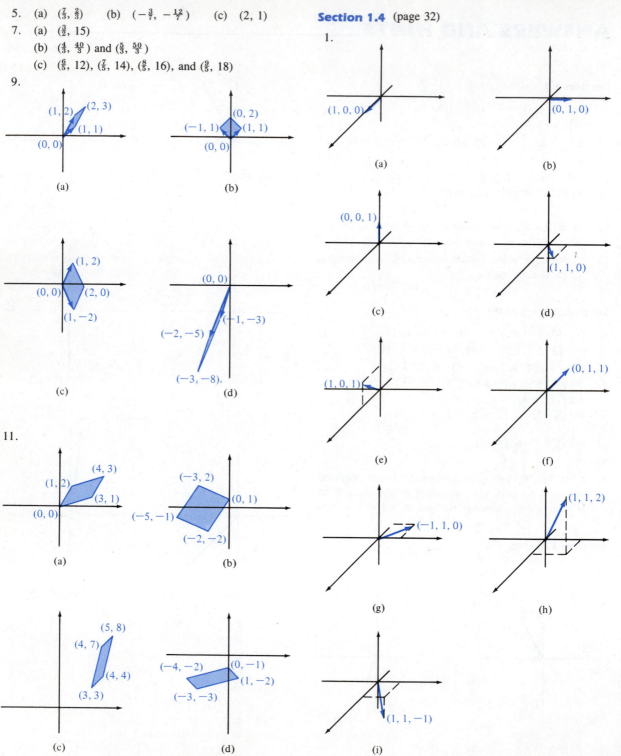

3. (a) and (d)

5. (a) $\sqrt{3}$ (b) $\sqrt{38}$ (c) $\sqrt{2}$
   (d) 3   (e) 1

7. (a) $\sqrt{29}$
   (b) $\mathbf{x} = (-1, 3, 7) + t(3, -2, -4)$
   (c) $(\frac{1}{2}, 2, 5)$
   (d) $(0, \frac{7}{3}, \frac{17}{3})$ and $(1, \frac{5}{3}, \frac{13}{3})$

9. (a) $\sqrt{14}$ (b) $\sqrt{78}$ (c) $\sqrt{6}$
   (d) $\sqrt{62}$ (e) $2\sqrt{19}$ (f) $\sqrt{6}$

11. (a) $(0, -1, 2)$ and $(1, -4, 5)$
   (b) $(-\frac{1}{4}, -\frac{1}{4}, \frac{5}{4}), (\frac{1}{2}, -\frac{5}{2}, \frac{7}{2}), (\frac{5}{4}, -\frac{19}{4}, \frac{23}{4})$
   (c) $(-\frac{2}{5}, \frac{1}{5}, \frac{4}{5})$ (d) $(\frac{4}{5}, -\frac{17}{5}, \frac{22}{5})$

13. $\mathbf{x} = (-2, 1, 3) + t(0, 2, 2)$

15. (a) $(1, \frac{8}{3}, \frac{5}{3})$ (b) $(\frac{7}{3}, 2, 2)$ (c) $(\frac{11}{3}, \frac{7}{3}, \frac{17}{3})$

17. $\mathbf{x} = (1, 3, 5) + t(-1, 1, 1)$

19. no (their trajectories intersect, but they do not collide.)

Section 1.5 (page 41)

1. (a) $\mathbf{a} \cdot \mathbf{b} = 4$, $\cos \theta = \dfrac{4}{\sqrt{65}}$, θ is acute, $\theta \approx 1.05$

   (b) $\mathbf{a} \cdot \mathbf{b} = 0$, $\cos \theta = 0$, $\theta = \dfrac{\pi}{2}$

   (c) $\mathbf{a} \cdot \mathbf{b} = 4$, $\cos \theta = 1$, $\theta = 0$

   (d) $\mathbf{a} \cdot \mathbf{b} = -10$, $\cos \theta = -1$, $\theta = \pi$

   (e) $\mathbf{a} \cdot \mathbf{b} = -5$, $\cos \theta = -\dfrac{1}{\sqrt{5}}$, θ is obtuse,
   $\theta \approx 2.03$

   (f) $\mathbf{a} \cdot \mathbf{b} = 2$, $\cos \theta = \dfrac{2}{3\sqrt{2}}$, θ is acute, $\theta \approx 1.08$

   (g) $\mathbf{a} \cdot \mathbf{b} = 4$, $\cos \theta = \dfrac{4}{\sqrt{66}}$, θ is acute, $\theta \approx 1.06$

   (h) $\mathbf{a} \cdot \mathbf{b} = -11$, $\cos \theta = -\dfrac{11}{7\sqrt{6}}$, θ is obtuse,
   $\theta \approx 2.27$

   (i) $\mathbf{a} \cdot \mathbf{b} = 0$, $\cos \theta = 0$, $\theta = \dfrac{\pi}{2}$

   (j) $\mathbf{a} \cdot \mathbf{b} = 8$, $\cos \theta = \frac{4}{5}$, θ is acute, $\theta \approx 0.64$

3. (a) $\mathbf{x} = (5, -2) + t(-1, 1)$
   (b) $\mathbf{x} = (5, -2) + t(-\frac{1}{2}, -\frac{1}{2})$
   (c) $\mathbf{x} = (5, -2) + t(-4, -3)$
   (d) $\mathbf{x} = (5, -2) + t(-3, 1)$
   (e) $\mathbf{x} = (5, -2) + t(-1, 2)$
   (f) $\mathbf{x} = (5, -2) + t(-1, -1)$

5. (a) $(\frac{1}{3}, -\frac{1}{3}, \frac{1}{3}), (\frac{2}{3}, \frac{1}{3}, -\frac{1}{3})$
   (b) $(-\frac{1}{3}, \frac{1}{3}, -\frac{1}{3}), (\frac{1}{3}, \frac{2}{3}, \frac{1}{3})$
   (c) $(\frac{1}{3}, -\frac{1}{3}, \frac{1}{3}), (-\frac{1}{3}, \frac{1}{3}, \frac{2}{3})$
   (d) $(-1, 1, -1), (0, 0, 0)$
   (e) $(0, 0, 0), (1, 2, 1)$
   (f) $(1, -1, 1), (1, 4, 3)$

7. (a) $\left(\dfrac{2}{\sqrt{13}}, \dfrac{3}{\sqrt{13}}\right)$ (b) $\left(-\dfrac{1}{\sqrt{2}}, \dfrac{1}{\sqrt{2}}\right)$
   (c) $(0, 1)$ (d) $(\frac{2}{3}, \frac{1}{3}, \frac{2}{3})$
   (e) $\left(\dfrac{1}{\sqrt{3}}, -\dfrac{1}{\sqrt{3}}, -\dfrac{1}{\sqrt{3}}\right)$
   (f) $\left(\dfrac{1}{\sqrt{3}}, \dfrac{1}{\sqrt{3}}, \dfrac{1}{\sqrt{3}}\right)$

9. *Hint:* Assume that the vertices of the triangle are $\mathbf{0}$, $\mathbf{a}$, and $\mathbf{b}$. Suppose the vector $\frac{1}{2}\mathbf{a} + \frac{1}{2}\mathbf{b}$ is an altitude and is therefore perpendicular to $\mathbf{b} - \mathbf{a}$. Compute the dot product of $\frac{1}{2}\mathbf{a} + \frac{1}{2}\mathbf{b}$ and $\mathbf{b} - \mathbf{a}$.

11. *Hint:* Expand the right-hand side using the fact that $\|\mathbf{a} + \mathbf{b}\|^2 = (\mathbf{a} + \mathbf{b}) \cdot (\mathbf{a} + \mathbf{b})$ and $\|\mathbf{a} - \mathbf{b}\|^2 = (\mathbf{a} - \mathbf{b}) \cdot (\mathbf{a} - \mathbf{b})$.

13. *Hint:* Compute $\|\mathbf{b} - \mathbf{a}\|^2 = (\mathbf{b} - \mathbf{a}) \cdot (\mathbf{b} - \mathbf{a})$.

Section 1.6 (page 49)

1. (a) $x_1 - x_2 = -1$ (b) $2x_1 - x_2 = 8$
   (c) $x_1 - x_2 = -6$ (d) $x_1 + 2x_2 = 0$
   (e) $x_2 = 4$

3. (a) $-x_1 + 2x_4 = 4$ (b) $3x_1 + 4x_2 = 18$
   (c) $x_2 = 3$ (d) $3x_1 + 2x_2 = 12$
   (e) $3x_1 + 4x_2 = 18$

5. (a) $-x_1 + x_2 = 0$ (b) $2x_1 + x_2 = 13$
   (c) $-4x_1 + 2x_2 = -4$ (d) $-4x_1 + 3x_2 = -2$
   (e) $x_2 = 0$

7. (a) $\dfrac{1}{\sqrt{2}}$ (b) $\dfrac{4}{\sqrt{5}}$ (c) 1
   (d) 0 (e) $\dfrac{2}{\sqrt{2}}$

9. (a) $2x_1 + 3x_2 - x_3 = 6$
   (b) $-x_1 + 2x_2 + 3x_3 = 10$
   (c) $x_1 + 6x_2 - 2x_3 = -5$
   (d) $-x_1 + 3x_2 + 2x_3 = 5$
   (e) $x_1 - x_3 = 0$

11. (a) not parallel, line of intersection:
   $$\mathbf{x} = (-\tfrac{2}{5}, -\tfrac{3}{5}, 0) + t(\tfrac{2}{5}, \tfrac{3}{5}, 1)$$

(b) not parallel, line of intersection:

$$\mathbf{x} = (7, -3, 0) + t(3, -1, 1)$$

(c) parallel

(d) not parallel, line of intersection:

$$\mathbf{x} = (1, 0, 0) + t(-1, 1, 0)$$

(e) parallel

13. (a) $(\frac{54}{21}, \frac{34}{21}, 3)$ (b) $(-\frac{6}{5}, \frac{7}{5}, \frac{1}{5})$ (c) $(\frac{1}{3}, \frac{1}{3}, \frac{1}{3})$

(d) $(5, 15, -5)$ (e) $(\frac{5}{3}, \frac{2}{3}, -\frac{5}{3})$

15. (a) $2x_1 - x_2 + 3x_3 + 4x_4 = -3$

(b) $6x_1 + 5x_2 + 5x_3 = 38$

(c) $-2x_1 - x_2 + 2x_4 + 5x_5 = 0$

(d) $2x_1 + 4x_2 + 6x_3 + 8x_4 + 10x_5 = 10$

(e) $2x_1 + 4x_2 + 6x_3 + \cdots + 2nx_n = 2n$

17. (a) $-x_1 - x_4 = -5$

(b) $x_1 + 2x_2 + x_3 + 2x_4 = -2$

(c) $2x_1 - x_2 + 2x_3 - x_4 + 2x_5 = 2$

19. (a) perpendicular   (b) parallel

(c) neither   (d) parallel

(e) perpendicular

21. *Hints:*

(a) If $\|\mathbf{b}\| \neq 1$, divide both sides of the equation $\mathbf{b} \cdot \mathbf{x} = c$ by $\|\mathbf{b}\|$. If $c < 0$, multiply both sides by -1.

(b) Use the formula

$$\text{distance} = \frac{|\mathbf{b} \cdot \mathbf{p} - c|}{\|\mathbf{b}\|}.$$

(c) An equation for ℓ is

$$(\cos\theta, \sin\theta) \cdot ((x_1, x_2) - (c\cos\theta, c\sin\theta))$$
$$= 0$$

where c is the distance from $\mathbf{0}$ to ℓ. Simplify.

23. (a) *Hint:* The hyperplane through $\mathbf{p}$ perpendicular to ℓ has equation $\mathbf{d} \cdot (\mathbf{x} - \mathbf{p}) = 0$. Let $\mathbf{x} = \mathbf{a} + t\mathbf{d}$ in the equation for the hyperplane, then solve for t.

(b) 3

Section 1.7 (page 61)

1. (a) $(-3, -1, 5)$ (b) $(-14, 2, -25)$

(c) $(0, 0, 0)$ (d) $(4, -10, 8)$

(e) $(-2, -2, 0)$ (f) $(2, -5, 6)$

3. (a) $3x_1 + 3x_2 - 3x_3 = 0$

(b) $x_1 = 1$

(c) $4x_1 + 2x_2 + x_3 = 9$

(d) $x_1 + x_2 + x_3 = 1$

(e) $6x_1 - 4x_2 - 2x_3 = 0$

(f) $2x_1 - 3x_2 - 11x_3 = -33$

5. (a) $\sqrt{14}$ (b) $4\sqrt{6}$ (c) 1

(d) $7\sqrt{3}$ (e) $\sqrt{89}$ (f) $\sqrt{419}$

7. (a) $3\sqrt{3}$ (b) $\sqrt{6}$ (c) $5\sqrt{74}$

(d) 1   (e) 3

9. (a) $\frac{1}{6}$ (b) 1   (c) 8   (d) 0

11. *Hint:* Let $\mathbf{a} = (a_1, a_2, a_3)$, $\mathbf{b} = (b_1, b_2, b_3)$. Calculate $\mathbf{a} \times \mathbf{b}$, then $(\mathbf{a} \times \mathbf{b}) \cdot \mathbf{a}$ and $(\mathbf{a} \times \mathbf{b}) \cdot \mathbf{b}$.

Section 1.8 (page 75)

1.

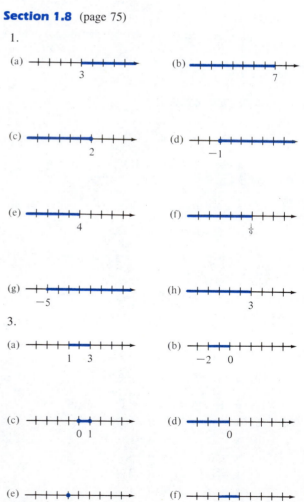

5.

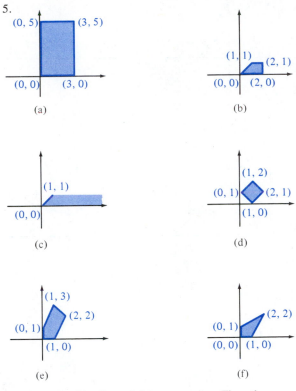

(a)

(b)

(c)

(d)

(e)

(f)

7. (b), (e), (h), (i), and (k) are convex. The others are not.

9. (a) Largest value, 20, at $(10, 0)$; smallest value, -10, at $(0, 10)$

  (b) Largest value, 20, at $(10, 0)$; smallest value, -8, at $(0, 8)$

  (c) Largest value, 14, at $(8, 2)$; smallest value, -8, at $(0, 8)$

11. (a) $\frac{5}{2}$ (b) $\frac{5}{2}$ (c) 4 (d) $\frac{13}{2}$
   (e) $\frac{13}{2}$ (f) $\frac{7}{2}$ (g) 1 (h) 1

13. $\dfrac{200}{3}$ pounds of regular, no deluxe

15. x_1 pounds of regular, x_2 pounds of deluxe, where (x_1, x_2) is any point on the line segment from $(0, 60)$ to $(40, 40)$.

17. 6 units of A, 0 units of B.

19. (b) *Hint:* Let a be the common value. Then (c_1, c_2) must satisfy the linear system

$$p_1 c_1 + p_2 c_2 = a$$

$$q_1 c_1 + q_2 c_2 = a$$

$$r_1 c_1 + r_2 c_2 = a.$$

Review Exercises (Chapter 1) (page 78)

1. (a) False   (b) True   (c) True
   (d) True   (e) False

3. $(-\frac{28}{37}, -\frac{69}{74}, -\frac{177}{74}, 0) + c(\frac{20}{37}, \frac{7}{74}, \frac{63}{74}, 1)$, where c is any real number

5. (a) $(3, -3, 4)$ (b) $\dfrac{5}{\sqrt{34}}$ (c) $\dfrac{3}{\sqrt{34}}$

   (d) $3x_1 - 3x_2 + 4x_3 = 3$
   (e) $x = (\frac{4}{3}, -\frac{1}{3}, 0) + t(-\frac{7}{6}, \frac{1}{6}, 1)$

7. (a) $(\frac{25}{17}, -\frac{11}{17})$
   (b) $(\frac{1}{2}, \frac{1}{2}, 0)$
   (c) none   (d) none
   (e) $(1, 0, 0) + c(1, 1, 1)$, where c is any real number

9. (a) $4x_1 + 2x_2 = 5$ (b) $8x_1 - 4x_2 + 2x_3 = 21$

11. *Hint:* $\|(\mathbf{a} - \mathbf{p}) \times \mathbf{d}\|$ is the area of the parallelogram spanned by $\mathbf{a} - \mathbf{p}$ and $\mathbf{d}$.

13. (a) Maximum 1 on the segment from $(1, 0)$ to $(3, 1)$, minimum -4 at $(0, 2)$
   (b) Maximum 2 at $(3, 1)$, minimum -2 at $(0, 2)$
   (c) Maximum 8 at $(2, 2)$, minimum 0 at $(0, 0)$
   (d) Maximum 10 at $(3, 1)$, minimum 0 at $(0, 0)$

Section 2.1 (page 90)

1. (a) $(8, -4)$ (b) $(0, 4)$ (c) $(7, -6)$
   (d) $(-1, 2)$ (e) $(3, -2)$

3. (a) $f(x_1, x_2) = (x_2, 0)$ (b) $f(x_1, x_2) = (0, x_1)$
   (c) $f(x_1, x_2) = (2x_1, 3x_2)$
   (d) $f(x_1, x_2) = (2x_2, 3x_1)$

5. (a) $\begin{pmatrix} 3 & -2 \\ -4 & 5 \end{pmatrix}$ (b) $\begin{pmatrix} \frac{1}{2} & 1 \\ 1 & \frac{1}{2} \end{pmatrix}$

   (c) $\begin{pmatrix} \sqrt{2} & -1 \\ 0 & 1 \end{pmatrix}$ (d) $\begin{pmatrix} 1 & -\sqrt{3} \\ -\sqrt{3} & 3 \end{pmatrix}$

7. (a) f reflects each vector in the x_2-axis
   (b) f reflects each vector in the origin
   (c) f multiplies each vector by the scalar 2
   (d) f projects each vector onto the x_1-axis
   (e) f rotates each vector counterclockwise through $\pi/2$
   (f) f rotates each vector clockwise through $\pi/2$
   (g) f multiplies each vector by the scalar $\frac{1}{2}$
   (h) f projects each vector onto the x_2-axis

9.

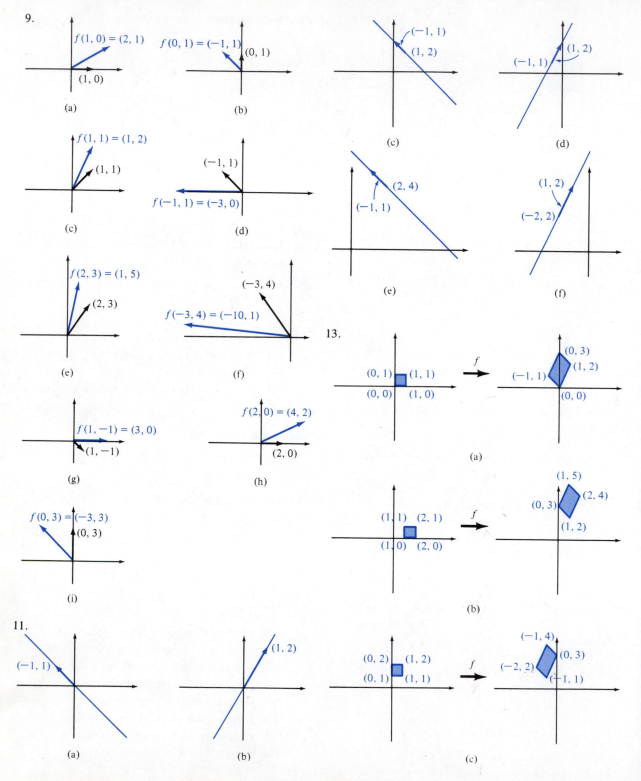

(a)   (b)   (c)   (d)

(c)   (d)   (e)   (f)

(e)   (f)

13.

(g)   (h)   (a)

(b)

(i)

11.

(a)   (b)   (c)

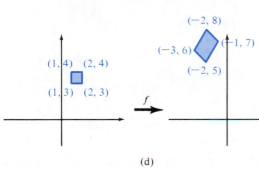

(d)

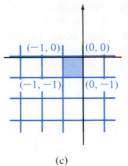

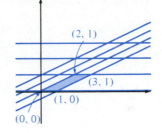

(c)

(d)

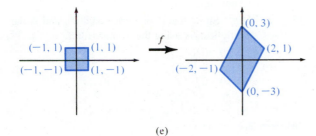

(e)

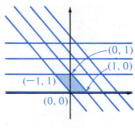

(e)

15.

3. (a) $\begin{pmatrix} 3 & 4 \\ -2 & 1 \end{pmatrix}$ (b) $\begin{pmatrix} 1 & -1 \\ 1 & 1 \end{pmatrix}$

   (c) $\begin{pmatrix} -1 & -\frac{1}{2} \\ 1 & \frac{1}{2} \end{pmatrix}$ (d) $\begin{pmatrix} 0 & 1 \\ 1 & 1 \end{pmatrix}$

   (e) $\begin{pmatrix} 1 & 0 \\ 1 & 1 \end{pmatrix}$

17. *Hint:* f maps line segments to line segments.

5. $\begin{pmatrix} \dfrac{1}{\sqrt{2}} & \dfrac{1}{\sqrt{2}} \\ \dfrac{1}{\sqrt{2}} & -\dfrac{1}{\sqrt{2}} \end{pmatrix}$

Section 2.2 (page 100)

7. (a) *Hint:* Check that f preserves addition and scalar multiplication. This can be done either algebraically or geometrically.

1. The images of the unit square and some coordinate lines are as follows:

   (b) $\begin{pmatrix} 1 & 0 \\ 0 & 0 \end{pmatrix}$

9. (a) *Hint:* Check that f preserves addition and scalar multiplication.

   (b) $\begin{pmatrix} u_1^2 & u_1 u_2 \\ u_1 u_2 & u_2^2 \end{pmatrix}$

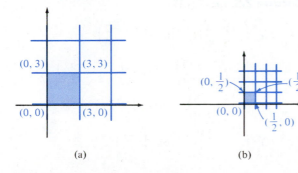

(a)   (b)

11. (a) *Hint:* The parallelogram $f(S)$ is spanned by the vectors (a, c) and (b, d). The area of this parallelogram is equal to the volume of the parallelepiped in $\mathbb{R}^3$ spanned by the vectors $(a, c, 0)$, $(b, d, 0)$, and $(0, 0, 1)$. Compute the triple scalar product.

(b) *Hint:* The matrix of a rotation is
$$\begin{pmatrix} \cos\theta & -\sin\theta \\ \sin\theta & \cos\theta \end{pmatrix}.$$

(c) *Hint:* The matrix of a reflection is
$$\begin{pmatrix} \cos 2\theta & \sin 2\theta \\ \sin 2\theta & -\cos 2\theta \end{pmatrix}.$$

(d) *Hint:* The matrix of a horizontal shear is
$$\begin{pmatrix} 1 & c \\ 0 & 1 \end{pmatrix}.$$

(e) c^2

Section 2.3 (page 106)

1. (a) $(1, 0)$ (b) $(0, 1)$ (c) $(1, -1)$
   (d) $(1, -2)$ (e) $(12, 1)$
   (f) $(x_1 + x_3, x_2 - x_3)$

3. (a) $(2, -1, 0)$ (b) $(1, 1, 3)$ (c) $(3, 2, 4)$
   (d) $(2, 4, 7)$ (e) $(2, 9, 14)$
   (f) $(2x_1 + x_2 + 3x_3, -x_1 + x_2 + 2x_3, 3x_2 + 4x_3)$

5. (a) $\begin{pmatrix} 3 & 2 & -1 \\ -1 & 4 & -5 \end{pmatrix}$ (b) $\begin{pmatrix} 2 & -3 \\ 7 & -6 \\ 1 & -1 \end{pmatrix}$

   (c) $\begin{pmatrix} 0 & 1 \\ 1 & 0 \\ 1 & 0 \\ 0 & 1 \end{pmatrix}$ (d) $(1 \;\; 1 \;\; 1)$ (e) $\begin{pmatrix} 1 \\ 2 \\ 3 \end{pmatrix}$

7. (a) $\begin{pmatrix} 2 & -1 & -1 \\ 3 & 1 & 2 \\ 1 & 0 & 1 \end{pmatrix}$,

   $f(x_1, x_2, x_3) =$
   $(2x_1 - x_2 - x_3, 3x_1 + x_2 + 2x_3, x_1 + x_3)$

   (b) $\begin{pmatrix} -1 & 3 & 0 \\ 1 & 4 & 0 \\ 1 & 5 & 1 \end{pmatrix}$,

   $f(x_1, x_2, x_3) =$
   $(-x_1 + 3x_2, x_1 + 4x_2, x_1 + 5x_2 + x_3$

   (c) $\begin{pmatrix} 0 & 0 & 1 \\ 1 & 0 & 0 \\ 0 & 1 & 0 \end{pmatrix}$, $f(x_1, x_2, x_3) = (x_3, x_1, x_2)$

(d) $\begin{pmatrix} a & b & c \\ b & c & a \\ c & a & b \end{pmatrix}$,

$f(x_1, x_2, x_3) =$
$(ax_1 + bx_2 + cx_3, bx_1 + cx_2 + ax_3, cx_1 + ax_2 + bx_3)$

9. (a) $(0, 0, -1)$ (b) $(-1, 0, 0)$
   (c) $(0, 1, 0)$ (d) $(-c, b, -a)$

11. *Hint:* Calculate $f(\mathbf{p} + s\mathbf{a} + t\mathbf{b} + u\mathbf{c})$ using the fact that f is a linear map.

13. *Hint:* Show that $f(\mathbf{0}) \neq \mathbf{0}$.

15. (a) *Hint:* Show that f preserves addition and scalar multiplication using the formula for f.

   (b) $\begin{pmatrix} 1 & 0 & 0 \\ 0 & 1 & 0 \\ 0 & 0 & 0 \end{pmatrix}$

Section 2.4 (page 114)

1. (a) $2x_1, 3x_1 + 3x_2, -x_1 + x_2$
   (b) $(0, 2x_2), (6x_1 + 3x_2, -3x_1 + 6x_2)$, $(2x_1 + x_2, -x_1)$
   (c) $(2x_1, 0, x_2), (3x_1, 3x_1, 3x_2), (-x_1, x_1, 0)$
   (d) $(x_1, x_2), (-3x_2, 3x_1), (-x_1 - x_2, x_1 - x_2)$

3. (a) $(0, x_2, x_3)$; $f - g$ is projection onto the x_2x_3-plane
   (b) $(-x_1, x_2, x_3)$; $f - 2g$ is reflection in the x_2x_3-plane

5. (a) $(0, 2)$ (b) $(1, 2)$ (c) $(1, 0)$
   (d) $(2, -2)$ (e) $(3, 4)$
   (f) $(x_2 + x_3, 2x_1 + 2x_2)$

7. (a) *Hint:* Find the formula for $g \circ f(x_1, x_2)$ by direct calculation.
   (b) *Hint:* Calculate $f \circ g(x_1, x_2)$ and compare with (a).

9. *Hint:* Calculate $f \circ f(\mathbf{x})$ from the formula for f.

Section 2.5 (page 121)

1. (a) $\begin{pmatrix} -1 & 4 \\ 6 & 6 \end{pmatrix}$ (b) $\begin{pmatrix} 2 & -2 \\ 4 & 6 \end{pmatrix}$

   (c) $\begin{pmatrix} 6 & -15 \\ -12 & -9 \end{pmatrix}$ (d) $\begin{pmatrix} 8 & -17 \\ -8 & -3 \end{pmatrix}$

   (e) $\begin{pmatrix} 3 & -6 \\ -2 & 0 \end{pmatrix}$ (f) $\begin{pmatrix} 5 & -2 \\ 18 & 24 \end{pmatrix}$

3. (a) $\begin{pmatrix} 2 & 1 \\ 6 & -7 \end{pmatrix}$ (b) $\begin{pmatrix} -1 \\ 7 \end{pmatrix}$

(c) $\begin{pmatrix} -3 & 4 & 1 \\ 5 & -5 & 0 \\ -9 & 12 & 3 \end{pmatrix}$ (d) $(5 \quad -6)$

(e) (5) (f) $\begin{pmatrix} 2 & -2 & 2 \\ -1 & 1 & -1 \\ 2 & -2 & 2 \end{pmatrix}$

5. (a) $\begin{pmatrix} 4 & -10 \\ 33 & 15 \end{pmatrix}$ (b) $\begin{pmatrix} 5 \\ -10 \end{pmatrix}$

(c) $\begin{pmatrix} 9 \\ 2 \\ 18 \end{pmatrix}$ (d) $\begin{pmatrix} -6 & -13 & -3 \\ -17 & 6 & 2 \\ -1 & 4 & 0 \end{pmatrix}$

(e) $(6 \quad 11 \quad 17)$ (f) $\begin{pmatrix} 1 & 0 & 0 \\ 0 & 1 & 0 \\ 0 & 0 & 1 \end{pmatrix}$

7. (a) $A^2 = \begin{pmatrix} 0 & 0 & ac \\ 0 & 0 & 0 \\ 0 & 0 & 0 \end{pmatrix}, A^3 = \begin{pmatrix} 0 & 0 & 0 \\ 0 & 0 & 0 \\ 0 & 0 & 0 \end{pmatrix}$

(b) $A^2 = \begin{pmatrix} 1 & 0 & 2a \\ 0 & 1 & 0 \\ 0 & 0 & 1 \end{pmatrix}, A^3 = \begin{pmatrix} 1 & 0 & 3a \\ 0 & 1 & 0 \\ 0 & 0 & 1 \end{pmatrix}$

9. (a) $A = \begin{pmatrix} 0 & 1 \\ 1 & 0 \end{pmatrix}, B = \begin{pmatrix} 1 & 1 \\ 1 & -1 \end{pmatrix}$

(b) $AB = \begin{pmatrix} 1 & -1 \\ 1 & 1 \end{pmatrix}$;

$g \circ f(x_1, x_2) = (x_1 - x_2, x_1 + x_2)$

(c) $BA = \begin{pmatrix} 1 & 1 \\ -1 & 1 \end{pmatrix}$;

$f \circ g(x_1, x_2) = (x_1 + x_2, -x_1 + x_2)$

11. (a) (i) $A = \begin{pmatrix} -3 & 1 \\ 1 & 1 \end{pmatrix}, B = \begin{pmatrix} 2 & -1 \\ 3 & 4 \end{pmatrix}$

(ii) $AB = \begin{pmatrix} -3 & 7 \\ 5 & 3 \end{pmatrix}$

(iii) $g \circ f(x_1, x_2) = (-3x_1 + 7x_2, 5x_1 + 3x_2)$

(b) (i) $A = \begin{pmatrix} 0 & 0 & 1 \\ 0 & 1 & 0 \\ 1 & 0 & 0 \end{pmatrix}, B = \begin{pmatrix} 1 & 1 \\ 1 & -1 \\ 2 & 0 \end{pmatrix}$

(ii) $AB = \begin{pmatrix} 2 & 0 \\ 1 & -1 \\ 1 & 1 \end{pmatrix}$

(iii) $g \circ f(x_1, x_2) = (2x_1, x_1 - x_2, x_1 + x_2)$

(c) (i) $A = \begin{pmatrix} 1 & -1 & 0 \\ 0 & 1 & -1 \\ -1 & 0 & 1 \end{pmatrix}$,

$B = \begin{pmatrix} 1 & 1 & 1 \\ 1 & 1 & 0 \\ 1 & 0 & 0 \end{pmatrix}$

(ii) $AB = \begin{pmatrix} 0 & 0 & 1 \\ 0 & 1 & 0 \\ 0 & -1 & -1 \end{pmatrix}$

(iii) $g \circ f(x_1, x_2, x_3) = (x_3, x_2, -x_2 - x_3)$

(d) (i) $A = (1 \quad -1), B = \begin{pmatrix} 1 & 1 & 0 \\ 0 & 1 & 1 \end{pmatrix}$

(ii) $AB = (1 \quad 0 \quad -1)$

(iii) $g \circ f(x_1, x_2, x_3) = x_1 - x_3$

(e) (i) $A = (1 \quad 1 \quad 1 \quad 1), B = \begin{pmatrix} 1 \\ -1 \\ 2 \\ 3 \end{pmatrix}$

(ii) $AB = (5)$

(iii) $g \circ f(x) = 5x$

13. *Hint:* Find the matrices of $g \circ f$ and $f \circ g$ by multiplying the matrix of f by the matrix of g in the proper order.

15. *Hint:* Write out the matrices in terms of their entries and use the definitions of matrix addition and scalar multiplication of matrices.

Section 2.6 (page 131)

1. (a) $g(\mathbf{x}) = \frac{1}{3}\mathbf{x}$ (b) $g(\mathbf{x}) = 5\mathbf{x}$

(c) $g(\mathbf{x}) = -\mathbf{x}$ (d) $g(x_1, x_2) = (\frac{1}{2}x_1, \frac{1}{3}x_2)$

(e) $g(x_1, x_2) = (x_2, x_1)$

(f) $g(x_1, x_2, x_3) = (x_3, x_1, x_2)$

3. (a) kernel $f = \{\mathbf{0}\}$; f is one-to-one

(b) The kernel of f is the line $\mathbf{x} = t(\frac{1}{2}, 1)$; f is not one-to-one.

(c) kernel $f = \{\mathbf{0}\}$; f is one-to-one

(d) The kernel of f is the line $\mathbf{x} = t(1, -2, 1)$; f is not one-to-one.

(e) The kernel of f is the line $\mathbf{x} = t(-1, 1, 1)$; f is not one-to-one.

5. (a) 2 (b) 1 (c) 2 (d) 2 (e) 2

7. (a) 2; invertible (b) 1; not invertible

(c) 2; not invertible (d) 3; invertible

(e) 4; invertible

9. *Hint:* Assume that $f(a) = f(b)$. Apply g to both sides of this equation and argue that $a = b$.

Section 2.7 (page 138)

1. (a) $\begin{pmatrix} \frac{1}{2} & \frac{1}{2} \\ \frac{1}{2} & -\frac{1}{2} \end{pmatrix}$ (b) $\begin{pmatrix} \frac{1}{4} & -\frac{3}{8} \\ \frac{1}{4} & \frac{1}{8} \end{pmatrix}$

(c) not invertible.   (d) $\begin{pmatrix} -1 & 0 & 1 \\ 1 & 1 & 0 \\ 3 & 1 & -1 \end{pmatrix}$

(e) $\begin{pmatrix} 3 & -\frac{5}{2} & \frac{1}{2} \\ -3 & 4 & -1 \\ 1 & -\frac{3}{2} & \frac{1}{2} \end{pmatrix}$ (f) $\begin{pmatrix} -7 & 5 & 3 \\ 3 & -2 & -2 \\ 3 & -2 & -1 \end{pmatrix}$

3. (a) $\begin{pmatrix} 1 & -a \\ 0 & 1 \end{pmatrix}$ (b) $\begin{pmatrix} 1/a & 0 \\ 0 & 1/b \end{pmatrix}$

(c) $\begin{pmatrix} 1 & 0 & -a \\ 0 & 1 & 0 \\ 0 & 0 & 1 \end{pmatrix}$ (d) $\begin{pmatrix} 1 & -a & ac-b \\ 0 & 1 & -c \\ 0 & 0 & 1 \end{pmatrix}$

(e) $\begin{pmatrix} 1 & 0 & 0 & -a \\ 0 & 1 & 0 & 0 \\ 0 & 0 & 1 & 0 \\ 0 & 0 & 0 & 1 \end{pmatrix}$

5. $g(x_1, x_2, x_3) = (\frac{1}{4}x_1 - \frac{3}{4}x_2 - \frac{1}{4}x_3, \frac{1}{4}x_1 + \frac{1}{4}x_2 - \frac{1}{4}x_3, -\frac{1}{4}x_1 + \frac{3}{4}x_2 + \frac{5}{4}x_3)$

7. (a) $g(x_1, x_2) = (\frac{2}{7}x_1 + \frac{5}{7}x_2, -\frac{1}{14}x_1 + \frac{5}{14}x_2)$

(b) not invertible

(c) not invertible

(d) $g(x_1, x_2, x_3) = (3x_1 - \frac{5}{2}x_2 + \frac{1}{2}x_3, -3x_1 + 4x_2 - x_3, x_1 - \frac{3}{2}x_2 + \frac{1}{2}x_3)$

(e) $g(x_1, x_2, x_3, x_4)$
    $= (x_1, -x_1 + x_2, -x_2 + x_3, -x_3 + x_4)$

9. (a) 1   (b) 3   (c) 2   (d) 2

11. (a) $\begin{pmatrix} 42 \\ -31 \end{pmatrix}$ (b) $\begin{pmatrix} -89 \\ 66 \end{pmatrix}$

(c) $\begin{pmatrix} 23\pi \\ -17\pi \end{pmatrix}$ (d) $\begin{pmatrix} -4 \\ 3 \end{pmatrix}$

13. *Hint:* Show that $(B^{-1}A^{-1})(AB) = I$ and $(AB)(B^{-1}A^{-1}) = I$.

15. *Hint:* If $AB = I$, then $f \circ g = i$, where f, g are the linear maps associated with A, B, respectively. Use Section 2.6 (Exercise 10 and Theorem 4) to conclude that rank $A = m$.

For the converse, assume that rank $A = m$. Construct the matrix $B = (\mathbf{b}_1, \dots, \mathbf{b}_m)$, where $\mathbf{b}_j =$ the jth column of B, by showing that the equations

$$A\mathbf{b}_1 = \mathbf{e}_1, \dots, A\mathbf{b}_m = \mathbf{e}_m$$

always have at least one, but possibly more than one, solution.

Section 2.8 (page 149)

1. (a) $\frac{1}{13}$ (b) $\frac{1}{52}$ (c) $\frac{1}{4}$ (d) $\frac{1}{26}$

3. (a) $\frac{3}{4}$ (b) $\frac{1}{3}$ (c) $\frac{2}{3}$

5. (c) and (d)

7. (a) $\frac{7}{16}$ (b) $\frac{11}{18}$ (c) $\frac{107}{216}$ (d) $\frac{1183}{2592}$

9. (a) 264   (b) 214   (c) 218

(d) 197 (approximately)

(e) 206 (approximately)

11. Approximately 229 in State A_1 and 203 in State A_2.

13. (a) $\mathbf{x} = (c, c)$, where c is any real number

(b) $\mathbf{x} = (c, c)$, where c is any real number

(c) $\mathbf{x} = (c, 0)$, where c is any real number

(d) $\mathbf{x} = (\frac{17}{29}c, \frac{18}{29}c, 0)$, where c is any real number

(e) $\mathbf{x} = (c_1, c_1, c_2)$, where c_1, c_2 are any two real numbers

15. (a) $T = \begin{pmatrix} 0 & \frac{2}{3} & \frac{1}{2} \\ \frac{2}{3} & 0 & \frac{1}{2} \\ \frac{1}{3} & \frac{1}{3} & 0 \end{pmatrix}$

(b) $\frac{2}{9}, \frac{14}{27}, \frac{7}{27}$

(c) about 30 or 31 in each of the first two compartments, and 20 in the third

(d) about 30 in each of the first two compartments, and 20 in the third

17. (a) $T = \begin{pmatrix} \frac{1}{2} & \frac{1}{3} & \frac{1}{3} \\ \frac{1}{4} & \frac{1}{2} & \frac{1}{6} \\ \frac{1}{4} & \frac{1}{6} & \frac{1}{2} \end{pmatrix}$, where the states A, B, C are ordered alphabetically.

(b) $\frac{5}{12}$

(c) We expect approximately 39 to buy Brand A, 36 to buy Brand B, and 25 to buy Brand C, two weeks hence.

(d) (40, 30, 30)

19. (a) $T = \begin{pmatrix} 1 & \frac{1}{2} & 0 & 0 \\ 0 & 0 & \frac{1}{2} & 0 \\ 0 & \frac{1}{2} & 0 & 0 \\ 0 & 0 & \frac{1}{2} & 1 \end{pmatrix}$.

(b) $\frac{1}{2}, \frac{5}{8}, \frac{5}{8}$

21. (a) *Hint:* $T^k\mathbf{e}_1$ is the first column vector of T^k.

Review Exercises (Chapter 2) (page 152)

1. (a) not linear   (b) $\begin{pmatrix} 2 & -2 \\ 2 & 2 \end{pmatrix}$

(c) $\begin{pmatrix} 1 & 1 & 0 & 0 & 0 & 0 \\ 0 & 0 & 1 & 1 & 0 & 0 \\ 0 & 0 & 0 & 0 & 1 & 1 \end{pmatrix}$ (d) not linear

(e) $\begin{pmatrix} 1 & 0 \\ 1 & 0 \\ 1 & 0 \\ 1 & 0 \end{pmatrix}$

3. (a) horizontal shear
   (b) expansion
   (c) reflection in $x_1 = x_2$
   (d) rotation counterclockwise through $\pi/2$
   (e) contraction
   (f) rotation counterclockwise through $\pi/6$

5. (a) (i) $A = \begin{pmatrix} 3 & 2 \\ -1 & 4 \end{pmatrix}, B = \begin{pmatrix} 4 & -5 \\ 2 & -3 \end{pmatrix}$

   (ii) $AB = \begin{pmatrix} 16 & -21 \\ 4 & -7 \end{pmatrix}$

   (iii) $f \circ g(x_1, x_2) = (16x_1 - 21x_2, 4x_1 - 7x_2)$

   (b) (i) $A = \begin{pmatrix} 1 & 0 \\ 1 & 1 \\ 1 & -1 \end{pmatrix}, B = \begin{pmatrix} 1 & -1 & 0 \\ 0 & 1 & -1 \end{pmatrix}$,

   (ii) $AB = \begin{pmatrix} 1 & -1 & 0 \\ 1 & 0 & -1 \\ 1 & -2 & 1 \end{pmatrix}$,

   (iii) $f \circ g(x_1, x_2, x_3)$
        $= (x_1 - x_2, x_1 - x_3, x_1 - 2x_2 + x_3)$

   (c) (i) $A = \begin{pmatrix} 1 & 0 \\ 0 & 1 \\ 0 & 1 \\ 1 & 0 \end{pmatrix}, B = \begin{pmatrix} 0 & 1 \\ 1 & 0 \end{pmatrix}$,

   (ii) $AB = \begin{pmatrix} 0 & 1 \\ 1 & 0 \\ 1 & 0 \\ 0 & 1 \end{pmatrix}$,

   (iii) $f \circ g(x_1, x_2) = (x_2, x_1, x_1, x_2)$

   (d) (i) $A = \begin{pmatrix} 1 & -1 & 0 \\ 0 & 1 & -1 \\ -1 & 0 & 1 \end{pmatrix}, B = \begin{pmatrix} 1 & 0 \\ 0 & 1 \\ 1 & 1 \end{pmatrix}$,

   (ii) $AB = \begin{pmatrix} 1 & -1 \\ -1 & 0 \\ 0 & 1 \end{pmatrix}$,

   (iii) $f \circ g(x_1, x_2) = (x_1 - x_2, -x_1, x_2)$

   (e) (i) $A = (1 \ \ 1 \ \ 1), B = \begin{pmatrix} 3 & 4 \\ -1 & 1 \\ 1 & 0 \end{pmatrix}$,

   (ii) $AB = (3 \ \ 5),$
   (iii) $f \circ g(x_1, x_2) = 3x_1 + 5x_2$

7. (a) (i) $A = \begin{pmatrix} 7 & 5 \\ 10 & 7 \end{pmatrix}$,

   (ii) $A^{-1} = \begin{pmatrix} -7 & 5 \\ 10 & -7 \end{pmatrix}$,

   (iii) $g(x_1, x_2) = (-7x_1 + 5x_2, 10x_1 - 7x_2)$

   (b) (i) $A = \begin{pmatrix} 4 & -3 \\ 6 & -5 \end{pmatrix}$, (ii) $A^{-1} = \begin{pmatrix} \frac{5}{2} & -\frac{3}{2} \\ 3 & -2 \end{pmatrix}$,

   (iii) $g(x_1, x_2) = (\frac{5}{2}x_1 - \frac{3}{2}x_2, 3x_1 - 2x_2)$

   (c) (i) $A = \begin{pmatrix} 1 & -1 & 0 \\ 3 & 4 & -1 \\ -9 & 3 & 1 \end{pmatrix}$,

   (ii) $A^{-1} = \begin{pmatrix} 7 & 1 & 1 \\ 6 & 1 & 1 \\ 45 & 6 & 7 \end{pmatrix}$,

   (iii) $g(x_1, x_2, x_3) = (7x_1 + x_2 + x_3, 6x_1 + x_2$
        $+ x_3, 45x_1 + 6x_2 + 7x_3)$

   (d) (i) $A = \begin{pmatrix} 5 & 4 & 3 \\ 2 & 1 & 0 \\ -1 & 1 & -1 \end{pmatrix}$,

   (ii) $A^{-1} = \begin{pmatrix} -\frac{1}{12} & \frac{7}{12} & -\frac{1}{4} \\ \frac{1}{6} & -\frac{1}{6} & \frac{1}{2} \\ \frac{1}{4} & -\frac{3}{4} & -\frac{1}{4} \end{pmatrix}$,

   (iii) $g(x_1, x_2, x_3) = (-\frac{1}{12}x_1 + \frac{7}{12}x_2 - \frac{1}{4}x_3,$
        $\frac{1}{6}x_1 - \frac{1}{6}x_2 + \frac{1}{2}x_3, \frac{1}{4}x_1 - \frac{3}{4}x_2 - \frac{1}{4}x_3)$

9. (a) True (b) False (c) False (d) True
   (e) True (f) True (g) False

11. (b) is one-to-one, the rest are not one-to-one; all but
    (d) are onto.

13. (b) .2 (c) .24 (d) (20, 50, 30)
    (e) $(\frac{110}{7}, \frac{410}{7}, \frac{180}{7})$

Section 3.1 (page 162)

1. (a) -10 (b) 5 (c) 6 (d) -32
   (e) -8 (f) 1
3. (a) -27 (b) -126 (c) -500
   (d) 10 (e) -6
5. (a) 1 (b) $(b - a)(c - a)(c - b)$
7. *Hint:* If the *j*th row of A is c times the *i*th row of A,
   then subtract c times the *i*th row from the *j*th row.
9. (a) -1 (b) -1 (c) 1 (d) 1
11. 1 and 2

13. *Hint:* Imitate the proof of Theorem 5, but work from the top down.

Section 3.2 (page 174)

1. (a) 58   (b) 5   (c) 69   (d) 60

    (e) 10   (f) -76

3. See the answers for Exercise 1.

7. (a) 0   (b) $2abc$

    (c) $a^2f^2 + b^2e^2 + c^2d^2 - 2abef + 2acdf - 2bcde$
$= (af - be + cd)^2$

Section 3.3 (page 182)

1. (1, 2, 3, 4) (2, 1, 3, 4) (3, 1, 2, 4) (4, 1, 2, 3)
(1, 2, 4, 3) (2, 1, 4, 3) (3, 1, 4, 2) (4, 1, 3, 2)
(1, 3, 2, 4) (2, 3, 1, 4) (3, 2, 1, 4) (4, 2, 1, 3)
(1, 3, 4, 2) (2, 3, 4, 1) (3, 2, 4, 1) (4, 2, 3, 1)
(1, 4, 2, 3) (2, 4, 1, 3) (3, 4, 1, 2) (4, 3, 1, 2)
(1, 4, 3, 2) (2, 4, 3, 1) (3, 4, 2, 1) (4, 3, 2, 1)

3. (a) 1   (b) 1   (c) 1   (d) -1

    (e) 1   (f) -1

5. (a) $\sigma = (3, 5, 4, 2, 1)$, sign $\sigma = 1$

    (b) $\sigma = (2, 3, 4, 5, 1)$, sign $\sigma = 1$

    (c) $\sigma = (5, 4, 1, 2, 3)$, sign $\sigma = -1$

    (d) $\sigma = (4, 3, 2, 1, 5)$, sign $\sigma = 1$

7. (a) $\begin{pmatrix} -1 & 2 \\ 1 & 3 \\ 0 & -1 \end{pmatrix}$ (b) $\begin{pmatrix} 1 & -1 & 1 \\ 3 & 0 & 1 \end{pmatrix}$

    (c) $\begin{pmatrix} 2 & 3 \\ 4 & 1 \end{pmatrix}$ (d) $\begin{pmatrix} 1 & 2 & 0 \\ -1 & 3 & -1 \\ 1 & 5 & 2 \end{pmatrix}$

    (e) $\begin{pmatrix} 0 & 2 \\ 2 & 3 \end{pmatrix}$ (f) $\begin{pmatrix} 0 & -1 & -2 \\ 1 & 0 & -3 \\ 2 & 3 & 0 \end{pmatrix}$

9. (b) and (c)

11. (a) 39   (b) -22

    (c) $ad - bc$ (d) $x^2 - 3x + 4$

13. *Hint:* (a) If B is obtained from A by multiplying the ith row by c, show that, for each permutation σ,

$$(\text{sign } \sigma)b_{1\sigma(1)} \cdots b_{n\sigma(n)} = c(\text{sign } \sigma)a_{1\sigma(1)} \cdots a_{n\sigma(n)}$$

    (b) If B is obtained from A by interchanging the first row with the second, show that, for each permutation σ,

$$(\text{sign } \sigma)b_{1\sigma(1)} \cdots b_{n\sigma(n)}$$
$$= -(\text{sign } \tau)a_{1\tau(1)} \cdots a_{n\tau(n)},$$

where τ is the permutation $\tau(1) = \sigma(2)$, $\tau(2) = \sigma(1)$, $\tau(k) = \sigma(k)$, $k = 3, 4, \ldots, n$.

15. *Hint:* (a) Using the formula for f_σ, read off the rows of the matrix of f_σ

    (b) Use the definition.

    (c) Use (a), (b), and Section 2.5, Theorem 2.

Section 3.4 (page 190)

1. (a) 0, not invertible   (b) 7, invertible

    (c) 1, invertible   (d) 1, invertible

3. (a) $\begin{pmatrix} -2 & 0 \\ 0 & -2 \end{pmatrix}$ (b) $\begin{pmatrix} 0 & 0 \\ 0 & 0 \end{pmatrix}$

    (c) $\begin{pmatrix} -17 & 0 & 0 \\ 0 & -17 & 0 \\ 0 & 0 & -17 \end{pmatrix}$ (d) $\begin{pmatrix} 0 & 0 & 0 \\ 0 & 0 & 0 \\ 0 & 0 & 0 \end{pmatrix}$

    (e) $\begin{pmatrix} 63 & 0 & 0 \\ 0 & 63 & 0 \\ 0 & 0 & 63 \end{pmatrix}$ (f) $\begin{pmatrix} 1 & 0 & 0 \\ 0 & 1 & 0 \\ 0 & 0 & 1 \end{pmatrix}$

5. (a) $\begin{pmatrix} -\frac{4}{7} & \frac{1}{7} \\ \frac{3}{7} & \frac{1}{7} \end{pmatrix}$ (b) $\begin{pmatrix} 1 & 0 \\ -a & 1 \end{pmatrix}$

    (c) $\begin{pmatrix} \frac{28}{84} & \frac{7}{84} & -\frac{7}{84} \\ -\frac{32}{84} & \frac{13}{84} & \frac{11}{84} \\ \frac{12}{84} & \frac{3}{84} & \frac{9}{84} \end{pmatrix}$

    (d) $\begin{pmatrix} 1 & 0 & 0 \\ -a & 1 & 0 \\ ac - b & -c & 1 \end{pmatrix}$

    (e) $\begin{pmatrix} \cos \theta & \sin \theta \\ -\sin \theta & \cos \theta \end{pmatrix}$

    (f) $\begin{pmatrix} 1 & 0 & 0 \\ 0 & \dfrac{d}{ad - bc} & -\dfrac{b}{ad - bc} \\ 0 & -\dfrac{c}{ad - bc} & \dfrac{a}{ad - bc} \end{pmatrix}$

7. *Hint:* (a) Show that whenever $i < j$, A_{ij} is upper triangular with at least one diagonal entry equal to zero.

    (b) If C is lower triangular, then C^t is upper triangular.

Section 3.5 (page 195)

1. (a) yes; add 3 times the second row to the first

    (b) no

    (c) yes; add -1 times the second row to the first

(d) yes; interchange the first and third rows

(e) yes; add -4 times the fourth row to the second

3. (a) $E_1 = \begin{pmatrix} -1 & 0 \\ 0 & 1 \end{pmatrix}, E_2 = \begin{pmatrix} 1 & 0 \\ -2 & 1 \end{pmatrix}$

(b) $E_1 = \begin{pmatrix} 1 & 0 \\ 1 & 1 \end{pmatrix}, E_2 = \begin{pmatrix} 1 & 0 \\ 0 & \frac{1}{6} \end{pmatrix}, E_3 = \begin{pmatrix} 1 & -3 \\ 0 & 1 \end{pmatrix}$

(c) $E_1 = \begin{pmatrix} 0 & 1 \\ 1 & 0 \end{pmatrix}, E_2 = \begin{pmatrix} 1 & -2 \\ 0 & 1 \end{pmatrix}$

(d) $E_1 = \begin{pmatrix} 0 & 0 & 1 \\ 0 & 1 & 0 \\ 1 & 0 & 0 \end{pmatrix}, E_2 = \begin{pmatrix} 1 & 0 & 0 \\ -2 & 1 & 0 \\ 0 & 0 & 1 \end{pmatrix},$

$E_3 = \begin{pmatrix} 1 & 0 & 0 \\ 0 & 1 & 0 \\ 1 & 0 & 1 \end{pmatrix}, E_4 = \begin{pmatrix} 1 & 0 & 0 \\ 0 & \frac{1}{7} & 0 \\ 0 & 0 & 1 \end{pmatrix},$

$E_5 = \begin{pmatrix} 1 & 2 & 0 \\ 0 & 1 & 0 \\ 0 & 0 & 1 \end{pmatrix}, E_6 = \begin{pmatrix} 1 & 0 & 0 \\ 0 & 1 & 0 \\ 0 & 1 & 1 \end{pmatrix},$

(e) $E_1 = \begin{pmatrix} 1 & 0 & 0 \\ -3 & 1 & 0 \\ 0 & 0 & 1 \end{pmatrix}, E_2 = \begin{pmatrix} 1 & 0 & 0 \\ 0 & 1 & 0 \\ 1 & 0 & 1 \end{pmatrix},$

$E_3 = \begin{pmatrix} 1 & 0 & 0 \\ 0 & -\frac{1}{7} & 0 \\ 0 & 0 & 1 \end{pmatrix}, E_4 = \begin{pmatrix} 1 & -2 & 0 \\ 0 & 1 & 0 \\ 0 & 0 & 1 \end{pmatrix},$

$E_5 = \begin{pmatrix} 1 & 0 & 0 \\ 0 & 1 & 0 \\ 0 & -7 & 1 \end{pmatrix},$

5. *Hint:* Apply Theorem 4 to the product $A^{-1}A = I$.

7. AE is obtained from A by applying a ''column'' operation.

9. *Hint:* Use Theorem 4 and Section 3.3 Exercise 15.

Section 3.6 (page 207)

1. (a) 1   (b) 1   (c) 3   (d) 1   (e) 3

3. 1

5. (a) 15   (b) 15   (c) 15   (d) 15

7. (a) 3   (b) 3   (c) 9   (d) 27   (e) 27

9. *Hint:* Let $f:\mathbb{R}^2 \rightarrow \mathbb{R}^2$ be the linear map such that $f(\mathbf{e}_1) = \mathbf{u}$ and $f(\mathbf{e}_2) = \mathbf{v}$. The parallelogram spanned by $\mathbf{u}$ and $\mathbf{v}$ is the image of the unit square under f.

11. *Hint:* Let $f:\mathbb{R}^3 \rightarrow \mathbb{R}^3$ be the linear map such that $f(\mathbf{e}_1) = \mathbf{u}$, $f(\mathbf{e}_2) = \mathbf{v}$, and $f(\mathbf{e}_3) = \mathbf{w}$. Then f sends the unit cube to the parallelepiped in question.

13. *Hint:* (a) Such an echelon matrix is either $\begin{pmatrix} 1 & c \\ 0 & 0 \end{pmatrix}$ or $\begin{pmatrix} 0 & 1 \\ 0 & 0 \end{pmatrix}$ or $\begin{pmatrix} 0 & 0 \\ 0 & 0 \end{pmatrix}$.

(b) Look at the image of the unit square under g.

Review Exercises (Chapter 3) (page 209)

1. (a) -1014 (b) -189

3. (a), (b), (c), (d), (e) are True; (f) False; (g), (h) True; (i) False

5. (a) 1   (b) 1   (c) -1 (d) 1

7. (a) $(x_1, x_2) = (-37, 24)$

(b) $(x_1, x_2) = (\frac{1}{25}, -\frac{3}{25})$

(c) $(x_1, x_2, x_3) = (0, -5, 5)$

9. (a) 1   (b) 1   (c) 2

11. $M_\sigma A$ is the matrix whose ith row is the $\sigma(i)$th row of A.

Section 4.1 (page 217)

1. (a), (d)

3. (b), (c), (d), (e), (f)

5. (a) $\mathbf{x} = t(-3, 2)$ (b) $\mathbf{x} = t(1, 1)$

(c) $\mathbf{x} = t(-5, -1, 1)$

(d) $\mathbf{x} = t(\frac{1}{2}, \frac{1}{2}, 1)$

9. *Hint:* If S is a subspace containing $a \neq 0$, then for every $b \in \mathbb{R}$, $b = (b/a)a$.

13. *Hint:* Since S is not empty, it must contain a vector $\mathbf{x}$. Notice that $\mathbf{0} = \mathbf{x} + (-1)\mathbf{x}$.

Section 4.2 (page 226)

1. $(1, 0) = \frac{1}{3}(1, -1) + \frac{1}{3}(2, 1)$

3. (a), (c), (e), (f)

5. (a) $x_1 - 3x_2 + 5x_3 = 0$

(b) $-9x_1 + 6x_2 - 4x_3 = 0$

(c) $x_1 + 2x_2 - x_3 = 0$

(d) $-12x_1 - 11x_2 + 5x_3 = 0$

(e) $-8x_1 - 7x_2 + 5x_3 = 0$ (f) $x_3 = 0$

7. (c), (d)

9. (a) $\{(2, -3, 1)\}$ (b) $\{(\frac{1}{5}, \frac{3}{5}, 1)\}$

(c) $\{(0, 0, 0)\}$ (d) $\{(1, 1, 1)\}$

(e) $\{(-3, 6, -2, 1, 0), (-1, 1, -5, 0, 1)\}$

11. (a) $\{(-1, 1, 0), (-1, 0, 1)\}$

(b) $\{(1, 0, 0), (0, -1, 1)\}$

(c) $\{(\frac{3}{2}, 1, 0,), (-2, 0, 1)\}$

(d) $\{(1, 0, 0)(0, 1, 0)\}$

13. (a) $\{(-1, 1)\}$ (b) $\{(-2, -1, 1)\}$

(c) $\{(-3, -1, 1)\}$ (d) $\{(\frac{1}{2}, 1, 0), (-\frac{3}{2}, 0, 1)\}$

(e) $\{(-1, 1, 0, 0), (0, 0, -1, 1)\}$

15. (b), (e)

17. $\{\mathbf{0}\}$

19. *Hint:* $\{f(\mathbf{e}_1), \ldots, f(\mathbf{e}_n)\}$ is a spanning set for the image of f.

Section 4.3 (page 235)

1. (a), (b), (e)

3. (a) independent   (b) independent

   (c) independent   (d) independent

   (e) dependent   (f) dependent

5. (a), (b), (e), (g)

7. (d), (e), (i)

9. (a) $\{(-1, 1, 0), (-1, 0, 1)\}$

(b) $\{(2, 1, 0, 0), (-3, 0, 1, 0), (-1, 0, 0, 1)\}$

(c) $\{(\frac{2}{3}, 1, 0, 0), (0, 0, 1, 0), (\frac{1}{3}, 0, 0, 1)\}$

(d) $\{(\frac{29}{12}, -\frac{11}{12}, 1)\}$

(e) $\{(\frac{3}{7}, -\frac{1}{7}, 1, 0), (-\frac{2}{7}, \frac{8}{7}, 0, 1)\}$

11. *Hint:* If either $\mathbf{v}$ or $\mathbf{w}$ equals zero, then there is an easy way to find a dependence relation. If $\mathbf{v}, \mathbf{w} \neq \mathbf{0}$, use Section 1.7, Theorem 1(g).

13. *Hint:* If there is a dependence relation $c_1\mathbf{v}_1 + \cdots + c_k\mathbf{v}_k = \mathbf{0}$, with $c_j \neq 0$, then solve for $\mathbf{v}_j$ to see that $\mathbf{v}_j$ is a linear combination of the other vectors. Conversely, if $\mathbf{v}_j$ is a linear combination of the vectors, it is easy to find a dependence relation.

15. *Hint:* Use Corollary 2 and Section 4.2, Theorem 2.

Section 4.4 (page 244)

1. (a) $(2, 1)$ (b) $(1, 1)$ (c) $(-1, -2)$

   (d) $(1, 0)$ (e) $(0, 1)$ (f) $(5, 4)$

3. (a) $(7, 4, 1)$ (b) $(19, 3, -2)$

   (c) $(-21, -2, 3)$ (d) $(13, 1, -2)$

   (e) $(7, -1, -2)$ (f) $(-29, 17, 16)$

5. (a) $(2, 3, 4)$ (b) $(3, 4, 2)$ (c) $(\frac{1}{2}, \frac{3}{2}, \frac{5}{2})$

   (d) $(1, 1, 0)$ (e) $(7, -11, 6)$ (f) $(0, 0, 1)$

7. (a) $(-\frac{5}{6}, -\frac{1}{6}, \frac{1}{3})$ (b) $(-\frac{1}{3}, \frac{1}{3}, \frac{1}{3})$

   (c) $(1, 0, 1)$ (d) $(-3, 0, 0)$

   (e) $(0, 1, 0)$ (f) $(\frac{4}{3}, -\frac{1}{3}, -\frac{4}{3})$

9. (a) $\frac{1}{2}$ (b) 1   (c) $-\frac{1}{2}$ (d) 3   (e) $\frac{7}{2}$

(f) 0

11. (a) $\{(1, 1, 0), (-1, 0, 1)\}, 2$ (b) $\{(\frac{1}{3}, \frac{1}{3}, 1)\}, 1$

(c) $\{\mathbf{e}_1, \mathbf{e}_2, \mathbf{e}_3\}, 3$

(d) $\{(1, 0, -1), (0, 1, -1)\}, 2$

(e) $\{(2, 1, 0)\}, 1$ (f) $\{(0, 1, 0), (0, 0, 1)\}, 2$

(g) $\{(1, 1, 1)\}, 1$ (h) $\{(1, 1, 1)\}, 1$

(i) $\{(1, 0, -1), (0, 1, 1)\}, 2$ (j) no basis, 0

(k) $\{\mathbf{e}_1, \mathbf{e}_2, \mathbf{e}_3\}, 3$

13. $n - 1$

15. *Hint:* (a) $\mathscr{L}(\mathbf{a}_1, \ldots, \overset{i}{\mathbf{a}_i}, \ldots, \overset{j}{\mathbf{a}_j}, \ldots, \mathbf{a}_n) = \mathscr{L}(\mathbf{a}_1, \ldots, \overset{i}{\mathbf{a}_j}, \ldots, \overset{j}{\mathbf{a}_i}, \ldots, \mathbf{a}_n)$

(b) $\mathscr{L}(\mathbf{a}_1, \ldots, c\mathbf{a}_i, \ldots, \mathbf{a}_n) = \mathscr{L}(\mathbf{a}_1, \ldots, \mathbf{a}_n)$

(c) $\mathscr{L}(\mathbf{a}_1, \ldots, \mathbf{a}_i + c\mathbf{a}_j, \ldots, \mathbf{a}_n) = \mathscr{L}(\mathbf{a}_1, \ldots, \mathbf{a}_n)$

17. *Hint:* Use Section 4.3, Theorem 1, and Section 4.2, Theorem 1.

19. (a) *Hint:* $A\mathbf{x} = \mathbf{b}$ has a solution $\mathbf{x} = \begin{pmatrix} x_1 \\ \vdots \\ x_n \end{pmatrix}$ if and only if $x_1\mathbf{c}_1 + \cdots + x_n\mathbf{c}_n = \mathbf{b}$, where $\mathbf{c}_1, \ldots, \mathbf{c}_n$ are the columns of A.

Section 4.5 (page 256)

1. (a) neither   (b) orthogonal   (c) orthonormal

   (d) orthonormal   (e) orthogonal

3. (a) $(0, 1, 0)$ (b) $(1, 0, \sqrt{2})$

(c) $\left(\frac{2}{3}, \frac{\sqrt{2}}{2}, \frac{\sqrt{2}}{6}\right)$ (d) $\left(\frac{2}{3}, \sqrt{2}, -\frac{4\sqrt{2}}{3}\right)$

(e) $(0, 1, 1)$

5. (a) $x_1 - x_3 = 0$ (d) $\left(\frac{2}{\sqrt{6}}, \frac{2}{\sqrt{3}}\right)$

7. (a) $\left\{\frac{1}{\sqrt{10}}(-1, 3, 0), \frac{1}{\sqrt{110}}(3, 1, 10)\right\}$

(b) $\left\{\frac{1}{\sqrt{5}}(1, 2, 0, 0), \frac{1}{\sqrt{70}}(-6, 3, 5, 0), \frac{1}{\sqrt{210}}(2, -1, 3, 14)\right\}$

(c) $\left\{\frac{1}{\sqrt{6}}(-1, -1, 2, 0), \frac{1}{\sqrt{66}}(1, -5, -2, 6)\right\}$

9. (a) $\left\{\dfrac{1}{\sqrt{3}}(1,-1,1),\dfrac{1}{\sqrt{2}}(1,1,0)\right\}$

(b) $(\frac{13}{3},\frac{23}{3},-\frac{5}{3})$ (c) $(\frac{2}{3},-\frac{2}{3},-\frac{4}{3})$

Section 4.6 (page 266)

1. (a) $y=x+\frac{4}{3}$ (b) $y=-\frac{5}{2}x+\frac{23}{3}$
   (c) $y=-\frac{1}{10}x-\frac{1}{5}$ (d) $y=2x-1$
   (e) $y=-\frac{17}{10}x+\frac{11}{10}$ (f) $y=-\frac{1}{5}x+\frac{11}{2}$
   (g) $y=\frac{39}{35}x-\frac{22}{7}$

3. (a) $y=4.13(1.72)^x$ (b) $y=10(5)^x$
   (c) $y=0.91(1.73)^x$ (d) $y=1.53(.601)^x$
   (e) $y=2^x$ (f) $y=7.41(2.93)^x$

5. $a=795.5$, $b=-0.49$. Approximately 83 worker-hours should be required to assemble the 100th automobile.

7. (a) $\mathbf{x}=\begin{pmatrix}\frac{19}{62}\\\frac{13}{31}\end{pmatrix}$ (b) $\mathbf{x}=\begin{pmatrix}\frac{82}{195}\\\frac{22}{195}\end{pmatrix}$

   (c) $\mathbf{x}=\begin{pmatrix}\frac{9}{14}\\-\frac{2}{14}\end{pmatrix}$

   (d) $\mathbf{x}=\begin{pmatrix}\frac{33}{18}\\0\end{pmatrix}+c\begin{pmatrix}\frac{4}{3}\\1\end{pmatrix}$, c any real number

   (e) $\mathbf{x}=\begin{pmatrix}\frac{1}{7}\\\frac{1}{7}\\\frac{1}{7}\end{pmatrix}$.

9. *Hint:* The least squares solutions of $A\mathbf{x}=\mathbf{b}$ are the solutions of $x_1\mathbf{a}_1+\cdots+x_n\mathbf{a}_n=\text{Proj}_S\mathbf{b}$, where $\mathbf{a}_1,\ldots,\mathbf{a}_n$ are the columns of A and $S=\mathcal{L}(\mathbf{a}_1,\ldots,\mathbf{a}_n)$.

11. *Hint:* The vector in $\mathcal{L}(\mathbf{x})$ closest to $\mathbf{y}$ is $\text{Proj}_\mathbf{x}\mathbf{y}$, so solve $m\mathbf{x}=\text{Proj}_\mathbf{x}\mathbf{y}$ for m by considering $(\mathbf{y}-m\mathbf{x})\cdot\mathbf{x}$.

13. *Hint:* (a) (a,b,c) is a least squares solution of $A\mathbf{x}=\mathbf{b}$ if and only if $\left\|A\begin{pmatrix}a\\b\\c\end{pmatrix}-\mathbf{b}\right\|^2$ is as small as possible.

(b) Take

$$A=\begin{pmatrix}x_1^k & x_1^{k-1} & \cdot & x_1 & 1\\x_2^k & x_2^{k-1} & \cdot & x_2 & 1\\\vdots & \vdots & \vdots & \vdots & \vdots\\x_n^k & x_n^{k-1} & \cdot & x_n & 1\end{pmatrix},\ \mathbf{b}=\begin{pmatrix}y_1\\y_2\\\vdots\\y_n\end{pmatrix},$$

and imitate the solution of (a).

Review Exercises (Chapter 4) (page 268)

1. (a), (d)

3. (a) $\{(0,1,1,0),(-1,-1,0,1)\}$ (b) 2
   (c) $\{(1,0,1),(0,1,1)\}$ (d) 2

5. (c) only

7. (c) none

9. (b) $\left(\dfrac{2+\sqrt{2}}{2},\dfrac{2-\sqrt{2}}{2},0\right)$

11. (a) $y=\frac{9}{5}x-\frac{11}{3}$ (b) $y=(1.04)x^{2.12}$
    (c) $y=(.37)(3.16)^x$

13. (a) True   (b) False   (c) True
    (d) True   (e) True   (f) False
    (g) False

Section 5.1 (page 278)

1. (a) no eigenvalues or eigenvectors
   (b) eigenvalue 3; eigenvectors $\{c(1,1)|c\neq0\}$
       eigenvalue 7; eigenvectors $\{c(-1,1)|c\neq0\}$
   (c) no eigenvalues or eigenvectors
   (d) eigenvalue -2; eigenvectors $\{c(\frac{2}{7},1)|c\neq0\}$
       eigenvalue 3; eigenvectors $\{c(\frac{1}{3},1)|c\neq0\}$
   (e) eigenvalue -4; eigenvectors $\{c(-2,1)|c\neq0\}$
       eigenvalue 3; eigenvectors $\{c(-1,1)|c\neq0\}$
   (f) eigenvalue -2; eigenvectors $\{c(1,1)|c\neq0\}$
       eigenvalue 2; eigenvectors $\{c(-1,1)|c\neq0\}$
   (g) eigenvalue -1; eigenvectors $\{c(\frac{2}{7},1)|c\neq0\}$
       eigenvalue 1; eigenvectors $\{c(\frac{1}{3},1)|c\neq0\}$
   (h) no eigenvalues or eigenvectors

3. (a) eigenvalue 1; eigenvectors $\{t(1,0)|t\neq0\}$
   (b) eigenvalue 1; eigenvectors $\{t(0,1)|t\neq0\}$
   (c) eigenvalue c; eigenvectors all nonzero vectors in $\mathbb{R}^2$
   (d) If θ is not an integer multiple of π, then f has no eigenvalues or eigenvectors; if θ is an even multiple of π, the eigenvalue is 1 and the eigenvectors are all nonzero vectors in $\mathbb{R}^2$; if θ is an odd multiple of π, the eigenvalue is -1 and the eigenvectors are all nonzero vectors in $\mathbb{R}^2$.
   (e) eigenvalue 1; eigenvectors $\{c(\cos\theta,\sin\theta)|c\neq0\}$
       eigenvalue -1;
       eigenvectors $\{c(-\sin\theta,\cos\theta)|c\neq0\}$

5. *Hint:* Apply g to the equation $f(\mathbf{v})=\lambda\mathbf{v}$. The eigenvalues are reciprocals of one another. (Why must λ be nonzero?)

Section 5.2 (page 287)

1. (a) $(13, -21)$ (b) $(-3, 5)$ (c) $(-10, 17)$
   (d) $(0, 1)$ (e) $(2, -3)$ (f) $(-1, 2)$

3. (a) $\left(\dfrac{\sqrt{3}}{2}, -\dfrac{1}{2}\right)$ (b) $\left(\dfrac{\sqrt{3}+1}{2}, \dfrac{\sqrt{3}-1}{2}\right)$

   (c) $\left(\dfrac{1-\sqrt{3}}{2}, \dfrac{1+\sqrt{3}}{2}\right)$ (d) $(2, 0)$

   (e) $(0, -2)$ (f) $(1-\sqrt{3}, \sqrt{3}-1)$

5. (a) $(-9, 3-11)$ (b) $(7, -2, 8)$
   (c) $(0, 1, 0)$ (d) $(0, 1, 1)$
   (e) $(-12, 4, -14)$ (f) $(-29, 10, -35)$

7. (a) $\begin{pmatrix} 0 & 2 \\ 1 & 0 \end{pmatrix}$ (b) $\begin{pmatrix} -1 & 0 \\ 1 & 1 \end{pmatrix}$

   (c) $\begin{pmatrix} 3 & -2 \\ -1 & 2 \end{pmatrix}$ (d) $\begin{pmatrix} 1 & 1 \\ 0 & 1 \end{pmatrix}$

9. (a) $\begin{pmatrix} 0 & 1 \\ 1 & 0 \end{pmatrix}$ (b) $\begin{pmatrix} 0 & 1 \\ 1 & 0 \end{pmatrix}$

   (c) $\begin{pmatrix} 1 & 0 \\ 0 & -1 \end{pmatrix}$ (d) $\begin{pmatrix} -1 & 0 \\ 0 & 1 \end{pmatrix}$

   (e) $\begin{pmatrix} -11 & 15 \\ -8 & 11 \end{pmatrix}$ (f) $\begin{pmatrix} -1 & 1 \\ 0 & 1 \end{pmatrix}$

11. (a) $((-1, 4), (1, 2)),\ \begin{pmatrix} -2 & 0 \\ 0 & 4 \end{pmatrix}$

   (b) $((1, 1), (1, -1)),\ \begin{pmatrix} 3 & 0 \\ 0 & -1 \end{pmatrix}$

   (c) $((1+\sqrt{2}, 1), (1-\sqrt{2}, 1)),\ \begin{pmatrix} \sqrt{2} & 0 \\ 0 & -\sqrt{2} \end{pmatrix}$

   (d) $((1, 1), (2, 1)),\ \begin{pmatrix} 1 & 0 \\ 0 & -1 \end{pmatrix}$

   (e) $((-\tfrac{3}{2}, 1), (-\tfrac{7}{5}, 1)),\ \begin{pmatrix} -3 & 0 \\ 0 & -4 \end{pmatrix}$

13. (a) $\begin{pmatrix} 2 \\ 3 \\ 1 \end{pmatrix}$ (b) $\begin{pmatrix} 0 \\ 3 \\ 3 \end{pmatrix}$ (c) $\begin{pmatrix} -14 \\ -8 \\ -26 \end{pmatrix}$

   (d) $\begin{pmatrix} 2 \\ 6 \\ 4 \end{pmatrix}$ (e) $\begin{pmatrix} 9 \\ 23 \\ 10 \end{pmatrix}$

15. *Hint:* Take **B** to be the ordered basis consisting of the column vectors of H, and see Theorem 3.

Section 5.3 (page 297)

1. (a) eigenvalue 5; eigenvectors $\{c(1, 1)|c \neq 0\}$
   eigenvalue -1; eigenvectors $\{c(1, -1)|c \neq 0\}$

   (b) eigenvalue 0; eigenvectors $\{c(1, 0)|c \neq 0\}$
   (c) eigenvalue 1; eigenvectors $\{c(1, 0)|c \neq 0\}$
   (d) eigenvalue 1; eigenvectors
   $\{c_1(-1, 1, 0) + c_2(-1, 0, 1)|c_1, c_2 \text{ not both } 0\}$
   eigenvalue 5; eigenvectors $\{c(1, 1, 2)|c \neq 0\}$

3. (a) $H = \begin{pmatrix} 1 & 2 \\ 1 & 1 \end{pmatrix}, D = \begin{pmatrix} 1 & 0 \\ 0 & -1 \end{pmatrix}$

   (b) $H = \begin{pmatrix} 1 & \frac{5}{4} \\ 1 & 1 \end{pmatrix}, D = \begin{pmatrix} -3 & 0 \\ 0 & 7 \end{pmatrix}$

   (c) $H \begin{pmatrix} -1 & -\frac{1}{2} \\ 1 & 1 \end{pmatrix}, D = \begin{pmatrix} 1 & 0 \\ 0 & 10 \end{pmatrix}$

   (d) $H = \begin{pmatrix} 1+\sqrt{2} & 1-\sqrt{2} \\ 1 & 1 \end{pmatrix},$
   $D = \begin{pmatrix} \sqrt{2} & 0 \\ 0 & -\sqrt{2} \end{pmatrix}$

   (e) $H = \begin{pmatrix} -\frac{3}{2} & -\frac{7}{5} \\ 1 & 1 \end{pmatrix}, D = \begin{pmatrix} 3 & 0 \\ 0 & 4 \end{pmatrix}$

5. (c), (d), (e)

7. (a) $H = \begin{pmatrix} \dfrac{1}{\sqrt{2}} & -\dfrac{1}{\sqrt{2}} & 0 \\ \dfrac{1}{\sqrt{2}} & \dfrac{1}{\sqrt{2}} & 0 \\ 0 & 0 & 1 \end{pmatrix},$

   $D = \begin{pmatrix} 1 & 0 & 0 \\ 0 & -1 & 0 \\ 0 & 0 & 2 \end{pmatrix}$

   (b) $H = \begin{pmatrix} -\frac{2}{3} & \frac{2}{3} & -\frac{1}{3} \\ \frac{2}{3} & \frac{1}{3} & -\frac{2}{3} \\ \frac{1}{3} & \frac{2}{3} & \frac{2}{3} \end{pmatrix}, D = \begin{pmatrix} 0 & 0 & 0 \\ 0 & 3 & 0 \\ 0 & 0 & -3 \end{pmatrix}$

   (c) $H = \begin{pmatrix} -\dfrac{2}{3} & \dfrac{1}{\sqrt{5}} & \dfrac{4}{3\sqrt{5}} \\ \dfrac{1}{3} & \dfrac{2}{\sqrt{5}} & -\dfrac{2}{3\sqrt{5}} \\ \dfrac{2}{3} & 0 & \dfrac{5}{3\sqrt{5}} \end{pmatrix}, D = \begin{pmatrix} 0 & 0 & 0 \\ 0 & 9 & 0 \\ 0 & 0 & 9 \end{pmatrix}$

   (d) $H = \begin{pmatrix} -\dfrac{1}{\sqrt{2}} & -\dfrac{1}{\sqrt{6}} & \dfrac{1}{\sqrt{3}} \\ \dfrac{1}{\sqrt{2}} & -\dfrac{1}{\sqrt{6}} & \dfrac{1}{\sqrt{3}} \\ 0 & \dfrac{2}{\sqrt{6}} & \dfrac{1}{\sqrt{3}} \end{pmatrix},$

   $D = \begin{pmatrix} -1 & 0 & 0 \\ 0 & -1 & 0 \\ 0 & 0 & 2 \end{pmatrix}$

11. *Hint:* Use Theorem 5 to verify that $(A^{-1})^{-1} = (A^{-1})'$.

13. *Hint:* The (i, j)-entry of the matrix A of f is $f(\mathbf{e}_j) \bullet \mathbf{e}_i$. What is the (j, i)-entry?

Section 5.4 (page 313)

1. (a) $s_{3\pi/8}$ (b) $s_{5\pi/8}$ (c) $r_{\pi/2}$ (d) $r_{\pi/2}$
   (e) $s_{\pi/2}$ (f) $r_{\pi/2}$ (g) r_0 (h) $s_{\pi/4}$

3. (a) r_0 (b) r_π (c) $s_{3\pi/8}$ (d) $r_{7\pi/4}$
   (e) $r_{\pi/4}$ (f) s_0

5. (a) *Hint:* Show that A is an orthogonal matrix and use Theorem 4.

   (b) *Hint:* Show that $\det A = 1$ and use Theorem 3.

   (c) $\mathbf{x} = t(1, 1, 0)$

   (d) $\mathbf{u} = \left(-\dfrac{1}{\sqrt{2}}, -\dfrac{1}{\sqrt{2}}, 0\right)$, $\theta = \dfrac{5\pi}{3}$

7. (a) $\mathbf{u} = (0, 0, 1)$, $\theta = \pi/4$

   (b) $\mathbf{u} = (0, 1, 0)$, $\theta = 5\pi/3$

   (c) $\mathbf{u} = (1, 0, 0)$, $\theta = \pi$

   (d) $\mathbf{u} = (0, 1, 0)$, $\theta = \pi/2$

9. (a) *Hint:* Use the result of Exercise 8 to find the matrices of g and h. Multiply them to find the matrix A of f. Then show that $\det A = 1$.

   (b) $\mathbf{x} = t(-1, -1, 1)$

   (c) $\mathbf{u} = \left(-\dfrac{1}{\sqrt{3}}, -\dfrac{1}{\sqrt{3}}, \dfrac{1}{\sqrt{3}}\right)$, $\theta = 4\pi/3$

11. (a) $g = t_{(-1,-2)}$, $h = r_\pi$

   (b) $g = t_{(-4,-3)}$, $h = r_{\pi/2}$

   (c) $g = t_{(-5,1)}$, $h = r_{3\pi/2}$

   (d) $g = t_{(2,1)}$, $h = s_{\pi/4}$

   (e) $g = t_{(2,-5)}$, $h = s_{\pi/2}$ (f) $g = t_{(1,4)}$, $h = s_0$

13. (a) $g = t_{(8,0)}$, $h = s_{\pi/2}$ (b) $g = t_{(0,14)}$, $h = s_0$

   (c) $g = t_{(1,1)}$, $h = s_{3\pi/4}$

   (d) $g = t_{(4,-4)}$, $h = s_{\pi/4}$

   (e) $g = t_{(2,4)}$, $h = s_{\ell_0}$, $\ell_0: x_1 + 2x_2 = 0$

   (f) $g = t_{(6,2)}$, $h = s_{\ell_0}$, $\ell_0: 3x_1 + x_2 = 0$

15. *Hint:* Find $r_\theta \circ t_\mathbf{v}(\mathbf{0})$ and $t_\mathbf{v} \circ r_\theta(\mathbf{0})$.

17. *Hint:* f is a rotation if and only if $\det f = 1$.

19. (a) *Hint:* Draw a figure.

   (b) *Hint:* Use Theorem 5(i).

   (d) *Hint:* Use (c).

21. *Hint:* Calculate $\|r_\theta(x_1, x_2) - r_\theta(y_1, y_2)\|$ and $\|(x_1, x_2) - (y_1, y_2)\|$.

Section 5.5 (page 333)

1. (a) ellipse   (b) hyperbola   (c) circle
   (d) ellipse   (e) circle   (f) hyperbola

3. (a) $\mathbf{B} = \left(\left(\dfrac{1}{\sqrt{2}}, \dfrac{1}{\sqrt{2}}\right), \left(-\dfrac{1}{\sqrt{2}}, \dfrac{1}{\sqrt{2}}\right)\right)$, $Q = 4y'^2$

   (b) $\mathbf{B} = \left(\left(\dfrac{1}{\sqrt{2}}, \dfrac{1}{\sqrt{2}}\right), \left(-\dfrac{1}{\sqrt{2}}, \dfrac{1}{\sqrt{2}}\right)\right)$,

     $Q = x'^2 - y'^2$

   (c) $\mathbf{B} = \left(\left(\tfrac{4}{5}, \tfrac{3}{5}\right), \left(-\tfrac{3}{5}, \tfrac{4}{5}\right)\right)$, $Q = 150y'^2$

   (d) $\mathbf{B} = \left(\left(\tfrac{4}{5}, \tfrac{3}{5}\right), \left(-\tfrac{3}{5}, \tfrac{4}{5}\right)\right)$, $Q = -25x'^2 + 100y'^2$

   (e) $\mathbf{B} = \left(\left(\dfrac{1}{2}, \dfrac{\sqrt{3}}{2}\right), \left(-\dfrac{\sqrt{3}}{2}, \dfrac{1}{2}\right)\right)$, $Q = x'^2 + 9y'^2$

   (f) $\mathbf{B} = \left(\left(\dfrac{\sqrt{3}}{2}, \dfrac{1}{2}\right), \left(-\dfrac{1}{2}, \dfrac{\sqrt{3}}{2}\right)\right)$, $Q = 2x'^2 - 2y'^2$

5. (a) parabola; the image of $x = 4y^2$ under $r_{\pi/4}$

   (b) hyperbola; the image of $-x^2 + y^2 = 1$ under $r_{\pi/4}$

   (c) parabola; the image of $x = 6y^2$ under r_θ, $\theta \approx .64$ radians

   (d) hyperbola; the image of $-x^2/4 + y^2 = 1$ under r_θ, $\theta \approx .64$ radians

   (e) ellipse; the image of $x^2/9 + y^2 = 1$ under $r_{\pi/3}$

   (f) hyperbola; the image of $-x^2 + y^2 = 1$ under $r_{\pi/6}$

7. (a) $\dfrac{(x - 2)^2}{4} + (y - 4)^2 = 1$; ellipse

   (b) $(x - 1)^2 - \dfrac{(y + 2)^2}{1/2} = 1$; hyperbola

   (c) $y = \tfrac{1}{4}(x + 5)^2$; parabola

   (d) $(x + 1)^2 - (y - 3)^2 = 4$; hyperbola

   (e) $\dfrac{(x + 2)^2}{25} + \dfrac{(y - 7)^2}{4} = 1$; ellipse

   (f) $x = \tfrac{1}{2}(y - \tfrac{1}{2})^2$; parabola

9. (a) parabola; image under translation by $\mathbf{v}$ of $x' = 4y'^2$, where $\mathbf{v}$ has $\mathbf{B}$-coordinates $\left(\dfrac{3\sqrt{2}}{2}, -\dfrac{1}{\sqrt{2}}\right)$, $\mathbf{B}$ as in 3(a).

   (b) hyperbola; image under translation by $\mathbf{v}$ of $-x'^2 + y'^2 = 1$, where $\mathbf{v}$ has $\mathbf{B}$-coordinates $(0, \sqrt{2})$, $\mathbf{B}$ as in 3(b).

   (c) parabola; image under translation by $\mathbf{v}$ of $x' = 6y'^2$, where $\mathbf{v}$ has $\mathbf{B}$-coordinates $(-\tfrac{8}{5}, \tfrac{6}{5})$, $\mathbf{B}$ as in 3(c)

   (d) hyperbola; image under translation by $\mathbf{v}$ of $-\dfrac{x'^2}{4} + y'^2 = 1$, where $\mathbf{v}$ has $\mathbf{B}$-coordinates $(\tfrac{27}{5}, \tfrac{11}{5})$, $\mathbf{B}$ as in 3(d)

(e) ellipse; image under translation by $\mathbf{v}$ of $\dfrac{x'^2}{9}$ $+ \; y'^2 \; = \; 1$, where $\mathbf{v}$ has $\mathbf{B}$-coordinates $\left(\dfrac{1 + \sqrt{3}}{2}, \dfrac{1 - \sqrt{3}}{2}\right)$, $\mathbf{B}$ as in 3(e)

(f) hyperbola; image under translation by $\mathbf{v}$ of $-x'^2 + y'^2 = 1$, where $\mathbf{v}$ has $\mathbf{B}$-coordinates $\left(\dfrac{3}{2}, \dfrac{3\sqrt{3}}{2}\right)$, $\mathbf{B}$ as in 3(f)

11. (a) $D = \frac{3}{4}$; ellipse   (b) $D = 0$; parabola
    (c) $D = -1$; hyperbola
    (d) $D = -4$; hyperbola
    (e) $D = -\frac{5}{4}$; hyperbola   (f) $D = 0$; parabola
    (g) $D = 4$; ellipse   (h) $D = 11$; ellipse

13. *Hint:* Show that if λ_1 and λ_2 are the eigenvalues of the matrix $\begin{pmatrix} a & b \\ b & c \end{pmatrix}$, then $\lambda_1 = \lambda_2$ if and only if $a = c$ and $b = 0$.

15. *Hints:* (a) Assume both circles are centered at $\mathbf{0}$, since translation is an isometry. If A has radius r and B has radius r', then $g(\mathbf{v}) = \dfrac{r'}{r}\mathbf{v}$ sends A onto B.

(b) Proceed as in (a)

(c) $4x^2 + y^2 = 1$ and $9x^2 + y^2 = 1$ are not similar.

(d) $x^2 - y^2 = 1$ and $x^2 - \dfrac{y^2}{4} = 1$ are not similar.

Review Exercises (Chapter 5) (page 335)

1. (a) $\lambda^2 - 6\lambda + 5$ (b) 1, 5

   (c) 1-eigenspace: $\left\{c\left(\dfrac{\sqrt{3}}{3}, 1\right)\middle| c \neq 0\right\}$

   5-eigenspace: $\{c(-\sqrt{3}, 1)|c \neq 0\}$

   (d) $\mathbf{B} = \left\{\left(\dfrac{1}{2}, \dfrac{\sqrt{3}}{2}\right), \left(-\dfrac{\sqrt{3}}{2}, \dfrac{1}{2}\right)\right\}$

   (e) $H = \begin{pmatrix} \dfrac{1}{2} & -\dfrac{\sqrt{3}}{2} \\ \dfrac{\sqrt{3}}{2} & \dfrac{1}{2} \end{pmatrix}$, $D = \begin{pmatrix} 1 & 0 \\ 0 & 5 \end{pmatrix}$

3. (a) $\begin{pmatrix} -3 & -2 \\ -4 & -3 \end{pmatrix}$, $(-11, -16)$

   (b) $\begin{pmatrix} -3 & 2 \\ 4 & -3 \end{pmatrix}$, $(5, -8)$

5. (a), (b) are not diagonable

   (c) $H = \begin{pmatrix} 1 & -1 \\ 1 & 1 \end{pmatrix}$, $D = \begin{pmatrix} 2 & 0 \\ 0 & -2 \end{pmatrix}$

7. (a) $f = r_{\pi/6}$ (b) f is not an isometry
   (c) $f = s_{\pi/12}$

9. Ellipse with equation $\left(x' - \dfrac{3}{\sqrt{2}}\right)^2 + \dfrac{\left(y' - \dfrac{1}{\sqrt{2}}\right)^2}{2}$ $= 1$, where

$$\begin{pmatrix} x \\ y \end{pmatrix} = \begin{pmatrix} \dfrac{1}{\sqrt{2}} & -\dfrac{1}{\sqrt{2}} \\ \dfrac{1}{\sqrt{2}} & \dfrac{1}{\sqrt{2}} \end{pmatrix} \begin{pmatrix} x' \\ y' \end{pmatrix}$$

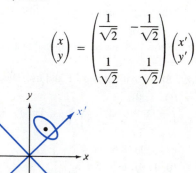

11. (a) False   (b) True   (c) True
    (d) False   (e) True   (f) True
    (g) True   (h) False   (i) True

Section 6.1 (page 344)

1. (a) $(1, 3, 5, 7, 9, \ldots)$ (b) $(1, 1, 1, 1, 1, \ldots)$
   (c) $(2, 4, 6, 8, 10, \ldots)$
   (d) $(\frac{1}{2}, -\frac{1}{2}, -\frac{3}{2}, -\frac{5}{2}, -\frac{7}{2}, \ldots)$
   (e) $(0, \pi, 2\pi, 3\pi, 4\pi, \ldots)$

3. (a) $2x^2 + 2x - 1$ (b) $4x - 9$
   (c) $3x^2 + 9x - 15$ (d) $x^2 + 11x - 23$
   (e) $9x^2 + 7x$

5. (a) yes
   (b) no; (i), (ii), (iii) not satisfied
   (c)   yes   (d) yes
   (e) no; (i), (ii), (iii) not satisfied
   (f) no; (iii) not satisfied
   (g) yes   (h) no; (iii) not satisfied
   (i)   yes   (j) yes

7. (a) no; (iii) not satisfied
   (b) no; (iii) not satisfied

(c) no; (i), (iii) not satisfied

(d) no; (ii) not satisfied

(e) no; (i), (ii), (iii) not satisfied

(f) no; (i), (ii), (iii) not satisfied

(g) yes   (h) yes   (i) yes

17. *Hint:* The verification is straightforward. $W_1 \cup W_2$ is not always a subspace.

21. yes

Section 6.2 (page 357)

1. (a), (d), (f)

3. (a), (c), (d), (f)

5. (a), (d)

7. (a), (c), (d)

9. *Hint:* Check that the set is a spanning set for $\mathcal{P}$ and that it is linearly independent.

11. (b)

13. (a) $\{x^2 + x + 1, x^2 - x + 1\}$, 2   (b) $\{1, x\}$, 2

(c) $\{x, x + x^2, x + x^3\}$, 3

(d) $\{x + 1, (x + 1)^2, (x + 1)^3\}$, 3

(e) $\{1, x, x^2, x^3 + x^2 + x + 1\}$, 4

15. (a) $\{\begin{pmatrix} 1 & 1 \\ 1 & 1 \end{pmatrix}, \begin{pmatrix} 1 & 0 \\ 0 & 1 \end{pmatrix}, \begin{pmatrix} 1 & 0 \\ 0 & 0 \end{pmatrix}, \begin{pmatrix} 0 & 1 \\ 0 & 0 \end{pmatrix}\}$

(b) $\{\begin{pmatrix} 1 & 0 \\ 0 & 1 \end{pmatrix}, \begin{pmatrix} 1 & 1 \\ 0 & 0 \end{pmatrix}, \begin{pmatrix} 1 & 0 \\ 0 & 0 \end{pmatrix}, \begin{pmatrix} 0 & 0 \\ 1 & 0 \end{pmatrix}\}$

(c) $\{\begin{pmatrix} 1 & 0 \\ 0 & 2 \end{pmatrix}, \begin{pmatrix} 1 & 2 \\ 0 & 1 \end{pmatrix}, \begin{pmatrix} 1 & 0 \\ 2 & 1 \end{pmatrix}, \begin{pmatrix} 1 & 0 \\ 0 & 0 \end{pmatrix}\}$

17. $\{f, f_1, f_2, f_3\}$, where f_1, f_2, f_3 have matrices $\begin{pmatrix} 1 & 0 \\ 0 & 0 \end{pmatrix}, \begin{pmatrix} 0 & 1 \\ 0 & 0 \end{pmatrix}, \begin{pmatrix} 0 & 0 \\ 1 & 0 \end{pmatrix}$, respectively.

19. (a) (1, 0, 0, 1)   (b) (0, 1, −1, 0)

(c) (2, 3, −1, 4)

21. mn

23. *Hint:* If $\{\mathbf{w}_1, \ldots, \mathbf{w}_m\}$ is a basis for W, we can enlarge it to a basis for V.

25. *Hint:* No vector in S can be a linear combination of other vectors in S.

27. *Hint:* The given equation implies that

$$(c_1 - d_1)\mathbf{v}_1 + \cdots + (c_n - d_n)\mathbf{v}_n = \mathbf{0}.$$

29. (a) *Hint:* If $c_1 p(x) + c_2 q(x) = 0$, then

$$c_1 p(0) + c_2 q(0) = 0$$

$$c_1 p(1) + c_2 q(1) = 0.$$

Section 6.3 (page 369)

1. (a) yes   (b) yes   (c) no   (d) no

(e) yes   (f) yes

3. (a) yes   (b) yes   (c) yes   (d) no

(e) yes   (f) yes   (g) yes   (h) no

5. (a) yes   (b) no   (c) yes   (d) yes

11. (a) $\ker f = \{0\}$, $\operatorname{im} f = \mathcal{P}^2$

(b) $\ker f = \mathcal{L}(x^2 - x - 1)$, $\operatorname{im} f = \mathcal{L}(x + 1, x^2)$

(c) $\ker f = \mathcal{L}(x - 1, x^2 - 1)$, $\operatorname{im} f = \mathcal{L}(x^2)$

13. $\dim(\ker D) = 1$, $\dim(\operatorname{im} D) = n$, $\dim \mathcal{P}^n = 1 + n$

15. (a) $2ax + b$

(c) *Hint:* Complete the square and notice that
$$a\left(x + \frac{b}{2a}\right)^2 \geq 0.$$

17. (a) $\frac{1}{4}x^4 + \frac{1}{3}x^3 + \frac{1}{2}x^2$ (b) $\frac{2}{5}x^5 - \frac{1}{4}x^4 + \frac{1}{2}x^2$

(c) $27x - \frac{27}{2}x^2 + 3x^3 - \frac{1}{4}x^4$

(d) $\frac{1}{5}x^5 - \frac{2}{3}x^3 + x$

(e) $x + \frac{1}{2}x^2 + \cdots + \dfrac{1}{n + 1}x^{n+1}$

(f) $x + x^2 + x^3 + \cdots + x^{n+1}$

19. (b) $\ker T$ is the set of all matrices with the sum of diagonal entries equal to zero

(c) $\mathbb{R}$ (d) 1, $n^2 - 1$

(e) $\{M_{ij} | i \neq j, i, j = 1, \ldots, n\} \cup \{D_i | i = 2, \ldots, n\}$, where M_{ij} has (i, j)-entry 1 and all other entries zero, D_i has (1, 1)-entry -1, (i, i)-entry 1, and all other entries zero.

27. (b) Assign to each mn-tuple $\mathbf{x}$ the matrix whose first row contains the first n entries in $\mathbf{x}$, whose second row contains the next n entries in $\mathbf{x}$, and so on.

Section 6.4 (page 378)

1. (a) $\begin{pmatrix} 1 & 0 & 0 \\ 0 & -1 & 0 \\ 0 & 0 & 1 \end{pmatrix}$ (b) $\begin{pmatrix} 1 & 0 & 0 \\ -2 & 1 & 0 \\ 1 & -1 & 1 \end{pmatrix}$

(c) $\begin{pmatrix} 1 & 0 & 0 \\ 4 & 1 & 0 \\ 4 & 2 & 1 \end{pmatrix}$ (d) $\begin{pmatrix} 2 & 0 & 0 \\ 0 & 1 & 0 \\ 0 & 0 & 0 \end{pmatrix}$

3. (a) $\begin{pmatrix} 1 & 1 & 1 \\ 0 & 1 & 2 \\ 0 & 0 & 1 \end{pmatrix}$ (b) $\begin{pmatrix} 0 & 0 & 1 \\ 0 & 1 & 2 \\ 1 & 1 & 1 \end{pmatrix}$

(c) $\begin{pmatrix} 1 & 1 & 1 \\ 2 & 1 & 0 \\ 1 & 0 & 0 \end{pmatrix}$ (d) $\begin{pmatrix} 1 & 1 & 1 \\ 0 & 1 & 2 \\ 0 & 0 & 1 \end{pmatrix}$

5. (a) $6x^2 + 3x$ (b) $9x^2 + 5x - 1$
    (c) $-3x^2 - 3x + 3$ (d) $2x^3 - x^2 + 3x + 1$
    (e) $-2x^3 + x^2 - 3x - 1$ (f) $-x + 2$

7. (a) $\begin{pmatrix} 0 & 0 & 0 & 0 & 0 \\ 4 & 0 & 0 & 0 & 0 \\ 0 & 3 & 0 & 0 & 0 \\ 0 & 0 & 2 & 0 & 0 \\ 0 & 0 & 0 & 1 & 0 \end{pmatrix}$

    (b) $4x^3 + 6x^2 - 4x + 1$

9. (a) $f \circ f(p(x)) = p(x + 2)$

    (b) $\begin{pmatrix} 1 & 0 & 0 & 0 \\ 6 & 1 & 0 & 0 \\ 12 & 4 & 1 & 0 \\ 8 & 4 & 2 & 1 \end{pmatrix}$

11. (a) $\begin{pmatrix} a_{22} & a_{21} & a_{23} \\ a_{12} & a_{11} & a_{13} \\ a_{32} & a_{31} & a_{33} \end{pmatrix}$ (b) $\begin{pmatrix} a_{22} & a_{23} & a_{21} \\ a_{32} & a_{33} & a_{31} \\ a_{12} & a_{13} & a_{11} \end{pmatrix}$

     (c) $\begin{pmatrix} a_{33} & a_{32} & a_{31} \\ a_{23} & a_{22} & a_{21} \\ a_{13} & a_{12} & a_{11} \end{pmatrix}$

13. (a) $\begin{pmatrix} -1 & 0 & 0 \\ -6 & 2 & 0 \\ -4 & 1 & 1 \end{pmatrix}$ (b) $\begin{pmatrix} 1 & 0 & 0 \\ 0 & 2 & 0 \\ 0 & 0 & -1 \end{pmatrix}$

     (c) $\begin{pmatrix} 0 & 0 & 1 \\ 0 & 1 & 2 \\ 1 & 1 & 1 \end{pmatrix}$

Section 6.5 (page 389)

1. (a) $\mathbf{x} = c_1(1, 1, 1, \ldots) + c_2(1, \frac{1}{5}, \frac{1}{25}, \ldots)$
        or $x_n = c_1 + c_2(\frac{1}{5})^{n-1}$
    (b) $\mathbf{x} = c_1(1, 1, 1, \ldots) + c_2(1, 5, 25, \ldots)$
        or $x_n = c_1 + c_2(5)^{n-1}$
    (c) $\mathbf{x} = c_1(1, -2, 4, -8, \ldots) + c_2(1, -4, 16, -64, \ldots)$ or $x_n = c_1(-2)^{n-1} + c_2(-4)^{n-1}$
    (d) $\mathbf{x} = c_1(1, 1 + \sqrt{3}, 4 + 2\sqrt{3}, \ldots)$
                $+ c_2(1, 1 - \sqrt{3}, 4 - 2\sqrt{3}, \ldots)$
        or $x_n = c_1(1 + \sqrt{3})^{n-1} + c_2(1 - \sqrt{3})^{n-1}$
    (e) $\mathbf{x} = c_1(1, -2, 4, -8, \ldots)$
                $+ c_2(0, 1, -4, 12, \ldots)$
        or $x_n = c_1(-2)^{n-1} + c_2(n - 1)(-2)^{n-2}$
    (f) $\mathbf{x} = c_1(1, -\frac{1}{2}, \frac{1}{4}, -\frac{1}{8}, \ldots)$
                $+ c_2(0, 1, -1, \frac{3}{4}, \ldots)$
        or $x_n = c_1(-\frac{1}{2})^{n-1} + c_2(n - 1)(-\frac{1}{2})^{n-2}$
    (g) $\mathbf{x} = c_1(1, -3, 9, -27, \ldots)$
                $+ c_2(0, 1, -6, 27, \ldots)$
        or $x_n = c_1(-3)^{n-1} + c_2(n - 1)(-3)^{n-2}$

    (h) $\mathbf{x} = c_1(1, -\frac{1}{2}, -\frac{1}{2}, 1, \ldots)$
                $+ c_2(0, \dfrac{\sqrt{3}}{2}, -\dfrac{\sqrt{3}}{2}, 0, \ldots)$
        or $x_n = c_1 \cos[(n - 1)2\pi/3]$
                $+ c_2 \sin[(n - 1)2\pi/3]$
    (i) $\mathbf{x} = c_1(1, -1, 0, 2, \ldots) + c_2(0, 1, -2, 2, \ldots)$
        or $x_n = c_1(\sqrt{2})^{n-1} \cos[(n - 1)3\pi/4]$
                $+ c_2(\sqrt{2})^{n-1} \sin[(n - 1)3\pi/4]$
    (j) $\mathbf{x} = c_1(1, 0, -1, 0, \ldots) + c_2(0, 1, 0, -1, \ldots)$
        or $x_n = c_1 \cos[(n - 1)\pi/2] + c_2 \sin[(n - 1)\pi/2]$

3. (a) $\mathbf{x} = (0, 1, 1, 0, -1, \ldots);$
        $x_n = \dfrac{2}{\sqrt{3}} \sin[(n - 1)\pi/3]$
    (b) $\mathbf{x} = (-1, 1, 2, -2, -4, \ldots);$
        $x_n = -(\sqrt{2})^{n-1} \cos[(n - 1)\pi/2]$
                $+ (\sqrt{2})^{n-2} \sin[(n - 1)\pi/2]$
    (c) $\mathbf{x} = (0, 1, 2, 2, 0, \ldots);$
        $x_n = (\sqrt{2})^{n-1} \sin[(n - 1)\pi/4]$
    (d) $\mathbf{x} = (0, 1, 4, 12, 32, \ldots); x_n = (n - 1)2^{n-2}$
    (e) $\mathbf{x} = (0, 1, 5, 20, 75, \ldots);$
        $x_n = \dfrac{1}{\sqrt{5}} [\left(\dfrac{5 + \sqrt{5}}{2}\right)^{n-1} - \left(\dfrac{5 - \sqrt{5}}{2}\right)^{n-1}]$

5. (b) $\Delta^2 \mathbf{x} = (x_3 - 2x_2 + x_1, x_4 - 2x_3 + x_2,$
        $x_5 - 2x_4 + x_3, \ldots)$
    (c) $p = a, q = 2a + b, r = a + b + c$

Review Exercises (Chapter 6) (page 391)

1. (c) and (e)
3. (a) Basis $\{x^3 + x^2 + x + 1, x^3 - x^2 + x - 1\}$,
        dimension 2
    (b) Basis $\{x^2 + 3x + 2, x^2 + 2x - 1\}$, dimension 2
    (c) Basis $\{x^3, (x + 1)^3, (x + 2)^3\}$, dimension 3
5. (b) and (d)
7. (a) ker f is the set of all $\mathbf{x} = (x_1, x_2, x_3, \ldots) \in \mathbb{R}^\infty$
        with $x_1 = x_2$. The image of f is the line $\mathbf{x} = t(-1, 1)$.
    (b) ker $f = \mathcal{L}(x(x - 1)(x + 1))$. The image of f is $\mathbb{R}^3$.
    (c) ker $f = \{\begin{pmatrix} a & a \\ b & b \end{pmatrix} | a, b \in \mathbb{R}\}$.
        im $f = \{\begin{pmatrix} a & -a \\ b & -b \end{pmatrix} | a, b \in \mathbb{R}\}$.
    (d) ker f consists of all constant polynomials.
        im $f = \mathcal{P}$.
9. $M_{\mathbf{B}'}(f) = \begin{pmatrix} 2 & -1 \\ 0 & 3 \end{pmatrix}$

11. (a) False   (b) False   (c) True   (d) False
     (e) True   (f) True   (g) True

13. $\mathbf{x} = c_1(1, \frac{1}{2}, \frac{1}{4}, \frac{1}{8}, \ldots) + c_2(1, -\frac{1}{3}, \frac{1}{9}, -\frac{1}{27}, \ldots)$,
where c_1 and c_2 are any real numbers.

Appendix (page 404)

1. (a) $\begin{pmatrix} 1 & 0 & 0 \\ 0 & 1 & 0 \\ 0 & 0 & 1 \end{pmatrix}$ (b) $\begin{pmatrix} 1 & 0 & .67 \\ 0 & 1 & 0 \\ 0 & 0 & 0 \end{pmatrix}$

    (c) $\begin{pmatrix} 1 & 0 & 0 \\ 0 & 1 & 0 \\ 0 & 0 & 1 \end{pmatrix}$

3. The exact solutions are:
    (a) $(\frac{1}{11}, -\frac{4}{11}) \approx (.091, -.364)$
    (b) $(\frac{29}{89}, \frac{2}{89}, -\frac{12}{89}) \approx (.326, .022, -.135)$
    (c) $(-\frac{8}{97}, -\frac{17}{97}, \frac{37}{97}) \approx (-.082, -.175, .381)$
    (d) $(\frac{82}{524}, -\frac{138}{524}, -\frac{169}{524}) \approx (.156, -.263, -.323)$

5. The exact dominant eigenvalues and corresponding eigenvectors are:
    (a) $\lambda = 10$, $\mathbf{v} = (-\frac{1}{2}, 1)$
    (b) $\lambda = 5$, $\mathbf{v} = (1, 1)$

    (c) $\lambda = 1 + \frac{1}{2}\sqrt{5} \approx 2.118$,
       $\mathbf{v} = (\frac{1}{2}\sqrt{5}, 1) \approx (.447, 1)$

    (d) $\lambda = 1 + \dfrac{\sqrt{2}}{2} \approx 1.707$,
       $\mathbf{v} = (1, 2 - \sqrt{2}) \approx (1, .586)$

7. (a) The unique solution is $(\frac{1}{2}, -\frac{1}{2})$
   (b) The Jacobi iterates are

$$\mathbf{x}_1 = \begin{pmatrix} 0 \\ -1 \end{pmatrix}, \mathbf{x}_2 = \begin{pmatrix} 1 \\ -1 \end{pmatrix},$$

$$\mathbf{x}_3 = \begin{pmatrix} 1 \\ 0 \end{pmatrix}, \mathbf{x}_4 = \begin{pmatrix} 0 \\ 0 \end{pmatrix} = \mathbf{x}_0,$$

$$\mathbf{x}_5 = \mathbf{x}_1, \mathbf{x}_6 = \mathbf{x}_2, \ldots$$

The iterates cycle forever through this set of four vectors.

   (c) The Gauss-Seidel iterates are

$$\mathbf{x}_1 = \begin{pmatrix} 0 \\ -1 \end{pmatrix}, \mathbf{x}_2 = \begin{pmatrix} 1 \\ 0 \end{pmatrix},$$
$$\mathbf{x}_3 = \mathbf{x}_1, \mathbf{x}_4 = \mathbf{x}_2, \ldots$$

The iterates cycle forever between these two vectors.

INDEX